GEOLOGY

AND THE ENVIRONMENT

SECOND EDITION

BERNARD W. PIPKIN
UNIVERSITY OF SOUTHERN CALIFORNIA

D. D. TRENT
CITRUS COLLEGE

West/Wadsworth

I(T)P® An International Thomson Publishing Company

Belmont, CA • Albany, NY • Bonn • Boston • Cincinnati • Detroit • Johannesburg
London • Los Angeles • Madrid • Melbourne • Mexico City • Minneapolis/St. Paul •
New York • Paris • Singapore • Tokyo • Toronto • Washington

PRODUCTION CREDITS

Production Editors
Steven P. Yaeger and Jayne A. Lindesmith

Copyeditor
Carole Bailey Reagle

Interior and Cover Designer
Diane Beasley

Cover Image
© 1997 Pete Kopischke/OUTBACK-IMAGES

Artwork
Precision Graphics, Carto-Graphics

Index
Northwind Editorial Services

Composition
Parkwood Composition

Prepress, Printing, and Binding
West Publishing Company

Figure Credits follow the Index.

A Note on the Type
This edition of Geology and the Environment is set in Bembo. Helvetica Light is used in the artwork.

COPYRIGHT © 1997 by Wadsworth Publishing Company
COPYRIGHT © 1994 By West Publishing Company
A Division of International Thomson Publishing Inc.
I(T)P®The ITP logo is a registered trademark under license.
Printed in the United States of America

04   03   02   01   00   99   98   8 7 6 5

For more information, contact Wadsworth Publishing Company, 10 Davis Drive, Belmont, CA 94002, or electronically at
http://www.thomson.com/wadsworth.html

International Thomson Publishing
  Europe
Berkshire House 168-173
High Holborn
London, WC1V 7AA, England

Thomas Nelson Australia
102 Dodds Street
South Melbourne 3205
Victoria, Australia

Nelson Canada
1120 Birchmount Road
Scarborough, Ontario
Canada M1K 5G4

International Thomson Publishing
  GmbH
Königswinterer Strasse 418
53227 Bonn, Germany

International Thomson Editores
Campos Eliseos 385, Piso 7
Col. Polanco
11560 México D.F. México

International Thomson Publishing
  Asia
221 Henderson Road
#05-10 Henderson Building
Singapore 0315

International Thomson Publishing
  Japan
Hirakawacho Kyowa Building, 3F
2-2-1 Hirakawacho
Chiyoda-ku, Tokyo 102, Japan

International Thomson Publishing
  Southern Africa
Building 18, Constantia Park
240 Old Pretoria Road
Halfway House, 1685 South Africa

LIBRARY OF CONGRESS CATALOGING-IN-PUBLICATION DATA

Pipkin, Bernard W.
    Geology and the environment / Bernard W. Pipkin, D. D. Trent.—
2nd ed.
      p. cm.
    Includes index.
    ISBN 0-314-09239-0 (pbk. : alk. paper)
    1. Environmental geology. I. Trent, D. D. II. Title.
QE38.P48   1997
    550—dc20   96-34500
                                                        CIP

ABOUT THE AUTHORS

DR. BERNARD W. PIPKIN is Professor Emeritus at the University of Southern California. He received his doctorate from the University of Arizona and is a licensed geologist and certified engineering geologist in the state of California. After graduation he worked for the U.S. Army Corps of Engineers on a variety of military and civil projects, from large dams to pioneer roads in rough terrain for microwave sites. He has been a consulting engineering geologist as well as a university teacher since 1965.

Dr. Pipkin is past president of the National Association of Geology Teachers and is a Fellow in the Geological Society of America. He hosted the PBS 30-part program *Oceanus* that won a local Emmy for Best Educational Television Series. He recently shared the Clare Coleridge Award (1995) with Richard Proctor from the Association of Engineering Geologists for their book *Engineering Geology Practice in Southern California.*

Dr. Pipkin is a private pilot with a flight instructor's rating; he took many of the aerial photos in the book. He and his wife Faye have three children and live south of Los Angeles in Palos Verdes.

DR. D. D. "DEE" TRENT has been working at, or teaching, geology since 1955. After graduating from college, he was employed by an oil company where his geological skills were sharpened with projects in Utah, Arizona, California, and Alaska. When the company decided to send him to Libya, he decided it was time to become a college geology teacher.

He has taught 28 years at Citrus Community College in Glendora, California, and along the way has worked with the National Park Service, picked up a Ph.D. at the University of Arizona, done field research on glaciers in Alaska and California, visited numerous mines in the United States and Germany, mapped deserts in California and Arizona, and served as an adjunct faculty member at the University of Southern California, where he taught Field Geology. When not involved with geology, he's either skiing or playing the banjo in a dixieland band.

Dr. Trent is married and has two grown children. He also claims to have the most intelligent and beautiful granddaughter born west of the Mississippi River in at least three decades.

BRIEF CONTENTS

CONTENTS

6 WEATHERING AND SOILS 152

7 LANDSLIDES AND MASS WASTING 182

8 SUBSIDENCE AND COLLAPSE 214

9 WATER ON LAND 236

13 ENERGY 364

14 MINERAL RESOURCES AND SOCIETY 410

15 WASTE MANAGEMENT AND GEOLOGY 456

PREFACE

 OUR CHANGING PLANET

Environmental geology is the study of the relationship between humans and their geologic environment. An underlying assumption is that this relationship is an interactive one. Not only do naturally occurring geologic phenomena affect the lives of people on a daily basis, but the activities of humans can induce or aggravate geological processes, with sometimes tragic consequences. Historian Will Durant noted that our physical environment ". . . exists by geological consent, subject to change without notice." We may interpret this to mean that during our lifetimes, we will probably be subject to a significant geologic hazard—an earthquake, a coastal or riverine flood, a landslide, volcanic activity, and so on. If for no other reason than this, we need to become informed about our geological environment and these hazards.

Environmental geology is dynamic—that is, it is the geology of "now" and not some distant past. During preparation of this edition, for example, unprecedented emissions of sulfur dioxide gas at Kilauea Volcano, Hawaii, forced short-term closure of Volcanoes National Park. Visibility was so reduced in the park that there was real concern for public safety, as well as for people with respiratory ailments. An earthquake was felt in several counties in Mississippi— an event which reminds us that the most widely felt and perhaps strongest earthquake in U.S. history was centered in the mid-continent region near New Madrid, Missouri. In Oregon and Washington torrential downpours yielded 76 cm (30 in) of rain in four days in 1996. The resulting floods left thousands homeless, and were aggravated by melting snowpack in the Cascade Range, triggering landslides and mudflows. This almost continuous and widespread activity is a reminder that no area is free from geological hazards.

In a society where science and technology are interwoven with economics and political action, an understanding of science takes on increasing importance. The National Science Foundation, the National Center for Earth Science Education, and several other prestigious earth-science organizations are promoting earth-science literacy in the United States for students from elementary school through college. The hope is that through education, today's students will become better stewards of our planet than their predecessors have been, working toward careful management of the land and its population, and controlling the reckless exploitation of earth resources that has been occurring. A fringe benefit is that by coming to understand the earth, our lives are enriched because we can better appreciate the earth's beauty and complexity as well as recognize its limitations. To achieve earth-science literacy goals, an understanding of the following concepts is essential:

- The planet earth is unique in our solar system.
- The Earth as a planet is old. It has existed for at least 4.6 billion years, and it has undergone many cycles of mountain building and erosion, ocean-basin formation and destruction, continental accretion and separation, and recycling of gases and liquids.
- Five earth systems—water, air, ice, solid earth, and life— all interact with each other in a complex manner.
- Earth's resources are limited, and their extraction and use must be controlled and judicious.

One fact is certain—the earth can sustain just so much life, and it is fast approaching that limit. The fragility of earth has been described very eloquently by *Apollo 14* astronaut Edgar Mitchell:

> "It is so incredibly impressive when you look back at our planet from out there in space and you realize so forcibly that it's a closed system—that we don't have unlimited resources, that there's only so much air and so much water. You get out there in space and you say to yourself 'That's home. That's the only home we have, and the only one we're going to have for a long time.' We had better take care of it, we don't get a second chance."

 ABOUT THIS BOOK

The authors, although longtime friends and former classmates, have concentrated on different aspects of geology. One of us (Pipkin) has specialized in the study of geologic hazards and their mitigation—more specifically, in the safe siting of engineering works ranging from dams and tunnels to single-family dwellings. The other (Trent) has focused upon geologic fieldwork, the exploration for natural resources, and glacier studies. As colleagues in the teaching profession, it seemed natural to combine our past experiences and mutual interest in teaching and environmental protection into a textbook.

Geology and the Environment is intended to fulfill the needs of a one-semester college course for students with little or no science background. It examines geologic principles, processes, and phenomena and relates them to humankind and human endeavors. What you learn in this course will be useful to you throughout your life. It will serve you as you form opinions about environmental issues and legislation, select a home site, or evaluate real property in a business venture. At the very least, when you understand how the earth "works," any fears and anxieties you have about such things as earthquakes, hurricanes, and volcanic eruptions should be lessened.

The emphasis in this textbook is to present geology that can be applied to improving the human endeavor. Chapter 1 presents an overview of the underlying cause of many of our present environmental problems. The stress of overpopulation on the environment has resulted in water and air pollution, land degradation, occupation of lands subject to geological hazards, and uncontrolled extraction of resources. The impact of overpopulation and the relatively new awareness of the environment are reasons that colleges are adding courses in environmentally related issues (such as environmental geology) to their curricula.

Because this textbook is intended for non-science majors, Chapters 2 and 3 offer some basic information useful for "getting around in geology." Although Chapter 2 presents a subset of information that will be useful in the subsequent examination of geological hazards, geologic processes, and mineral resources, the emphasis is upon the fact that the entire earth is an interacting "system." We will see that one cannot divorce atmospheric and oceanic processes from solid earth surface processes, nor separate moving tectonic plates (Chapter 3) from volcanic and seismic activity.

Following these groundwork chapters, the focus shifts to various types of geologic processes and hazards. Earthquakes and volcanoes are two of the most spectacular examples of the energy contained *within* the earth, and Chapters 4 and 5 examine how this energy is transferred to the earth's surface—where we live—and how we can learn to recognize the signals that "something's up" and work to minimize the damage when the inevitable occurs.

The next three chapters focus on geologic processes on or very near the surface of the earth. Much of the soil that we build our homes on and grow our food in today is the bedrock of the far distant past, chemically and physically altered through the processes of weathering (Chapter 6). Present-day rocks and soils continue to undergo such processes, in some cases assisted by means of human activities such as construction, mining, and water extraction from the earth. The results—mass wasting (landslides) and ground subsidence and collapse—cause loss of lives and millions of dollars of damage each year. These ground-failure problems are examined in detail in Chapters 7 and 8.

Earth is sometimes described as the water planet, and the presence or absence of water is essential in defining the geo-logical and biological environments in which we live. The next four chapters focus on the many ways in which this most essential of all substances affects us. Streams and rivers are vital resources that sustain life along their banks and can serve as important transportation systems, but they also pose hazards to those they serve through the threat of flooding (Chapter 9). Of course, not all water is on the surface of the earth, and in fact, in many areas the most important source of water come from underground. Ground water is a finite resource, however, and one that can easily be wasted through excessive use or contamination. These problems are examined in Chapter 10.

We next look at the effects that the largest bodies of water on our planet—the oceans—have on human life (Chapter 11). These include coastal processes and the creation and destruction of beaches, and the impact of the most life-threatening of all natural hazards—hurricanes and typhoons. However, the oceans are most important because they play a central role in producing world climate and thus establish the ground rules for the continued existence of life as we know it. There follows logically in Chapter 12 an examination of the geological aspects of extreme climates, arid and glacial landscapes, and their effects on the human scene. This, in turn, leads into the contemplation of global climate change that currently is so much in the popular press.

Geology and the Environment closes with a look at all sources of energy that can be utilized by humans (Chapter 13), mineral resources, their extraction, and the reclamation of mining sites (Chapter 14), and how wastes generated by affluent industrial societies are disposed of (Chapter 15). Along with overpopulation, the safe disposal of solid and liquid waste products may be one of the greatest problems facing humankind. The discussion illustrates that a "no deposit, no return" society is no longer tenable.

Distinctive Features of this Text

It is reasonable that the college course in which this textbook is used falls into the category of a general education course. General education, whether it be science, humanities, or social science, is meant to enrich our lives and minds beyond the specialization of our major interest. To this end, there are galleries of photos at the end of each chapter that illustrate geologic wonders and some geologic "oddities." For example, at the end of Chapter 5 (Volcanoes and the Environment), there are photographs of often-visited scenic volcanic features with explanations of how they formed. It is hoped that this kind of presentation will stimulate the reader to learn more and give him or her a greater appreciation of natural geologic wonders.

Case Studies appear throughout the text to highlight the events or topics in the text discussion. These cover a broad spectrum of topics and geographic areas, with many focusing on the causes and aftereffects of bad environmental and geological planning. Also located within each chapter are provocative Consider This questions, new to this edition. They appear shortly after or at the end of a discussion and are worded such that students must consider the information just presented in the text in order to answer the questions. It is hoped that in this way the important points just discussed will be reinforced in the student's mind, and that the questions will stimulate classroom discussion.

At the end of each chapter can be found a list of Key Terms introduced in the chapter, a Summary of the chapter in outline form. Study Questions geared to test understanding of the key concepts of the chapter, and a list of books, articles, and videos for Further Information.

This Edition

New Case Studies, photographs, and recent geologic events (for instance, the Kobe, Japan earthquake of 1995; the Mt. Ruapehu, New Zealand, eruption of 1995–96) are to be found within the second edition. The Galleries, which were well received in the first edition, have been expanded to be included in all chapters. As noted above, Consider This questions are now included within the chapters to help reinforce important points, and to create opportunities for discussion.

We have attempted to employ a systems approach to the study of our geological environment. The scientific emphasis today is on the idea that all of the earth's spheres (atmosphere, hydrosphere, lithosphere, et al.) and the processes active within them do not operate independently but are interconnected. There are five elements or "geospheres" that make up the complete earth system, and it is within these that the hazards and processes discussed in the book are generated and take place.

Some users of the first edition suggested that more material on environmental law should be included in the text. To this end a table of the most important environmental legislation is presented in Chapter 1, and each of these laws is cited in the text where it is applicable. It was felt that this would be a better approach than writing a separate chapter on environmental law which included case citations and histories. We believe the student can better appreciate the need for legislation when presented with the associated geologic hazard or human activity that led to the need for government intervention.

Of course, the excellent quality and number of illustrations and figures was maintained and even increased in some chapters. First-edition adopters particularly praised how well the art and photos supported the text discussions.

In keeping with the volume and high quality of information available to those with access to modern computers, Appendix V lists World Wide Web (www) sites that can be easily accessed and have a wealth of information on geologic hazards, mineral resources, population problems, and environmental law. The appendix may be doubly useful as the Web can provide new information or data useful to instructors in both their research and lectures.

SUPPLEMENTS

For Instructors

- **Instructor's Manual and Test Bank** includes test questions in multiple-choice, true/false, and essay format. Media resources for each chapter are also listed. Also available on the WESTEST 3.0 computerized testing program.
- **Author's Full-Color slide set** (40) from the authors' collection are available to qualified adopters.
- **Full-Color Transparency Acetates** show approximately 40 key figures and illustrations from the text. Available to qualified adopters.
- **West's Great Ideas for Teaching Geology** offers dozens of ideas, demonstrations, and analogies gathered from teachers around the world.

For Students

- **Student Study Guide,** by Joan Licari (Cerritos College) summarizes the key terms and concepts from each chapter section, followed by a question-and-answer review. Each chapter concludes with a practice test.
- **Study Skills for Science Students,** by Daniel Chiras (University of Colorado–Denver) is designed to accompany any introductory science text. It offers tips on how to improve memory, learn more quickly, get the most out of lectures, prepare for tests, produce top-notch term papers and improve critical thinking skills.
- **Current Issues in Geology—1997 Edition,** edited by Michael McKinney and Robert Tolliver of the University of Tennessee–Knoxville, offer a collection of approximately 70 very current articles to supplement material that students will encounter in their coursework. The articles are collected from a number of general interest and science magazines.

Multimedia Products

- **In-TERRA-Active, West's Physical Geology Interactive CD-ROM** by Philip E. Brown (University of Wisconsin–Madison) and Jeremy Dunning (Indiana University) provides your students with an interactive study environment that enhances class material by providing visual and mental stimulus that static images cannot provide.
- **GeoTutor Software** (Macintosh) by Vicki Harder, serves as an entertaining review program for introductory courses. Categories, points, and time limits make the review more interesting and encourage discussion of the questions and answers. Classroom review is possible with the aid of a projection pad.
- **Interactive Geoscience Tutorials** (Macintosh) by Vicky Harder reinforce text concepts with interactive visual aids and review questions. Available tutorials

include: Introduction to Geology, Igneous Rocks, Mass Wasting, Minerals, Plate Tectonics, and Weathering, Erosion, Rock Cycle, Earthquakes, Geologic Time, Groundwater, Sedimentary Rocks, History of the Universe, Solar system, and Planet.

- **West's Earth Science Video Library** consists of 21 videotapes on topics in geology, oceanography, and meteorology. Adopters can select and exchange from this list. Standard policy.
- **WESTEST™ 3.2 computerized testing**
- **Geology and the Environment: An Internet-Based Resource Guide,** by John Butler (University of Houston), is a disk-based (Mac and IBM) introduction to environmental geology-related sites throughout the Internet, suitable for use with most major web browsers. Most chapters contain exercises involving exploration of specific sites (links to sites are embedded within the exercise). Several appendices offer additional geological background material that supplements the discussions in chapters 2 and 3 of the text.
- **Earth Online,** by Michael Ritter (University of Wisconsin–Stevens Point), is an inexpensive, hands-on Internet guide written for the Internet novice. It provides opportunities for students to participate in Internet homework exercises, lab exercises, or just to explore the vast resources on the Internet. Earth Online provides a tool to get students "up and running" on the Internet while providing guidance and exercises to immediately apply text material, or information students encounter by visiting the Earth Online homepage. The Earth Online home page accessed via (http://www.wadsworth.com/geo) is created and maintained by the text author, and will house exercises, tips, author chats, new links, and constant updates of the exercises and reference sites.

ACKNOWLEDGEMENTS

We gratefully acknowledge the photographs, published works, unpublished information, and field trip guides contributed by the following individuals. These added appreciably to the substance and appearance of the second edition of this book.

Hobart King, *Mansfield University, Pennsylvania*
Albert J. Robb, III, *Mobil Exploration and Producing U.S., Inc., Liberal, Kansas*
Joan Van Velsor, *California Department of Transportation*
Lydia K. Fox, *University of the Pacific*
Bill Kane, *University of the Pacific*
Randall Jibson, *U.S.G.S.*
Edwin Harp, *U.S.G.S.*
Lynn Highland, *U.S.G.S.*
Robert Schuster, *U.S.G.S.*
Pam Irvine, *California Division of Mines and Geology*
Siang Tan, *California Division of Mines and Geology*

Douglas J. Lathwell, *Cornell University*
Jack A. Muncy, *Tennessee Valley Authority*
T. K. Buntzen, *Alaska Department of Natural Resources*
Chuck Meyers, *U.S. Dept. of the Interior, Surface Mining Reclamation and Enforcement*
Ed Nuhfer, *University of Colorado, Denver*
Peter W. Weigand, *California State University, Northridge*
Joe Snowden, *University of Southeastern Missouri*
James L. Shrack, *Arco, Anaconda, Montana*
Marcie Kerner, *Arco, Anaconda, Montana*
John Gamble and Ed Belcher, *Wellington, New Zealand*
Rick Giardino, *Texas A&M University*
Kenneth Ashton, *West Virginia Geological Survey*
Peter Martini, *University of Guelph*
Ward Chesworth, *University of Guelph*

Prior to preparation of the second edition, the following first-edition reviewers offered comments and suggestions that were helpful in planning this revision.

Herbert G. Adams, *California State University, Northridge*
Lisa DuBois, *San Diego State University*
Edward B. Evenson, *Lehigh University*
John Field, *Western Washington University*
Robert B. Furlong, *Wayne State University*
Raymond W. Grant, *Mesa Community College*
Douglas W. Haywick, *University of South Alabama*
Roger D. Hoggan, *Ricks College*
Steve Kenaga, *Grand Valley State University*
Barbara Hill, *Onondaga Community College*
Hobart M. King, *Mansfield University*
Robert Kuhlman, *Montgomery County Community College*
Rita Leafgren, *University of Northern Colorado*
Joan Licari, *Cerritos College*
Larry Mayer, *Miami University*
David McConnell, *University of Akron*
James Neiheisel, *George Mason University*
Geoffrey Seltzer, *Syracuse University*

The following reviewers were kind enough to read and review the first edition, upon which this revision was built. Some were very helpful in pointing out typographic errors or discrepancies that were corrected in subsequent printings.

James L. Baer, *Brigham Young University*
William B. N. Berry, *University of California, Berkeley*
Lynn A. Brant, *University of Northern Iowa*
Don W. Byerly, *University of Tennessee, Knoxville*
Susan M. Cashman, *Humboldt State University*
Robert D. Cody, *Iowa State University*
Mark W. Evans, *Emory University*
Robert B. Furlong, *Wayne State University*
Bryan Gregor, *Wright State University*
Douglas W. Haywick, *University of South Alabama*
Roger D. Hoggan, *Ricks College*
Alan C. Hurt, *San Bernardino Valley College*
Michael Lyle, *Tidewater Community College*
Lawrence Lundgren, *University of Rochester*

Garry McKenzie, *Ohio State University*
Robert Meade, *California State University, Los Angeles*
William N. Mode, *University of Wisconsin–Oshkosh*
Marie Morisawa, *State University of New York at Binghampton*
William J. Neal, *Grand Valley State University*
Michael J. Nelson, *University of Alabama at Birmingham*
June A. Oberdorfer, *San Jose State University*
Charles Rovey, *Southwest Missouri State University*
Susan C. Slaymaker, *California State University, Sacramento*
Frederick M. Soster, *DePauw University*

Special thanks are due to Douglas W. Haywick, University of South Alabama, and Hobart King, Mansfield University, who worked closely with us on the second edition. Many of our colleagues at the University of Southern California, particularly Doug Hammond, Steve Lund, and Greg Davis, gave generously of their knowledge when we fell short or our memories lapsed. Carole Bailey Reagle, the copy editor for both editions, performed in her usual magnificent way, at times creating a "silk purse" where, before she unleashed her pen, there was something much less. Steve Yaeger and Jayne Lindesmith, the Production Editors, kept order in the presence of chaos and we think their efforts yielded a very attractive textbook. Our Editor, Rick Mixter was the driving force behind the second edition and we thank him for his faith in the project and for committing the resources of West Educational Publishing Company. We owe special thanks to Keith Dodson, the Developmental Editor and Lauren Fogel for working with us throughout the project. We are especially grateful to Keith for his careful attention to detail and his many worthy suggestions.

And finally to our wives, Faye Pipkin and Patricia Trent—our traveling companions, pick bearers, and best friends—our loving thanks. Although these great ladies have been "widowed" for long spells between 1993 and 1996, they gave us nothing but encouragement during the long workdays on the project.

BERNARD W. PIPKIN
Palos Verdes Estates, California

D. D. TRENT
Glendora, California

Every 96 hours, the human population grows by one million people. We create the equivalent of a new Germany every month, a new Mexico every year . . . and the equivalent of a new state of Maine every four days.

GEORGE MOFFET, INSTITUTE FOR GLOBAL ETHICS (1930–)

Geology is the study of the earth, and **environmental geology** is the subdiscipline that focuses on the relationship between humans and their earth environment. A moment's reflection reveals that environmental geology encompasses many essentials of our daily lives: the resources used to manufacture our material goods, the water we drink, the soil that grows our food, and the very ground on which we

OPENING PHOTO
The Statue of Liberty with New York City and Manhattan Island in the background. The towers of the World Trade Center can be seen in the center of the photograph. Beneath the skyscrapers is a maze of subway and sewer tunnels. This is possible because the city is built upon competent metamorphic bedrock that is capable of supporting enormous loads. New York City has one of the largest concentrations of people in the world and, as a result, is beset with problems. Many of the problems are geologic in nature, such as safe disposal of solid and liquid wastes, provision of an adequate supply of fresh water, and mitigation of pollution of the Hudson River Estuary and Long Island Sound.

stand and travel. Many of our life choices have geological ramifications. The decision of where to live, for example, should involve geological questions. Is the homesite stable against landslides? What are the risks of disastrous floods? Could an earthquake damage this dwelling?

Yes, environmental geology is the study of human interaction with the land, with all its sociological, economic, and political ramifications. It is the application of the science of geology to solve the problems that arise from the complex interactions of the earth's five systems: water, air, soil, solid earth, and life. As such, environmental geology encompasses all of the traditional branches of geological study: mineralogy and economic geology (mineral resources), petrology (the study of rocks), engineering geology (the building of structures on land), sedimentology (the origin and interpretation of sediments), structural geology (how rocks deform), geophysics and geochemistry (the physics and chemistry of the earth), and hydrology (the study of water). ◉ Figure 1.1 illustrates the components of environmental geology. Environmental geology applies all that is known about geology to the problems of human existence, particularly to the problems of geologic hazards and how the earth's resources can best be utilized.

Although the world's peoples have been applying the general principles of environmental geology without thinking very much about them for several thousand years, the designation "environmental" is a very recent addition to the formal study of geology. In Europe and elsewhere, many old monasteries, cathedrals, castles, and fortresses are sited on the

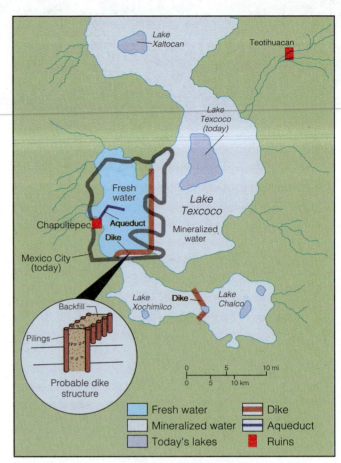

◉ **FIGURE 1.2 An ancient Aztec solution to ground water problems. About 1000 B.C. ancient Lake Texcoco was drained, leaving Lakes Xochimilco and Chalco as remnants in the Mexico City area.**

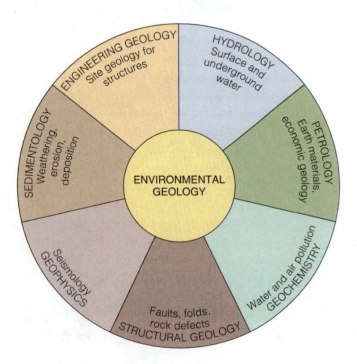

◉ **FIGURE 1.1 The relationship between the geological sciences and environmental geology. Environmental geology incorporates many skills and specialties of the discipline of geology.**

soundest locations for miles around. Until about 1000 B.C., large Lake Texcoco occupied the Valley of Mexico, the home of the Aztec civilization. The Aztecs drained the lake in order to improve foundation conditions and placed impermeable walls into the ground to impede sulfur-contaminated water from mixing with fresh underground water. The dry bed of ancient Lake Texcoco is now occupied by Mexico City, one of the largest cities in the world (◉ Figure 1.2). In ancient Persia, present-day Iran, underground water was collected for daily use by digging tunnels, known as *kanats*, many kilometers into hillsides (◉ Figure 1.3). Although the practice started in Persia, it spread rapidly, and by about 500 B.C. kanats were being used in Afghanistan, China, and Egypt. Many of them are still operative, and ten new ones were constructed in the People's Republic of China in 1949. Even years ago, however, some unfavorable or hazardous geological conditions were ignored or were not recognized. This explains, for example, why some towers in Italy now tilt due to the underlying soil conditions (see Chapter 8). Many ancient structures that were built in concert with the environment are still serving human needs today; others, built to overwhelm or to command the environment, are in ruins.

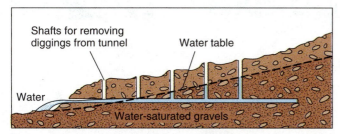

● FIGURE 1.3 An ancient water-collecting tunnel, or *kanat,* driven into a gravel deposit. Kanats are still used today.

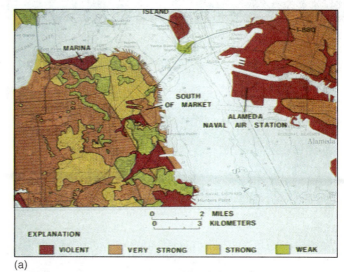

EXPLANATION

| VIOLENT | VERY STRONG | STRONG | WEAK |

(a)

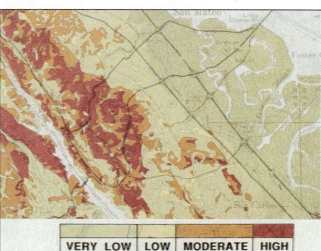

| VERY LOW | LOW | MODERATE | HIGH |

(b)

● FIGURE 1.4 Hazard maps of the San Francisco Bay region. *(a)* Maximum predicted ground shaking due to strong earthquake motion. Ground shaking is a function of the geologic foundation at the site, such as soft bay muds in the Marina district (violent shaking) or hard bedrock on Nob Hill (weak shaking). *(b)* Risk of earthquake-induced landslides along the San Andreas fault (dotted line) west of San Francisco Bay.

A major goal of those who work in environmental geology is to identify adverse geologic conditions in advance by collecting surface data from rock outcrops and subsurface data from drilled boreholes and by using geophysical techniques (see Case Study 1.1). This need to identify potential geological problems was recognized in the 1950s in the United States because of burgeoning urban growth. Marginally suitable lands that are subject to landslides, earthquake damage, and flooding were being developed for human use. In addition, the demand for resources increased as cities grew, which required advance planning for local sourcing of building materials, providing adequate drinking water of good quality, and deciding what would be the best use for given parcels of land. Most of the planning problems then, as now, ultimately became engineering or geological ones. Not surprisingly, many city planners are civil engineers, and these engineers were among the first to recognize the need for input from professional geologists. Thus the field of engineering geology expanded into "urban geology," and as communities became larger and more complex, "environmental geology" was born. Although the original definition of **engineering geology** was restricted to "the application of geological data and principles to the design and construction of engineering works," it has been expanded to include "the development, protection, and remediation (cleanup) of underground water resources." Such environmental applications of geology as ridding waters and soils of toxic substances remaining from past industrial and commercial activities are now major activities of geological firms that once dealt only in heavy construction and land development.

Multipurpose maps of areas being considered for development are essential to effective planning, and these maps must be available to planners long before the bulldozers arrive. Maps (such as the ones in ● Figure 1.4) can be generated that show areas of unstable ground, locations of active earthquake faults, areas with high flood potential, the depth to underground water of a particular quality, and possible mineral resources. Such maps can also indicate areas that are unsuitable for residential or commercial development but which could be used for parks and recreation. The generation of a multipurpose map usually begins with the drawing of a geologic map that shows the distribution of rocks and their spatial relationships at the surface and below, and

drainage patterns. The National Geologic Mapping Act of 1992 authorized the federal government to spend $184 million over four years for geologic mapping and support activities. The act recognizes that "geologic mapping is a fundamental part of most earth-science studies, with applications in such fields as applied-geology investigations; water-resource planning; and land-use planning for waste dumps, coastal housing development, and transportation." Geologic maps are used extensively in the planning process and beyond. They can be adapted to many special requirements of a given project or a long-term plan.

Nature is a good teacher, and many significant advances in environmental geology have resulted from

CASE STUDY 1.1

A Day in the Life of an Environmental Geologist

I conjure you, my brethren, to remain faithful to earth . . .

NIETZSCHE

The public perception of geologists has been that of rugged outdoor types who do field work in beautiful natural surroundings, wear Pendleton shirts, carry rock hammers, look for fossils, and dig holes in the ground for mining and oil companies. True, many professional geologists' work is in remote field areas and connected with extracting natural resources such as minerals and oil. However, most of the new work for geologists today is in urban settings and is related to protecting the existing environment or correcting past transgressions. The majority of non-teaching geologists in the United States and Canada carry out a broad spectrum of activities aimed at helping people to develop land and exploit resources in a sensible, environmentally safe manner. Many others study geologic hazards, such as earthquakes, landslides, and volcanic activity, processes that take more than 100,000 lives each year and wrench billions of dollars from the world economy. Environmental geologic studies in areas of high geologic-hazard potential establish bases for assessing hazard risks. This work significantly reduces deaths and economic losses. Similar studies assess the pollution potential of particular actions or projects.

Environmental geologists work with planners and decision makers in long-range land-use planning. They gather field data in order to make sound recommendations that are useful in the planning process. They collect the data by mapping rock strata, digging trenches across active earthquake faults, drilling exploratory holes to collect samples, and in some cases, drilling boreholes large enough to permit them to go down into the

(a)

(b)

ground for direct observations. In addition, geologists may make geophysical surveys using sound-wave (shallow seismic) transmission, measures of electrical conductivity of the ground, and even radar-wave transmission to locate buried objects or underground caverns. Typical projects of environmental geologists include:

- Water resource development and protection projects (hydrogeology), including locating and developing underground water resources, remediation of contaminated underground water supplies, and sanitizing contaminated soils that have potential for polluting underground water supplies. Such projects include studies of surface waters and commonly require the services of a geochemist.
- Waste disposal projects for solid wastes (landfills) and hazardous wastes, both solid and liquid.
- Hill-slope stability studies for proposed structures of all types, including dams, power plants, tunnels, and private dwellings (◉ Figure 1, parts a and b).
- Seismic and volcanic risk analysis (Figure 1, part c).
- Environmental geophysicists are called upon to locate and map such things as toxic or radioactive materials that have been buried in large steel drums, underground openings such as caves and caverns in limestone terrane, underground water that has been contaminated by sea-water intrusion or by dissolved ionic contaminants, and buried or concealed faults.

(c)

◉ FIGURE 1 Environmental geologists at work. (a) Split-tube core samplers help geologists to test soil for hydro-carbon contamination, run percolation tests for seepage-line evaluation, and perform strength tests for foundation design. (b) Geologists use tilt-up drilling rigs on hillsides and in close quarters such as people's back-yards. At this residence a landslide is encroaching upon the swimming pool. (c) Geologists carefully log the sidewalls of a trench across the San Andreas fault near Wrightwood, California, to find and date the youngest rupture (an 1857 earthquake) and then older breaks. The data allow earthquake recurrence intervals for the fault to be determined.

In the past, such work fell strictly within the realm of engineering geologists. Today, however, the distinction between geologic work on engineered structures and that that is strictly environmental has all but disappeared. In fact, the Association of Engineering Geologists has recently renamed their journal *The Bulletin of Environmental and Engineering Geosciences*. By whatever label, all geologists must have proper academic and on-the-job training. Many states now require that environmental and engineering geologists who sign consulting reports be licensed, and educational and experience requirements for licensing are stringent.

(a)

(b)

(c)

⊙ FIGURE 1.5 *(a)* A hillside home in Studio City, California, was separated from its foundation during the Northridge earthquake and rolled onto its side, leaving *(b)* an automobile perched precariously on the stable sidewalk above. *(c)* Fires after the Kobe, Japan, earthquake in 1995 destroyed the equivalent of 70 U.S. city blocks.

studying environmental geology "case histories"—serious attempts to describe and analyze events that have occurred. For example, recent geological disasters have forced reevaluation of what had been considered state-of-the-art earthquake-resistant design of freeways, high-rise buildings, and low-rise buildings such as parking structures. They were the earthquakes in San Francisco (1989) and Northridge, California (1994), and Kobe, Japan (1995), (⊙ Figure 1.5). Although none of these is considered a *great* earthquake as defined by the Richter magnitude scale (> magnitude 8), the one at Kobe (7.2 on the Japanese earthquake scale) resulted in the largest monetary loss for a natural disaster, $100 billion. This is much greater than the estimated $20 billion spent to repair the damage of Hurricane Andrew (1992) in Florida and the adjacent Gulf Coast region. The two California earthquakes are not far behind Hurricane Andrew in their overall dollar losses. Mount Pinatubo in the Philippines is another example. It was considered a dormant volcano when the U.S. Air Force built the immense Clark Air

Force Base nearby. Pinatubo's eruption in 1991 forced this facility to be closed and abandoned (⊙ Figure 1.6 and Chapter 5). Although people normally attribute great loss of life and huge economic losses to natural disasters, we must recognize that societal problems have also had catastrophic impacts. ■ Table 1.1 makes this point.

Other advances in environmental geology are developed in response to problems that become apparent less catastrophically. U.S. laws now require underground gasoline-storage tanks to be examined for leaks. If any are found, nearby underground water supplies must be examined for possible contamination. New technologies for sanitizing the gasoline-contaminated water and soils discovered by these mandatory inspections have resulted (see Case Study 1.1). Disposal of human-generated solid waste is a major problem, which requires identifying geologically acceptable disposal sites—including sites that are safe for sequestering hazardous substances. (The toxic-waste tragedy of Love Canal and its remediation are discussed in Chapter 15.) Extracting water from beneath the land surface for human use may seem

■ TABLE 1.1 Comparison of Deaths Due to Geologic and Societal Catastrophes

CATASTROPHE	DEATHS
Wars versus Earthquakes (Chapter 4)	
U.S. casualties in World War II	292,000
Atomic bomb, Hiroshima, Japan, 1945	80,000–200,000
T'ang-shan, China, 1976	655,000
Kobe, Japan, 1995	5,200
Murders versus Volcanic Eruptions (Chapter 5)	
Murders in the United States, 1995	20,045
Mt. Pelée, Martinique, 1902	30,000–40,000
Armero, Colombia, 1985	20,000–22,000
Plagues versus Landslide Events (Chapter 7)	
The Black Death, Europe, 14th century	25,000,000
World plague, 1894–1914	10,000,000+
U.S. AIDS deaths to 1 January 1995	267,500
Kansu, China, 1920	200,000
Mameyes, Puerto Rico, 1985 (most disastrous U.S. landslide)	129
Atrocity versus Flood Events (Chapter 9)	
The Holocaust, Europe, 1939–1945	6,000,000
Yellow River, China, 1887	900,000–6,000,000
Yangtze River, China, 1931	3,700,000

SOURCES: Various, including Edward B. Nuhfer, Richard J. Proctor, and Paul H. Moser, *The Citizen's Guide to Geologic Hazards* (Arvada, Colo.: American Institute of Professional Geologists, 1993); Center for Disease Control and Prevention, *HIV-AIDS Surveillance Report, 1996;* and *Encyclopaedia Britannica.*

● FIGURE 1.6 Volcanic ash rises several kilometers above Mount Pinatubo, the Philippine Islands, June 1992.

harmless enough, but in Venice, Italy, this is causing the land beneath the city to subside. This sinking causes periodic flooding, with serious consequences for the residents and their largest source of revenue, tourism. Similarly, subsidence due to extraction of underground water has altered the flow of surface water and storm drains over large areas of the U.S. Gulf Coast and the Great Valley of California.

CONSIDER THIS . . .

Do societal hazards or geological ones present greater risks to your health and well being? What is the greatest risk to life and limb? (Hint: Consider societal hazards not listed in Table 1.1.) What factors influence your amount of exposure to geological hazards?

At the core of these and many other environmental problems is a burgeoning world population of more than 5.76 billion people. Every one of these persons wishes to share in the good life, but this growing population exists on a planet with limited air, land, water, and mineral resources.

 ## TOO MANY PEOPLE

The concept of **carrying capacity**—the idea that a given amount of land or other critical resource can support a limited number of people or animals—has intrigued scientists, economists, and philosophers for centuries. Thomas Malthus postulated in 1812 that populations grow geometrically, whereas food supplies increase only linearly, which leads to a gap between demand (population) and supply (food) that eventually results in famine. Human life is inextricably linked to resource utilization, environmental quality, and economic well-being. Thus, a good starting point for any environmental study is to learn how populations grow, a study referred to as **population dynamics**.

World population has increased dramatically this century. In 1900, the population of the earth was estimated at 1.6 billion people. By 1995 this figure had more than tripled, to 5.76 billion. Stated another way, it took about 10,000 generations for humankind to reach the first two billion in population, but only one generation to add the most recent two billion. Pogo, a popular comic strip character of the 1960s, once said, "Yep, son, we have met the enemy, and he is us," and this applies to the global population dilemma. Much of the recent dramatic increase in population can be explained by a reduction in the death rate due to overall better health globally.

Exponential Growth and the Population Bomb

Growth and increases are familiar processes to all of us; we see them daily in our national debt, taxes, rents, food costs,

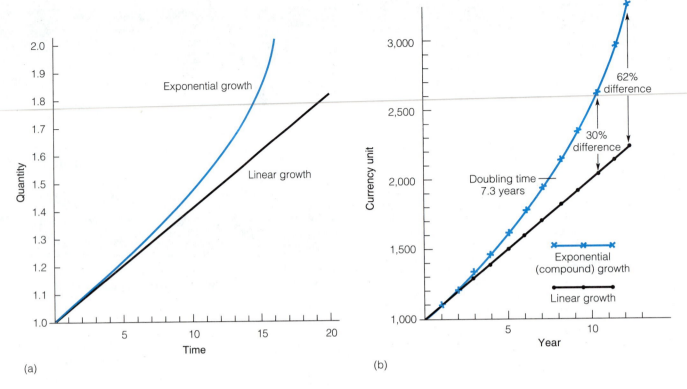

(a)

(b)

⦿ FIGURE 1.7 Linear versus exponential growth. *(a)* The principle. *(b)* The growth of a savings account at 10 percent interest per year. Although the text example uses the dollar as the currency unit, the unit could be pounds, yen, or rupees—and the principle is the same for population growth. Note that the doubling time for exponential growth at a 10 percent annual rate is about 7 years.

and so on. Increases that occur in nearly equal increments are described as *linear*. Examples include your electric bill variation with usage of electricity and growth of the grass in your lawn between weekly cuttings. Some other things grow by increments that become larger with time, and this kind of growth is described as *geometric* or *exponential*. Examples include the expansion of weeds in a neglected garden and increases in the national debt. The growth of a savings account also is exponential, assuming no withdrawals. A thousand dollars invested at 10 percent annual interest rate earns $100 the first year and becomes $1,100. During the second year $1,100 at 10 percent earns $110, and the value of the account grows to $1,210, and so on in each succeeding year. Thus, interest is earned on interest (compounded) so that at the end of ten years, for instance, the account balance is not just $1,000 *plus* 10 percent interest *times* 10 years, or $2,000, but $2,593.74.

It may seem incongruous that to our well-being and way of life the most important increases are not interest rates or the national debt, but the explosive growth of world population and natural-resource consumption. The graphs of ⦿ Figure 1.7 contrast linear (straight) and exponential (J-shaped) growth curves. Although a population growth rate of 2 percent per year or a similar increase in the consumption of a particular natural resource may seem trivial, when we realize that this growth is exponential, our assessment changes. Valid concerns for the future of the human

condition arise. As world population increases exponentially, natural resources are rapidly being depleted; even if geological exploration can double or triple known oil or mineral reserves, it will make little difference in the useful lifetime of those resources (see Case Study 1.2).

Population growth rate is determined by subtracting a population's death rate from its birthrate, with death and birth rates expressed as deaths and live births per 1,000 people per year, respectively. For example, a birthrate of 20 *minus* a death rate of 10 would yield a growth rate of 10 per 1,000 persons, or 1.0 percent. World growth rates were low at the turn of the century (about 0.8%), and they have fluctuated with time, reaching a high of about 2.06 percent during the 1960s; they have slowly declined since the 1960s to 1.68 during the 1990–95 period. At first glance you might think that a growth rate of 1 percent would yield a population **doubling time** of 100 years (since 1% per year × 100 years = 100%). This is not the case, however. Populations grow like savings accounts with compounded interest. Just as you can earn interest on interest, the people that are added produce more people, resulting in a shorter doubling time. A very close approximation of doubling time can be obtained by the "rule of 70"; that is,

Population doubling time = 70 ÷ growth rate (%)

Thus, the doubling time for a 1 percent growth rate would be 70 years; that for a 2 percent growth rate (close to the

CASE STUDY 1.2

The United Nations Conference on Population—Ten Billion Big Macs, Please!

The United Nations Conference on Population convened in Cairo, Egypt, late in 1994. After extensive debate the four main blocs at the conference—Muslim, Roman Catholic, emerging nations, and developed Western nations—adopted a carefully worded consensus "program." The thrust of the program is to hold down population growth by family planning and by giving women the power to make decisions about their own lives. The program states that women have the right to decide "freely and *responsibly*" (author's emphasis) how many children they have and when they have them. The program is not binding and will probably be interpreted differently in different countries. The Vatican and many Muslim nations were unhappy with the section on reproductive rights because they believe those rights belong to couples, rather than individuals, and because they object to abortion.

According to a United Nations spokesperson, if family planning succeeds, the world population will rise to 7.27 billion by the year 2015. If it fails, the population will grow to 7.92 billion by 2015 and explode to 12.5 billion by 2050.

However, the axiom that global population control is essential because world population is about to exceed or has exceeded earth's carrying capacity is being challenged by, of all groups, the Economic Research Service (ERS) of the U.S. Department of Agriculture. ERS has gone on record saying that the world is not near its limit of food-production capacity and that the earth could easily support 10 billion people. Food production is not limited by the amount of soil and water, they say, but by human ingenuity, citing corn yields in Iowa that are almost five times the world average. These USDA experts argue that crop yields can increase at the same rate as population if sufficient money is allocated to research and development of new agricultural technologies. Acknowledging that there are 700 million undernourished

people in the world, they allege that the problem is not that the world does not produce enough food; rather, it is that poor nations cannot afford to buy enough food.

Some of us will be able to monitor such issues, because it appears that population momentum, even with the most successful birth-control efforts, will generate 10 billion people by the year 2050. This dramatic growth is driven by increases in longevity and by the great increase in the number of women of child-bearing age in the next few decades (■ Table 1).

■ **TABLE 1** Estimated World Population by Age Group, 1995 and 2050

AGE GROUP	POPULATION, BILLIONS	
	1995	*2050*
70+	.221	.857
60–69	.314	.941
50–59	.425	1.18
40–49	.611	1.25
30–39	.805	1.31
20–29★	.999	1.34
10–19★	1.080	1.32
0–9★	1.270	1.34
TOTAL	5.7 billion	9.5 billion

★Future parents
SOURCE: *U.S. News and World Report,* 12 September 1994. Data from World Bank, World Population Projections.

present world growth rate) would be 35 years, and so forth. ■ Table 1.2 provides some examples. The global population growth and the average annual growth rate for 1950 to 2005 are shown in ■ Table 1.3. Although some progress has been made in slowing the growth rate, the number of people added each year continues to grow larger. The world population is projected to grow by 959 million in the decade 1990–2000, the largest increment ever for a single decade.

U.S. Population Growth and Immigration

In 1995 an estimated 263 million people were living in the United States. They generated 15.9 live births per thousand people and a death rate of 9.0/1,000, for a growth rate of 6.9/1,000, or 0.69 percent. At first glance these figures appear to be real progress toward **zero population growth,**

a condition that exists when, on average, as many people die as are born each year. However, even at this low rate the population would double in 100 years, without accounting for immigration—which adds significantly to the U.S. population, particularly in Texas, Florida, and California. From the 1920s through the 1960s about 200,000 immigrants came to the United States each year. Since then this number has grown dramatically for a variety of political and economic reasons. By 1993 there were 810,000 legal immigrants *plus* an estimated 200,000 illegal ones, for a total of about a million new persons—five times the annual immigration of 30 years earlier. From the year 2000 through the end of the next century, it is estimated that 90 percent of the population growth in the United States will be immigrants and their progeny.

Another factor in population dynamics is the **total fertility rate (TFR),** the average number of children born to

■ TABLE 1.2 How Populations Grow	
GROWTH RATE, %	DOUBLING TIME, YEARS★
1.0	70.0
2.0	35.0
3.0	23.3
4.0	17.5
5.0	14.0
6.0	11.7
7.0	10.0

★Calculated by using the formula 70 ÷ growth rate (%), which yields a close approximation up to a growth rate of 10%.

■ TABLE 1.3 World Population and Growth Rate, 1950–2005		
YEAR	POPULATION, BILLIONS	GROWTH RATE, %★
1950	2.52	
1955	2.75	1.77
1960	3.03	1.95
1965	3.34	1.99
1970	3.77	1.90
1975	4.08	1.84
1980	4.45	1.81
1985	4.85	1.75
1990	5.30	1.70
1995	5.76	1.68
2000	6.26	1.56
2005	6.65	1.43

★Average annual rate for the previous five-year period.

SOURCE: Estimates in United Nations, *Demographics Yearbooks* for 1985 and 1990 and in *World Resources, 1994–95* (New York: Oxford, University Press).

women of child-bearing age. A TFR of 2.0 would be ideal today, because it would stabilize the population. This rate is the **replacement rate,** the rate at which just enough babies are born to replace their parents. If the two-child family were to become a reality in the United States, the TFR would drop below 2.0, however, because not all women of child-bearing age have children, particularly in developed countries, where women have other options for their lives. Curve A (◉Figure 1.8) shows where the U.S. population is headed at the present fertility rate of 2.1 and high level of immigration. Curve B shows what could occur if we chose to adopt the two-child-family ideal and to hold immigration to levels that prevailed before 1980. Population would actually decrease, and people might need to be encouraged to have more children. Although the practices that would generate curve B are probably not realistic for this country, the curve shows that population can be stabilized by con-

trolling family size and immigration, both of which may eventually be forced upon us because of population pressures. It should be noted that immigration is not a problem for the earth as a whole—yet!

World Population Growth and the Environment

In 1990 about 90 million people were added to a world population of 5.2 billion people. If this rate of growth continues (1.68 percent in 1995, down from 1.81 percent in 1980), our planet will need to support 10 billion people by the year 2030. Slower growth has been reported in all parts of the world except Africa. Growth-rate reduction measures, largely family planning, are being implemented or encouraged throughout the world. The growth rate of China, whose population has broken the billion mark, has decreased more than 30 percent, but it was still 1.42 percent for 1990–95.

Due to variations in growth-rate momentum, world population projections for the middle of the next century range from a low of 9 to 10 billion to a high of 15 billion people (see ◉Figure 1.9). The population growth curve in the figure is the typical **J-curve** for exponential growth of any type. In such curves, growth is slow at first (the flat part of the curve) and then becomes rapid (the steep part of the curve). As the population has more than tripled this century and demanded more from the earth, 20 percent of arable lands have suffered serious topsoil erosion, 20 percent of tropical rain forests have been obliterated, and many thousands of plant and animal species have disappeared. As population increases, marginal lands are tilled because of the

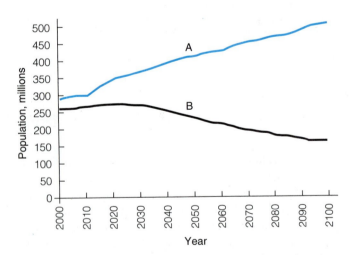

◉ FIGURE 1.8 Two U.S. population projections to the year 2100. Curve A projects population at the current levels of fertility and immigration to nearly a half-billion people by 2100. Curve B shows what could be achieved with the two-child family and immigration at levels that prevailed before 1980.

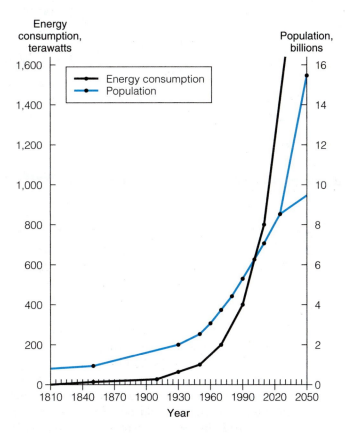

Energy consumption, terawatts

Population, billions

— Energy consumption
— Population

Year

⊙ FIGURE 1.9 Projected world population and energy consumption to the year 2050. The two projections for population are based on minimum and maximum birthrate estimates. A terawatt is a thousand billion (trillion) watts, or a billion kilowatts.

need for more food, and thus, poor agricultural practices become exaggerated. In addition, between 1940 and 1990, atmospheric carbon dioxide (CO_2) levels increased 13 percent, which may have implications for global warming, and the protective stratospheric ozone layer decreased 2 percent worldwide (and more than that over Antarctica; see Chapter 12). None of these problems is insurmountable, but human ingenuity will certainly be taxed to overcome the challenges they present.

Overpopulation Scenario One: The Sky Is Falling!

We know that catastrophic extinctions have occurred at particular geologic periods in the earth's history. Fossil evidence indicates that during an interval of catastrophic extinction in the late stages of the Cretaceous Period, the extinction rate averaged one species per 25 years. In contrast, it is estimated that more than 1,700 species were pushed to extinction in 1991 alone, either from loss of habitat or human predation (that's 4.6 species per day). Granted that our biological inventory of the Cretaceous is fragmental and that modern estimates are biased by our better

knowledge of present-day species, but such estimates clearly indicate the malady: too many people and too few resources. We must face the facts that the present world population is stressing the environment and that the carrying capacity of the earth has either been reached or is fast approaching. Furthermore, students of population dynamics, economics, and environmental science generally agree that the earth cannot sustain the 9–10 billion people projected for 2050, assuming the lower future growth rates. Drastic action to stem population growth must be implemented now, for if nothing is done, future population reductions will occur by increased fatalities from natural disasters, mass starvation, ecological degradation, and social catastrophes.

The population problem has been attributed to what is loosely called the "north–south disparity," the economic disparity between developed nations, largely in the Northern Hemisphere, and lesser developed or emerging nations, many of which are in the Southern Hemisphere—India being an important exception. Increasingly, the world's people are being divided into two societies, one rich and one poor. Twenty percent of the world's people live in developed countries, and they use 10 times the energy and produce 20 times the waste of the eighty percent who live in the underdeveloped or emerging nations.

Regardless of life-styles, the earth is overpopulated both north and south. The burning of oil, gas, and coal—the so-called **fossil fuels**—doubled every 20 years between 1890 and 1970, mostly to support the enhanced quality of life in the north (⊙ Figure 1.10). Burning fossil fuels depletes finite resources and creates problems that adversely affect air and water quality, most importantly by producing the "greenhouse" gas carbon dioxide. If we are in fact at the threshold of human-induced global warming—not all scientists agree this is so—sea level could rise and inundate coastal areas, including all or parts of many large cities. If we desire a healthy and sustainable habitat for future generations, we must stop uncontrolled growth, runaway resource consumption, and degradation of the environment. This will require many changes in our way of life that most of us "northerners" are likely to perceive as sacrifices.

CONSIDER THIS . . .

Thomas Malthus (1776–1834) is best known for his theory that population growth will always tend to outrun food supply, and that famine, plagues, and even wars are the inevitable results of uncontrolled growth. Why can't food supplies grow exponentially as do populations, thereby preserving human existence as we know it?

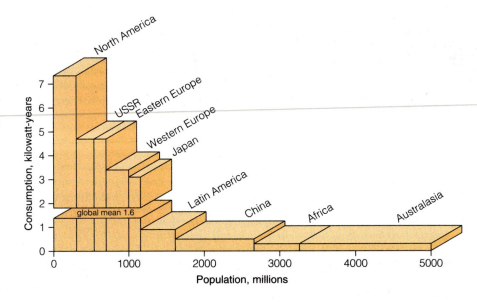

● FIGURE 1.10 Per capita energy consumption in kilowatt-years (kwy) and population by geographic or political region for 1987. Note that the industrialized nations consumed 2–4 times the global per capita mean and 10–20 times that of such areas as Australasia and Africa.

Overpopulation Scenario Two: The Gaia Hypothesis

British physicist James Lovelock and his work with atmospheric processes and chemistry are respected in the scientific community. He developed an apparatus for measuring trace (parts per trillion) amounts of atmospheric gases and was the first person to establish that chlorofluorocarbons (CFCs) were persistent and long-lived in the atmosphere. This led to speculations about the impact of CFCs on the ozone layer, which has become an area of intense scientific research and debate.

The name of Gaia, the Greek goddess of the earth, has been applied by Lovelock and U.S. biologist Lynn Margulis to their hypothesis that the earth is a superorganism whose environment is controlled by the plants and animals that inhabit it, rather than vice-versa. According to this hypothesis, the planet is "self-adjusting," so to speak. The superorganism regulates its environment with a complex system of mechanisms and buffers, just as an animal adjusts to varying temperatures, chemistry, and other environmental factors. For example, carbon dioxide, CO_2, a gas that is required for photosynthesis and therefore all of life, also acts as an insulating blanket in that it holds the sun's heat close to the earth. Terrestrial and marine plants act as regulators that "pump out" excess CO_2 from the atmosphere when great amounts are added by volcanic activity, which, if it were to accumulate, would smother humans and other animals. Were it not for the development early on in geologic time of bacteria and algae, which "pump" carbon dioxide from the air, the earth's atmosphere would have evolved much differently and would be poisonous to life as we know it—much like the atmospheres of Mars and Venus.

According to the Gaia hypothesis, the last ice age ended abruptly when atmospheric CO_2 doubled due to a sudden failure of the "pumps," not by the slow processes of geochemistry as invoked by conventional scientific theory. The amount of carbon dioxide being added to the atmosphere by humans today is comparable to the natural rise that terminated the last ice age and resulted in global warming. (■ Table 1.4 presents 1989 estimates of atmospheric carbon emissions for selected countries.) Hypothetically, the earth, being responsive to change, should be able to absorb the excess carbon dioxide in the oceans, soils, and plants—if it is left alone. Unfortunately, it may be that the natural system is being prevented from establishing a new, tolerable equilibrium, mostly because deforestation is removing the plant life that is necessary to counteract the change. Similar arguments are made for other, less crucial constituents of the

■ TABLE 1.4 Estimated Annual Atmospheric Carbon Emissions from Industrial Processes; Selected Countries, 1989

Country	ESTIMATED CARBON EMISSION	
	Total Amount, millions of tons	Tons per Capita
U.S.A.	4,869	19.7
Canada	456	17.3
Australia	257	15.5
Soviet Union	3,804	13.3
Japan	1,040	8.5
China	2,389	2.2
Brazil	207	1.4
India	652	0.8
Zaire	3.8	0.1

SOURCE: World Resources Institute in collaboration with the United Nations Environment Programme and the United Nations Development Programme, *World Resources 1992–1993* (New York: Oxford University Press, 1992).

atmosphere that have remained constant for long periods of geologic time.

The Gaia hypothesis is appealing because it offers a solution for environmental problems. However, even its strongest supporters admit that by sometime between 2050 and 2100, the amount of carbon dioxide in the atmosphere will be doubled. According to Lovelock, "It is a near certainty that the new state will be less favorable for humans than the one we enjoy now."

 ## TOWARD A SUSTAINABLE SOCIETY

A **sustainable society** is one that satisfies its needs without jeopardizing the needs of future generations. The current generation *should* strive to pass along its legacy of food, fuel, clean air and water, and mineral resources for material needs to the generations of the twenty-first century. We know what needs to be done to accomplish this, but knowing what to do and implementing a long-range plan for carrying it out are entirely different things. Some optimism is found in the birthrates of the United States and Western Europe; they are approaching the replacement rate. Nonetheless, these populations are still growing because of the momentum of population growth and—in the United States—because of net immigration. Also encouraging are indications of changing attitudes about fertility among people in less-developed countries such as India, Bangladesh, Indonesia, Egypt, and China. A promising development in industrialized nations is the growing respect for basing an economy on steady-state conditions, rather than on continual growth. The conceptual ideal is sometimes referred to as the *spaceship economy*, since all resources must be conserved, shared, and recycled in space travel. This, together with population control, could lead in time to a sustainable society (see Case Study 1.3).

Do efforts toward a sustainable society threaten "the American way of life"? Alma Paty of the American Mining Congress says *yes*. With only 5 percent of the world's population, the United States has been consuming more than a third of its raw materials. Our high standard of living has been supporting two cars, a nice big home in the "burbs," beryllium–copper golf clubs, personal computers, and odor-killing cat litter. We have gizmos galore, and they all require the use of mined raw materials. Unfortunately, most of us seem to have the mentality that cars and microwaves come from factories, not from mines, and that production and consumption can continue at their present levels forever. The truth, Paty points out, is that resources are limited, and sustainable development requires a life-style of "sophisticated modesty," characterized by more use of public transportation, smaller houses, and fewer gadgets and adult toys.

Energy and Global Warming

After population growth, global warming is potentially the most serious long-term environmental problem facing

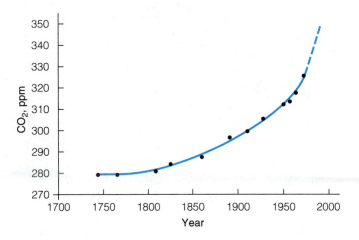

⊙ **FIGURE 1.11** Estimated atmospheric carbon dioxide changes from 1700 to 1980 based on data from air bubbles trapped in Antarctic ice cores and on atmospheric measurements.

humankind. It is closely associated with the increase in atmospheric carbon dioxide produced by the burning of coal, oil, and natural gas (⊙ Figure 1.11). Other greenhouse gases, such as methane, nitrous oxide, ozone, and chlorofluorocarbons (CFCs), are also increasing in the atmosphere, and this, too, is closely associated with population growth. It is estimated that these gases may contribute to about half of the global warming projected for the next century.

Analysis of air bubbles trapped in deep cores of glacial ice shows that CO_2 in the atmosphere has increased 25 percent since the eighteenth century. Calculations suggest that by reducing world CO_2 emissions from the 1992 level of 22 billion tons/year to 8 billion tons/year, the atmosphere (and possibly the climate?) would stabilize at acceptable CO_2 levels. In the developed nations this would require an 85 percent reduction per capita in consumption of fossil fuels, which translates to the personal sacrifice of conventional fossil-fuel-driven automobiles and other machines. In the early 1990s 500 million motor vehicles were registered in the world—each consuming an average of 2 gallons of gasoline per day—and that number was expected to quadruple by 2025. Thus, alternative fuels *must* be developed. Perhaps these will be liquid fuels derived from coal, biomass (e.g., alcohol), or solar energy.

As developing nations raise their standards of living, their energy consumption will also increase, and competition between the developed and the developing nations for energy supplies will increase. This competition will surely increase as the world's coal and oil reserves decrease. In the decade 1982–1992 energy consumption increased more than 50 percent in Asia, 40 percent in Africa, and 30 percent in Latin America, compared to increases of 8 percent or less in the United States and Europe. Per capita oil consumption actually decreased 16 percent during the 1980s in the United States. The developing nations were using 30 percent of the world's energy in the early 1990s, a usage that

Earth Day 1995—Assessing 25 Years of "Thinking Green"

In April 1995 the United States celebrated the 25th anniversary of the first Earth Day with much fanfare. Looking back at the achievements of those years, all that fuss seems to have been warranted.

During the 1960s a relatively small group of environmentalists, conservationists, and scientists, all widely regarded at the time as silly do-gooders and "tree huggers," began raising the public awareness that there is such a thing as "the environment" and that it needs protection. One of them was an outspoken biologist named Barry Commoner. In his 1971 book, *The Closing Circle*, Commoner advanced four laws of ecology that have had great impact on our thinking:

1. Everything is connected to everything else.
2. Everything must go somewhere.
3. Nature knows best.
4. There is no such thing as a free lunch.

Today, most of us appreciate Commoner's first law, the concept that all elements of an ecosystem are related and that disrupting any of them may adversely impact the entire system. The second law stresses that humans do not destroy matter but simply alter it into forms that may be detrimental to health and the environment. Consider toxic waste and greenhouse gases, for example. Number three, "Nature knows best," seems simple enough, but it presents an enormous educational problem because it conflicts with the normal human tendency to attempt to control nature. This dangerous tendency has been reinforced in our Judeo-Christian culture by the biblical

> Be fruitful and multiply and replenish the earth, and subdue it; and have dominion over the fish of the sea, the fowl of the air, and over every living thing that moveth upon the earth (Genesis 1:28).

Too many times we have not "listened" to nature, even after carrying out exhaustive scientific investigations, with disastrous results, as illustrated by many examples detailed in this book. We are learning that we cannot force nature to absorb our wastes and poisons beyond its capacity to do so without damaging the ecosphere. The last law puts the teeth in the first three; it assures us that we will pay for any ill-gotten benefits. The cost of polluting streams, oceans, or soil or of overpopulating unfit lands is eventual remediation expense, resource depletion, health problems, natural disasters, global warming, or the like.

Before 1970 little serious thought was given to environmental education. Today, however, one would be hard pressed to find a college, university, law school, or graduate program that does not offer courses dealing with environmental issues. "Green" thinking is now ingrained in policies, practices, and activities of industry, government, and the public. Great progress has been made. Some successes of the 25 years since the first Earth Day are these:

- Hydrocarbon emissions from automobiles have decreased by almost half.
- The number of large cities violating clean-air standards has dropped from 40 to 9.
- The use of unleaded gasoline has reduced lead emissions by 98 percent.
- Industrial toxic-waste spills have declined 43 percent.
- The California gray whale, the bald eagle, and the American alligator have been removed from the endangered species list.
- Mandatory recycling programs have been implemented in 6,600 U.S. cities. The amount of waste that is recycled increased from 17 percent in 1990 to 22 percent in 1995.
- 100 million acres have been set aside as wilderness areas.
- Since 1979 the country has invoked energy savings valued at almost five times all the new sources of energy combined. Some people call these savings "negawatts."
- The number of environmental groups on college and university campuses grew from 50 in 1989 to more than 2,000 in 1995.

Underlying these achievements is a large body of federal and state laws enacted since the 1960s to clean up and protect the environment. Some of them are listed in ■ Table 1.

On the negative side we still have tons of highly radioactive spent nuclear fuel (see Chapter 13) and stockpiles of nuclear

was predicted to double by 2015. It is in the best interest of the industrialized nations to undertake research and development of alternative energy sources and to expand their current methods of reducing energy usage.

The only options for large-scale power production sufficient to drive economic growth and human requirements in the near future are solar-energy conversion and nuclear energy, including both fission and fusion. Solar-derived electricity at present costs one-third less than electricity from a nuclear power plant and it lacks the hazards associated with radioactive substances. On the negative side, with present technology, solar-energy conversion requires vast areas of land and an effective storage system. Alternative and more sophisticated sources of energy will be developed or expanded—sources such as biomass, geothermal, wind, ocean thermal gradient, and perhaps tidal energy—but these are expected to supply only a small percentage of the total need (see Chapter 13).

TABLE 1 Some Important U.S. Environmental Laws, 1963–1994

1963	Clean Air Act
1964	Wilderness Act—sets aside lands permanently for public use, restricts resource development activities
1970	National Environmental Policy Act (NEPA)★
	Environmental Protection Agency (EPA) established
1971	Vermont Bottle Bill—"Recycle or pay."
1972	Water Pollution Control Act
	Marine Mammal Protection Act
	Pesticide Act
	Coastal Zone Management Act
1973	Endangered Species Act★
	Oregon Bottle Bill
1974	Safe Drinking Water Act★—protects drinking water, establishes EPA water-quality standards
1976	Resource Conservation and Recovery Act (RCRA)★—controls hazardous and toxic wastes
1977	Clean Air Act amendment
	Ocean Dumping Act amendment
1978	Energy Policy and Conservation Act
1980	Comprehensive Environmental Response, Compensation, and Liability Act (CERCLA)★—established Superfund to clean up most severely polluted sites
	Low-Level Radioactive Waste Act
1982	Nuclear Waste Policy Act★
1984	Resource Conservation and Recovery Act amendment
1986	Safe Drinking Water Act amendment
	Superfund amendment
1987	Montreal Protocol—bans chlorofluorocarbons (CFCs)
	Marine Plastic Pollution Control Act
1988	Alternative Motor Fuels Act
1990	National Environmental Education Act
	Clean Air Act amendment
1994	California Desert Protection Act—protects environmentally sensitive desert areas of Western U.S.

★Landmark legislation.
SOURCES: Various.

weapons awaiting the development of safe disposal facilities. The government estimates this task may cost future generations $300 billion. There were 108 million cars in the United States on the first Earth Day. Today there are more than 200 million, and the number of miles traveled has nearly doubled. Although per capita energy use has decreased, total energy use is up significantly (see Chapter 13). The period from March to December in 1994 was the warmest on record in the United States, and sea level rose about three millimeters on many stable coasts. If that is shown to be the start of a trend, it could be strong evidence for global warming (see Chapter 12).

In the mid 1990s the policies and practices of the Environmental Protection Agency have received severe criticism by members of Congress and some sectors of the public. Complying with some EPA requirements has proven to be so expensive that some industries are unable to compete in world markets, thus costing jobs for U.S. workers. Many legislators want the states to take over the policing role, thereby reducing federal interference with business. Some studies have shown that states with the best environmental protection have the healthiest economies, which may lead other states to conclude that it would be good business to "clean up their acts." Americans have become "lite green" say some institute and university researchers. "Lite" green or "dark" green, the record of the first 25 years of Earth Days is pretty good.

The Land

The amount of cultivatable soil limits the number of humans earth can support, because soil determines grain production and therefore food supplies. Arable soils worldwide suffered slight to severe degradation between 1945 and 1990 due to poor agricultural practices, natural erosion, and erosion accelerated by deforestation (■ Table 1.5). The world's farmers lost an estimated 500 million tons of topsoil in the 1970s and 80s, an amount equal to the tillable area of India and France combined. Waterlogging and salt-contamination of soils are reducing the productivity of at least a fourth of the world's cropland, and human-generated smog (ozone) and acid rain also are taking their toll on crops. It is the lack of soil, sometimes combined with drought, that is the cause of famine, not soil infertility. (Soils and degradation of the atmosphere are discussed in Chapters 6 and 13.)

Soil fertility can be improved by applying manufactured

	DEGRADED LAND AREA AS A PERCENT OF VEGETATED LAND		
Region	Total	Light Erosion*	Moderate to Extreme Erosion**
World	17	7	10
Asia	20	7	13
South America	14	6	8
Europe	23	6	17
Africa	22	8	14
North America (U.S., Canada)	5	1	4
Central America, Mexico	25	1	24

* *Light:* crop yields reduced less than 10%.

***Moderate:* crop yields reduced 10%–50%. *Severe:* crop yields reduced more than 50%. *Extreme:* no crop growth possible. About 9 million hectares worldwide exhibit extreme erosion, less than 0.5 percent of all degraded lands.

SOURCE: Various sources compiled by World Resources Institute and the United Nations, *World Resources 1992–93* (New York: Oxford University Press, 1992).

chemical and biochemical fertilizers, whereas drought or a lack of soil is difficult to remedy. China, for example, has developed and encourages such environmentally sound agricultural practices as fertilizing rice paddies with ferns that have nitrogen-fixing capability. This practice has spread throughout Southeast Asia.

As populations grow, humans are forced to occupy and to attempt to cultivate less-desirable lands and lands that may be subject to various life-threatening geologic hazards. Scientific approaches are now available for assessing the risks of flooding, massive landslides, earthquakes, volcanic eruptions, and land subsidence. This book explains these phenomena as they are presently understood and the means that have been developed for alleviating the risks that they pose to human life.

Forests

Deforestation is one of the hottest environmental issues of the day. Although most deforestation is occurring in the tropics and remains largely unmeasured, the Food and Agricultural Organization (FAO) of the United Nations reported in 1991 that 17 million hectares (42.5 million acres) per year of tropical rain forest were being deforested, up 50 percent from the early 1980s rate. Deforestation leads to loss of habitat and decreased biodiversity, contributes to climate change by adding CO_2 to the atmosphere, and often results in soil degradation. ● Figure 1.12 shows late 1980s estimates of annual rates of deforestation in Asia, Latin America, and parts of Africa. Deforestation estimates do not include forests that are logged out or clear-cut and then replanted or simply left to rejuvenate.

One reason for clearing forests is to convert land to agricultural use. In Brazil, an estimated 10-year annual average of 1.7 million hectares (4.3 million acres) is converted. At one time the Brazilian government offered tax credits to landowners who cleared their forest, but these credits were stopped in 1987. Since then, the rate has slowed from an estimated peak of 8 million hectares (20 million acres). More than half of the tropical-forest lands destroyed each year are in Brazil. It should be noted, however, that Brazil plans to set aside 1.6 million square kilometers (618,000 mi^2), about 20 percent of its land area, as public reserves and parks. The Indonesian government, in a full-page advertisement in a U.S. magazine in 1989, explained that the aspirations of its 170 million people were the same as those of citizens in developed nations, and that in order to better their lives, 20 percent of Indonesian forests must be converted to plantations of teak, rubber, rice, and coffee. This conversion has been accomplished successfully. A poor African farmer, on the other hand, has little choice but to clear land to grow crops just for subsistence. How a country uses its land is a complex political and sociological issue, and there are no simple ways to reconcile the conflicting human needs that the issue involves.

Logging is another cause of deforestation, but it has been demonstrated that selective logging in tropical forests is sustainable if the logged lands are allowed to rejuvenate. In 1988 Thailand was inundated with 40 inches of rain in five days. Thousands of logs left on hillsides to dry floated off in mud and water that engulfed entire villages, killing 350 people. The Thai government banned logging, and was caught between the companies that want compensation for lost business and the landless poor who were given farms in the previously logged areas.

Area Cleared Annually, thousands of hectares

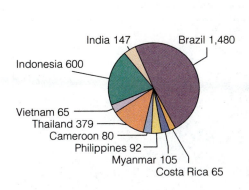

1981–85 Estimates
Total: 3,013

India 147   Brazil 1,480
Indonesia 600
Vietnam 65
Thailand 379
Cameroon 80
Philippines 92
Myanmar 105
Costa Rica 65

(a)

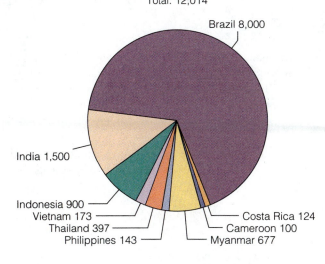

1987–88 Estimates
Total: 12,014

Brazil 8,000

India 1,500

Indonesia 900
Vietnam 173
Thailand 397
Philippines 143
Costa Rica 124
Cameroon 100
Myanmar 677

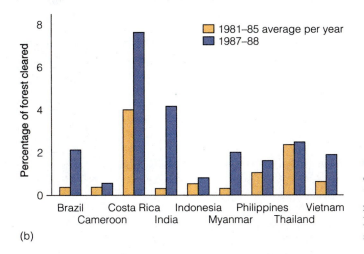

□ 1981–85 average per year
■ 1987–88

Percentage of forest cleared

Brazil   Costa Rica   Indonesia   Philippines   Vietnam
   Cameroon   India   Myanmar   Thailand

(b)

● **FIGURE 1.12** *(a)* **Estimated area lost to deforestation each year in nine tropical countries, 1981–85 versus 1987–88.** *(b)* **Annual percentage of forest cleared in the nine countries, 1981–85 versus 1987–88. Note that Costa Rica has the largest percentage lost but the second-smallest area lost.**

A third cause of deforestation is the demand for wood fuel and forest products. Many of the forests in India and the Sahel of West Africa have been decimated for use as domestic fuel and in light industry.

Deforestation is second only to the burning of fossil fuels as a source of atmospheric CO_2. Forests store 450 billion metric tons of carbon. When cleared, they can no longer sequester the carbon, and it is released rapidly to the atmosphere when trees are burned, or slowly if they are left to decay.

Resources

Our throw-away society, though certainly a convenient one, has become a recycling society as natural resources dwindle and waste-disposal sites become scarce and expensive to operate. Most of what we used to discard after one use—paper, plastic, glass, aluminum, and even steel, is now recycled. Trash disposal has become so expensive, at least in urban settings where most trash is generated, that recycling is close to being cost-effective. Unless alternative energy sources are maximally utilized, oil reserves will be exhausted in your lifetime or in your children's. United States coal reserves are estimated at a 400-year reserve, but such pollution considerations as smog, acid rain, and global warming will continue to limit its use. World reserves of metallic and nonmetallic resources, many of which are in abundant supply at present, can certainly be extended by recycling. The geological and environmental consequences of energy and mining and the problems associated with waste disposal are considered in Chapters 13 through 15.

Overpopulation Problems

As emphasized in this chapter, overpopulation is causing environmental damage, societal problems, and human suffering. In some countries population growth is utterly out of control, and too little is being done to educate the world's people on the long-term dangers of this uncontrolled growth. In other countries—China, for instance—measures and incentives are in place for curbing the growth and achieving a sustainable population. This photo gallery highlights the dilemma.

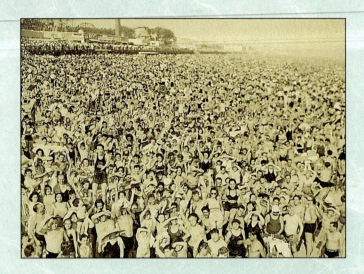

◉ FIGURE 1.13 (right) Coney Island in the 1930s, when the U.S. population was a fraction of what it is today. Imagine this beach on a national holiday today.

◉ FIGURE 1.14 India's population in 1990 was 853 million and it will exceed one billion early in the twenty-first century. The waters of the Ganges River are used for drinking, irrigation, and bathing; Vārānasi (Benares), Uttar Pradesh, India.

(a)

(b)

(c)

● FIGURE 1.15 (at left) Deforestation near the city of Manaus in Amazonia, Brazil's largest state. (*a*) False-color satellite photo of the confluence of the Rio Negro *(black)* and the sediment-laden Amazon River *(light green)*. Healthy rain forest is shown in red, and deforested areas appear in light colors. Manaus is on the point of land that extends into the Rio Negro. The waters of the Rio Negro are recognizable for more than 80 kilometers (48 mi) downstream in the Amazon. (*b*) The Rio Negro and the natural Amazon rain forest at Manaus. (*c*) Adjacent rain forests have been removed as the population of the area has grown. Manaus's population is now more than a million people, and the city is rapidly expanding into areas that were once sparsely populated.

● FIGURE 1.16 (above) A model Chinese family in front of a billboard encouraging the one-child family; Beijing. Although the People's Republic of China is pushing for population control, its population of more than a billion people makes continued growth inevitable.

SUMMARY

Environmental Geology

DEFINED The interaction of humans and the geological environment.

UTILIZES Most of the traditional geologic specialties such as petrology, structural geology, sedimentology, economic geology, engineering geology, geophysics, and geochemistry.

AREAS OF INTEREST Geologic hazards such as volcanism, earthquakes, and flooding; mineral and water resources; and land-use planning—should I build my house on this hillside?

GOALS To predict or anticipate geologic problems by collecting data in the field and analyzing it in the laboratory. Data may be portrayed on special-use maps and utilized by planners and public officials.

CASE HISTORIES Nature is a good teacher, and geologists learn much by studying case histories of geologic events.

Population Dynamics

CARRYING CAPACITY Defined for our purposes as the number of people the earth can support. This number is uncertain but lies somewhere between the 1995 population of 5.76 billion and the 10–15 billion projected for the year 2050.

POPULATION GROWTH RATE Defined as live births per 1,000 persons (that is, the birthrate) *minus* the death rate per 1,000. A birthrate of 20 *minus* a death rate of ten *equals* a growth rate of 10 per 1,000 persons, or 1%.

DOUBLING TIME The number of years in which a population is doubled. It can be projected closely by dividing 70 by the percentage birthrate. The 1995–96 world growth rate of 1.68 percent yields a doubling time of 40 years.

Overpopulation Scenarios

UNCONTROLLED GROWTH, WORST-CASE SCENARIO The earth becomes uninhabitable as resources dwindle, sea-level rises due to global warming and inundates coastal cities, and hazardous ultraviolet radiation strikes the earth's surface due to depletion of the ozone layer.

GAIA HYPOTHESIS Life is an important regulator of the earth environment. If left alone, natural systems are self-regulating; for example, as carbon dioxide builds up in the atmosphere, plant growth and the oceans will remove it, thereby reducing global warming. Human intervention into natural systems (by deforestation, for example) has rendered a basic premise of this hypothesis untenable.

Sustainable Society

DEFINED Society's current needs are satisfied without jeopardizing future generations. To accomplish this, population control is needed.

ENERGY IMPLICATIONS Increasing energy demands dictate that we seek alternatives to burning the earth's reserves of limited fossil fuels. Alternatives include solar, wind, geothermal, and nuclear energy.

LAND IMPLICATIONS Soil erosion and the occupation of marginal lands subject to geologic hazards are problems that must be addressed.

FORESTS IMPLICATIONS Deforestation is the second-greatest contributor to the problem of human-induced carbon dioxide accumulating in the atmosphere. It results in loss of habitat for humans and animals and contributes to desertification.

RESOURCES IMPLICATIONS We must invest our efforts in recycling and conserving resources.

KEY TERMS

carrying capacity	population dynamics
doubling time	population growth rate
engineering geology	replacement rate
environmental geology	sustainable society
fossil fuels	total fertility rate (TFR)
J-curve	zero population growth

STUDY QUESTIONS

1. Physical geology is the study of earth materials, the processes that act upon them, and the resulting products. How does environmental geology differ from physical geology? In order to answer this question you might compare textbooks in the two subject areas.
2. Distinguish between engineering geology and environmental geology.
3. Which natural resources are most important in determining the carrying capacity of the earth?
4. The United States has a current growth rate of 0.7 percent. At this rate how much time is required for a population to double? Discuss the impact of a doubling of your community's population in that length of time.
5. Of what value are case histories in solving environmental geological problems?
6. What is the impact of overpopulation on energy use and global warming? How does this bode for the well-being of future generations? In what parts of the world will the largest percentage increases in energy consumption probably occur in the next decade, and why?
7. What are some of the adverse consequences of clearing tropical rain forests?

FURTHER INFORMATION

BOOKS AND PERIODICALS

Brown, L., and others. 1990. *The state of the world, 1990*. New York: W. W. Norton and Co.

Blue Planet Group. 1991. *Blue planet: Now or never,* case study 9734. Ottawa, Ontario, KIG 5J4, Canada: Blue Planet Group.

Ehrlich, Paul R., and Anne Ehrlich. 1970. *Population, resources, and environment: Issues in human ecology.* New York: W. H. Freeman and Co.

_____. 1987. *Earth.* New York: Franklin Watts.

Gore, Al. 1992. *Earth in balance: Ecology and the human spirit.* Boston: Houghton Mifflin Co.

Joseph, Lawrence E. 1990. *Gaia: The growth of an idea.* New York: St. Martin's Press.

Knickerbocker, Brad. 1995. Earth Day at 25: A U.S. environmental report card. *Christian Science monitor,* April 18.

Lovelock, James. 1988. *The ages of Gaia: A biography of the living earth.* New York: W. W. Norton and Co.

Moffet, George. 1994. *Critical masses: The global population challenge.* New York: Viking Press.

Paty, Alma Hale. 1993. Sustainable development: End of the American dream? *Geotimes,* July: 6.

Silver, Cheryl S., and Ruth S. DeFries. 1990. *One earth one future: Our changing global environment.* Washington, D.C.: National Academy of Sciences, National Academy Press.

Sowers, G. F. 1981. *There were giants on the earth in those days,* 15th Terzhagi lecture. Geotechnical Division of the American Society of Civil Engineers 107, no. 6T4: 383–419.

United Nations. 1980. *World population trends and policies, 1979 monitoring report, Volume 1: Population trends.* New York: United Nations.

World Resources Institute and the United Nations Development and Environment Programs. 1992–93. *Toward sustainable development: A guide to the global environment.* New York: Oxford University Press.

World Resources Institute and the United Nations Development and Environment Programs. 1994–95. *People and the environment: Resource consumption, population growth, and women.* New York: Oxford University Press.

*Man makes a great fuss about this planet, which is only
a ball bearing in the hub of the universe.*

CHRISTOPHER MORLEY, AMERICAN WRITER (1890–1957)

Geologic materials such as rocks, valuable mineral deposits, and soils are an integral part of the human geologic environment. Rocks form the geologic foundations and provide building material for many of our structures. In addition, rocks are involved in many surface earth processes that are hazardous to humans, such as landslides and rockfalls. In this chapter we will discuss briefly how the planet earth formed and how its rocky crust and atmosphere arrived at their present condition. Environmental geologists are often asked about future geologic events, and in order to better understand the future a geologist must learn as much as possible about the past. An important concept in geology

OPENING PHOTO
View down Lone Pine Canyon toward Valhalla, Sequoia National Park, California. Granites and country rocks of the Sierra Nevada have been sculpted into broad valleys and pinnacles by glaciers that existed in the region less than 20,000 years ago. Bare rock and thin soils testify to the erosive power of ice.

known as the Law of Uniformity is that all geologic processes active today have been active in the past, only at differing rates. The law can be paraphrased as "the present is the key to the past." For example, by studying the geologically recent past, we have learned that North America was subject to multiple glacial advances in the past 1.6 million years, a period of time known as the Ice Ages. As a result, scientists monitor and computer-model weather and climate in order to anticipate long-term trends and perhaps even the next ice age (see Chapter 12).

THE BEGINNING

The universe is mostly space, a void, nothingness. Even so, it contains immense quantities of the simplest elements known, hydrogen and helium, which fuel the stars and our sun. This is explained by the widely accepted model that the universe began with a colossal **big bang** when an incredibly dense "cosmic baseball" exploded between 10 and 15 billion years ago, sending matter and energy outward in all directions. The primordial universe was composed of an expanding cloud of gas containing about 75 percent hydrogen, 23 percent helium, and 2 percent other elements. After about a billion years had passed, the clouds segregated into smaller clouds that eventually formed galaxies and groups of galaxies. Within these protogalactic gas clouds were many far smaller clouds, each held together by gravity. Eventually, as the small clouds, or nebulae, became more condensed, gravitational attraction caused them to collapse inward, generating tremendous heat in the core of each cloud (about 10 million degrees centigrade) and initiating thermonuclear reactions whereby hydrogen atoms fused to form helium atoms. In this manner the individual stars composing the galaxies were formed (◉ Figure 2.1). It is the tremendous energy released by the fusing of hydrogen atoms to form helium that causes stars to shine and our sun to give off heat and light. There are as many as 50 billion galaxies of tens or even hundreds of billions of stars each, which gives the galaxies a milky appearance in the sky (Greek *galaktos*, "milk"). It has been speculated that there are more stars in the universe than grains of sand on all the beaches on earth! Many of the galaxies are spiral in shape and so large that it takes 100,000 years for light to travel their width (at the speed of light, 186,281 miles per second). Our solar system

is in the galaxy we call the *Milky Way*, and our sun is an average-sized star in our galaxy.

Few stars that formed relatively soon after the big bang exist today because they had very high core temperatures and they consumed hydrogen at a prodigious rate. Elements as heavy as iron formed in these large stars, which eventually collapsed into their depleted cores and exploded as **supernovas**. During the explosions, elements heavier than iron formed and were spread about the universe. These were the raw materials for a later generation of stars and for our planetary system.

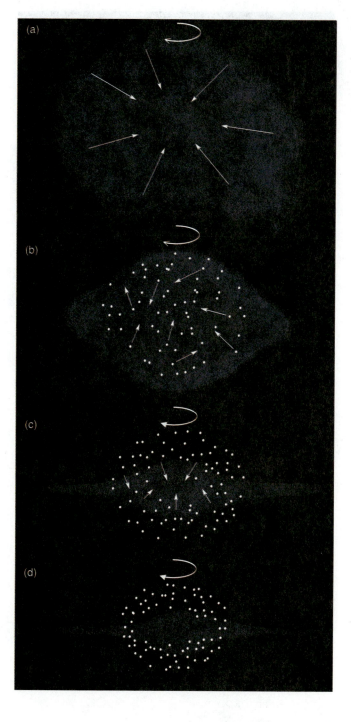

◉ FIGURE 2.1 (right) Formation of the galaxy. *(a)* The pregalactic cloud, rotating slowly, begins to collapse. *(b)* Stars formed before and during collapse have a spherical distribution and noncircular, randomly oriented orbits. *(c)* The collapse leads to a disk with a large central bulge, surrounded by a spherical halo of old stars and clusters. *(d)* The disk flattens further and eventually forms spiral arms. Stars in the disk have relatively high concentrations of heavy elements because they formed from material that had been through stellar nuclear processing.

OUR SOLAR SYSTEM

Nebulae, or interstellar gas clouds, are usually stable; they do not collapse in on themselves unless they are pushed by an outside event. One such event, a supernova explosion sometimes called the *bing bang,* as opposed to the earlier big bang, may have occurred about five billion years ago. This explosion caused the cloud that was to become our solar system to spin and to start collapsing. As the inner part of the nebula contracted, it was heated by the fusion of hydrogen atoms, and the outer part of the gas cloud was forced into a disc shape by rotation—a condition suggested some 200

years ago by Immanuel Kant and Pierre de Laplace (●Figure 2.2). At this stage the hot protosun was surrounded by a cooler, disc-shaped, rotating cloud—the solar nebula. Volatile (low-boiling-point) compounds such as methane and ammonia condensed in the cold outer reaches of the nebula to form the large outer planets, whereas metallic and rocky substances with high melting points condensed closer to the sun and formed the dense inner planets. In time the disc was composed of asteroid-sized lumps of solid matter known as **planetesimals,** which collided and grew into planets by a process called *accretion.* The inner, *terrestrial planets* are relatively small, dense, and composed mainly of rock, whereas the outer, *Jovian planets* are large and gaseous (●Figure 2.3). Pluto, the planet farthest from the sun, is an exception in that it is both small and dense and has an inclined orbit. The planets rotate in the same direction as the overall rotation of the nebular disc with two exceptions. It is postulated that the slow opposite rotation of Venus and the exaggerated tilt of Uranus are accounted for by glancing collisions with large planetesimals. Such an encounter is also thought to have formed the earth's moon. Heat generated by colliding planetesimals and radioactive decay caused the earth to melt and differentiate into layers according to their density. Nickel and elemental iron, being heaviest, sank to form the **core** of the earth, and a lighter, crystalline mush of rock-forming silicates, the **mantle,** formed around the core (●Figure 2.4). The earth as a planetary body formed about 4.6 billion years ago, at which time it was too hot to hold an atmosphere. As it cooled, however, gases leaked from the interior—mainly hydrogen, ammonia, methane, and water vapor—which eventually evolved into our oxygen- and nitrogen-rich atmosphere. Thus, the earth's early atmosphere is the result of "outgassing" from the interior. Finally, lighter minerals formed the outer shell of the earth, the **crust**—which includes rocks of both the continents and the ocean basins—about 4.0–4.3 billion years ago. This is the age of the oldest known minerals in continental rocks. (■Table 2.5, later in the chapter, summarizes age-dating methods.)

THE EARTH SYSTEM

If you were to ask the earth, if the earth could speak, "How is your geology today," or, "How is your biosphere," the

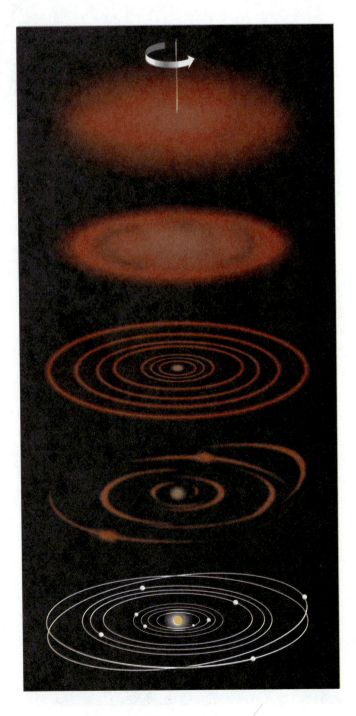

● FIGURE 2.2 (left) Formation of the solar system. An interstellar cloud, initially very extended and rotating very slowly, collapses under its own gravitation. This happens most quickly at the center. As the collapse occurs, the internal temperature rises and the rotation rate increases. Eventually the central condensation becomes hot and dense enough to be a star, with nuclear reactions at its core. The hot protosun is surrounded by a cooler, disk-shaped rotating cloud (the solar nebula) from which the planets eventually are formed.

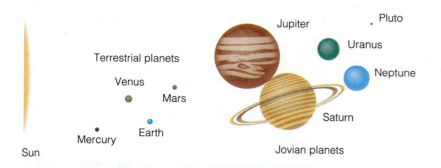

reply in all likelihood would be, "How's my *what?*" Just as the elements of human physiological systems are connected and interdependent, so on earth there are linkages and interactions between the rocky surface and the atmosphere and oceans, and between these systems and all plant and animal life. Our interest extends from the ecosphere, where life evolves, to the reactions between the crust and its fluid and gaseous envelopes, and even to the earth's core, where magnetic fields that protect most organisms from lethal radiation are generated. An example of a little-known connection is the one that exists between dust storms in the Gobi and Takla Makan deserts in China and the North Pacific Ocean. Airborne silt from those deserts is carried all the way to the North Pacific, where it increases the nutrient content and fertility of this otherwise barren oceanic realm. The periodic weather phenomenon known as *El Niño,* which heavily impacts the west coasts of South and North America, has its origins in ocean–atmosphere interactions called the *Southern Oscillation* in the western Pacific thousands of miles away. We now know that ENSO—the acronym for *El Niño–Southern Oscillation*—is a global weather event whose effects are not limited to floods along the west coasts of the Americas.

The five interconnected reservoirs, the so-called geospheres, that make up the complete earth system are

- the *atmosphere,* the gaseous envelope surrounding the earth;
- the *pedosphere,* soil or any other loose material on the earth's surface above bedrock that supports plant life;
- the *lithosphere,* the solid portion of the earth's surface;
- the *hydrosphere,* all water on and in the earth, including oceans, lakes, rivers, clouds, water underground, and glaciers; and
- the *biosphere,* all of the living and dead organic components of the planet (● Figure 2.5). The *ecosphere* is a subsystem that includes the biosphere and its interactions with the other geospheres.

The solid earth sciences attempt to understand the past, present, and future behavior of the whole earth system; that is, they take a planetary approach to the study of global change by studying the present and the past. The spheres of the earth system are all linked, and separate disciplines such as oceanography, geology, hydrology, and atmospheric physics have simply developed in order to organize information and award academic degrees. Over the past century the sciences have become increasingly segmented and specialized, going in the exact opposite direction of what is needed for solving today's environmental problems. Strong disciplinary studies are needed, but we also need a systems approach for understanding how the earth works and how a sustainable environment can be provided for future generations.

The surface of the lithosphere is a major boundary in the earth system, as this is where energy from the sun is transferred to the planet and its inhabitants. It is also at this boundary that we find landslides, volcanic activity, earthquakes, severe weather, and long-term global weather changes that impact humans. "The whole of humankind," wrote noted historian Norman Pounds, "has been in the shadow of an ice age." When Columbus was a teenager, the earth fell into a period of severe cold called the "Little Ice Age," which lasted 400 years to the mid 1800s. The earth is a dynamic system, so there is nothing new about global change. What *is* new is how *fast* change seems to be taking place today. Temperatures changed at a rate of about 0.5° C

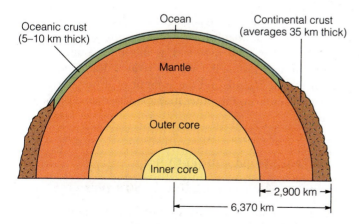

● FIGURE 2.4 Structure of the earth showing oceanic and continental crust, mantle, and outer and inner cores. Not to scale.

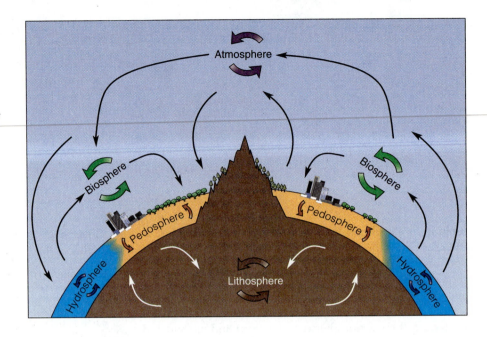

● FIGURE 2.5 The five interconnected spheres of the earth system. Each sphere plays a role in the mobilization, transfer, and deposition of the chemical elements vital to the human existence. How many geologic cycles can you identify in this artist's rendering?

per century during the Little Ice Age as best as we can determine from written records and ice cores. The rate of temperature change today is about ten times faster, 0.5° C per decade, even in the most conservative models. We must learn how much of this increase is anthropogenic (due to human activities). This is the first time humans have had the luxury of being able to look into the future and predict 10 years in advance what the climate will be like—and humans today have the intellect, technology, and obligation to mitigate or at least to slow adverse anthropogenic changes.

The geospheres represent closed systems in which the mass movements of material that we call *geologic cycles* (rock and hydrologic cycles) and *biogeochemical cycles* (cycling of life-forming elements such as carbon and nitrogen) occur. The most readily understood of these cycles is probably the hydrologic cycle, by which water on land and at the sea surface is evaporated to the atmosphere and transported as clouds, which then redistribute the moisture around the globe (see Figure 9.1). It is a short-term cycle. The rock cycle, on the other hand, is a long-term cycle involving slow changes, some of them taking millions of years (see Figure 2.13). Less obvious is the recycling of the elements hydro-

gen, oxygen, carbon, nitrogen, and potassium, the main players in biogeochemical cycles. The atmosphere and hydrosphere get carbon, nitrogen, and oxygen by interacting with the solid earth and oceans, by weathering, and by volcanic action, and the cycle is completed when animals and plants give up these elements to the atmosphere, lithosphere, and hydrosphere by their life processes or after death by decomposition. Current knowledge of the carbon cycle is woefully inadequate, as scientists cannot account for 25 percent of sequestered carbon (carbon that is isolated from the environment by burial or other means). The assumption is that it is buried beneath sediment in the oceans and on land. Knowledge of the carbon cycle is of major importance in predicting future global climate and, specifically, the amount of global warming that can be caused by anthropogenic carbon dioxide (CO_2) and methane (CH_4), so-called greenhouse gases.

The time scale and areal extent of selected earth-system processes are summarized in ● Figure 2.6. We cannot do anything about such processes as mountain building and petroleum generation, but we can reduce death and injury from geologic hazards and the amount of atmospheric pollution, soil erosion, and other human-caused environmental problems.

CONSIDER THIS . . .

Some people question the reality and the seriousness of global warming and depletion of natural resources. Are scientists and futurists who warn of the possible consequences of uncontrolled emissions and consumption making "mountains out of mole-hills"?

 EARTH MATERIALS

Elements, Atoms, and Atomic Structure

Elements are substances that cannot be changed into other substances by normal chemical methods. They are composed of infinitesimal particles called **atoms**. In 1870, Lothar Meyer published a table of the 57 then-known elements arranged in the order of their atomic weights. Meyer left blank spaces in the table wherever elements of particu-

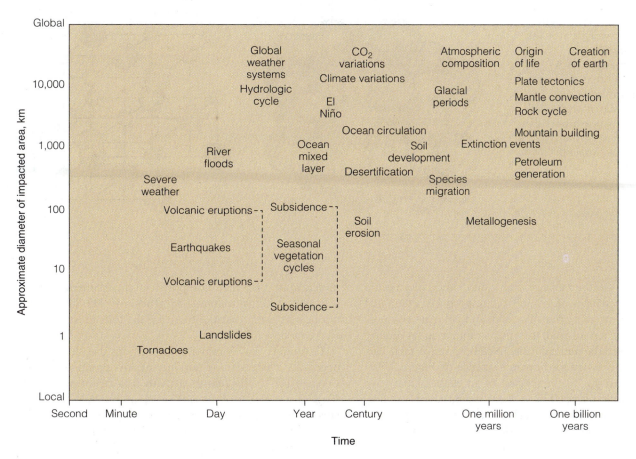

● FIGURE 2.6 Approximate space and time scales of earth-system processes. Space scale is the approximate diameter of the region impacted by the earth-system process.

lar weights were not known. About the same time Dmitry Mendeleyev had developed a similar table, but his table was based upon the similar chemical properties of the particular known elements. Where an element's placement based on weight did not group it with elements of similar properties, Mendeleyev did not hesitate to suggest that its weight had been measured incorrectly. He also predicted the properties of the "missing" elements, and very shortly his predictions proved correct with the discovery of the elements gallium, scandium, and germanium (Appendix 1).

The weight of an atom of an element is contained almost entirely in its **nucleus,** which contains protons (atomic weight = 1, electrical charge = 1^+) and neutrons (atomic weight = 1, electrical charge = 0). An element's **atomic number** is the number of protons in its nucleus, and this number is unique to that element. The weight of an atom—its **atomic mass**—is the sum of its nuclear protons and neutrons. For example, $_2^4$He denotes the element helium, which has 2 protons (denoted by the subscript) and two neutrons, for an atomic mass of 4 (superscript). Electrical neutrality of the atom is provided by balancing the positive proton charges by an equal number of negatively charged electrons orbiting in shells around the nucleus (● Figure 2.7). **Ions** are atoms that are positively or negatively charged owing to a loss or gain of electrons (e^-) in the outer electron shell.

Thus, sodium (Na) may lose an electron in its outer shell to become sodium ion (Na^+, a *cation*), and chlorine (Cl) may gain an electron to become chloride ion (Cl^-, an *anion*).

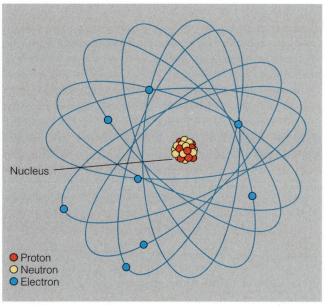

● FIGURE 2.7 The structure of an atom showing the nucleus and its surrounding electron cloud.

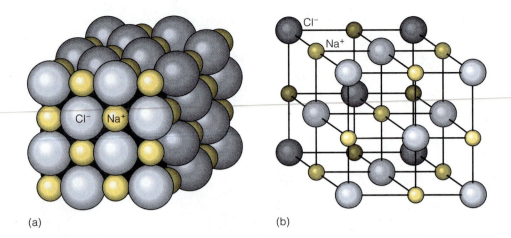

● FIGURE 2.8 Two models of the atomic structure of the mineral halite (NaCl), common table salt. *(a)* A packing model that shows the location and relative sizes of the sodium and chloride ions. *(b)* A ball-and-stick model that shows the cubic crystal structure of the mineral.

Valence refers to ionic charge, and sodium and chlorine are said to have valences of 1^+ and 1^-, respectively. Elemental sodium is an unstable metallic solid, and chlorine is a poisonous gas. When ionized, they may combine in an orderly fashion to form the mineral halite (NaCl), common table salt, a substance necessary for human existence (● Figure 2.8).

Isotopes are forms (or species) of an element that have different atomic masses. The element uranium, for instance, always has 92 protons in its nucleus, but varying numbers of neutrons define its isotopes. For example, Uranium-238 (^{238}U) is the common naturally occurring isotope of uranium. It weighs 238 atomic mass units and contains 92 protons and 146 (238 *minus* 92) neutrons. Uranium-235 is the rare isotope (0.7% of all uranium) that has 143 neutrons. Similarly, ^{12}C is the common isotope of carbon, but ^{13}C and ^{14}C also exist. A modern periodic table of the elements that shows their atomic number and mass appears in Appendix 1.

Minerals

The earth's crust is composed of rocks, and rocks are made up of one or more minerals. Although most people seldom think about minerals that do not have nutritional importance or great beauty and value such as gold and silver, the mineral kingdom encompasses a broad spectrum of chemical compositions. **Minerals** are defined as *naturally occurring, inorganic, crystalline* substances, each with a narrow range of *chemical compositions* and characteristic *physical properties*. Thus, neither artificial gems such as zirconia nor organic deposits such as coal and oil are minerals by definition. The crystal form of a mineral reflects an orderly atomic structure, which in turn determines the crystal shape. Solids that have random or noncrystalline atomic structures—glass and opal, for example—are described as **amorphous**. Minerals composed of a single element, such as copper and carbon (as in graphite and diamonds), are known as *native elements*. Most minerals, however, are composed of combinations of two or more elements. Although more than 3,000 minerals have been named and described, only about 20 make up the bulk of the earth's crustal rocks. The chemistry of minerals is the basis of their classification, and the most common

minerals are primarily composed of the elements oxygen (O), silicon (Si), aluminum (Al), and iron (Fe). This should not be surprising, because these are the four most abundant elements on earth.

Geologists classify minerals into groups that share similar negatively charged ions (anions) or ion groups (radicals). Oxygen (O^{-2}) may combine with iron to form hematite (Fe_2O_3), a member of the *oxide group* of minerals. The *sulfide minerals* typically are combinations of a metal with sulfur. Examples include galena (lead sulfide, PbS), chalcopyrite (copper-iron sulfide, $CuFeS_2$), and pyrite (iron sulfide, FeS_2), the "fool's gold" of inexperienced prospectors. *Carbonate minerals* contain the negatively charged $(CO_3)^{-2}$ ion. Calcite ($CaCO_3$), the principal mineral of limestone and marble, is an example.

The most common and important minerals are the *silicates,* which are composed of combinations of oxygen and silicon with or without metallic elements. The basic building block of a silicate mineral is the silica tetrahedron (● Figure 2.9). It consists of one silicon atom surrounded by four oxygen atoms at the corners of a four-faced tetrahedron. The manner in which the tetrahedra are packed or arranged in the mineral structure is the basis for classifying silicates. Some of the ways the tetrahedra may be arranged are in layers or sheets (as in mica and chrysotile asbestos; see Case Study 2.1), in long chains (as in pyroxene, hornblende amphibole, and crocidolite asbestos), and in three-dimensional networks (as in quartz and feldspar). The *feldspar group* of silicate minerals is the largest and most significant mineral group in igneous rocks. The feldspars range in composition from the potassium-rich orthoclase that is common in granites, to the sodium- and calcium-rich plagioclase found in basalt and gabbro (Figures 2.9 and ● 2.10). The relative abundances of rock-forming silicate and nonsilicate minerals are shown in ■ Table 2.1.

Serious students of mineralogy are able to identify several hundred minerals without destructive testing. They are able to do this because minerals have distinctive physical properties, most of which are easily determined and associated with particular mineral species (Appendix 2). With practice, one can build up a mental catalog of minerals, just

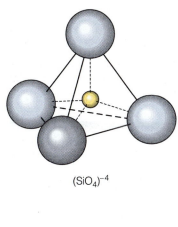

$(SiO_4)^{-4}$

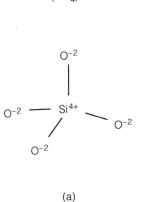

(a)

		Formula of negatively charged ion group	Example
Isolated tetrahedra	▲	$(SiO_4)^{-4}$	Olivine
Continuous chains of tetrahedra	← ⬧⬧⬧ →	$(SiO_3)^{-2}$	Pyroxene group (augite)
	← ⬧⬧⬧ →	$(Si_4O_{11})^{-6}$	Amphibole group (hornblende)
Continuous sheets	↑ ← ⬧⬧⬧ → ↓	$(Si_4O_{10})^{-4}$	Micas (muscovite)
Three-dimensional networks	Too complex to be shown by a simple two-dimensional drawing	$(SiO_2)^0$ $(Si_3AlO_8)^{-1}$ $(Si_2Al_2O_8)^{-2}$	Quartz Orthoclase feldspars Plagioclase feldspars

(b)

● FIGURE 2.9 *(a)* Model of the silica tetrahedron, showing the unsatisfied negative charge at each oxygen that allows it to form *(b)* chains, sheets, and networks.

like building a vocabulary in a foreign language. The most useful physical properties are hardness, cleavage, crystal form, and to lesser degrees, color (which is variable) and luster. Mineralogists use the **Mohs hardness scale,** developed by Friedrich Mohs in 1812 (■ Table 2.2, page 32). Mohs assigned a hardness value *(H)* of 10 to diamond and of 1 to talc, a very soft mineral (Table 2.2). Diamond scratches everything, including other diamonds; quartz ($H = 7$) and feldspar ($H = 6$) can scratch glass ($H = 5½-6$) and calcite ($H = 3$), and your fingernail can scratch gypsum ($H = 2$).

■ TABLE 2.1 Common Rock-Forming Minerals

MINERAL	ABUNDANCE IN CRUST, %	ROCK IN WHICH FOUND
Plagioclase*	39	Igneous rocks mostly
Quartz	12	Detrital sedimentary rocks, granites
Orthoclase**	12	Granites, detrital sedimentary rocks
Pyroxenes	11	Dark-colored igneous rocks
Micas	5	All rock types as accessory minerals
Amphiboles	5	Granites and other igneous rocks
Clay minerals	5	Shales, slates, decomposed granites
Olivine	3	Iron-rich igneous rocks, basalt
Others	11	Rock salt, gypsum, limestone, etc.

*A series of six minerals within the feldspar group from albite ($NaAlSi_3O_8$) to anorthite ($CaAl_2Si_2O_8$).
**Feldspar group of minerals.

Minerals, Cancer, and OSHA—Fact and Fiction

When an international health organization designated quartz as "probably" carcinogenic (cancer-causing) to humans, the U.S. Occupational Safety and Health Agency (OSHA) immediately went into action. It issued a mandate requiring products containing more than one tenth of one percent (0.1%) of quartz or "free silica" (SiO_2) to display warning signs. Because quartz is chemically inert and because it is the most common mineral species, questions arose as to whether this requirement extended to sandpaper, beach sand, and unpaved roads that may produce clouds of fine-grained quartz dust. Questions—and more—arose when some Delaware truck drivers were "ticketed" for not displaying the required "quartz on board" signs during transport of crushed rock to construction sites. Such well-meaning OSHA actions have been perceived as ridiculous. They illustrate what happens when a government authority establishes health regulations without understanding the substances involved. In this case OSHA did not understand the where, how, and chemistry of the mineral—its mineralogy. Unless one has had a long-term craving to inhale beach sand or crushed rock, there are no health hazards in the examples cited. Nonetheless, because exposure to very large quantities of any substance, even table salt or vitamins, can be harmful to health, it is certain that long-term workplace exposure to dusts from some minerals can pose health problems.

The World Health Organization's Group I classification of "known" carcinogenic minerals includes asbestos minerals (there are six), erionite (a zeolite), and minerals containing arsenic, chromium, and nickel. Its Group 2 of "probable" carcinogenic minerals includes all radioactive minerals and minerals containing lead, beryllium, and silica. The classification is based upon suspected relationships between specific diseases and workplace exposure to the identified substances and on laboratory experiments with animals.

A major human health concern is the relationship between lung cancer and inhaling the fine fibers of asbestos minerals. The long, fibrous crystal forms of asbestos belonging to the amphibole group (see hornblende structure, Figure 2.11) are considered the most carcinogenic. Of these, blue asbestos, crocidolite (pronounced cro-cid´-o-lite; ◉ Figure 1), is thought to be most dan-

gerous and is believed to cause mesothelioma, a relatively rare cancer of the lining of the heart and lungs. Although crocidolite constitutes only 5 percent of all industrial asbestos, its morbidity statistics are grim indeed. For example, a 1990 survey of 33 men who, in 1953, worked in a factory where crocidolite was utilized in manufacturing cigarette filters found that 19 of them had died of asbestos-related diseases.

Common white asbestos, chrysotile (pronounced chrys-o-tile´; ◉ Figure 2), is a sheet-structure mineral similar to mica. It appears under high magnification as long fibers, which are in reality rolled sheets, much like long rolls of gift-wrapping paper. Chrysotile constitutes 95 percent of all industrial asbestos and has never been demonstrated to cause cancer at the levels found in most schools and public buildings. A study conducted in Thetford, Quebec, Canada, evaluated persons who had been exposed to high concentrations of white asbestos in their workplace and found no deaths attributable to it. Many of the diagnosed chrysotile-related lung problems have been traced back to

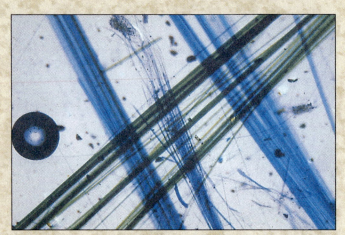

◉ FIGURE 1 Deadly crocidolite, "blue asbestos," magnified about 100 times. The mineral appears in its characteristic blue and green colors when viewed through a petrographic microscope. It was used in only a small percentage of commercial products.

Cleavage refers to the characteristic way particular minerals split along definite planes as determined by their crystal structure. Mica has perfect cleavage in 1 direction; this is called *basal* cleavage, because it is parallel to the basal plane of the crystal structure. Feldspars split in 2 directions; halite, which has *cubic* structure, in 3 directions; and so on, as shown in ◉ Figure 2.11, page 33. Minerals that have perfect cleavage can be split readily by a tap with a rock hammer or even peeled apart, as in the case of mica. Some minerals do not

cleave but have distinctive **fracture** patterns that help one to identify them. The common crystal forms shown in ◉ Figure 2.12, page 33, can be useful in identifying a particular mineral. Color tends to vary within a given mineral species, so it is not a reliable identifying property. **Luster,** how a mineral reflects light, is useful, however. We recognize metallic and nonmetallic lusters, the latter being divided into types such as glassy, oily, greasy, and earthy. (Case Study 2.2 discusses some gem minerals.)

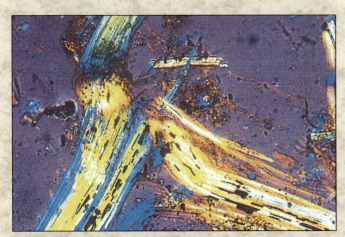

● FIGURE 2 Chrysotile asbestos (× 100) has many uses in homes, offices, and public buildings. Although it is the target of removal in thousands of structures, at low levels it poses no documented risk. Note the difference between chrysotile and crocidolite's crystal structure. One can see in the structure why crocidolite is a potent carcinogen.

improper handling of the material and inadequate respiratory protection during asbestos removal procedures.

Inhaling finely divided airborne asbestos fibers in large quantities is definitely a health hazard. The fibers stay in the lungs and after some time, perhaps several decades, can cause cancerous growths. Under the Toxic Substance Control Act of 1972, the Asbestos Hazard Emergency Response (AHER) of 1986 required the EPA to develop regulations addressing the health risks posed by asbestos in schools and other public buildings. In response, the EPA mandated that *all* asbestos minerals be treated as identical health hazards, regardless of their mineralogy and ambient concentration in a specific structure, and that they be removed. By the mid 1990s the cost of removal had amounted to hundreds of millions of dollars, and the total cumulative cost for remediation of rental and commercial buildings and litigation and enforcement was estimated at $100 billion. The EPA requirements were

sweeping. They applied to buildings containing the relatively benign white asbestos and to those where airborne concentrations were so low that they could not be measured. The fear of asbestos and the EPA regulations are based on the idea that there is no safe "threshold" concentration; that even one fiber can kill. No doctor or public health official would agree with this. A New York school district's administration was besieged by irate parents when it admitted that it could not *guarantee* that *all* asbestos had been removed from its buildings. The district had spent millions of dollars to remove it.

Industry and government can ill afford to squander capital and tax dollars on poorly conceived hazard-mitigation requirements based upon the "no threshold" theory of health risk; that is, that one molecule of pesticide or one asbestos fiber can cause health problems. Many government regulations regarding "toxic substances" address risks to humans that are no greater than those incurred by routinely drinking several cups of coffee each day or eating a peanut butter sandwich for lunch. The "no threshold" criterion forces "carcinogen" classification upon otherwise benign minerals and many other useful substances and needlessly increases public anxiety. (Remember when synthetic sweeteners and cellular telephones were labeled carcinogenic?) The costs of removal and liability protection in such cases are horrendous, and a fiscal crisis in the environmental field is acknowledged. Liability extends to building owners, real estate agents who sell the buildings, banks that lend money for property purchases, and purchasers of land upon which any material classified as hazardous is found.

A person's chances of being struck by a lightning bolt are about 35 per 1 million lifetimes, and the risk of a nonsmoker's dying from asbestos exposure is about 1 in 100,000—about a third of the chance of being struck by lightning. Incidentally, the risk of dying from cigarette-related diseases is about 1 in 5 for smokers. Certainly, some substances in the environment pose dangers, and the key to risk reduction and longevity is awareness. Just as a reasonable person would not play golf during a violent thunderstorm because of the risk of being struck by lightning on the course, one should not handle hazardous substances without wearing protective gear.

Rocks

Rocks are defined as consolidated or poorly consolidated aggregates of one or more minerals, glass, or solidified organic matter (such as coal) that cover a significant part of the earth's crust. There are three classes of rocks, based upon their origin: igneous, sedimentary, and metamorphic. **Igneous** (Latin *ignis,* "fire") rock is crystallized from molten or partly molten material. **Sedimentary** rocks include both

lithified (that is, turned to stone) fragments of preexisting rock and rocks that were formed from chemical or biological action. **Metamorphic** rocks are those that have been changed, essentially in the solid state, by heat, fluids, and/or pressure within the earth.

The **rock cycle** is one of many natural cycles on earth. The illustration of it in ● Figure 2.13 shows the interactions of energy, earth materials, and geologic processes that form and destroy rocks and minerals. The rock cycle is essentially

● FIGURE 2.10 *(a)* Coarse-grained granite, in which feldspar, quartz, and minor amounts of mica and hornblende are visible. *(b)* A photomicrograph of a thinly sliced section of granite reveals large interlocking grains. *(c)* Minerals in a hand sample of fine-grained basalt are not identifiable. *(d)* A photomicrograph of basalt reveals its interlocking fine-grained texture.

■ TABLE 2.2 Mohs Hardness Scale

HARDNESS	MINERAL	COMMON EXAMPLE
1	Talc	Pencil lead 1–2
2	Gypsum	
		Fingernail 2½
3	Calcite	Copper penny 3
		Brass
4	Fluorite	Iron
5	Apatite	Tooth enamel
		Knife blade
		Glass 5½–6
6	Orthoclase	
		Steel file 6½
7	Quartz	
8	Topaz	
9	Corundum	Sapphire, ruby
10	Diamond	Synthetic diamond

a *closed system;* "what goes around comes around," so to speak. In a simple cycle there might be a sequence of formation, destruction, and alteration of rocks by earth processes. For example, an igneous rock may be eroded to form sediment that subsequently becomes a sedimentary rock, which may then be metamorphosed by heat and pressure to become a metamorphic rock, and then melted to become an igneous rock.

Throughout this book we emphasize those rocks which, because of their composition or structure, are involved in geologic events that endanger human life or well-being or that are important as resources.

IGNEOUS ROCKS. Igneous rocks are classified according to their texture and their mineral composition. A rock's texture is a function of the size and shape of its mineral grains, and for igneous rocks is determined by how fast or slowly a melted mass cools (see Figure 2.10). If **magma**—molten rock within the earth—cools slowly, large crystals develop and a rock with coarse-grained, **phaneritic texture** such as granite is formed (see Figure 2.10). The resulting large rock mass formed within the earth is known as a **batholith.** Greater than 100 square kilometers (39 mi^2) in area by definition, batholiths are mostly granitic in composition and show evidence of having invaded and pushed aside the *country rock* into which they intruded. Also, many batholiths have formed in the cores of mountain ranges during mountain-building episodes. Batholiths are a type of igneous mass referred to as *plutons,* and rocks of batholiths are described as **plutonic** (after Pluto, the Greek god of the underworld), because they were formed at great depth. The Sierra Nevada of California is an example of an uplifted and eroded mountain range with an exposed core composed of many plutons (⦿Figure 2.14).

Lava is molten material at the earth's surface produced by volcanic activity. Lava cools rapidly, resulting in restricted crystal growth and a fine-grained, or **aphanitic,** texture. An example of a rock with aphanitic texture is rhyolite, which has about the same composition as granite but a much finer texture. The grain size of an igneous rock can tell us if it is an **intrusive** (cooled within the earth) or an **extrusive** (cooled at the surface of the earth) rock. The classification by texture and composition of igneous rocks is shown in ⦿Figure 2.15. Note that for each composition there are pairs of rocks that are distinguished by their texture. Granite and rhyolite, diorite and andesite, and gabbro and basalt are the most common pairs found in nature. It should be noted that mineral composition ranges over a continuum from light-colored granites to dark gabbros, with many identified intermediate rock types between them. The end member shown in Figure 2.15, peridotite, is rarely found because it originates deep within the earth. Peridotite is described as *ultramafic* (*ma* for magnesium, *fic* for iron) because it contains mostly *ferromagnesian* minerals—silicates that are rich in iron (Fe) and magnesium (Mg), such as certain pyroxenes and olivine—that give it a dark or greenish color. Three important and relatively common glassy (composed of amorphous or finely crystalline SiO$_2$) volcanic rocks not included in Figure 2.15 are obsidian, which looks like black glass, and pumice, which has the composition of glass but is not really glassy-

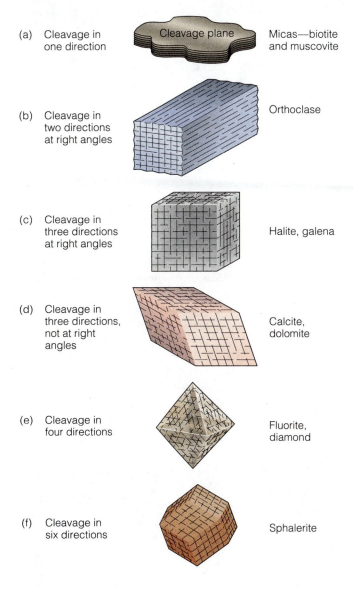

(a) Cleavage in one direction — Cleavage plane — Micas—biotite and muscovite

(b) Cleavage in two directions at right angles — Orthoclase

(c) Cleavage in three directions at right angles — Halite, galena

(d) Cleavage in three directions, not at right angles — Calcite, dolomite

(e) Cleavage in four directions — Fluorite, diamond

(f) Cleavage in six directions — Sphalerite

⦿ **FIGURE 2.11 Types of cleavage and typical minerals in which they occur.**

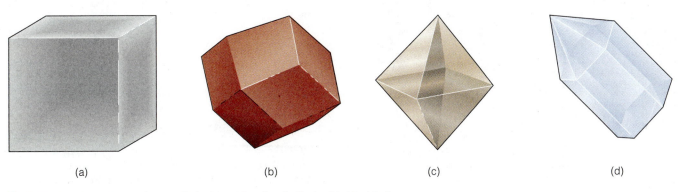

(a)   (b)   (c)   (d)

⦿ **FIGURE 2.12 Mineral crystal shapes:** *(a)* cube (halite), *(b)* 12-sided dodecahedron (garnet), *(c)* 8-sided octahedron (diamond, fluorite), and *(d)* hexagonal prism (quartz).

Gems—Nature's Artwork

Rarity and beauty are what make precious and semiprecious gems so desirable. Those shown in ⊙ Figure 1 are but a sample of several hundred valued gem minerals and their varieties. The largest uncut diamond in the world, the Cullinan diamond (Figure 1, part a), was recovered in 1905 from the Premier Mine at Kimberley, Transvaal, South Africa; the huge, colorless stone weighed 3,106 carats (1.3 pounds). The smooth face shown in the photograph is a natural cleavage surface, which indicates that this is not the entire stone. Unfortunately, the remainder of it lies undiscovered in a deep volcanic pipe—which is where all diamonds originate. The Transvaal government bought the diamond for $1.6 million and gave it uncut to King Edward VII of England. After months of study, an Amsterdam diamond cutter's decisive blow yielded nine large stones and about 100 smaller ones, all of them flawless. All are now in the British crown jewels collection.

Diamonds are the high-pressure form of carbon and are the hardest natural substance known. Curiously, graphite, the high-temperature, low-pressure form of carbon, is among the softest minerals known. Diamonds are also found in Australia, India, South America, and near Murfreesboro, Arkansas, where the initial discovery was made by a farmer in 1906. Thousands of diamonds have been recovered near Murfreesboro since then, one weighing 40 carats, but the deposit has not been found to be commercially viable. For the price of admission, one may dig for diamonds at Murfreesboro in the "blue ground" of the eroded diamond-bearing volcanic pipe. One lucky hunter uncovered a 15-carat stone. Little known to the general public are the diamond finds in California gold-bearing gravels and in the soils and glacial deposits of the Midwest and the East (⊙ Figure 2). The sources of these diamonds are unknown, but those found in ice-age deposits originated somewhere near the boundary between the United States and Canada.

Precious opal (Sanskrit *upala,* "precious stone"; Figure 1, part b) occurs in shallow surface diggings, and because it lacks crystalline form, it actually is not a mineral. It is amorphous hydrated silica and gets its "fire" from an orderly arrangement of silica spheres interspersed with water molecules. Black and fire opals are the most desirable, and each stone is priced individually.

Emerald, the most valuable mineral carat-for-carat, is the precious variety of the silicate mineral beryl (Figure 1, part c). It is found in coarse-grained granites called **pegmatites** (see Chapter 14) and in mica schist.

Quartz, plain old SiO_2, one of the most common minerals, has a host of semiprecious varieties such as purple amethyst (Figure 1, part d), banded agate, yellow citrine, and smoky and rose quartz. Searching for polishing-quality quartz varieties is the passion of many "rock hounds," and good crystal specimens also are cherished.

(a)

(b)

(c)

(d)

⊙ **FIGURE 1** *(a)* An exact replica of the Cullinan diamond with the two largest stones cut from it, the Great Star of Africa (530.2 carats) and the Second Star of Africa (317 carats), which now reside in the British crown jewels collection. *(b)* Uncut precious opal, *(c)* emerald, and *(d)* quartz.

⊙ **FIGURE 2** Reported diamond finds in glacial drift in the Great Lakes region. Numerals indicate the number of diamonds found at the given location; single finds have been reported at the other locations.

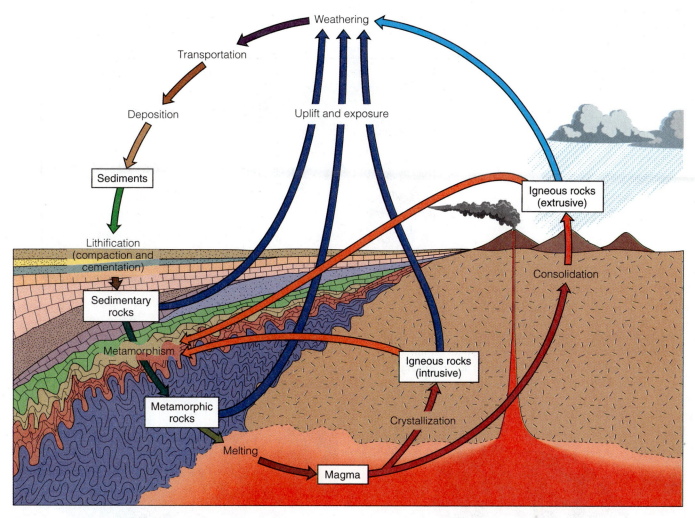

● **FIGURE 2.13 The rock cycle. The three rock types are interrelated by internal and external processes involving the atmosphere, ocean, biosphere, crust, and upper mantle.**

looking (● Figure 2.16, parts a and b). The third rock not included in Figure 2.15, tuff, is consolidated volcanic ash or cinders (Figure 2.16, part c). Tuff has a **pyroclastic** (literally "fire-broken") texture resulting from fragmentation during violent volcanic eruptions.

SEDIMENTARY ROCKS. Sediment is particulate matter derived from the physical or chemical weathering of materials of the earth's crust and by certain organic processes. It may be transported and redeposited by streams, glaciers, wind, or waves. Sedimentary rock is sediment that has become **lithified**—turned to stone—by pressure from deep burial, by cementation, or by both of these processes. **Detrital sedimentary rocks** are those that are composed of *detritus,* fragments of preexisting rocks and minerals. Their texture, which makes this composition quite obvious, is described as **clastic.** Clastic sedimentary rocks are classified according to their grain size. Thus sand-size sediment lithifies to sandstone, clay to shale, and gravel to conglomerate (● Figures 2.17 and ● 2.18, parts a, b, and c). Other sediments result from chemical and biological

● **FIGURE 2.14 The Sierra Nevada batholith and Mount Whitney, looking west from Owens Valley; east-central California.**

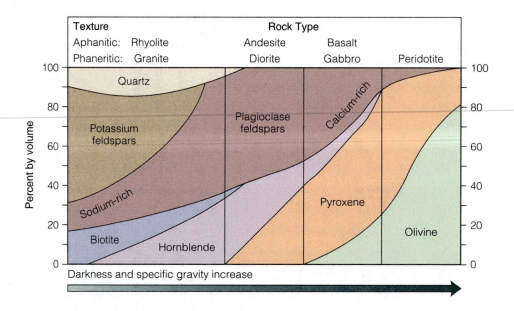

activity. **Chemical sedimentary rocks,** which may be clastic or nonclastic, include chemically precipitated limestone, rock salt, and gypsum. **Biogenic sedimentary rocks** are produced directly by biological activity, such as coal (lithified plant debris), some limestone and chalk ($CaCO_3$, shell material), and chert (siliceous shells) (Figure 2.18, parts d and e). ■ Table 2.3 presents the classification of sedimentary rocks.

(a)

(b)

● FIGURE 2.16 Extrusive igneous rocks: *(a)* glassy obsidian; *(b)* pumice, which is formed from gas-charged magmas; and *(c)* tuff. The lavas that form obsidian and pumice cool so rapidly that crystals do not grow to any size. The reference scale in each photo is 5 centimeters long.

(c)

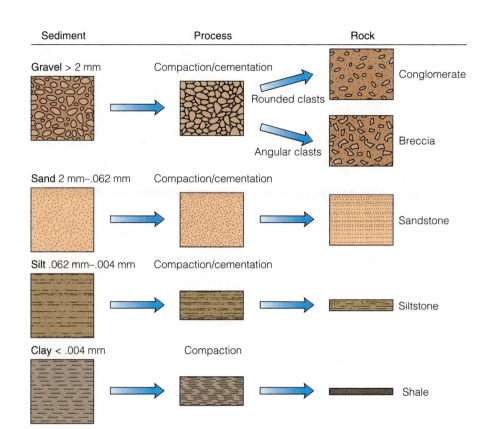

● FIGURE 2.17 How sediments are transformed into clastic sedimentary rocks. The sequence of lithification is deposition to compaction to cementation to hard rock.

(a)

(b)

(c)

(d)

(e)

● FIGURE 2.18 Common clastic sedimentary rocks: *(a)* shale, *(b)* sandstone, *(c)* conglomerate. Biogenic sedimentary rocks: *(d)* limestone composed entirely of shell materials (called *coquina*), *(e)* coal.

DETRITAL SEDIMENTARY ROCKS (CLASTIC TEXTURE)

Sediment	Description	Rock Name
Gravel (>2.0 mm)	Rounded rock fragments	Conglomerate
	Angular rock fragments	Breccia
Sand (0.062–2.0 mm)	Quartz predominant	Quartz sandstone
	>25% feldspars	Arkose
Silt (0.004–0.062 mm)	Quartz predominant, gritty feel	Siltstone
Clay, mud (<0.004 mm)	Laminated, splits into thin sheets	Shale
	Thick beds, blocky	Mudstone

CHEMICAL SEDIMENTARY ROCKS

Texture	Composition	Rock Name
Clastic	Calcite ($CaCO_3$)	Limestone
	Dolomite [$CaMg(CO_3)_2$]	Dolostone
Crystalline	Halite (NaCl)	Rock salt
	Gypsum ($CaSO_4 \cdot 2H_2O$)	Rock gypsum

BIOGENIC SEDIMENTARY ROCKS

Texture	Composition	Rock Name
Clastic	Shell calcite, skeletons, broken shells	Limestone, coquina
	Microscopic shells ($CaCO_3$)	Chalk
Nonclastic (altered)	Microscopic shells (SiO_2), recrystallized silica	Chert
	Consolidated plant remains (largely carbon)	Coal

A distinguishing characteristic of most sedimentary rock is bedding, or **stratification** into layers. Because shale is very thinly stratified, or *laminated,* it splits into thin sheets. Some sedimentary rocks, such as sandstone and limestone, may occur in beds that are several feet thick. Other structures in these rocks give hints as to how the rock formed. *Cross-bedding*—stratification that is inclined at an angle to the main stratification—indicates the influence of wind or water currents. Thick cross-beds and frosted quartz grains in sandstone indicate an ancient desert sand-dune environment (◉ Figure 2.19). Some shale and claystone show polygonal *mud cracks* on bedding planes, similar to those

◉ FIGURE 2.19 Wind-blown sandstone exhibiting large cross-beds; Zion National Park, Utah.

◉ FIGURE 2.20 Mud cracks in an old clay mine; Ione, California.

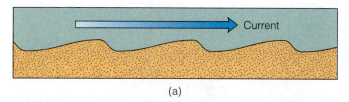

(a)

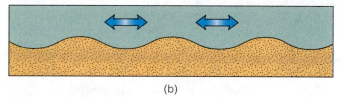

(b)

⊙ FIGURE 2.21 Ripple marks found in sedimentary rocks. *(a)* Asymmetrical ripple marks are common in streambeds and are also found on bedding planes in sedimentary rocks. *(b)* Symmetrical ripple marks are due to oscillating water motion; that is, wave action.

found on the surface of modern dry lakes; these indicate desiccation in a subaerial environment (⊙ Figure 2.20). *Ripple marks,* low, parallel ridges in deposits of fine sand and silt, may be asymmetrical or symmetrical (⊙ Figure 2.21). Asymmetrical ripples indicate a unidirectional current; symmetrical ones, the back-and-forth motion produced by waves in shallow water. Only in sedimentary rock do we find abundant fossils, the remains or traces of life.

METAMORPHIC ROCKS. Rocks that have been changed from preexisting rocks by heat, pressure, or chemical processes are classified as metamorphic rocks. The process of metamorphism results in new structures, textures, and minerals. **Foliation** (Latin *folium,* "leaf") is the flattening and layering of minerals by nonuniform stresses. Foliated metamorphic rocks are classified by the development of this structure. Slate, schist, and gneiss are foliated, for example, and they are identified by the thickness or crudeness of their foliation (⊙ Figure 2.22). Metamorphic rocks may also form by **recrystallization.** This occurs when a rock is heated and strained by uniform stresses so that larger, more perfect grains result or new minerals form. In this manner a limestone may recrystallize to marble or a quartz sandstone to quartzite, one of the most resistant of all rocks (⊙ Figure 2.23). ▪ Table 2.4 summarizes the characteristics of common metamorphic rocks.

CONSIDER THIS . . .

Geology can be thought of as a form of detective work in which events and physical settings from millions of years ago can be reconstructed from "clues" found in layers of sedimentary rocks. What are some of these "clues" in the sedimentary rock record that tell us about the environments in which they formed?

(a)

(b)

(c)

⊙ FIGURE 2.22 Metamorphic rocks: *(a)* coarsely foliated gneiss, *(b)* mica schist showing wavy or crinkly foliation, *(c)* finely foliated slate showing slaty cleavage.

Rock Defects

Some rocks have structures that geologists view as "defects"; that is, surfaces along which landslides or rockfalls may occur. Almost any planar structure, such as a bedding plane in sedimentary rock or a foliation plane in metamorphic rock, holds potential for rock slides or falls. The orientation of a plane in space, such as a stratification plane, a fault, or a

(a)

(b)

◉ FIGURE 2.23 *(a)* Limestone recrystallizes to form white marble. The streaks are due to disseminated minerals, mostly the green mineral epidote. *(b)* Quartzite, the hardest and most durable common rock, is metamorphosed sandstone.

TABLE 2.4 Common Metamorphic Rocks

ROCK	PARENT ROCK	CHARACTERISTICS
Foliated or Layered		
Slate	Shale and mudstone	Splits into thin sheets
Schist	Fine-grained rocks, siltstone, shale, tuff	Mica minerals often crinkled
Gneiss	Coarse-grained rocks	Dark and light layers of aligned minerals
Nonfoliated or Recrystallized		
Marble	Limestone	Interlocking crystals
Quartzite	Sandstone	Interlocking, almost fused quartz grains

joint, may be defined by its **dip** and **strike** (Appendix 3). **Joints** are rock fractures without displacement and they occur in all rock types. They are commonly found in paral-

lel sets spaced several feet apart (◉ Figure 2.24). **Faults** are also fractures in crustal rocks, but they differ from joints in that some movement or displacement has occurred along the fault surface. Faults also are found in all rock types and are potential surfaces of "failure." We will examine the relationships of rock defects to geologic hazards when we discuss landslides, subsidence, and earthquakes.

◉ GEOLOGIC TIME

Interest in extremely long periods of time sets geology and astronomy apart from other sciences. Geologists think in terms of billions of years for the age of the earth and its oldest rocks, numbers that, like the national debt, are not easily comprehended. Time is such an important element in earth science that it has its own designation: *geologic time*. Why is the time of ancient events relevant to environmental geology? Ongoing research is being directed toward determining recurrence intervals for large earthquakes and volcanic

CONSIDER THIS . . .

In a famous scene from the 1959 film "North by Northwest," the protagonists (played by Cary Grant and Eva Marie Saint) try to avoid the villains by scrambling down the faces carved in the granite of Mount Rushmore (see Figure 2.32). In real life, rock climbers are often found scaling near-vertical cliffs of granite, which they prefer over cliffs composed of schist or gneiss. Why should this be the case?

◉ FIGURE 2.24 Joints in basalt flows near Chico, California.

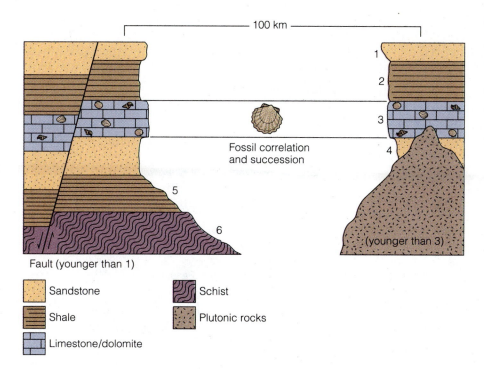

100 km

Fossil correlation
and succession

(younger than 3)

Fault (younger than 1)

Sandstone

Shale

Limestone/dolomite

Schist

Plutonic rocks

eruptions. It is postulated that if we know the average time interval between catastrophic events over the past several thousand years, we can predict more reliably when the next such events will occur.

Geologists use two different means of dating rocks and geologic events. In the field we normally use **relative age dating,** which establishes the sequence of geologic events. For example, we can establish that "these" rock strata are oldest, or that "this" igneous rock is younger than that rock. In the laboratory, **absolute age dating** can be carried out. This yields an age in years before the present for an event, rock, or artifact.

Relative age is determined by applying geologic laws based upon similar reasoning. For instance, the Law of Superposition tells us that in a stack of undeformed sedimentary rocks, the stratum (layer) at the top is the youngest. The Law of Cross-cutting Relationships tells us that a fault is younger than the youngest rocks it displaces or cuts. Similarly, we know that a pluton is younger than the rocks it intrudes (● Figure 2.25).

Using these laws, geologists arranged a great thickness of sedimentary rocks and their contained fossils representing an immense span of geologic time. The geologic age of a particular sequence of rock was then determined by applying the Law of Fossil Succession, the observed chronologic sequence of life forms through geologic time. This allows fossiliferous rocks from two widely separated areas to be correlated by matching key fossils or groups of fossils found in the rocks of the two areas (Figure 2.25). Using such indicator fossils and radioactive dating methods, geologists have developed the geologic time scale to chronicle the documented events of earth history (● Figure 2.26). Note that the scale is divided into units of time during which rocks were deposited, life evolved, and significant geologic events

such as mountain building occurred. Eons are the longest time intervals, followed, respectively, by eras, periods, and epochs. The Phanerozoic ("revealed life") Eon began 570 million years ago with the Cambrian Period, the rocks of which contain the first extensive fossils of organisms with hard skeletons. Because of the significance of the Cambrian Period, the informal term *Precambrian* is widely used to denote the time before it, which extends back to the formation of the earth 4.6 billion years ago. Note that the Precambrian is divided into the Archean and Proterozoic Eons, with the Archean Eon extending back to the formation of the oldest known in-place rocks about 3.9 billion years ago. Precambrian time accounts for 88 percent of geologic time, and the Phanerozoic for a mere 12 percent. The eras of geologic time correspond to the relative complexity of life forms: Paleozoic (oldest life), Mesozoic (middle life), and Cenozoic (most recent life). Environmental geologists are most interested in the events of the past few million years, a mere heartbeat in the history of the earth.

Absolute age-dating requires some kind of natural clock. The ticks of the clock may be the annual growth rings of trees or established rates of disintegration of radioactive elements to form other elements. At the turn of the twentieth century, American chemist and physicist Bertram Borden Boltwood (1870–1927) discovered that the ratio of lead to uranium in uranium-bearing rocks increases as the rocks' ages increase. He developed a process for determining the age of ancient geologic events that is unaffected by heat or pressure—**radiometric dating.** The "ticks" of the radioactive clocks are radioactive decay processes—spontaneous disintegrations of the nuclei of heavier elements such as uranium and thorium to lead. A radioactive element may decay to another element, or to an isotope of the same element. This decay occurs at a precise rate that can be deter-

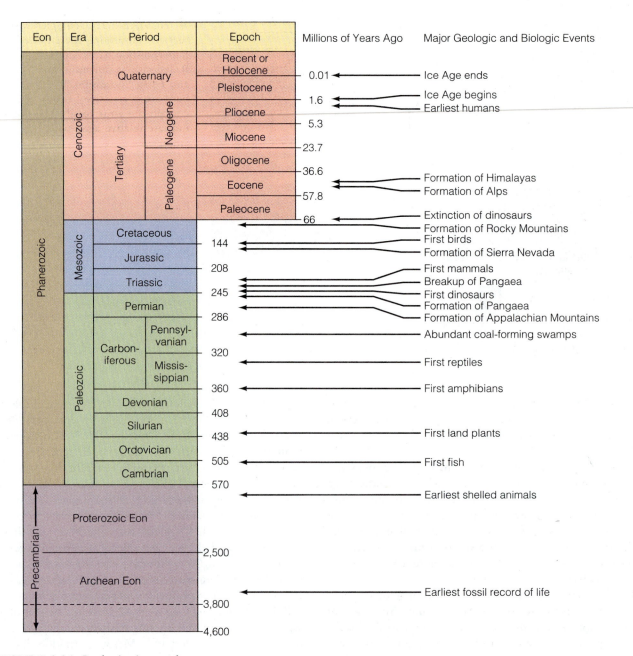

⊙ **FIGURE 2.26 Geologic time scale.**

mined experimentally. The most common emissions are alpha particles ($_2^4\alpha$), which are helium atoms, and beta particles (β^-), which are nuclear electrons. New *radiogenic* "daughter" elements or isotopes result from this **alpha decay** and **beta decay** (⊙ Figure 2.27).

For example, of the three isotopes of carbon, ^{12}C, ^{13}C, and ^{14}C, only ^{14}C is radioactive, and this radioactivity can be used to date events between a few hundred and a few tens-of-thousands of years ago. Carbon-14 is formed continually in the upper atmosphere by neutron bombardment of nitrogen, and it exists in a fixed ratio to the common isotope, ^{12}C. All plants and animals contain radioactive ^{14}C in equilibrium with the atmospheric abundance until they die, at which time ^{14}C begins to decrease in abundance, and along with it, the object's radioactivity. Thus by measuring the radioactiv-

ity of an ancient parchment, log, or piece of charcoal and comparing the measurement with the activity of a modern standard, the age of archaeological materials and geological events can be determined (⊙ Figure 2.28). Carbon-14 is formed by the collision of cosmic neutrons with ^{14}N in the atmosphere, and then it decays back to ^{14}N by emitting a nuclear electron (β^-).

$$^{14}_{7}N + neutron \longrightarrow {}^{14}_{6}C + proton$$
$$^{14}_{6}C \longrightarrow {}^{14}_{7}N + \beta^-$$

Both carbon—the radioactive "parent" element—and nitrogen—the radiogenic "daughter"—have 14 atomic mass units. However, one of ^{14}C's neutrons is converted to a proton by the emission of a beta particle, and the carbon

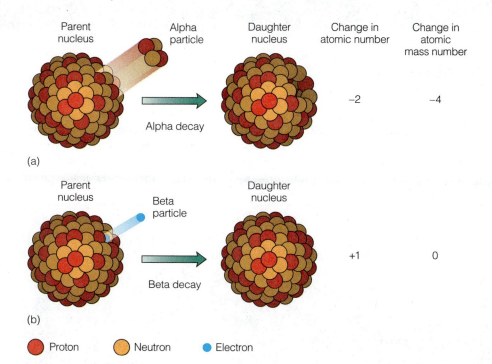

(a)

(b)

Proton   Neutron   Electron

changes (or *transmutes*) to nitrogen. This process proceeds at a set rate that can be expressed as a **half-life,** the time required for half of a population of radioactive atoms to decay. For ^{14}C this is about 5,730 years.

Radioactive elements decay exponentially; that is, in two half-lives one fourth of the original number of atoms remain, in three half-lives one eighth remain, and so on (⊙ Figure 2.29). So few parent atoms remain after seven or eight half-lives (less than 1 percent) that experimental uncertainty creates limits for the various radiometric dating methods.

Whereas the practical age limit for dating carbon-bearing materials such as wood, paper, and cloth is about 40,000 to 50,000 years, ^{238}U disintegrates to ^{206}Pb and has a half-life of 4.5×10^9 years. Uranium–238 emits alpha particles (helium atoms). Since both alpha and beta disintegrations can be measured with a Geiger counter, we have a means of determining the half-lives of a geological or archaeological sample. ■ Table 2.5 shows radioactive parents, daughters, and half-lives commonly used in age-dating.

 AGE OF THE EARTH

Before the advent of radiometric dating, determining the age of the earth was a source of controversy between established religious interpretations and early scientists. Archbishop Ussher (1585–1656), the Archbishop of Armagh and a professor at Trinity College in Dublin, declared that the

⊙ FIGURE 2.28 (right) Carbon-14 is formed from nitrogen-14 by neutron capture and subsequent proton emission. Carbon-14 decays back to nitrogen-14 by emission of a nuclear electron (β⁻).

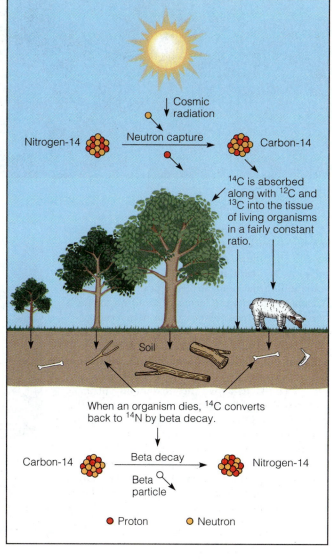

Cosmic radiation

Nitrogen-14   Neutron capture   Carbon-14

^{14}C is absorbed along with ^{12}C and ^{13}C into the tissue of living organisms in a fairly constant ratio.

Soil

When an organism dies, ^{14}C converts back to ^{14}N by beta decay.

Carbon-14   Beta decay   Nitrogen-14
Beta particle

● Proton   ● Neutron

GALLERY

A Rock Collection

S pecific rocks have had historical, aesthetic, and practical significance to humans; Plymouth Rock, the Rock of Gibraltar, the biblical Rock of Ages, and the Rosetta Stone are examples. Various geologic agents of erosion have shaped the rocks of the Grand Canyon, Bryce Canyon, Yosemite, and other national parks into areas that we admire for their great beauty. Early humans lived in rock caves and threw rocks to defend themselves against their

(a)

◉ FIGURE 2.31 Ayers Rock, the world's largest sandstone monolith, is one of Australia's leading tourist attractions. An erosional remnant of a previous highland, it rises abruptly above the surrounding flat land. In geological terminology it is called an *inselberg* (German "island mountain") because of its resemblance to an island rising from the sea. Aborigines (native Australians) call Ayers Rock *Uluru.* In 1958 Ayers Rock and the rocks of the Olgas nearby were incorporated into Uluru National Park. *(a)* Ayers Rock rises 348 meters (1,141 ft) above the surrounding plain, and is scaled by 80 percent of the visitors to the park. *(b)* The Olgas, a circular grouping of more than 30 red conglomerate domes, is a subdued but still spectacular sandstone inselberg in Uluru National Park. Its highest point, Mt. Olga, rises 460 meters (1,500 ft) above the plain. The Olgas' Aboriginal name is *katajuta,* "many heads." Ayers Rock is in the background.

(b)

enemies and to kill animals for food. Over time humans came to build crude rock shelters, and then rock fences, and eventually beautiful castles and abbeys. Today, rock and rock products are important mineral resources, used mostly as building materials. The beauty and usefulness of rocks are celebrated in ◉ Figures 2.31 to 2.34.

◉ FIGURE 2.32 Sixty-foot-high representations of the heads of four presidents were carved into Mount Rushmore's Harney Peak Granite in southwestern South Dakota between 1927 and 1941. The great work was possible because the Precambrian granite is relatively homogeneous (massive) and not riddled with defects. Why would such a project not have been considered if Mount Rushmore were composed of slate, shale, or schist? Can you identify the presidents honored here?

◉ FIGURE 2.33 President Theodore Roosevelt (left) and naturalist John Muir at Yosemite Valley in 1903. The rock on which they were photographed is Cathedral Peak Granite, one of the many granite intrusions that make up the Sierra Nevada batholith. Note the rounded exfoliation domes in the background (see Chapter 6).

◉ FIGURE 2.34 A skillfully assembled sandstone wall in Yorkshire, England. Yorkshire is the largest county (shire) in England, and literally thousands of kilometers of these walls are found there. This commonly occurring sandstone breaks readily into platy and blocky pieces, which makes it perfect for wall building.

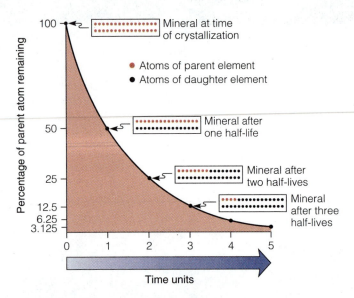

100 — Mineral at time of crystallization

● Atoms of parent element
● Atoms of daughter element

50 — Mineral after one half-life

25 — Mineral after two half-lives

12.5 — Mineral after three half-lives
6.25
3.125

Percentage of parent atom remaining

0 1 2 3 4 5

Time units

⊙ **FIGURE 2.29** Decay of a radioactive parent element with time. Each time unit is one half-life. Note that after two half-lives one fourth of the parent element remains, and that after three half-lives one eighth remains.

earth was formed in the year 4004 B.C. Not only did Ussher provide the year but also the day, October 23, and the time, 9:00 A.M. Ussher has many detractors, but Stephen J. Gould of Harvard University, though not proposing acceptance of Ussher's date, is not one of them. Gould points out that Ussher's work was good scholarship for its time, because other religious scholars had extrapolated from Greek and Hebrew scriptures that the earth was formed in 5500 B.C. and 3761 B.C., respectively. Ussher based his age determination on the verse in the Bible that says "one day is with

the Lord as a thousand years" (2 Peter 3:8). Because the Bible also says that God created heaven and earth in 6 days, Ussher arrived at 4004 years B.C.—the extra four years because he believed Christ's birth year was wrong by that amount of time. Ussher's age for the earth was accepted by many as "gospel" for almost 200 years.

By the late 1800s geologists believed that the earth was on the order of 100 million years old. They reached their estimates by dividing the total thickness of sedimentary rocks (tens of kilometers) by an assumed annual rate of deposition (mm/year). Evolutionists such as Charles Darwin thought that geologic time must be almost limitless in order that minute changes in organisms could eventually produce the present diversity of species. Both geologists and evolutionists were embarrassed when the British physicist William Thomson (later Lord Kelvin) demonstrated with elegant mathematics how the earth could be no older than 400 million years, and maybe as young as 20 million years. Thomson based this on the rate of cooling of an initially molten earth and the assumption that the material composing the earth was incapable of creating new heat through time. He did not know about radioactivity, which adds heat to rocks in the crust and mantle.

Bertram Boltwood postulated that older uranium-bearing minerals should carry a higher proportion of lead than younger samples. He analyzed a number of specimens of known relative age, and the absolute ages he came up with ranged from 410 million to 2.2 billion years old. These ages put Lord Kelvin's dates based on cooling rates to rest and ushered in the new radiometric dating technique. By extrapolating backward to the time when no radiogenic lead had been produced on earth, we arrive at an age of 4.6 billion years for the earth. This corresponds to the dates obtained from meteorites and lunar

■ **TABLE 2.5 Age-Dating Methods and Some Important Ages**

	METHOD		
Parent Isotope	Daughter Isotope	Half-life, years	Range, years
$^{238}_{92}$U $\longrightarrow$	$^{206}_{82}$Pb $+ 8^4_2\alpha + 6\beta^- + Q\star$	4.5×10^9	$10^7 - 4.5 \times 10^9$
$^{232}_{90}$Th $\longrightarrow$	$^{208}_{82}$Pb $+ 6^4_2\alpha + 4\beta^- + Q$	1.5×10^{10}	$10^7 - 4.5 \times 10^9$
$^{235}_{92}$U $\longrightarrow$	$^{207}_{82}$Pb $+ 7^4_2\alpha + 4\beta^- + Q$	7.1×10^8	$10^7 - 4.5 \times 10^9$
$^{87}_{37}$Rb $\longrightarrow$	$^{87}_{38}$Sr $+ \beta^- + Q$	5.0×10^{10}	$10^7 - 4.5 \times 10^9$
$^{40}_{19}$K $\longrightarrow$	$^{40}_{18}$Ar $+ \beta^- + Q$	1.3×10^9	$10^6 - 4.5 \times 10^9$
$^{14}_6$C $\longrightarrow$	$^{14}_7$N $+ \beta^- + Q$	5,730	100 to 50,000

Oldest known rocks: zircon minerals in the Acasta gneisses of Canada have been dated as 3.96 billion years old.

Oldest known crust: zircon minerals in rocks of western Australia derived from older crust have been dated as 4.0–4.3 billion years old.

Oldest known fossils: algae and bacteria have been dated as 3.5 billion years old.

★Q = heat

rocks, which are part of our solar system. The oldest known intact terrestrial rocks are found in the Acasta gneisses of the Slave geological province of Canada's Northwest Territories. Analyses of lead to uranium ratios on the gneisses' zircon minerals indicate that the rocks are 3.96 billion years old. However, older detrital zircons on the order of 4.0–4.3 billion years old have been found in

western Australia, indicating that some stable continental crust was present as early as 4.3 billion years ago (Table 2.5). Suffice it to say that the earth is very old and that there has been abundant time to produce the features we see today. (● Figure 2.30)

Environmental geology deals mostly with the present and the recent past—the time interval known as the *Holocene*

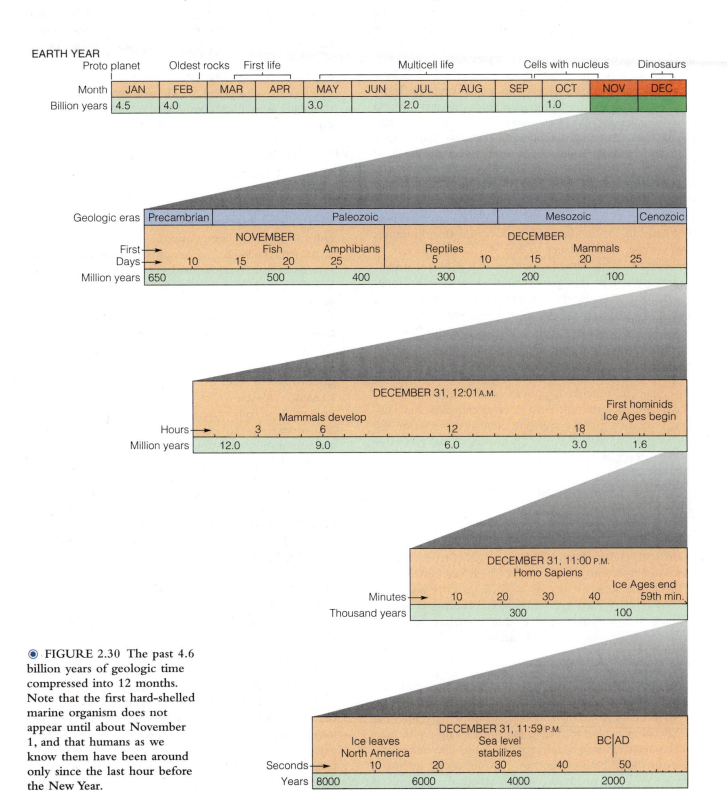

● FIGURE 2.30 The past 4.6 billion years of geologic time compressed into 12 months. Note that the first hard-shelled marine organism does not appear until about November 1, and that humans as we know them have been around only since the last hour before the New Year.

Epoch (10,000 years ago to the present) of the *Quaternary Period*. However, because the Great Ice Age advances of the Pleistocene Epoch of the Quaternary Period (which preceded our own Holocene Epoch) had such a tremendous impact on the landscape of today, we will investigate the probable causes of the "ice ages" in order to speculate a bit about what the future might bring. In addition, Pleistocene glacial deposits are valuable sources of underground water, and they have economic value as sources of building materials. We will see that "ice ages" have occurred several times throughout geologic history (see Chapter 12).

What impresses one most about geologic time is how short the period of human life on earth has been. If we could compress the 4.6 billion years since the earth formed into one calendar year, *Homo sapiens* would appear about 30 minutes before midnight on December 31 (see Figure 2.30, page 47). The last ice-age glaciers would begin wasting away a bit more than 2 minutes before midnight, and written history would exist for only the last 30 seconds of the year. Perhaps we should keep this calendar in mind when we hear that the dinosaurs were an unsuccessful group of reptiles. After all, they endured 50 times longer than hominids have existed on the planet to date. At the present rate of population growth, there is some doubt that humankind as we know it will be able to survive anywhere near that long.

The earth we see today is the product of earth processes acting on earth materials over the immensity of geologic time. The response of the materials to the processes has been the formation of mountain ranges, plateaus, and the multitude of other physical features found on the planet. Perhaps the greatest of these processes has been the formation and movement of platelike segments of the earth's surface, which influence the distribution of earthquakes and volcanic eruptions, global geography, mineral deposits, mountain ranges, and even climate. These *plate tectonic* processes are the subject of the next chapter.

 # SUMMARY

Big Bang

DESCRIPTION Rapid expansion of dense matter, mostly hydrogen and helium, which segregated into smaller "clouds" of matter to form galaxies and stars 10–15 billion years ago. Gravitational collapse of clouds raised temperatures so that thermonuclear reactions occurred, forming heavier elements.

Solar System

DESCRIPTION Formed 5 billion years ago from an interstellar cloud (nebula).

PROBABLE CAUSE The nebula collapsed as a result of a supernova explosion (called the "bing bang" by some). The nebula started to spin and condense. The inner part of the cloud became the hot protosun and it was surrounded by a cooler, disc-shaped cloud (the solar neb-

ula). Volatile substances in the outer reaches of the nebula formed the large outer planets; rocky substances closest to the sun formed the dense inner planets.

EARTH Probably formed by accretion of asteroid-sized masses (planetesimals).

Earth System

The earth can be thought of as having five interconnected regions or spheres of chemical, physical, and/or biological activity, each of which plays a role in the mobilization, transport, and deposition of the chemical elements. The spheres are the atmosphere (air), pedosphere (soil), lithosphere (rock), hydrosphere (water), and biosphere (life). They continually recycle and are vital to the dynamic earth system.

Earth Materials

CRUST Outermost rocky layer of earth composed of rocks that are aggregates of minerals.

MINERALS Naturally occurring inorganic substances with a definite set of physical properties and a narrow range of chemical compositions.

CLASSIFICATION BY Chemistry of anions (negative ions or radicals), such as oxides, sulfides, carbonates, etc. Silicates are the most common and important mineral group and are composed of silica tetrahedra $(SiO_4)^{-4}$ units that are joined to form chains, sheets, and networks.

IDENTIFICATION Physical properties such as hardness, cleavage, crystal structure, fracture pattern, and luster.

ROCKS Consolidated or poorly consolidated aggregates of one or more minerals or organic matter.

CLASSES *Igneous*—rock formed by crystallization of molten or partially molten material.

Sedimentary—layered rock resulting from consolidation and lithification of sediment.

Metamorphic—preexisting rock that has been changed by heat, pressure, or chemically active fluids.

CLASSIFICATION BY Texture and composition.

STRUCTURES Many planar rock structures may be viewed as defects along which landslides, rock falls, or other potentially hazardous events may occur. Stratification in sedimentary rocks and foliation in metamorphic rocks are common planar features. Faults and joints occur in all classes of rock.

ROCK CYCLE A sequence of events by which rocks are formed, altered, destroyed, and reformed as a result of internal and external earth processes.

Geologic Time

RELATIVE TIME The sequential order of geologic events established by using basic geologic principles or laws.

GEOLOGIC TIME SCALE The division of geologic history into eons, eras, periods, and epochs. The earliest 88 percent of geologic time is known informally as *Precambrian* time. Precambrian rocks are generally not fossiliferous. Rocks of the Cambrian Period and younger contain good fossil records of shelled and skeletonized organisms. Environmental geology is concerned mainly with geologic events of the present epoch, the Holocene, and the one that preceded it, the Pleistocene.

ABSOLUTE AGE Absolute dating methods use some natural "clock." For long periods of time, such clocks are known rates of radioactive decay. A parent radioactive isotope decays to a daughter isotope at a rate that can be determined experimentally. This established rate, expressed as a half-life, enables us to determine the age of a sample of material.

AGE OF THE EARTH Comparing present U/Pb ratios with those in iron meteorites, we can extrapolate backward and obtain an earth age of 4.6 billion years old.

ROCKS AND CRUST The oldest known in-place rocks are 3.96 billion years old. Dating of minerals derived from older crust yields an age of 4.0–4.3 billion years for stable continental crust.

KEY TERMS

absolute age dating	ion
alpha decay	isotope
amorphous	joint
aphanitic	lava
atom	lithified
atomic mass	luster
atomic number	magma
batholith	mantle
beta decay	metamorphic rocks
big bang	mineral
biogenic sedimentary rocks	Mohs hardness scale
chemical sedimentary rocks	nucleus
clastic	pegmatite
cleavage	phaneritic texture
core	planetesimal
crust	plutonic
detrital sedimentary rocks	pyroclastic
dip	radiometric dating
element	recrystallization
extrusive	relative age dating
fault	rock
foliation	rock cycle
fracture	sedimentary rocks
half-life	stratification
igneous rocks	strike
intrusive	supernova

STUDY QUESTIONS

1. Explain the "big bang" theory of how the universe formed. Then explain how our planetary system may have come into existence.
2. What physical and chemical factors are the bases of the rock-classification system? What are the three classes of rocks, and how does each form? Identify one characteristic of each class that usually makes it readily distinguishable from the other rock classes.
3. How and where do batholiths form? What type of rock most commonly forms in batholiths?
4. What are some of the planar surfaces in rocks that may be weak and thus lead to various types of slope failure?
5. How may structures in sedimentary rocks be used to reconstruct past environments?
6. How may absolute age-dating techniques be used to the betterment of human existence?
7. The earth is 4.6 billion years old, and humans have been on earth for only a few hundred thousand years—but a blink of an eye in the great length of geologic time. What changes of a global nature have humans invoked on the earth's natural systems (water, air, ice, the solid earth, and biology) in this short length of time? Which impacts are reversible, and which ones cannot be reversed or mitigated?
8. What is the most common mineral species? Name several rocks in which it is a prominent constituent.
9. What is cleavage? How can it serve as an aid in identifying minerals?
10. The most common intrusive igneous rock is composed mostly of (1) the most common mineral and (2) a mineral of the most common mineral group. Name the rock and its constituent minerals.
11. What mineral is found in both limestone and marble?

FURTHER INFORMATION

BOOKS AND PERIODICALS

Albritton, C. C. 1984. Geologic time. *Journal of Geological Education* 32, no. 1: 29–47.

Bowring, S. A.; I. S. Williams; and W. Compston. 1989. 3.96 Ga gneisses from the Slave province, Northwest Territories, Canada. *Geology* 17: 971–975.

Brown, V. M., and J. A. Harrell. 1991. Megascopic classification of rocks. *Journal of Geological Education* 39: 379.

Cameron, A. G. W. 1975. The origin and evolution of the solar system. *Scientific American* 233, no. 3: 32.

Dietrich, R. V., and Brian Skinner. 1979. *Rocks and rock minerals.* New York: John Wiley and Sons.

Eicher, D. L. 1976. *Geologic time,* 2d ed. Englewood Cliffs, N.J.: Prentice-Hall, Inc.

Gunter, Mickey Eugene. 1994. Asbestos as a metaphor for teaching risk perception. *Journal of Geological Education* 42: 17.

Hurley, Patrick. 1959. *How old is the earth?* New York: Anchor Books.

Libby, W. F. 1955. *Radiocarbon dating.* Chicago: University of Chicago Press.

Mackenzie, Fred T. G., and Judith A. Mackenzie. 1995. *Our changing planet: An introduction to earth system science and global environmental change.* Englewood Cliffs, N.J.: Prentice-Hall, Inc.

National Research Council, National Academy of Sciences. 1993. *Solid-earth sciences and society.* Washington, D.C.: National Research Council Commission on Geosciences, Environment, and Resources, National Academy of Sciences Press.

Nuhfer, E. B.; R. J. Proctor; and Paul H. Moser. 1993. *The citizen's guide to geologic hazards.* Arvada, Colo.: American Institute of Professional Geologists.

Sagan, Carl. 1975. The solar system. *Scientific American* 233, no. 4: 98.

Silk, Joseph. 1989. *The big bang,* 2d ed. New York: W. H. Freeman and Co.

Skinner, H. Catherine W., and Malcom Ross. 1994. Geology and health. *Geotimes* (Washington, D.C.: American Geological Institute), January: 11–12.

_____. 1994. Minerals and cancer. *Geotimes,* January: 13–15.

Snow, T. P. 1993, *Essentials of the dynamic universe.* St. Paul, Minn.: West Publishing Company.

Wicander, Reed, and James Monroe. 1993. *Historical geology,* 2d ed. St. Paul, Minn.: West Publishing Company.

No man but feels more of a man in the world if he can have a bit of ground that he can call his own. However small it is on the surface, it is 4,000 miles deep; and that is a very handsome property.

CHARLES DUDLEY WARNER, AMERICAN EDITOR AND ESSAYIST (1829–1900)

During the 1960s our concepts of the architecture of the earth's crust and how the earth "works" were completely revolutionized by the theory of **plate tectonics** (Greek *tektonikos,* "builder"). The theory maintains that the lithosphere, the outer rocky shell of the earth, consists of a number of rigid plates, 70–150 kilometers (40–90 mi) thick, that slide under, past, and away from one another. Continents, although parts of the moving plates, are but passive passengers. Where plates interact with each other, earthquakes and volcanic activity occur, mountain ranges may form, and internal heat is released to the earth's surface. Plate tectonics integrates two older explanations for the distribution of the

OPENING PHOTO
This is the view southward across the Thingvellir graben on the crest of the Mid-Atlantic ridge, where it passes through part of Iceland. The photo was taken from the American plate, looking across the graben toward the Atlantic plate. Lake Thingvallavatn, in the background, occupies part of the graben.

earth's physical and geophysical features—the theories of continental drift and sea-floor spreading.

CONTINENTAL DRIFT

What ultimately was to become the theory of plate tectonics began in 1910 when Alfred Wegener (1880–1930), a German meteorologist, presented a paper to the Frankfurt Geological Society in which he expressed his opinion that the continents have moved great horizontal distances in the past 100 million years. He called his hypothesis *die Verschiebung der Kontinente,* meaning "continental displacement," which English-speaking scientists translated to **continental drift,** the name that stuck. Wegener called the late Paleozoic supercontinent—a large landmass consisting of all the continents we know today—**Pangaea** (Greek "all land"; ◉ Figure 3.1). He theorized that Pangaea split apart during Mesozoic time and that the continents then moved steadily in all directions to their present positions. Wegener explained that as the continents drifted westward *(Westwanderung),* they plowed up mountain ranges such as the Andes and the North American Coast Ranges. Australia became geographically isolated, and the Atlantic Ocean became larger at the expense of a diminishing Pacific Ocean.

Strong evidence in favor of the hypothesis, according to Wegener and many Southern Hemisphere geologists, is the distribution of late Paleozoic (Permian Period) glacial deposits and plant fossils known collectively as the *Gondwana Succession.* These distinctive fossil land plants (◉ Figure 3.2) and their associated glacial deposits are found only in South America, Africa, India, Australia, and Antarctica (◉ Figure 3.3). The only explanation for this, it was argued, is that these continents were once connected in the southern part of Pangaea. Some years earlier this hypothetical former landmass with the distinct Permian glaciation pattern had been named **Gondwanaland.** (The northern landmass had been named **Laurasia.**)

Additional paleontological evidence supporting continental drift is supplied by the amphibian and reptile fossils currently found on widely separated Gondwana continents. In particular are the fossil remains of the Permian freshwater reptile *Mesosaurus* found in Brazil and South Africa. It is inconceivable that these freshwater creatures swam or island-hopped across the salty Atlantic Ocean to an identical freshwater habitat on the opposite side. It seems much more reasonable that *Mesosaurus* inhabited large Permian lakes and/or rivers spanning parts of both continents when they were joined, and that the rocks containing their fossils were separated when Pangaea split apart.

Wegener believed that the ultimate proof of drifting was the absolute motion of Greenland relative to Europe that had been found. He relied upon surveys made between 1823 and 1870—later proved in error—indicating that Greenland had moved 420 meters (¼ mi) to the west in 47

(a)

(b)

Late Carboniferous Period (250 million years ago)

◉ **FIGURE 3.1** The earth of 250 million years ago according to Alfred Wegener. The supercontinent Pangaea began to break up about 100 million years ago, and the individual continents, roughly outlined as we know them today, have since then "drifted" to their present positions. The stippled areas represent widespread shallow seas over the continents in which sediments and fossils accumulated.

◉ **FIGURE 3.2** Plant fossils of the Gondwana Succession. These fossils are found only in Gondwana; that is, the present-day continents of South America, Africa, India, Australia, and Antarctica. Shown are *Glossopteris* leaves from the Upper Permian *(a)* Dunedoo Formation and *(b)* Illawarra Coal Measures in Australia.

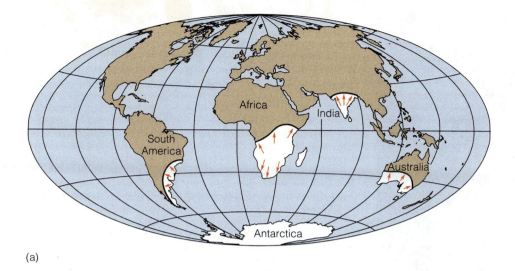

(a)

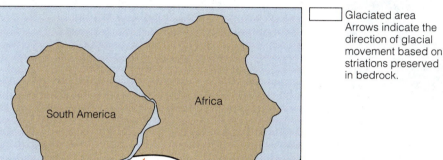

Glaciated area
Arrows indicate the direction of glacial movement based on striations preserved in bedrock.

(b)

(c)

● FIGURE 3.3 (a) The directions of glacial grooves preserved in bedrock on continents as they are now positioned show the direction of glacier movement during Late Paleozoic (Gondwana Succession) time. For a large glacier or a number of glaciers to produce the directions of the observed grooves would otherwise require a source area in the Indian and South Atlantic Oceans. (b) With Gondwana continents reunited and an ice sheet over a south pole in South Africa, the directions of glacier motion are resolved. (c) Gondwana glacial grooves at Hallet's Cove, Australia. Formed in Permian time, these grooves are more than 200 million years old. The width of the area shown in the photo is approximately 2 meters.

years (an astounding 8.9 m or 29 ft per year). Nonetheless, persuasive arguments against drifting were founded on the lack of a known mechanism for driving the system and on the rather simple logic that if the oceans were sufficiently rigid to cause mountain ranges to crumple up along the forward edges of the moving continents, then how could they *move*. Wegener spent his entire life defending his hypothesis in scientific papers and five editions of his classic book *The Origin of Continents and Oceans*. He died on an expedition to Greenland, where, it is said, his theory was

born 30 years earlier when he saw a giant iceberg float away from a calving glacier. Unfortunately, he was never able to reconcile the forces necessary to drive continental drift, and his hypothesis gained relatively few adherents until thirty years after his death. Nevertheless, there persisted the nagging correlations of fossils, mountain ranges, and ages of rocks and geologic provinces across the Atlantic and Indian Oceans. Wegener deserves our admiration for his dogged determination, scientific insight, and ability to synthesize seemingly disparate data. To evaluate Wegener's attractive

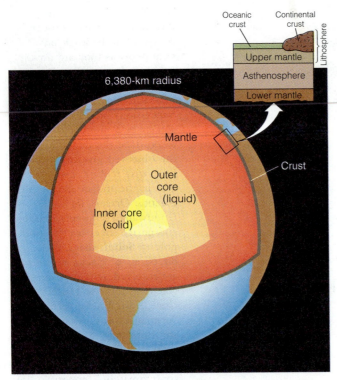

FIGURE 3.4 The internal structure of the earth. The lower part of the figure shows the layers that are defined by contrasts in density, as deduced from observations of seismic waves. The upper part of the figure illustrates the relative thickness of the crust in continental areas as compared with the sea floor. The lithosphere composes the rigid tectonic plates moving on a plastic asthenosphere.

Structure of the Earth

The interior of the earth is differentiated into distinct layers separated by boundaries called **discontinuities**: see ◉ Figure 3.4. The model of the earth's interior in the figure is not speculation. Our "X-ray" vision is provided by earthquake (seismic) waves, which give us clues to the physical and chemical nature of the earth's interior. Although seismic waves cannot tell us exactly what composes the earth hundreds of kilometers below its surface, their velocities and travel paths give us information about the physical properties (density, elasticity, etc.) of the materials, from which we can make deductions about their composition. The **lithosphere** is the solid, rigid part of the earth comprising the crust and the upper mantle. It extends 100–150 kilometers (60–90 mi) below the surface of the continents and is 70–80 kilometers (40–50 mi) thick beneath the oceans. What we regard as the plates in plate tectonic theory are composed of lithosphere. The crust, or upper part of the lithosphere, consists of relatively light (granitic) continental crust and somewhat denser (basaltic) oceanic crust. Continental crust has a density of about 2.7 grams per cubic centimeter (gm/cm^3) and stands topographically above the denser oceanic (basalt) crust with a density of 3.0 gm/cm^3. This "floating" of lighter (less dense) lithospheric blocks on denser lithosphere is known as **isostasy**. At the base of the crust is the **Mohorovičić discontinuity** (usually called simply "M" or *Moho),* which is indicated by a drastic change in the behavior of seismic waves due to a change in the chemical or physical nature of the material (◉ Figure 3.5). The thickness of the crust and depth to the "M" varies, but it is about 35 kilometers (22 mi) beneath the continents and 10 kilometers (6 mi) beneath the sea floor. The crust is thickest under mountain ranges and thinnest under ocean basins (Figure 3.5).

but controversial concept requires some knowledge of the structure and composition of the earth's interior. We examine this subject next.

◉ FIGURE 3.5 Geologic cross section of the earth's crust, showing continental crust, oceanic crust, and the "M" discontinuity. This section typifies the west coast of South America, where an oceanic plate descends below a continental plate, and an andesitic volcanic mountain range, the Andes, has formed. Note that the rigid lithospheric plates, consisting of crust and the upper mantle, overlie the asthenosphere.

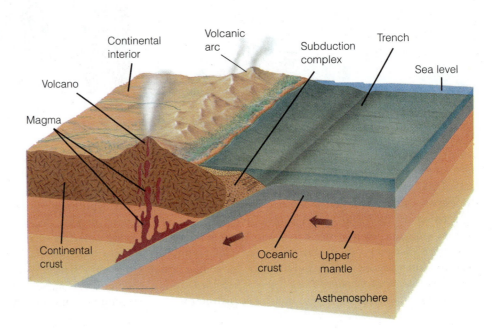

At the base of the lithosphere, at a depth of about 100 kilometers (60 mi), is a zone of low seismic velocities known as the *plastic layer,* or the **asthenosphere.** It is here that the rigid lithosphere uncouples, or detaches itself from the underlying mantle, and moves about. What drives the plates is poorly understood, but it is most certainly a combination of gravity acting on the lithosphere and thermal differences within the mantle. Thermal convection—analogous to warm air rising and cool air sinking in closed loops in our atmosphere—has been suggested. (We can produce a similar, observable convective phenomenon by carefully pouring cold milk into hot coffee or by gently heating tomato soup on the kitchen stove.) Although thermal convection does not adequately explain all the nuances of plate tectonics, it serves as a first approximation of the driving force. To complete the picture of the earth's interior, we find a very dense molten core and a solid inner core, both believed to be composed of a nickel–iron alloy. This composition is suggested by theory and by analogy to nickel–iron meteorites that strike the earth, which may represent chunks of a disintegrated planetary body.

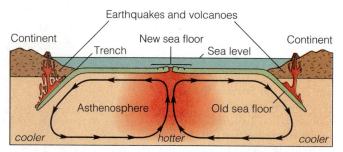

● **FIGURE 3.6** Sea-floor spreading driven by convection currents. New crust forms at ridges and is consumed at trenches. Volcanic activity and earthquakes are generated along the sinking oceanic crust.

 SEA-FLOOR SPREADING

U.S. geologists Harry Hess and Robert Dietz independently concluded in 1960 that new sea floor is being created at mid-ocean ridges by volcanic activity. Inasmuch as the earth is not growing larger, they said, an equal amount of oceanic crust is probably being lost at trenches (● Figure 3.6). They envisioned the driving force to be convection currents in the mantle caused by heat differences within the earth and gravity, an explanation that had been proposed 15 years earlier by British geologist Arthur Holmes. According to this theory, convection carries the rigid new crust away from the mid-ocean ridges like a conveyor belt and drives it into the mantle at trenches. This new theory, called **sea-floor spreading** by Dietz, explained the distribution of earthquakes and volcanoes on earth quite nicely.

 PLATE TECTONICS:
THE GLOBAL THEORY

Plate tectonics is a unifying concept in that it provides a coherent model of the outer part of the earth and how it works. According to the theory, the outer part of the earth is composed of seven large and a number of small interacting plates. There are only three ways these plates can interact and, therefore, three types of boundaries: divergent, convergent, and transform. Each of the three boundaries is produced by different stresses:

■ At **divergent boundaries** there is *tension;* stresses acting in opposite directions away from bodies tend to pull them apart.

■ At **convergent boundaries** there is *compression;* stresses converge in the same plane on bodies and tend to shorten or compress them.

■ At **transform boundaries** *shear* forces acting in opposite directions in different planes cause adjacent parts of a body to slide relative to each other in a direction parallel to a fracture or plane of contact (● Figure 3.7).

Where plates move apart, openings are created and molten material fills them, forming new oceanic lithosphere. These divergent boundaries, also known as **spreading centers,** occur at mid-ocean ridges. Here we find high heat flow, usually mild volcanic activity, and shallow (< 40 km or 25 mi deep) earthquakes.

Where plates come together, or collide, at convergent boundaries, crust is destroyed. One form of convergence occurs where an oceanic plate is driven beneath another oceanic plate or a continental plate; this is called **subduction.** Where two continental plates collide, folded mountain ranges may form. The Himalayas originated this way when the sediments forming them were caught in a "vice" between the Indian plate and the Eurasian plate. There we find strong earthquakes but no true subduction, because the lighter continental crust cannot sink into denser lithosphere. Where there is subduction, the descending plates may be identified by a zone of earthquake **foci** (points within the earth where the earthquakes originate) inclined at an angle of about 60 degrees to the earth's surface. This zone was discovered by Japanese seismologists in 1927 and is called the **Benioff zone,** in honor of Hugo Benioff who mapped the zones worldwide. Only at subduction zones do we find earthquake foci greater than 400 kilometers (250 mi) deep. This fact is cited as evidence that one rigid plate is sinking beneath another, thereby creating earthquakes at great depth. Subduction is most commonly associated with volcanic island arcs, such as the Japanese and Aleutian Islands, and the most destructive earthquakes and explosive volcanoes occur at these boundaries. (Earthquakes and volcanoes are the subjects of Chapters 4 and 5.)

At transform boundaries, plates slide past one another along faults, and crust is neither created nor destroyed. Also known as **transform faults,** we find that portions of mid-ocean ridges and trenches are offset or that the fault con-

● FIGURE 3.7 *(a)* Divergent, convergent, and transform plate boundaries. Each type of boundary is characterized by a particular type of plate motion and stress. *(b)* Stresses found at plate boundaries *(1)* Compression. Two forces acting in opposite directions in the same plane results in shortening of crust by folding and faulting. *(2)* Tension. Extension, or pulling the crust apart, causes faulting. (Rifting is shown.) *(3)* Shear. Two forces acting in opposite directions on different planes causes faulting and displacement along closely spaced planes.

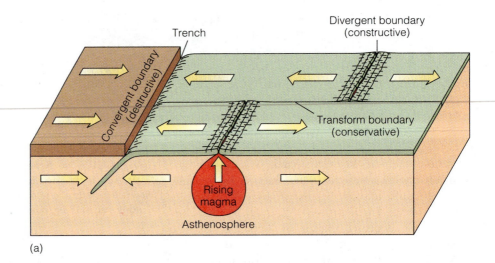

(a)

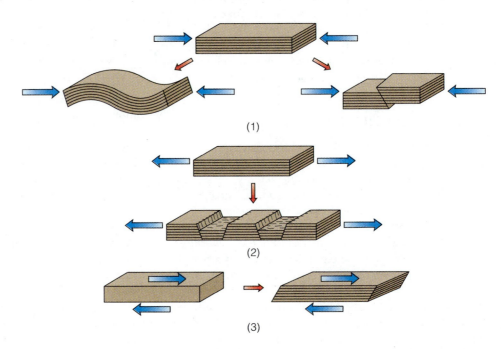

(1)

(2)

(3)

● FIGURE 3.8 The San Andreas fault is a transform fault, one of the plate boundary types, between the North American and Pacific plates.

nects a ridge and a trench. At the San Andreas fault, an example of a transform boundary, we find only earthquakes and no volcanic activity directly related to the boundary (●Figure 3.8). Regardless of the type of boundary, earthquakes and their associated epicenters (the points at the surface of the earth that are directly above earthquake foci) outline the major plates (●Figure 3.9).

Plate boundaries may also be described by what happens to the crust at that plate margin. Where plates diverge at a spreading center, new ocean crust is created, and we have a *constructive* plate boundary. Where an oceanic plate sinks and disappears at a subduction zone, the plate boundary is described as *destructive*. Finally, at transform boundaries crust is neither created nor destroyed, and we have a *conservative* boundary. Because these designations describe how the margin is modified, they also serve as a memory aid for what processes are taking place there.

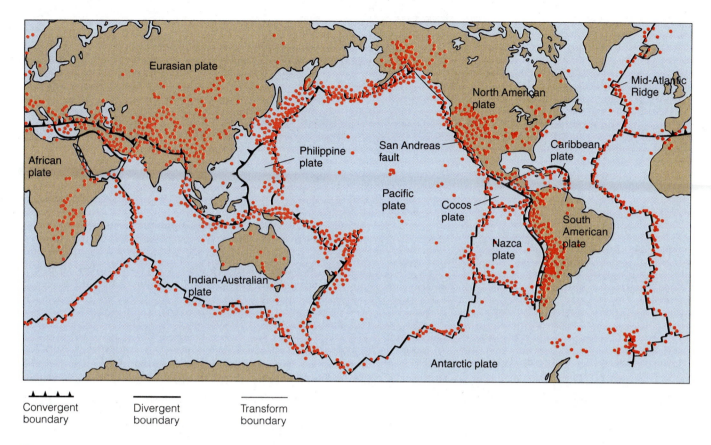

Convergent boundary

Divergent boundary

Transform boundary

⦿ **FIGURE 3.9** The earth's major tectonic plates and several smaller ones. Note that the plates are outlined by earthquake epicenters, each dot on the map representing an epicenter.

Proof of Plate Tectonics

The essential evidence for sea-floor spreading and plate tectonics is found by studying a totally unrelated phenomenon, the magnetic forces within the earth. This study determines the direction and strength of the earth's ancient "fossil" magnetic fields—its **paleomagnetism**—by establishing the **polarity** of magnetic minerals contained in ancient rocks, preferably rocks of known age. You see, the earth is essentially a magnetic dipole, like a bar magnet, that has a positive (north) pole and negative (south) pole. *Polarity* refers to the orientation of a magnet's positive or negative pole that is exhibited when the earth's north magnetic pole attracts the north-seeking end of a compass needle. Studies in the mid 1960s indicated that the earth's magnetic field was reversed periodically throughout geologic time; that is, the north magnetic pole became the south magnetic pole, and vice versa. Rocks that formed when the earth's magnetic field was the same as today's field yield a strong magnetic signal, or *positive anomaly,* whereas those that formed when the field was the opposite of today's field (reversed) yield a weak signal, a *negative anomaly.* An **anomaly** is a departure from what is expected or from the mean. These anomalous signals can be mapped one sample at a

time in a laboratory or on a large scale with an airborne or shipboard instrument called a *magnetometer.* We now know that during the past five million years the earth's magnetic poles have flip-flopped many times. The long periods of normal and reversed polarity are called *magnetic epochs,* and within the epochs there were shorter reversals, called *magnetic events* (⦿ Figure 3.10).

Oceanographers and geologists had long been puzzled by the striped magnetic-anomaly patterns mapped in sea-floor rocks that parallel mid-ocean ridges (⦿ Figure 3.11). Based on U.S. geologists' findings on the timing of magnetic reversals, British researchers reasoned that new sea floor created at mid-ocean ridges should preserve the magnetic field of the time it was formed and that, if sea-floor spreading had occurred, the pattern should change from normal to reversed as one travels away from the ridge. Repeated traverses by ships over several spreading centers were made, and just such a parallel, symmetric pattern of normally and reversely magnetized rocks was confirmed (Figure 3.11). In fact, at least 171 reversals over the past 76 million years are now known. The rate at which the sea floor is moving away from a ridge axis may be calculated by measuring the distance from the ridge to a magnetic anomaly of known age, and then simply dividing the distance by

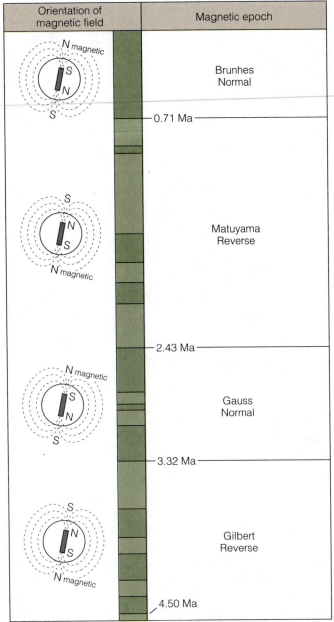

Orientation of magnetic field	Magnetic epoch
N magnetic / S / N / S	Brunhes Normal
	— 0.71 Ma —
S / N / S / N magnetic	Matuyama Reverse
	— 2.43 Ma —
N magnetic / S / N / S	Gauss Normal
	— 3.32 Ma —
S / N / S / N magnetic	Gilbert Reverse
	— 4.50 Ma —

Ma = millions of years ago

■ Normal

■ Reverse

● **FIGURE 3.10** The earth's magnetic field is much like the field that would be generated by a bar magnet inclined at 11 degrees from the earth's rotational axis. Reversals of the field have occurred periodically, leaving "fossil" magnetism in rocks that can be dated. Each epoch is named after an important student of earth magnetism.

the age of the anomaly. The rate at which the ocean is opening up, becoming larger, is then determined as twice the spreading rate calculated for one side of the ridge, because the spreading movement is presumed to be sym-

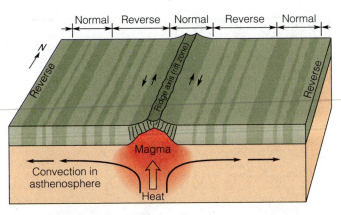

Normal | Reverse | Normal | Reverse | Normal

● **FIGURE 3.11** Magnetic anomalies form "stripes" parallel to a mid-ocean ridge. As lava forms at the ridge axis, it assumes the magnetic polarity of the earth at that time. Normal polarity episodes are shown as dark stripes; reversed polarity episodes are represented by light stripes.

metrical. Plates move at rates on the order of centimeters per year (about the same rate as human fingernail growth), not meters per year as Wegener believed. The East Pacific Rise is spreading at the fastest rate, with a rate up to 18 centimeters (7 in) per year, whereas the northern Mid-Atlantic Ridge is slowest, at about 2 centimeters (¾ in) per year (● Figure 3.12). Techniques using very-long-baseline radio interferometry (VLBI), satellite laser ranging (SLR), and global positioning system (GPS) are now providing direct measurements that confirm the theory of plate tectonics. As these techniques' use expands, it will be possible to monitor surface deformation directly. This will contribute to our understanding of earthquakes as well as other earth processes.

How Plate Tectonics Works

As can be seen from the model of plate-tectonic movement in ● Figure 3.13, plate boundaries determine the distribution and type of volcanic and earthquake activity. This relatively new global theory also explains other things that have puzzled scientists in the past. For instance, in the mid twentieth century many geologists believed that ocean basins and continents were stable, semipermanent features of the earth's crust. This belief was refuted when ancient sediments could not be found on the sea floor. Only young deposits, no more than 200 million years old (Jurassic age; representing just five percent of geologic time), have been found. Plate tectonics resolves this problem, because it allows that ocean basins have opened and closed many times throughout geologic time as the plates have moved about; therefore, no sediments older than the last opening cycle would be expected. In fact, if we multiply the average spreading rate by the length of the mid-ocean-ridge system, we find that new oceanic crust is formed at a rate of 2.8 square kilometers

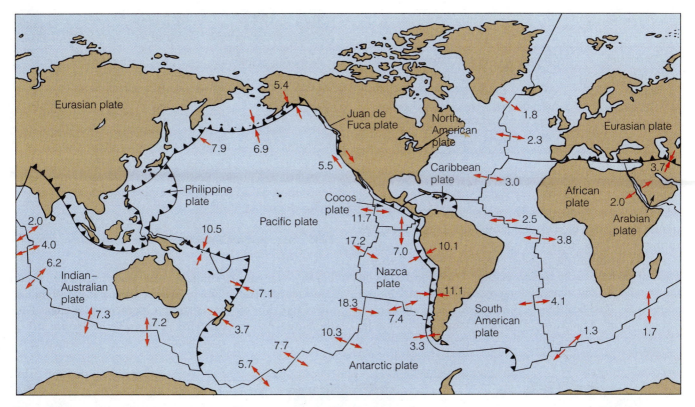

● FIGURE 3.12 Magnetic anomalies of known age are plotted against the distance from the respective mid-ocean ridge to determine the spreading rate. This map shows the average movement rate (cm/year) and the relative motion of the tectonic plates.

(1.1 mi²) per year. Because our oceans cover an area of 310 million square kilometers (120 million mi²), they could have formed in as little as 110 million years, and over the past two billion years, as many as 20 oceans may have been created and destroyed! Land geology also sup-

ports at least two and as many as five complete openings and closings of all ocean basins.

The *Wilson cycle* is an evolutionary sequence for oceans named for its developer, J. Tuzo Wilson, a Canadian geophysicist and a major contributor to plate tectonic theory:

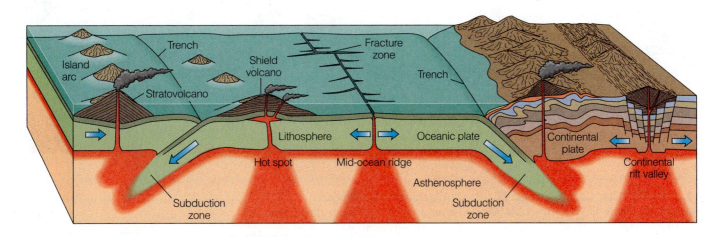

● FIGURE 3.13 Schematic cross section of the plate tectonic model. Divergent and convergent boundaries and their relationships to sea-floor and continental volcanoes are shown. Fracture zones on the sea floor are transform boundaries, and the mid-ocean ridge is offset along these features.

EVOLUTIONARY STAGE	EXAMPLE
Embryonic	Rift valleys of East Africa
Youthful	Gulf of California, Red Sea
Mature	Atlantic Ocean (growing larger)
Declining	Pacific Ocean (becoming smaller)
Terminal	Mediterranean Sea (closing, almost extinct)

The embryonic-stage rift valleys of East Africa exemplify down-dropped fault features called **grabens** that result when the crust is being pulled apart, "rifted," by convection in the mantle below the continent (◉ Figure 3.14). As rifting progresses, the crust becomes thinner, volcanism occurs, and eventually a new ocean basin forms with a floor of oceanic crust and an oceanic spreading center in the middle. In this manner long, narrow seas, such as the present-day Red Sea (Figure 3.14, part d) and Gulf of California, form. The South Atlantic Ocean, for example,

CONSIDER THIS . . .

A stockbroker might speculate on what the market will do the next day, a businessperson might try to forecast six-month trends, but only a geologist (or a geology student) will hazard a guess on the face of our planet millions of years from now. Here's your chance. What will be the fate of (1) the Mediterranean Sea, (2) the Atlantic and Pacific Ocean basins, and (3) the west coast of the United States exclusive of Alaska if present plate motions continue for the next 50 million years?

began as a rift in Pangaea between present-day South America and Africa in late Triassic time about 180 million years ago. The North Atlantic Ocean opened later. Thus the entire Atlantic basin has matured to its present size in

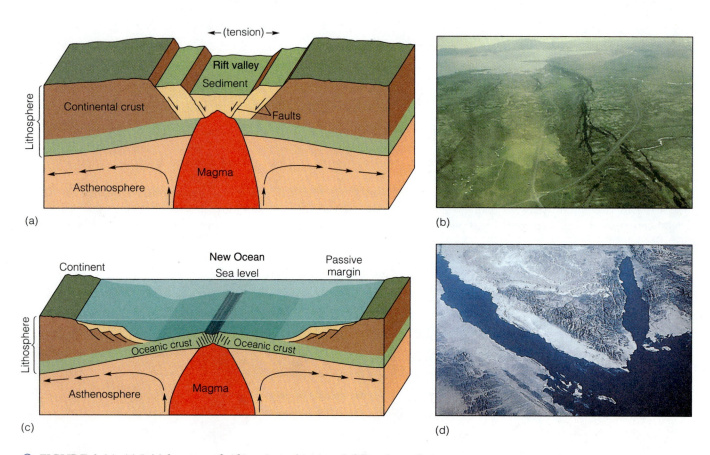

◉ FIGURE 3.14 *(a)* Initial stages of rifting (stretching) and formation of a graben like those in East Africa. *(b)* Western boundary of the Thingvellir graben, Iceland, formed by stretching (tension) along the Mid-Atlantic Ridge. This is similar to the mechanism that is forming the rift valleys in East Africa. *(c)* An ocean is born as the rift widens and new oceanic crust forms at the spreading center. *(d)* The Red Sea, the Gulf of Suez, and the Gulf of Aqaba are on a divergent plate boundary. A youthful ocean basin, the Red Sea looks today much like the South Atlantic Ocean must have appeared about 170 million years ago.

the last 180 million years (● Figure 3.15). It has done so at the expense of the declining-stage Pacific Ocean basin, which is growing smaller as North and South America move westward. The Mediterranean Sea, on the other hand, is slowly being "terminated" as the African and Eurasian plates converge on the basin. Evidence for this closure is found around the Mediterranean basin, where we find strong earthquakes, active volcanoes, and the east–west-trending Atlas and Alps mountain ranges.

A more dramatic evolution is apparent in Asia. There the Indian plate moved thousands of kilometers northward at about five centimeters (2 in) per year and collided with the Eurasian plate, folding, faulting, and uplifting the sediments of the ancient Tethys Sea to form the Himalaya Mountains. Where continent–continent collisions occur, mountain ranges form and catastrophic earthquakes occur. This is because both colliding plates are composed of low-density granitic crust, and thus, neither can subduct completely beneath the other into the heavier mantle (Case Study 3.1).

Within some ocean basins there are chains of aligned volcanic islands that become progressively older in one direction. The Hawaiian Islands, Line Islands, and Tuamotus are such chains in the Pacific Ocean. Although the Galápagos Islands off the coast of Ecuador—made famous by the writings of Charles Darwin—are only roughly aligned, they are of similar origin. These lines-of-islands are not associated with plate boundaries, but have formed over rising plumes of lava, or **hot spots,** in the mantle. These rising intraplate plumes penetrate the general convective circulation in the mantle that drives plate motion and are independent of spreading centers. As a tectonic plate moves over one of these plumes, it carries with it the volcano that formed there from the plume, and a new volcano forms on the sea floor in its place. A line of extinct volcanoes forms over time, with their ages increasing with distance from the hot spot. In the case of the Hawaiian chain, each island formed over the stationary hot spot and was then carried northwestward on the moving lithospheric plate (● Figure 3.16, page 64, parts a and b). The Island of Hawaii is the youngest of the Hawaiian hot-spot volcanoes and is the only one with active volcanism today. A line drawn from the Island of Hawaii to Midway Island, the oldest of the

(a)

(b)

(c)

● FIGURE 3.15 Paleogeographic reconstruction of continents' movements over the past 180 million years during *(a)* the early Triassic Period, *(b)* late in the Cretaceous Period, and *(c)* as it is today. Note that Panthalassa was the "one" ocean dominating the globe, as Pangaea was the "one" continent. These maps show present-day coastlines; ancient coastlines differed from these. Compare these maps with Figure 3.1, which was originally published in 1915 and revised by Wegener in 1929 just before his death.

CONSIDER THIS . . .

Southern California is a highly populated area that is known for being subject to a wide variety of geologic hazards. The Gulf of California immediately to the south is a divergent plate boundary with high heat flow and high potential for volcanic activity. Why needn't Southern Californians worry about this particular geologic hazard?

chain, would represent an arrow marking the direction and rate of movement of the Pacific plate in this region. The track of the Hawaiian Islands forms a "lazy *L*" with the Emperor Seamount chain, a line of submerged volcanoes that extends northward from Midway, tracing a different direction of plate movement prior to 45 million years ago. Of greater environmental concern because of population

Exotic Terranes—A Continental Mosaic

Because continental plates divide, collide, and slip along faults, it is not surprising that bits and pieces of continental crust existing within the oceans ultimately collide with continents at subduction zones and become stuck there. Because of their low density, they stand high above oceanic crust (due to isostatic equilibrium), and these microplates or microcontinents become plastered, *accreted,* to larger continental plates when they collide with them. The accreted plates are known as **terranes,** defined as fault-bounded blocks of rock with histories quite different from those of adjacent rocks or terranes.★ In size they may be several thousand square kilometers or just a few tens of square kilometers, and they may become part of a continent composed of many terranes. Oceanic crust and the sediment resting on it may be scraped up onto the continents. Terranes may consist of almost any type of rock, but each one is fault-bounded, has a paleomagnetic signature indicating a distant origin, and has little geology in common with adjacent terranes or with the continental **craton**—the continent's stable core.

The terrane concept developed in Alaska when geologic mapping for land use planning revealed that the usually predictable pattern of rocks and structures was not valid for any distance. In fact, the rocks the geologists found a few kilometers away were almost always of a "wrong" composition and age. Further studies revealed that Alaska is a collection of microplates (terranes)—tectonic flotsam and jetsam, if you wish—that have mashed together over the past 160 million years and that are still arriving from the south (◉ Figure 1). One block, the Wrangellia terrane, was an island during Triassic time and it has a paleo-

magnetic signature of rocks that formed 16° from the equator. It is not known whether Wrangellia formed north or south of the

---

★*Terrane,* as noted, is a geological term describing the area or surface over which a particular rock or group of rocks is prevalent. *Terrain* is a geographical term referring to the topography or physical features of a tract of land.

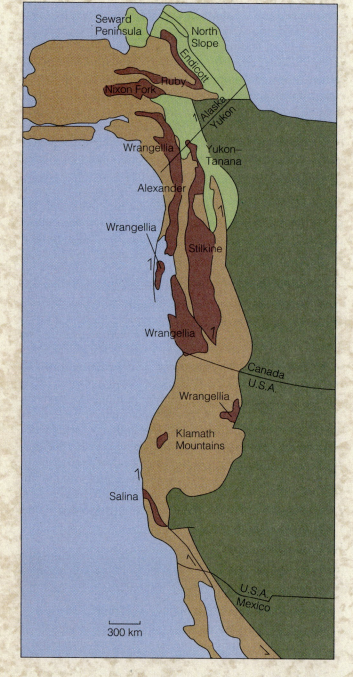

◉ **FIGURE 1** Accreted terranes of the west coast of the United States and Canada. The Salina, Wrangellia, Alexander, Stikine, Klamath Mountains, Ruby, and Nixon Fork blocks (dark brown) were probably once parts of other continents and have been displaced long distances. The terranes shown as light-green areas are probably displaced parts of North America, the stable North American craton (darker green). The lighter-brown areas represent rocks that have not traveled great distances from their places of origin.

equator, because it is not known whether the magnetic field at the time was normal or reversed. In either case, the terrane traveled a long distance to become part of present-day Alaska (⦿Figure 2). It now appears that about 25 percent of the western edge of North America, from Alaska to Baja California, formed in this way; that is, by bits and pieces being grafted onto the core of the continent and thus enlarging it. In the distant future, part of California may become an exotic terrane of Alaska.

If you've deduced that terrane accretion onto the continents (also known as *docking*) is a characteristic of active continental margins, you are right (⦿Figure 3). The east coast of North America is a rifted (pulled-apart), or *passive*, margin. Material is added to it by river sediment forming flat-lying sedimentary rocks that become part of the continent. Such rocks are not subject to the mountain-building forces of active margins and they accumulate in thick, undisturbed sequences.

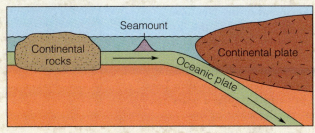
(a)

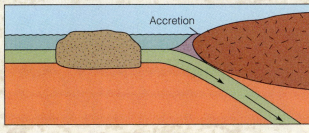
(b)

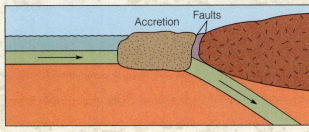
(c)

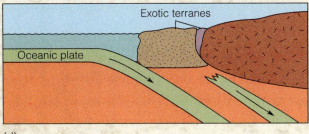
(d)

⦿ FIGURE 3 Formation of exotic terranes by a small continental mass and a submarine volcano being scraped off from the subducting plate and onto the continental mass. Much of the west coasts of the U.S. and Canada have grown in this manner.

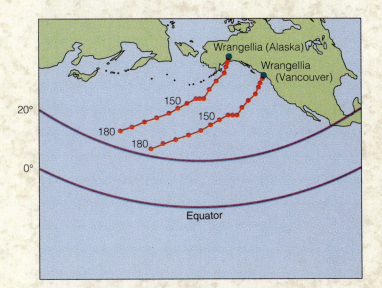

⦿ FIGURE 2 An analysis of Wrangellia terrane trajectories for the period 180 Ma to 100 Ma, assuming initial position in the Northern Hemisphere. In this interpretation both Vancouver Island and the Alaskan terranes (see Figure 1) rode with the oceanic plate until they collided (docked) at their present positions on the North American continent about 100 million years ago. The data points (the dots) represent paleomagnetic evidence at the respective locations.

FIGURE 3.16 (a) Cross section of six islands formed and forming over the Hawaiian "hot spot." The Hawaiian Islands become systematically older in a northwest direction. The new volcano Loihi is forming underwater off the southeast end of the Hawaiian chain. (b) The bend in the line at the beginning of the Emperor Seamount chain indicates a more northerly direction of plate movement over the Hawaiian hot spot prior to 45 million years ago. Note the alignment of the Marquesas, Tuamotu–Pitcairn, and Cook–Austral Islands that have formed over other hot spots.

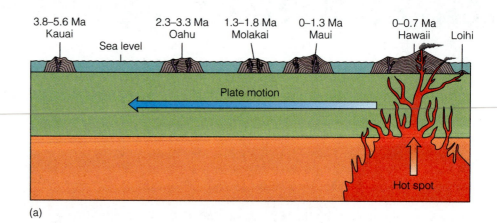

(a)

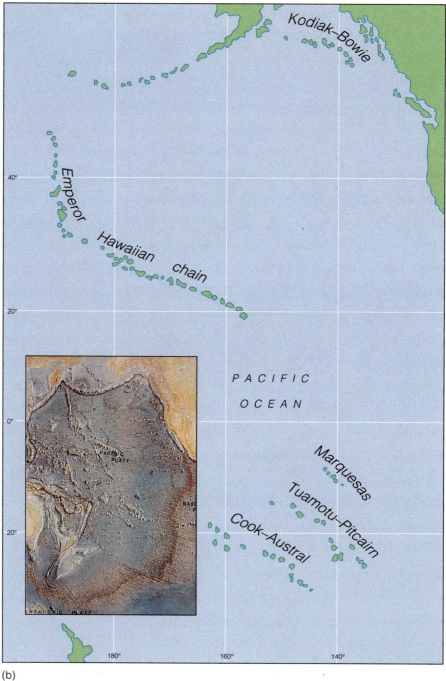

(b)

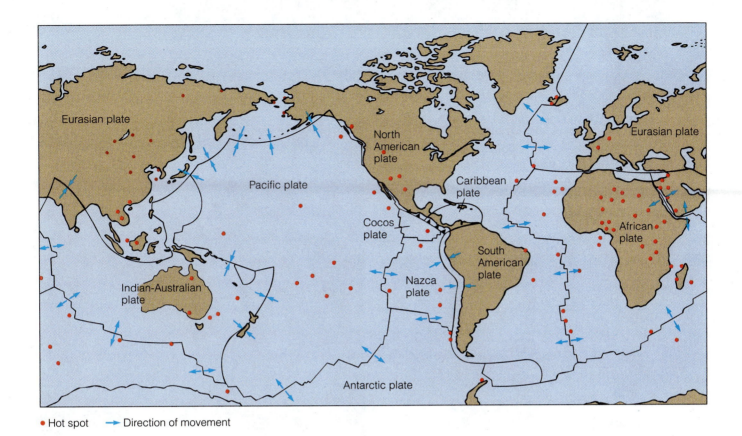

● Hot spot → Direction of movement

◉ FIGURE 3.17 The earth's hot spots and their relationships to plates and the
directions of plate movement.

densities are hot spots that occur beneath the continents. Yellowstone National Park, for example, is situated above a midplate hot spot with a potential for destructive volcanic activity. About a third of the more than 120 identified hot spots that have been active within the last ten million years penetrate continental areas (◉ Figure 3.17). Almost 40 volcanic plumes underlie the African plate, suggesting that it is either moving very slowly or is stationary.

Much of the sea-floor evidence that supports the plate tectonics theory comes from the Deep-Sea Drilling Project funded by the National Science Foundation and implemented aboard the research drilling ship *Glomar Challenger*. During a series of cruises in the Mediterranean Sea between 1970 and 1975, the researchers discovered thick salt deposits beneath the sea floor, indicating that about ten million years ago (Miocene time) the sea was isolated from the Atlantic Ocean and that the sea totally evaporated. Access of Atlantic Ocean water to the Mediterranean through the Strait of Gibraltar was cut off by uplift as the African plate moved northward, and also by a global lowering of sea level at the time. Further drilling yielded sediments indicating that the floors of the two Mediterranean basins once resembled the great dry lakes of present-day deserts in the western United States. Occasionally they were occupied by water when the connection with the

Atlantic was periodically reestablished (see ◉ Figure 3.20, part a, on page 67), but generally they were lifeless salt flats. The Mediterranean must have looked as Death Valley does today from a high vantage point, except that major rivers such as the Rhone, Danube, and Nile cascaded down the continental slopes to the flats below. About five million years ago the Strait of Gibraltar reopened, and the Atlantic Ocean flowed into the Mediterranean Sea as a waterfall or cascade of great magnitude (Figure 3.20, part b). This inflow filled the Mediterranean in a few thousand years. The sediments on top of the five-million-year-old salt-flat deposits contain fossils of marine organisms, indicating

CONSIDER THIS . . .

Keeping in mind that the "earth system" is composed of different earth cycles (such as the Wilson cycle), predict the general arrangement of oceanic and continental plates in, say, 75 million years. In particular, will the present continents be more consolidated or less consolidated at this time?

Visions of How the Earth Works

The evolution of modern plate-tectonic theory started with Alfred Wegener's 1912 hypothesis of Continental Drift (◉ Figure 3.18). Fifty years later the idea of sea-floor spreading was proposed. Although scientists were skeptical at first, this far-reaching concept led to the unifying theory of plate tectonics (◉ Figure 3.19).

(a)

◉ FIGURE 3.18 Alfred Wegener formulated his hypothesis that continents move about on the face of the earth while sitting out a long Arctic winter in Greenland. His ideas were considered so outrageous that more than 50 years passed before scientists realized he was on the right track.

(b)

◉ FIGURE 3.19 Evidence of colliding and sliding tectonic plates. *(a)* Collision of the Indian and Asian plates produced the great Himalaya Mountains. The rocks of Nepal's Dhaulāgiri I (8,172 m, 26,810 ft), one of the highest peaks in the world, were uplifted and contorted by the continent–continent collision. The steep-sided mountain's south wall is 15,000 feet high. *(b)* Contorted sedimentary rocks are exposed in a roadcut adjacent to the San Andreas fault near Palmdale, California. Since the rocks formed, they have been stressed and folded by many kilometers of movement on the fault, a transform boundary between the Pacific and the North American plates.

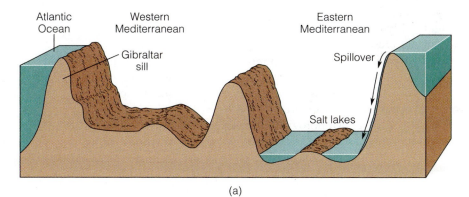

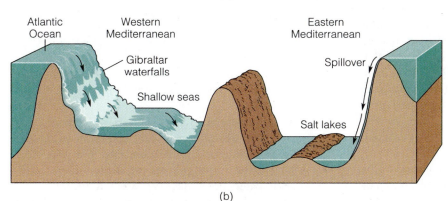

FIGURE 3.20 Desiccation of the eastern and western Mediterranean basins about ten million years ago (Miocene time) when its connection with the Atlantic Ocean through the Strait of Gibraltar was blocked. *(a)* Total evaporation took as little as 1,000 years, during which salt was deposited and dry lakes or salt lakes existed in the basins. Some spillover from the Red Sea occurred in the eastern Mediterranean. *(b)* Two-kilometer-thick deposits of sediments and evaporites (especially gypsum and halite) below the floors of these basins indicate that the cycle of filling, evaporation, and refilling was repeated many times.

rapid filling with normal sea water, and no abnormal salinity, which would have occurred if the filling had been slow.

Plate tectonics theory unifies and incorporates the basic concepts of continental drift and sea-floor spreading. In one way or another, plate tectonics explains why nearly all geologic phenomena occur, such as why volcanoes are either explosive or quiet at given locations, why mountain ranges are located where they are, and why large faults such as the San Andreas in California cut across the landscape and occasionally generate damaging earthquakes. The theory has revolutionized our understanding of how the earth works and has provided geologists with a useful model in their searches for new mineral and oil deposits.

CONSIDER THIS...

It is said that all knowledge of the natural world is potentially beneficial to humanity, even knowledge about past conditions that have subsequently changed. How might our knowledge of past and present plate motions be applied to benefit the human race?

SUMMARY

Continental Drift

DESCRIPTION Alfred Wegener's theory that during the Permian Period there was a single supercontinent, Pangaea, which split apart during Jurassic time into individual continents that have since migrated to their present positions.

CAUSE The theory provided no explanation for a driving force.

EVIDENCE

1. Similar glacial deposits and plant fossils in Permian rocks (the Gondwana Succession) in Africa, South America, India, Australia, and Antarctica.
2. Similarities between fossils, ages of rocks, and geologic provinces on opposite sides of the Atlantic Ocean.
3. The jigsaw fit of the continents if the Atlantic and Indian Oceans were closed.

Sea-Floor Spreading

DESCRIPTION Theory independently proposed by both Harry Hess and Robert Dietz that new oceanic crust is created at mid-ocean ridges and moves laterally across the ocean basins to trenches, where it sinks and returns to the mantle.

CAUSE Convection cells in the mantle generated by thermal and density differences carry the rigid overlying crust (lithosphere) away from the ridges until the crust cools and descends toward the mantle at oceanic trenches.

EVIDENCE
1. Distribution of deep-focus earthquakes at trenches.
2. Lack of ancient sediments on the sea floor.
3. Increasing age of sea-floor sediments away from mid-ocean ridges.

Plate Tectonics

DESCRIPTION Theory that unifies and incorporates the earlier theories of continental drift and sea-floor spreading. According to it, the lithosphere is composed of seven large plates and a number of smaller ones that move about on the surface of the earth. Plates interact in only three ways: they *collide* at convergent boundaries marked by trenches or mountain ranges; they *move apart,* forming new sea floor, at divergent boundaries marked by mid-ocean ridges; or they *slip by one another* along transform boundaries.

CAUSE Convection currents in the asthenosphere (the plastic layer of the upper mantle) carry the plates as passive passengers from divergent boundaries to convergent and conservative boundaries.

EVIDENCE
1. Land geology, ocean drilling, and geophysical data support past openings and closings of the ocean basins (Wilson cycle).
2. Magnetic anomaly patterns ("stripes") on both sides of mid-ocean ridges.
3. Chains of oceanic volcanoes and seamounts that systematically increase in age with distance from the present locations of active volcanism (hot spots) indicate that plates are moving over fixed plumes of lava coming from deep in the mantle.
4. The geological and geophysical evidence for continental drift and sea-floor spreading also support plate tectonics.

EXPLAINS
1. The thick salt deposits beneath the Mediterranean Sea and its desiccation 10 million years ago.
2. The existence of narrow seas and gulfs such as the Gulf of California and the Red Sea as spreading centers.
3. The location and geology of the Himalaya Mountains.
4. The location of violent earthquakes and volcanoes at convergent margins.

KEY TERMS

anomaly	convergent boundary
asthenosphere	craton
Benioff zone	discontinuity
continental drift	divergent boundary

focus (pl. *foci*)	Pangaea
Gondwanaland	plate tectonics
graben	polarity
hot spot (plume)	sea-floor spreading
isostasy	spreading center
Laurasia	subduction
lithosphere	terrane
Mohorovičić discontinuity ("M" or Moho)	transform boundary transform fault
paleomagnetism	

STUDY QUESTIONS

1. What geologic and geographic evidence did Alfred Wegener cite for continental drift?
2. How do we know that the interior of the earth is hot, and why has it not cooled entirely in the last 4.5 billion years?
3. How do density differences cause movement in fluids? In your own words describe heat transfer by convection currents, giving some illustrations from everyday life.
4. What are the major spheres, or shells, of the earth's interior based upon density? What is meant by the term *lithosphere?* Describe the lithosphere and tell how it is separated from the deep mantle.
5. What is a tectonic "plate"? Give an example (exact geographic location) of each type of plate boundary. How fast do plates move?
6. What geophysical observations and measurements support the model of sea-floor spreading and plate tectonics?
7. How can the change in direction of a line of oceanic islands—such as that between the Hawaiian Islands chain and the Emperor Seamount chain—be explained?
8. Draw a cross section of a typical convergent plate boundary (a subduction zone) and label the following: trench, volcanic areas, Benioff zone, lithosphere, asthenosphere.
9. How might the distribution of the Permian Gondwana Succession be explained without invoking plate tectonics (ancient landmass, land bridges, etc.)? Might an alternative explanation be more reasonable than drifting continents and plate tectonics?
10. How have the narrow bodies of the Red Sea and the Gulf of California formed? Study a map, seeking an obvious relationship between the Red Sea and the Dead Sea.

FURTHER INFORMATION

BOOKS AND PERIODICALS

Ballard, Robert. 1983. *Exploring our living planet.* Washington, D.C.: National Geographic Society.

Cox, Allan. 1986. *Plate tectonics: How it works.* Palo Alto, Calif.: Blackwell Scientific Publications.

Dalziel, Ian W. D. 1995. Earth before Pangaea. *Scientific American,* January: 58.

Debiche, M. G.; Allan Cox; and D. Engebretson. 1987. *The motion of allochthonous terranes across the north Pacific basin.* Geological Society of America Special Paper 207, 49 pp.

Hallam, A. 1973. *A revolution in the earth sciences.* Oxford, U.K.: Oxford University Press.

Hsu, Ken. 1983. *The Mediterranean was a desert.* Princeton, N.J.: Princeton University Press.

Larson, Robert. 1995. The mid-Cretaceous superplume episode. *Scientific American,* January: 82–86.

Moores, E. M., ed. 1990. Shaping the earth: tectonics of continents and oceans. *Scientific American readings.* New York: W. H. Freeman and Co.

Sullivan, Walter. 1974. *Continents in motion: The new earth debate.* New York: McGraw-Hill Book Co.

Tarling, D., and Maureen Tarling. 1970. *Continental drift: A study of the earth's moving surface.* New York: Anchor Books.

Uyeda, Seiya. 1977. *The new view of the earth.* New York: W. H. Freeman.

Wegener, Alfred. 1929, 1966 translation. *The origin of continents and oceans.* New York: Dover Publications.

Wilson, J. Tuzo. 1972. Continents adrift. *Scientific American readings.* San Francisco: W. H. Freeman.

VIDEOS

National Geographic Society. 1980. *Dive to the edge of creation.* Emmy Award-winner; 60 minutes.

National Geographic Society with WQED, Pittsburgh. 1983. *Born of fire.* 60 minutes.

EARTHQUAKES AND HUMAN ACTIVITIES

4

The traffic has stopped, but the freeways are moving.

L.A. RADIO ANNOUNCER, 17 JANUARY 1994

It was early Monday morning and still dark out. The weather was clear and warm for January, and many of the residents of the greater Los Angeles area had planned to sleep late because it was the holiday honoring Martin Luther King, Jr. At 4:30 A.M. a previously unrecognized fault beneath the San Fernando Valley suddenly ruptured. The block of crust overlying the gently sloping earth fracture moved upward 2.0 meters (6.6 ft), producing such violent ground motion in the densely populated San Fernando Valley that several thousand one- and two-story structures collapsed or were badly damaged. Damage was pervasive in the Northridge vicinity; hardly a house escaped without some kind of damage. Tens of kilometers distant occasional severe damage was found, seemingly capricious and without explanation. Sixty people were killed. The exact monetary cost of the event will never be known, but

OPENING PHOTO

One of many scenes resulting from the Northridge earthquake of 17 January 1994. Failure of a high overpass on the Golden State freeway stranded a motor home, a 65-foot truck-trailer (note skid marks), and a pickup truck. Amazingly, occupants of two cars that fell through the gap to the road survived the fall.

estimates ranged from $13 billion to $20 billion, making it the most costly earthquake in U.S. history.

The event is called the Northridge earthquake, even though the epicenter was in the Reseda zip code area immediately south of Northridge near the intersection of Roscoe and Reseda Boulevards (● Figure 4.1). Some Resedans wanted the earthquake to be renamed the "Reseda earthquake," which others believed was taking civic pride too far. Had the earthquake occurred a few hours later, the death toll would have been much greater. Because of the early hour, large commercial buildings and

(b)

(c)

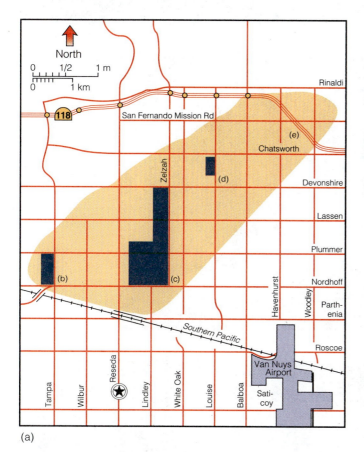
(a)

(d)

● FIGURE 4.1 *(a)* Map of the northern San Fernando Valley showing the location of the Northridge earthquake's epicenter. Note the locations of photos (b) to (e) on the map. *(b)* Collapse of the facade of Bullock's department store. Three multistory parking structures at this shopping center collapsed and had to be demolished. *(c)* Typical geology professor's office at California State University, Northridge. It is uncertain whether the picture was taken before or after the earthquake. *(d)* Mid-level collapse of this building resulted in its total destruction. The third floor shifted and dropped onto the second floor. Had this occurred during business hours, there would have been a large death toll. *(e)* Aerial view of the 118 Freeway about 11 kilometers (7 mi) northeast of the epicenter. Note that the freeway decks separated from the column caps on which they rested and dropped like tipped dominoes (see Case Study 4.4).

(e)

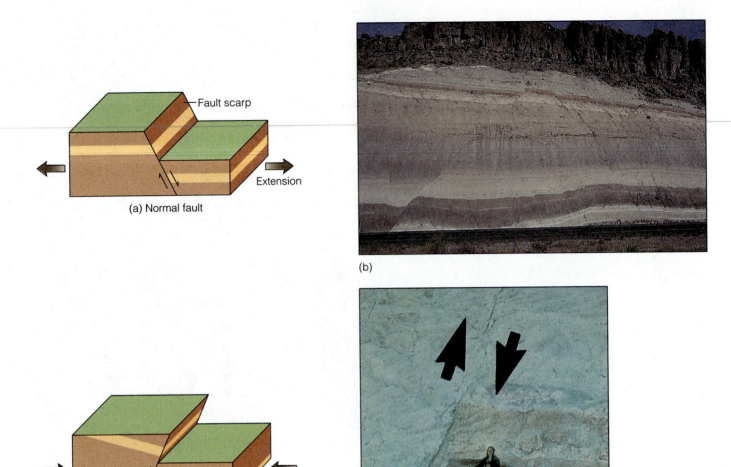

(a) Normal fault

Fault scarp

Extension

(c) Reverse fault

Compression

(e) Right-lateral strike-slip fault

Shear

(b)

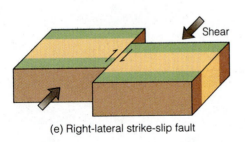

(d)

(f)

◉ FIGURE 4.2 *(a)* Normal fault geometry and *(b)* examples in a road cut on I-40 east of Kingman, Arizona. *(c)* Reverse fault geometry and *(d)* example in Death Valley, California. *(e)* Right-lateral strike-slip fault geometry and *(f)* right-lateral offset of resistant sedimentary rock ridges, Frontier Formation, Wyoming.

parking structures were practically vacant, and freeway traffic was light. In comparison with earthquakes of similar size in communities lacking seismic building codes, the number of deaths is extremely low. For instance, the Armenian earthquake of 1988 had 25,000 fatalities, and the one in India in 1993 had 11,000. The reason the 1995 Kobe, Japan, earthquake took 5,500 lives in a country noted for advanced seismic design requirements is discussed later in this section.

The Northridge earthquake may also be compared with the Great Alaskan earthquake of 1964, but for a different reason. The latter produced 100 times the ground shaking of the Northridge earthquake, but only a fraction (2 percent to 3 percent) of the dollar loss. This is because Alaska's population density is so low; it is less than 1 person per square kilometer. As many as 2,000 persons live within a square kilometer of the San Fernando Valley and vicinity. Thus, the force of an earthquake is only one factor that determines its impact on humans. A small earthquake may cause more human misery than a big one if its effects are concentrated in a densely populated area. Fortunately, neither New York City (10,000 people/km²) nor Hong Kong (55,000 people/km²) is in earthquake country.

It has long been known that earthquakes are not randomly distributed over the earth. Plate tectonic theory explains that most earthquakes occur at or near plate boundaries and that the strength of an earthquake is closely related to the type of boundary on which it occurs.

 ## THE NATURE OF EARTHQUAKES

Earthquakes are the result of abrupt movements on **faults**—fractures in the earth's lithosphere. The types of faults and the earth forces that cause them are shown in ◉ Figure 4.2, (at left). The movements occur as the earth's crustal plates slip past, under, and away from one another. Because the **stress** (force per unit area) that produces **strain** (deformation) can be transmitted long distances in rocks, active faults do not necessarily occur exactly on a plate boundary, but they generally occur in the vicinity of one. The mechanism by which stressed rocks store up strain energy along a fault to produce an earthquake was explained by Harold F. Reid after the great San Francisco earthquake of 1906. Reid proposed a mechanism to explain the shaking that resulted from movement on the San Andreas fault, known as the **elastic rebound theory** (see ◉ Figure 4.3). According to this theory, when sufficient strain energy has accumulated in rocks, they may rupture rapidly—just as a rubber band breaks when it is stretched too far—and the stored strain energy is released as vibrations that radiate outward in all directions (◉ Figure 4.4).

Most earthquakes are generated by movements on faults within the crust and upper mantle that do not produce ruptures at the ground surface. We can thus recognize a point within the earth where the fault rupture starts, the **focus,** and the **epicenter,** the point on the earth surface directly above the focus (Figure 4.4). Elastic waves—vibrations—

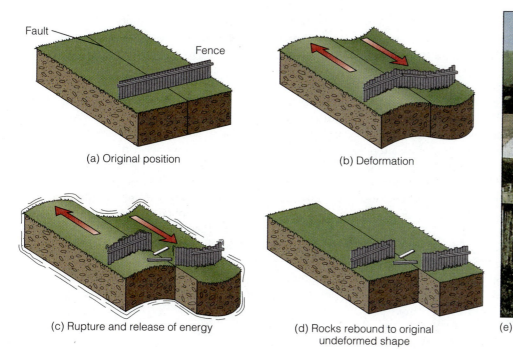

(a) Original position

(b) Deformation

(c) Rupture and release of energy

(d) Rocks rebound to original undeformed shape

(e)

◉ **FIGURE 4.3** *(a–d)* **The cycle of elastic-strain buildup and release for a right-lateral strike-slip fault according to Reid's elastic rebound theory of earthquakes. At the instant of rupture,** *(c),* **energy is released in the form of earthquake waves that radiate out in all directions.** *(e)* **Right-lateral offset of a fence by 2.5 meters (8 ft) by displacement on the San Andreas fault in 1906; Marin County, California.**

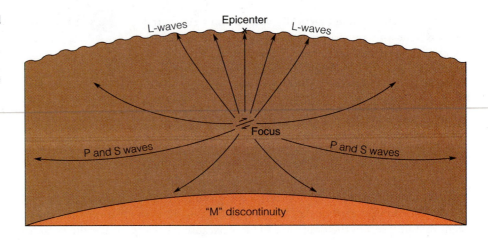

● FIGURE 4.4 Paths of body (P and S) and surface (L) waves generated at an earthquake focus by a buried fault. The epicenter is the point on the earth surface that is directly above the focus.

move out spherically in all directions from the focus and strike the surface of the earth. Damaging earthquake foci are generally within a few kilometers of the earth's surface. Deep-focus earthquakes, on the other hand, those whose foci are 300–700 kilometers (190–440 mi) below the surface, do little or no damage. Earthquake foci are not known below 700 kilometers. This indicates that the mantle at that depth behaves plastically due to high temperatures and confining pressures, deforming continuously as ductile substances do, rather than storing up strain energy.

The vibration produced by an earthquake is complex, but it can be described as three distinctly different types of waves (● Figure 4.5). Primary waves, **P-waves,** and secondary waves, **S-waves,** are generated at the focus and travel through the interior of the earth; thus, they are known as **body waves.** They are designated P- and S-waves because they are the first (*primary*) and second (*secondary*) waves to arrive from distant earthquakes. As these body waves strike the earth's surface, they generate long waves, **L-waves,** which travel only at the surface, being analogous to water ripples on a pond.

P-waves (Figure 4.5, part a) are **longitudinal waves;** the solids, liquids, and gases through which they travel are alternately compressed and expanded in the same direction the waves move. Their velocity depends upon the resistance to changes in volume (compressibility) and shape of the material through which they travel. P-waves travel about 300 meters (1,000 ft) per second in air, 300–1,000 meters (1,000–3,000 ft) per second in soil, and faster than 5 kilometers (3 mi) per second in solid rock at the surface. P-wave velocity increases with depth in the earth, because the materials composing the mantle and core increase in rigidity with depth. This increasing rigidity causes the waves' travel paths to be bowed downward as they move through the earth. P-waves speed through the earth with a velocity of about 10 kilometers (6 mi) per second. At the ground surface they have very small amplitudes (motion) and cause little property damage. Although they are physically identical to sound waves, they vibrate at frequencies

below what the human ear can detect. The earthquake noise that has been reported could be P-waves of a slightly higher frequency or some other vibrations whose frequencies are in the audible range.

S-waves (Figure 4.5, part b) are **transverse (shear) waves;** they produce ground motion "normal" (perpendicular) to their direction of travel. This causes the rocks through which they travel to be twisted and sheared. These waves have the shape produced when one end of a garden hose or a loosely hanging rope is given a vigorous flip. They can travel only through material that resists shearing; that is, material that resists two forces acting in opposite directions in different planes. Thus, S-waves travel only through solids, because liquids and gases have no shear strength (try piling water in a mound). S-wave ground motion may be largely in the horizontal plane and can result in considerable property damage. S-wave velocity is approximately 5.2 kilometers (3.2 mi) per second from distant earthquakes; hence, they arrive after the P-waves.

Unexpended P- and S-wave energy bouncing off the earth's surface generates the complex surface waves known as L-waves (Figure 4.5, part c). These waves produce a rolling motion at the ground surface. (In fact, slight motion sickness is a common response to L-waves of long duration.) The waves are the result of a complex interaction of several wave types, the most important of which are *Love waves,* which exhibit horizontal motion normal to the direction of travel, and *Rayleigh waves,* which exhibit retrograde (opposite to the direction of travel) elliptical motion in a plane perpendicular to the ground surface. L-waves may produce large displacements of the ground surface and they are the most destructive of the earthquake waves.

Locating the Epicenter

A **seismograph** is an instrument designed specifically to detect, measure, and record vibrations in the earth's crust (● Figure 4.6). It is relatively simple in concept, but some very sophisticated electronic systems have been developed

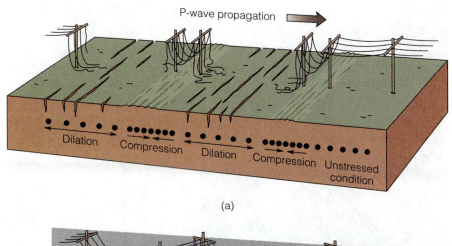

P-wave propagation

Dilation   Compression   Dilation   Compression   Unstressed condition

(a)

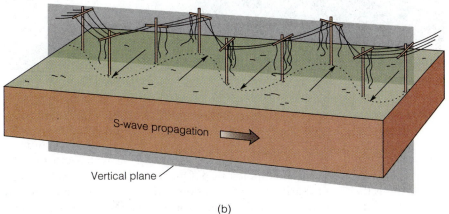

S-wave propagation

Vertical plane

(b)

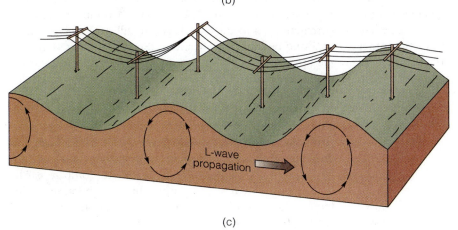

L-wave propagation

(c)

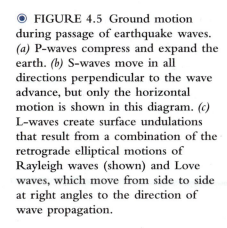

● FIGURE 4.5 Ground motion during passage of earthquake waves. *(a)* P-waves compress and expand the earth. *(b)* S-waves move in all directions perpendicular to the wave advance, but only the horizontal motion is shown in this diagram. *(c)* L-waves create surface undulations that result from a combination of the retrograde elliptical motions of Rayleigh waves (shown) and Love waves, which move from side to side at right angles to the direction of wave propagation.

for transmitting earthquake records to seismological stations by radio transmission from remote areas and by telephone from developed areas. Seismic data are recorded onto a **seismogram** (● Figure 4.7) and on magnetic tape for later playback and analysis. Because seismographs are extremely sensitive to vibrations of any kind, they are installed in quiet areas such as abandoned oil and water wells, cemeteries, and parks.

When an earthquake occurs, the distance to its epicenter can be approximated by computing the difference in P- and S-wave arrival times at various seismograph stations. Although the method actually used by seismologists today

is more accurate and determines the depth and location of the quake's focus and its epicenter, what is described here serves to show how seismic-wave arrival times can be used to determine the distance to an epicenter. Because it is known that the two kinds of waves are generated simultaneously at the earthquake focus and that P-waves travel faster than S-waves, it is possible to calculate where the waves started.

Imagine that two trains leave a station at the same time, one traveling at 60 kilometers per hour and the other at 30 km/h, and that the second train passes your house one hour after the first train goes by. If you know their speeds, you

● FIGURE 4.6 Seismometer (left) and seismogram (right) used to demonstrate method of detecting and recording earthquakes. The record on the seismogram is not an earthquake, but represents the seismometer's detection of vibrations from students' footsteps amplified about 2,000 times.

can readily calculate your distance from the train station as 60 kilometers. The relationship between distance to an epicenter and the arrival times of P- and S-waves is illustrated in ● Figure 4.8. The epicenter will lie somewhere on a circle whose center is the seismograph and whose radius is the distance from the seismograph to the epicenter. The problem is then to determine where on the circumference of the circle the epicenter is. To learn this, more data are needed; specifically, the distances to the epicenter from two other seismograph stations. Then the intersection of the circles drawn around each of the three stations—a method of map location called *triangulation*—specifies the epicenter of the earthquake (● Figure 4.9).

Earthquake Measurement

INTENSITY SCALES. The reactions of people (geologists included) to an earthquake, or an earthquake prediction, typically range from mild curiosity to outright panic. However, a sampling of the reactions of people who have been

subjected to an earthquake can be put to good use. Numerical values can be assigned to the individuals' perceptions of earthquake shaking and local damage, which can then be contoured upon a map. One **intensity scale** developed for measuring these perceptions is the 1931 **Modified Mercalli scale (MM).** The scale's values range from *MM* = I (denoting not felt at all) to *MM* = XII (denoting widespread destruction), and they are keyed to specific U.S. architectural and building specifications. (See Case Study 4.1.) People's perceptions and responses are compiled from returned questionnaires, and then lines of earthquake intensity, called **isoseismals,** are plotted on maps. Isoseismals enclose areas of equal earthquake damage and can indicate areas of weak rock or soil as well as areas of substandard building construction. Such maps have proved useful to planners and building officials in developing building codes and standards (● Figure 4.10, page 79). It is difficult to compare earthquakes in different regions using this scale because of differences in construction practices, population density, and sociological factors.

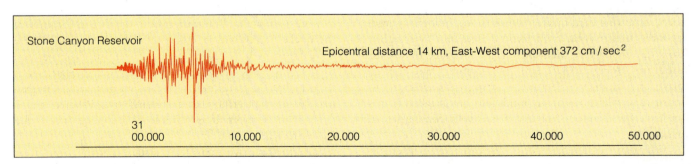

Stone Canyon Reservoir   Epicentral distance 14 km, East-West component 372 cm / sec^2

31
00.000   10.000   20.000   30.000   40.000   50.000

● FIGURE 4.7 Seismogram of the main shock of the Northridge earthquake, 17 January 1994. The time in minutes and seconds after 4:00 A.M. appears at the bottom. Distance from the epicenter and direction and ground acceleration at the recording site is also indicated. (Acceleration is explained in the next section.)

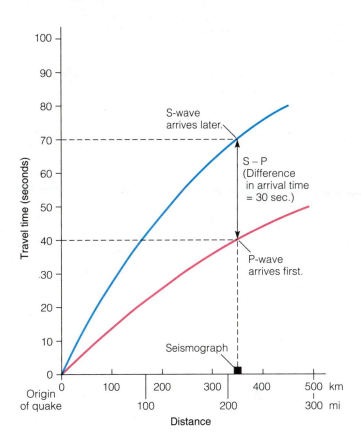

● FIGURE 4.8 (right) Generalized graph of distance versus travel time for P- and S-waves. Note that the P-wave has traveled farther from the origin of the earthquake than the S-wave at any given elapsed time.

RICHTER MAGNITUDE SCALE—THE BEST-KNOWN SCALE. The best-known measure of earthquake strength is the **Richter magnitude scale,** which was introduced in 1935 by Charles Richter and Beno Gutenberg at the California Institute of Technology. It is a scale of the energy released by an earthquake and thus, in contrast to the intensity scale, may be used to compare earthquakes in widely separated geographic areas. It is calculated by measuring the maximum amplitude of the ground motion as shown on the seismogram using a specified seismic wave, usually the surface wave. Next, the seismologist "corrects" the measured amplitude (in microns) to what a "standard" seismograph would record at the station. After an additional correction for distance from the epicenter, the Richter **magnitude** is the common logarithm of that ground motion in microns. For example, a magnitude-4 earthquake is specified as having a corrected ground motion of 10,000 microns ($\log_{10}$ of 10,000 = 4) and thus can be compared to any other earthquake for which the

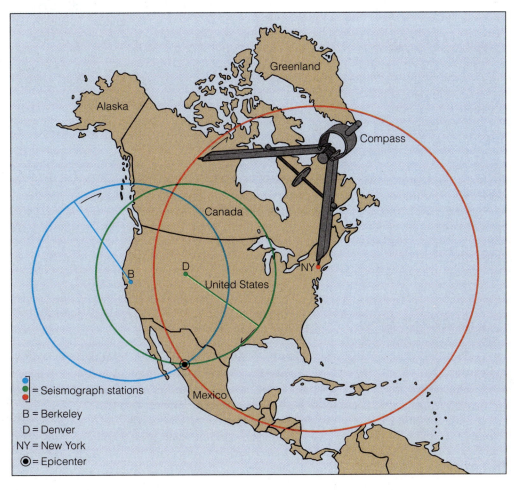

● FIGURE 4.9 An earthquake epicenter can be closely approximated by triangulation from three seismic stations.

● = Seismograph stations
B = Berkeley
D = Denver
NY = New York
◉ = Epicenter

Shake, Rattle, and Roll

The Rossi–Forel intensity scale was the first scale developed for evaluating the effects of an earthquake on structures and humans. It was developed in Europe in 1878 and assigned Roman numerals from *I* (for *barely felt*) to *X* (for *total destruction*) for varying levels of intensity. This scale was first modified by Father Giuseppi Mercalli in Italy and later in 1931 by Frank Neumann and Harry O. Wood in the United States. The Modified Mercalli scale is the scale most widely used today for evaluating the effects of earthquakes in the field. An abbreviated version is given here.

I. Not felt except by a very few persons under especially favorable circumstances.

II. Felt by a few persons at rest, especially by persons on upper floors of multistory buildings, and by nervous or sensitive persons on lower floors.

III. Felt quite noticeably indoors, especially on upper floors, but many people do not recognize it as an earthquake; vibrations resemble those made by a passing truck.

IV. Felt indoors by many persons and outdoors by only a few. Dishes, windows, and doors are disturbed, and walls make creaking sounds as though a heavy truck had struck the building. Standing cars are rocked noticeably.

V. Felt by nearly everyone. Many sleeping people are awakened. Some dishes and windows are broken, and some plaster is cracked. Disturbances of trees, poles, and other tall objects may be noticed. Some persons run outdoors.

VI. Felt by all. Many people are frightened and run outdoors.

Some heavy furniture is moved, and some plaster and chimneys fall. Damage is slight, but humans are disturbed.

VII. General fright and alarm. Everyone runs outdoors. Damage is negligible in buildings of good design and construction, considerable in those that are poorly built. Noticeable in moving cars.

VIII. General fright approaching panic. Damage is slight in specially designed structures; considerable in ordinary buildings, with partial collapse; great in older or poorly built structures. Chimneys, smokestacks, columns, and walls fall. Sand and mud are ejected from ground openings (liquefaction).

IX. General panic. Damage is considerable in specially designed structures, well-designed frame structures being deformed; great in substantial buildings, with partial collapse. Ground is cracked; underground pipes are broken.

X. General panic. Some well-built wooden structures and most masonry and frame structures are destroyed. Ground is badly cracked; railway rails are bent; underground pipes are torn apart. Considerable landsliding on riverbanks and steep slopes.

XI. General panic. Ground is greatly disturbed, with cracks and landslides common. Sea waves of significant height may be seen. Few if any structures remain standing; bridges are destroyed.

XII. Total panic. Total damage to human engineering works. Waves seen on the ground surface. Lines of sight and level are distorted. Objects are thrown upward into the air.

---

same corrections have been made. It should be noted that because the scale is logarithmic, each whole number represents a ground shaking (at the seismograph site) 10 times greater than the next-lower number. Thus a magnitude-7 produces 10 times greater shaking than a magnitude-6; 100 times that of a magnitude-5; and 1,000 times that of a magnitude-4. Total energy released, on the other hand, varies

CONSIDER THIS . . .

You are shopping for a house in earthquake country. How could an isoseismal map of an earthquake whose epicenter was near a house you would like to buy be useful? What patterns would you look for when examining the isoseismal map?

logarithmically as some exponent of 30. Compared to the energy released by a magnitude-5 earthquake, a $M_L = 6$ releases 30 times (30^1) more energy; a $M_L = 7$ releases 900 times (30^2) more; and a $M_L = 8$ releases 27,000 (30^3) times more energy (⊙ Figure 4.11, part a).

The Richter scale is open-ended; that is, theoretically it has no upper limit. However, rocks in nature do have a limited ability to store strain energy without rupturing, and no earthquake has been observed with a Richter magnitude greater than 8.9, yet.

MOMENT MAGNITUDE—THE MOST WIDELY USED SCALE. Seismologists have abandoned Richter magnitudes in favor of **moment magnitudes** *(M_w or M)* for describing earthquakes. The reason is that Richter magnitudes do not accurately portray the energy released by large earthquakes on faults with great rupture lengths. The seismic waves used to determine the Richter magnitude come from only a small part of the fault rupture and, hence,

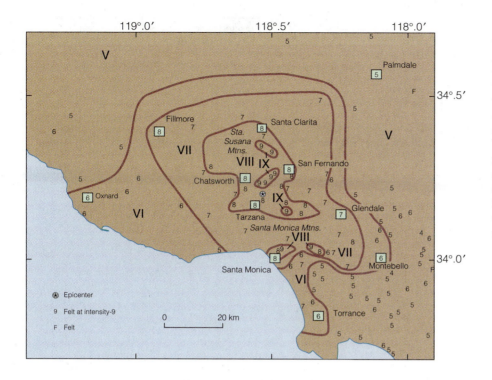

● FIGURE 4.10 Isoseismal map of the near-field (close to the epicenter) Modified Mercalli scale intensities of the Northridge earthquake. Intensity-9 values are assigned to locations where there are spectacular partial collapses of modern store buildings, destroyed wood-frame apartment buildings, and collapses of elevated freeways. Intensity-6 values are assigned to locations where there are some broken windows, a few instances of fallen plaster or damaged unreinforced masonry chimneys, and many items fallen from shelves.

cannot provide an accurate measure of the total seismic energy released by a very large event.

Moment magnitude is derived from seismic moment, M_0 (in dyne centimeters), which is proportional to the average displacement (slip) on the fault *times* the rupture area on the fault surface *times* the rigidity of the faulted rock. The amount of seismic energy (in ergs) released from the ruptured fault surface is linearly related to seismic moment by a simple factor, whereas Richter magnitude is logarithmically related to energy. Because of the linear relationship of seismic moment to energy released, the equivalent energy released by other natural and human-caused phenomena can be conveniently compared to earthquakes' moment magnitudes on a graph (Figure 4.11, part b).

Moment magnitudes (M_w) are derived from seismic moments (M_0) by the formula: $M_w = (\frac{2}{3} \log M_0 - 10.7)$. This table compares the two most commonly used scales for selected significant earthquakes:

EARTHQUAKE	RICHTER MAGNITUDE	MOMENT MAGNITUDE
Chile, 1960	8.3	9.5
Alaska, 1964	8.4	9.2
New Madrid, 1812	8.7 (est.)	8.1
Mexico City, 1985	8.1	8.1
San Francisco, 1906	8.3 (est.)	7.7
Loma Prieta, 1989	7.1	7.0
San Fernando, 1971	6.4	6.7
Northridge, 1994	6.4	6.7
Kobe, Japan, 1995	7.2 JMA	6.9

CONSIDER THIS . . .

"Then the Lord rained upon Sodom and upon Gomorrha brimstone and fire. . . . And he overthrew those cities, and all the plain about . . ." (Genesis 19:24–25). Sodom and Gomorrha are believed to have been located, respectively, near the modern industrial city of Sedom in Israel at the south end of the Dead Sea and at the north end of the Dead Sea. Could some geological event have caused the destruction of these towns? If so, what events might have been possible?

The Richter scale is not used in Japan. The official earthquake scale there is the scale developed by the Japanese Meteorological Agency (JMA). The Chilean earthquake of 1960 $(M_w = 9.5)$ has the greatest seismic moment and energy release ever measured; that is, it had the longest fault rupture and the greatest displacement. The Alaskan earthquake of 1964 has the highest recent Richter magnitude and theoretically more severe ground motion than the event in Chile.

Fault Creep, the "Nonearthquake"

Some faults move almost continuously, or in short spurts, and do not produce detectable earthquakes. This type of

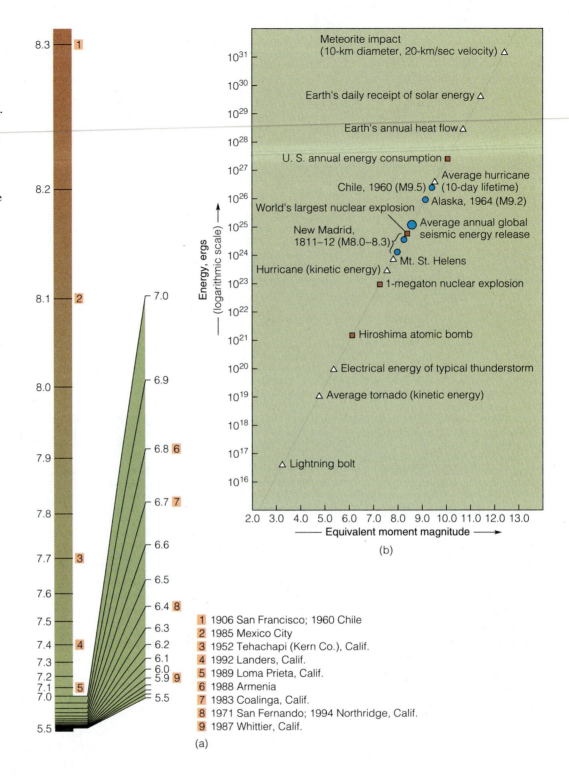

● **FIGURE 4.11**
Earthquake magnitude and energy.
(a) Richter magnitudes of eleven selected earthquakes, 1906–1995.
(b) Equivalent moment magnitudes of energetic human-caused events *(squares)*, natural events *(triangles)*, and large earthquakes *(circles)*. The erg is a unit of energy, or work, in the metric (cgs) system. To lift a one-pound weight one foot requires 1.4×10^7 ergs.

1 1906 San Francisco; 1960 Chile
2 1985 Mexico City
3 1952 Tehachapi (Kern Co.), Calif.
4 1992 Landers, Calif.
5 1989 Loma Prieta, Calif.
6 1988 Armenia
7 1983 Coalinga, Calif.
8 1971 San Fernando; 1994 Northridge, Calif.
9 1987 Whittier, Calif.

(a)

movement, called **fault creep,** is well known but poorly understood. The Hayward fault, which is part of the San Andreas fault system, is an example of a creeping fault. It is just east of the San Andreas fault and runs through the cities of Hayward and Berkeley, California. Creep along this fault causes displacements of millimeters per year, and is one way that this plate boundary accommodates motion between the Pacific and North American plates. An alternative "accommodation" of plate motion on the San Andreas system is the storage of the strain energy and periodic release as a large displacement on a fault that generates a damaging earthquake. The obvious benefit of a creeping fault does not come without a price, as the residents of Hollister, California, can testify. The "creeping" Calaveras fault runs through their town and, as in Hayward, it gradually displaces curbs, sidewalks, and even residences (● Figure 4.12).

FIGURE 4.12 Fault creep along the Hayward fault at Hayward, California. Creep of a few millimeters per year has resulted in about 18 centimeters (7 in) of offset during this curb's lifetime.

A good strategy in areas subject to fault creep is to map the surface trace of the fault precisely so that it can be avoided in future construction. Land adjacent to such known faults can be designated as limited-use areas. It may provide recreation space for local residents, for example.

An interesting lack of geological foresight is seen at the University of California–Berkeley Memorial Stadium. The huge structure was built on the Hayward fault before the hazard of fault creep was recognized, and creep-caused damage to the stadium's drainage system requires periodic maintenance. It is ironic that creep damage occurs at an institution that is highly regarded for seismological research and that installed the first seismometer in the United States.

SEISMIC DESIGN CONSIDERATIONS
Ground Shaking

Movements, particularly rapid back-and-forth displacements, are the real "killers" during an earthquake. Shear and surface waves are the culprits, and the potential for them must be evaluated when establishing design specifications. The design objective for earthquake-resistant buildings is relatively straightforward: structures should be designed to withstand the maximum potential horizontal ground acceleration expected in the particular region. Engineers call this acceleration *base shear,* and it is usually expressed as a percentage of the *acceleration of gravity (g).* On earth, g is the acceleration of a falling object in a vacuum (9.8 m/sec², or

32 ft/sec²). In your car it is equivalent to accelerating from a dead stop through 100 meters in 4.5 seconds. An acceleration of 1 *g* downward produces weightlessness, and a fraction of 1 *g* in the horizontal direction can cause buildings to separate from their foundations or to collapse completely. An analogy is to imagine rapidly pulling a carpet on which a person is standing; most assuredly the person will topple.

The effect of high horizontal acceleration on poorly constructed buildings is twofold. Flexible-frame structures may be deformed from cube-shaped to rhomb-shaped, or they may be knocked off their foundations (◉ Figure 4.13). More rigid multistory buildings may suffer "story shift" if floors and walls are not adequately tied together (◉ Figure 4.14). The result is a shifting of floor levels and the collapse of one floor upon another like a stack of pancakes. The terrible death toll (over 25,000 fatalities) in Armenia in December 1988 was largely due to this kind of collapse. There, prefabricated multistory buildings simply came apart because their design could not accommodate the shearing force (◉ Figure 4.15). Such structural failures are not survivable by inhabitants and they clearly illustrate the adage that "earthquakes don't kill people; buildings do."

Damage due to shearing forces can be mitigated by bolting frame houses to their foundation and by *shear walls.* An example of a shear wall is plywood sheeting nailed in place over a wood frame, which makes the structure highly resistant to deformation. Wall framing, usually two-by-fours, should be nailed very securely to a wooden sill that is bolted to the foundation. Diagonal bracing and blocking also provide shear resistance (◉ Figure 4.16). L-shaped structures may suffer damage where they join, as each wing of the structure vibrates independently. Such damage can be minimized by designing *seismic joints* between the building wings or between adjacent buildings of different heights. These joints are filled with a compressible substance such as felt, plastic, or rubber that will accommodate movement between the structures (Figure 4.16).

Wave period is the time interval between arrivals of successive wave crests, or of equivalent points of waves, and it is expressed as *T* in seconds. It is an important consideration when assessing a structure's potential for seismic damage, because if a building's natural period of vibration is equal to that of seismic waves, a condition of resonance exists. **Resonance** occurs when a building sways in step with an oscillatory seismic wave. As a structure sways back and forth under resonant conditions, it gets a push in its direction of sway with the passage of each seismic wave. This causes the sway to increase, just as pushing a child's swing at the proper moments makes it go higher with each push. Resonance also may cause a glass tumbler to shatter when an operatic soprano sings just the right note or frequency.

Low-rise buildings have short natural wave periods (0.05–0.1 seconds), and high-rise buildings have long natural periods (1–2 seconds). Therefore, high-frequency (short-

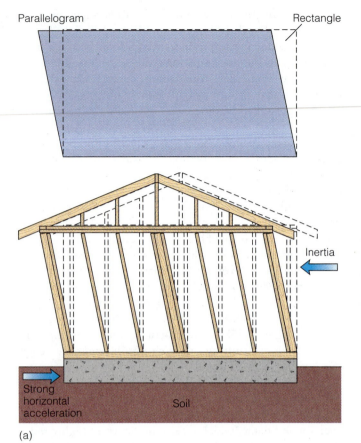

Parallelogram   Rectangle

Inertia

Strong
horizontal
acceleration   Soil

(a)

(b)

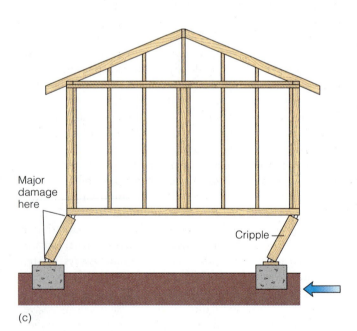

Major
damage
here

Cripple

(c)

(d)

● FIGURE 4.13 *(a)* Strong horizontal motion may deform a house from a cube to a rhomboid or knock it from its foundation completely. *(b)* A Coalinga, California, frame house that was deformed by the magnitude-6.3 earthquake in 1983. *(c)* A cripple-wall consists of short vertical members that connect the floor of the house to the foundation. Cripple-walls are common in older construction. *(d)* A Watsonville, California, house that was knocked off its cripple-wall base during the Loma Prieta earthquake of 1989. Without exception, cripple-walls bent or folded over to the north relative to their foundations.

● FIGURE 4.14 Total vertical collapse as a result of "story shift"; Mexico City, 1985. Such structural failures are not survivable.

● FIGURE 4.15 Wreckage of an older three-story building in Leninakan, Armenia, 1988. A collapsed nine-story building is at the right. Membrane (floor) failures in precast-concrete-frame structures were common here; 132 of these buildings "pancaked," leaving little space for occupant survival.

period) waves affect single-family dwellings and low-rise buildings, and long-period (low-frequency) waves affect tall structures. Close to an earthquake epicenter, high-frequency waves dominate, and thus more extensive home and low-rise damage can be expected. With distance from the epicenter, the short-period wave energy is absorbed or dissipated, resulting in the domination of longer-period waves. (The differential effects of long and short wave periods were evident after the 1985 Mexico City earthquake, as discussed and illustrated in the next section.)

Landslides

Thousands of landslides are triggered by earthquakes in mountainous or hilly terrain; there were an estimated 17,000 during the Northridge earthquake alone (see Case

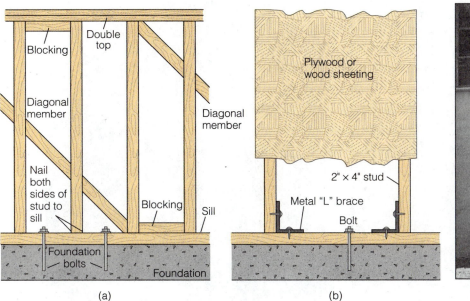

(a)   (b)   (c)

● FIGURE 4.16 Methods of reinforcing structures against base shear. *(a)* Diagonal cross-members and blocks resist horizontal earthquake motion (shear). *(b)* Plywood sheeting forms a competent shear wall, and metal "L" braces and bolts tie the structure to the foundation. *(c)* Seismic joint between two wings of a classroom building; San Bernardino Valley College, California. The joint is the vertical, dark path in the center of the photo. The active San Jacinto fault runs beneath the campus of this community college.

Study 4.2). The 1989 Loma Prieta earthquake caused landsliding in the Santa Cruz Mountains and adjacent parts of the California Coast Ranges. Such areas are slide prone under the best of conditions, and only a small earthquake will trigger many slope failures. In the greater San Francisco Bay area there was an estimated $10 million damage to homes, utilities, and transportation systems because of landslides and surficial ground failures resulting from the Loma Prieta earthquake.

One of the worst earthquake-triggered tragedies occurred in Peru in 1970. An earthquake-initiated avalanche consisting of a mixture of snow, ice, and rock gave way at an elevation of 6,000 meters (20,000 ft) on the western slopes of Nevado Huascarán. A kilometer wide and 1.5 kilometers long, it accumulated water as it plummeted downhill at more than 160 kilometers per hour (100 mph), jumping a natural barrier, and burying 25,000 residents and the villages of Yungay and Ranrahirca. The inundated areas remain completely covered with mud and boulders, leaving little indication of what lies below (● Figure 4.17).

Ground or Foundation Failure

Liquefaction is the sudden loss of strength of water-saturated, sandy soils resulting from shaking during an earthquake. Sometimes called **spontaneous liquefaction,** it can cause large ground cracks to open, lending support to the ancient myth that the earth opens up to swallow people and animals during earthquakes. Shaking can cause saturated sands to consolidate and thus to occupy a smaller volume. If the water is slow in draining from the consolidated

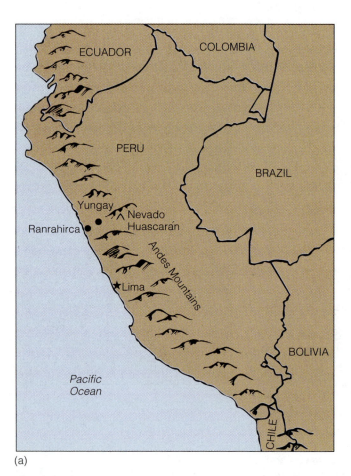

(a)

(b)

● FIGURE 4.17 *(a)* and *(b)*. A 1970 earthquake triggered a rock and snow avalanche down the slopes of Nevado Huascarán in Peru that buried the towns of Yungay and Ranrahirca. The depth of mud at Yungay was as much as 4.6 meters (15 ft).

Earthquakes, Landslides, and Disease

The January 1994 Northridge earthquake caused 17,000 land-slides over an area of about 10,000 square kilometers (4,000 mi^2), according to E. L. Harp and Randall Jibson of the U.S. Geological Survey (● Figure 1). The ground failures occurred up to 70 kilometers (43 mi) from the epicenter, with the slides and falls concentrated in the dry, weakly cemented Neogene sedimentary rocks of the Santa Susana Mountains. Most failures were thin, less than 5 meters (16 ft) thick, although failures more than 10 meters (32 ft) thick were not uncommon. The largest landslide involved a volume greater than 200,000 cubic meters. The Santa Susana Mountains are near a plate-tectonic boundary (the San Andreas fault) and are being rapidly uplifted and deeply eroded by running water. The soft Neogene marine and nonmarine sediments there are among the most susceptible in the world to seismically induced ground failures. The toes of some of the landslides advanced as much as 200 meters (650 ft) from their point of origin, a condition that could be disastrous if it occurred in developed areas of the Santa Susana Mountains.

Dynamically induced (caused by earthquake shaking) land-slides are not particularly newsworthy, but these slides caused an outbreak of coccidiodomycosis (CM), commonly known as "valley fever." Endemic to the desert Southwest, the disease causes flulike symptoms and, in severe cases, can be fatal. It is caused by the victim inhaling airborne *Coccidioides immitis* spores that reside in the top 20 centimeters of the soil. From January 24 to March 15, 166 people were diagnosed with valley fever symptoms in Ventura County, up from only 53 cases in all of 1993. Most of the cases were reported from the Simi Valley, an area in which only 14 percent of the county's population resides (● Figure 2).

Large clouds of dust hung over the Santa Susana Mountains for several days after the earthquake, promoted by the lack of winter rains preceding the quake. During this period, pressure-gradient winds known locally as "Santa Ana" winds of 10–15 knots (11–17 mph) blew into the Simi Valley, carrying in spore-

● FIGURE 1 Thousands of landslides in the Santa Susana Mountains triggered by the Northridge earthquake created a tremendous dust cloud that remained over the mountains for several days.

laden dust from the Santa Susana Mountains to the northeast. Researchers believe that the more "competent" metamorphic rocks of the San Gabriel Mountains northeast of the San Fernando Valley probably account for the lack of cases reported in the epicentral region.

The U.S. Center for Disease Control and Prevention in Atlanta, Georgia, concluded that there is epidemiological, geological, and meteorological evidence to establish a direct association between the Northridge earthquake and the outbreak of CM in the Simi Valley. Earthquake ground motion triggered many landslides, which generated the spore-laden dust that was carried into the Simi Valley. This is the first report of an earthquake-associated outbreak of valley fever, even though many earthquakes have occurred in CM endemic areas.

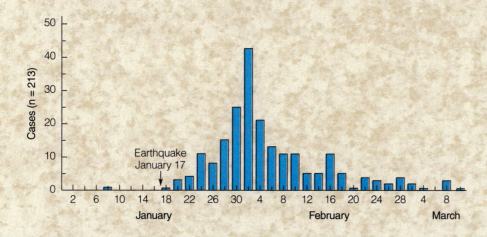

● FIGURE 2 Histogram of the almost-epidemic outbreak of coccidioidomycosis (valley fever) in Ventura County in early 1994, following the Northridge earthquake. The affected area was downwind from the dust clouds shown in Figure 1.

(a)

(b)

 FIGURE 4.18 *(a)* These apartment buildings tilted as a result of soil liquefaction in Niigata, Japan, in 1964. Many residents of the building in the center exited by walking down the side of the structure. *(b)* "Sand blows" in a lettuce field near Castroville, California, caused by liquefaction during the Loma Prieta earthquake.

material, the overlying soil comes to be supported only by pore water, which has no resistance to shearing. This may cause buildings to settle, earth dams to fail, and sand below the ground surface to blow out through openings at the surface much like small volcanoes (Figure 4.18). Lique-faction at shallow depth may result in extensive lateral movement or spreading of the ground, leaving great cracks and openings.

Ground areas most susceptible to liquefaction are those that are underlain at shallow depth—usually less than 30 feet—by layers of water-saturated fine sand. With subsurface geologic data obtained from water wells and foundation borings, liquefaction-susceptibility maps have been prepared for many seismically active areas in the United States.

Similar failures occur in certain clays that lose their strength when they are shaken or remolded. Such clays are called *quick clays* and they are natural aggregations of fine-

 FIGURE 4.19 "House of cards" collapse of quick clay structure in Turnagain Heights, Anchorage, Alaska. Total destruction occurred within the slide area, which is now called Earthquake Park.

grained clays and water. They have the peculiar property of turning from a solid (actually a gel-like state) to a liquid when they are agitated by an earthquake, an explosion, or even vibrations from pile driving. They occur in deposits of glacial-marine or glacial-lake origin and are therefore found mostly in northern latitudes, particularly in Scandinavia, Canada, and the New England states. Failure of quick clays underlying Anchorage, Alaska, produced extensive lateral spreading throughout the city in the 1964 earthquake (Figure 4.19).

The physics of failure in spontaneous liquefaction and in quick clays is similar. When the earth materials are water-saturated and the earth shakes, the loosely packed sand consolidates or the clay collapses like a house of cards. The pore-water pressure pushing the grains apart becomes greater than the grain-to-grain friction, and the material becomes "quick" or "liquefies" (Figure 4.20). The potential for such geologic conditions is not easily recognized. In many cases it can be determined only by information gained from bore holes drilled to depths of 30 meters (100 feet). Because of this expense, many site investigations do not include deep drilling, and the condition can be unsuspected until an earthquake occurs.

Ground Rupture and Changes in Ground Level

Structures that straddle an active fault may be destroyed by actual ground shifting and the formation of a fault scarp (Figure 4.21, part a). By excavating trenches across fault zones, geologists can usually locate potentially active rupture surfaces, which reduces the possibility of construction across an active fault trace. In 1970 the California legislature enacted the Alquist–Priolo Special Study Zones Act, which was renamed the *Earthquake Fault Zones Act* in 1995. It mandates that all known active faults in the state be accurately mapped and zoned for seismic safety. The act provides funds for state and private geologists to locate the

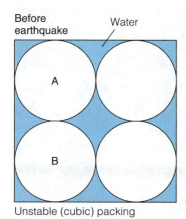
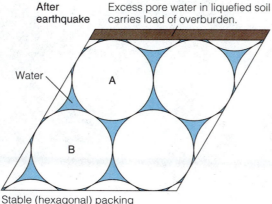

● FIGURE 4.20 Liquefaction (lateral spreading) due to repacking of spheres (idealized grains of sand) during an earthquake. The earthquake's shaking causes the solids to become packed more efficiently and thus to occupy less volume. A part of the overburden load is supported by water, which has no resistance to lateral motion.

youngest fault ruptures within a zone and requires city and county governments to limit land use adjacent to identified faults within their jurisdictions. Ironically, the faults responsible for the 1992 Landers earthquake ($M_w = 7.5$) were designated as Earthquake Fault Zones just prior to the June 28 event.

Changes in ground level as a result of faulting may have an impact, particularly in coastal areas that are uplifted or down-dropped. For instance, during the 1964 Alaskan event, parts of the Gulf of Alaska thrust upward 11 meters (36 ft), exposing vast tracts of former tidelands on island and mainland coasts (Figure 4.21, part b).

Fires

Fires caused by ruptured gas mains or fallen electric power lines can add considerably to the damage caused by an earthquake. In fact, most damage attributed to the San Francisco earthquake of 1906 and much of that in Kobe, Japan, in 1995 was due to the uncontrolled fires that followed the earth-

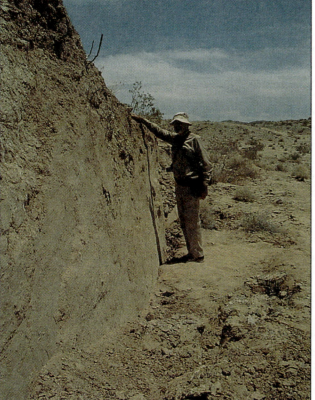

(a)

(b)

● FIGURE 4.21 (a) The 1992 Landers earthquake (*M*$_w$ = 7.5) in the Colorado Desert of southern California was felt from Phoenix Arizona, to Reno, Nevada. Right-lateral offset of 4.27 meters (14 ft) on the Johnson Valley fault created a 2-meter (6.5-ft) vertical scarp due to lateral offset of a ridge. Providing a scale for this photograph, the geologist is 185 cm tall. (b) Fault scarp on Montague Island in Prince William Sound, Alaska. In places, the scarp is 4.8 meters (16 ft) high between the white (shell-covered) uplifted sea floor (*right*) and the landward area.

⊙ FIGURE 4.22 Fire in San Francisco's Marina district after the 1989 earthquake. Liquefaction caused gas mains to break, and fire ensued. The structure at the far right collapsed due to liquefaction and foundation failure.

quakes. The Kobe quake hit at breakfast time. In neighborhoods crowded with wooden structures, fires erupted when natural-gas lines broke and falling debris tipped over kerosene stoves. Broken water mains made fire-fighting efforts futile.

Many gas pipelines that cross active faults in California today are above ground in order to facilitate repair in case of fault rupture. Fire is always a threat in urban areas after an earthquake (⊙ Figure 4.22). One of the principal "do's" for citizens immediately after a quake is to shut off the gas supply to homes and other buildings in order to prevent gas leaks into the structure from damaged lines. This in itself will save many lives (see Figure 4.57).

Tsunamis

The most myth-ridden hazard associated with earthquakes are **tsunamis** (pronounced "soo-nah´-meez"), or seismic sea waves. *Tsunami* is a Japanese word meaning "great wave in harbor" and it is appropriate, because these waves most commonly wreak death and destruction along shorelines in bays and harbors. The Japanese record of tsunamis goes back 2,000 years, and their power is dramatically displayed in the well-known print by Hokusai (⊙ Figure 4.23). Tsunamis are impulsively generated waves that are pro-

duced when the sea floor is disrupted by faulting, a volcanic eruption, or a landslide. This motion displaces the overlying water column, giving rise to waves that move outward from the source in all directions—waves similar to those produced by throwing a rock into a pond. Tsunamis may have wavelengths in excess of 160 kilometers (100 mi) and velocities on the order of 800 kilometers per hour (480 mph), but curiously, the waves are only a few feet high in the open ocean (⊙ Figure 4.24). As the wave moves into shallow water, a transformation takes place as the velocity and wavelength decrease. This in turn causes the wave height to increase, creating a higher and more dangerous wave.

Tsunamis resulting from major (magnitude >7) earthquakes may be 15 meters (50 ft) high and travel 50–60 kilometers per hour (30–36 mph) on land, like the one that struck Hawaii in 1946. The time between crests is about 20 minutes, and the first crest is not always the largest. The trough of the initial wave may arrive first, and many lives have been lost by curious persons who wandered into tidal regions exposed as water withdrew. In Lisbon, Portugal, in 1755, many people were in church when an earthquake struck on All Soul's Day. The worshipers ran outside to escape falling debris and fire, many of them joining a group seeking safety on the waterfront. There was a quiet with-

drawal of water followed by a huge wave minutes later. Sixty thousand people died. The length of time of strong shaking has been estimated from historical accounts of how many *Ave Marias* and *Pater Nosters* were recited by the churchgoers during their fearful experience.

Water withdrawal before the first high wave also contributed to fatalities in the 1946 Hawaiian tsunami; many people walked offshore to collect exposed mollusks just before the first wave crest arrived. A good rule of thumb is to head for high ground at once should you witness a sudden recession of water along a beach or coastline.

Tsunamis are most common within the Pacific Ocean, where they are generated mostly by large-scale dip-slip faulting in subduction zones that ring the basin. Japanese studies have shown that strike-slip faulting seldom produces tsunamis. They are not limited to the Pacific, however, as demonstrated by the 1883 eruption of the volcano Krakatoa in the Indian Ocean. The eruption created a tsunami 35 meters (115 ft) high that struck the west coasts of Java and Sumatra, killing 36,000 people. It then traveled across the Pacific Ocean and arrived at South America, still a meter high. In recent history tsunamis generated by earthquakes on the Pacific Rim have damaged Chile (1960), Alaska (1964), and Hawaii (⊙ Figure 4.25). In fact, Hawaii has been struck by 85 damaging tsunamis in the past 150 years, about once every two years.

The potential for tsunami damage at a given locality depends on the source mechanism, the distance of travel to the shoreline, the topography of the continental shelf, and the configuration of the coast. Tsunamis are known that have risen to heights of more than 35 meters (114 ft), whereas others have appeared only as unusually high tides on the shoreline. The 1964 Alaska earthquake caused a landslide at the head of Valdez fiord that generated a tsunami within the long, narrow bay (⊙ Figure 4.26). The

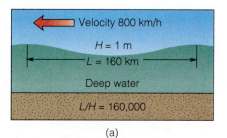

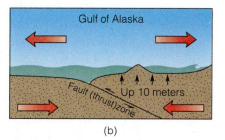

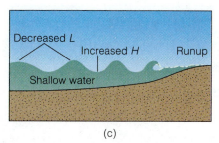

⊙ FIGURE 4.24 Evolution of a tsunami from deep to shallow water. *(a)* In deep water, the tsunami wave has a low height, a very long length, and a high velocity. *(b)* The wave moves out in all directions from an area of uplift. This example is of the 1964 Gulf of Alaska tsunami. *(c)* Upon arriving in shallow water, the wave's length decreases and its height increases until eventually the wave breaks and runs up onto the land, where it causes damage.

● FIGURE 4.25 Tsunami sources and travel paths to the Hawaiian Islands, 1900–1983.

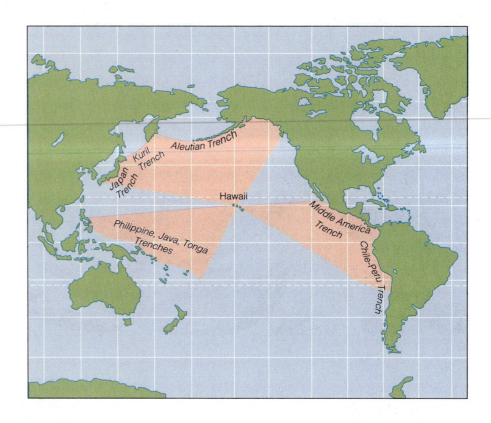

Valdez pier and waterfront slid into the bay, and where the water was 10 meters (33 ft) deep, it is now 30 meters (100 ft) deep. Thirty-five people disappeared who were standing in the waterfront area watching a ship unload. The rise and fall of the water continued throughout the night as the sea sloshed from one end of the fiord to the other. In Prince William Sound outside the entrance to the fiord, little or no tsunami activity was observed. Uplift of the sea floor in the sound, however, generated a tsunami that traveled in all directions, devastating Seward and other coastal towns. Much of the wave energy had dissipated by the time it reached northern California. However, the sea–

floor topography off the little town of Crescent City just south of the Oregon border served to concentrate wave energy on a narrow stretch of shoreline. Lives were lost there, although no damage was reported to the north and south. All other things being equal, a flat coastal plain is more susceptible to damage than is a steep-cliffed shoreline. The reason for this is that tsunami *runup*, the elevation above sea level to which a wave rises over the land, covers a much larger area on a low-lying coast than on a steep one.

The best defense at present is an early warning system. Soon after 159 people lost their lives in 1946 in Hawaii, the

(a)

(b)

● FIGURE 4.26 *(a)* Graphic depiction of the landslide in Valdez fiord that caused the wave and the fatalities at Valdez. *(b)* Valdez, Alaska, three weeks after the 1964 earthquake. The extent of the tsunami runup is visible in the photo.

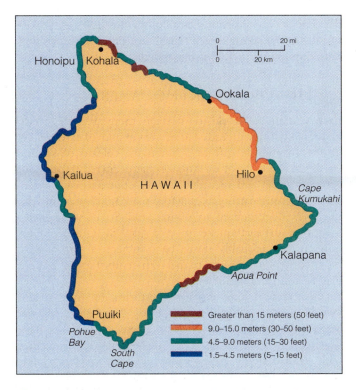

● FIGURE 4.28 Parking meters bent by the 1960 tsunami at Hilo, Hawaii. The meters resemble arrows aligned with the wave runup.

● FIGURE 4.27 Forecast tsunami heights along the coasts of the Island of Hawaii. Note that the greatest wave heights occur along the northeast shore. This shore faces the Aleutian trench, a common source of Hawaiian tsunamis.

U.S. Coast and Geodetic Survey set up a network of seismographs to aid in predicting arrival times of waves from tsunami-generating earthquakes. Indeed, the Honolulu, Hawaii, telephone book shows predicted tsunami runup heights and areas to be avoided during an alert. Predicted maximum tsunami heights for the big island of Hawaii are shown in ● Figure 4.27. Note that the city of Hilo and its bay, which faces northeast toward the Aleutian trench, is a coastal section with high expected wave heights. This is due

to the coastline's orientation and the bay's funnel shape, which concentrates tsunami energy. An engineering study in the aftermath of the 1960 tsunami that struck Hilo resulted in some recommendations for mitigating tsunami damage in that city. Parking meters in the runup area were bent flat in the direction of wave travel (● Figure 4.28), with the direction varying throughout the affected area. It is now recommended that new structures be situated with their narrowest dimension oriented toward the wave. Several waterfront structures had open fronts that allowed water to pass through them, demolishing the solid bearing walls at the rear holding up the roof. Waterfront buildings are now designed with open parking or glassed-in space on the lower level, so that much damage from water impact can be avoided as wave runup flows through the open part of the structure.

Finally, people living in a coastal area subject to tsunamis should heed civil defense officials and their tsunami warnings (● Figure 4.29). When a warning was issued for the U.S. West Coast after the 1960 Chilean earthquake, Los

● FIGURE 4.29 Residents of Hilo, Hawaii, head for high ground as a wall of water moves over the land. This 1946 tsunami killed 159 people. It was generated by an earthquake in the Aleutian Trench more than 2,200 miles to the northeast. (See also Figures 4.25 and 4.27).

(a)

(b)

(c)

● FIGURE 4.30 *(a–c)* Sequential photographs of the 1957 tsunami at Laie Point, Oahu. The photos show the arrival of a major wave and its progressive runup. This tsunami caused 57 deaths and $300,000 worth of damage.

Angeles Harbor was lined with people hoping to witness a "tidal wave"—an erroneous synonym for *tsunami*. Although it is probably human nature to want to witness such an event, be forewarned that "the big one" could ruin your day. For obvious reasons, few photographs have been taken of tsunamis breaking directly on shore. Thus, the series presented in ● Figure 4.30 is remarkable.

 ## HISTORIC EARTHQUAKES

Each year in the world, on average, there are at least two earthquakes of magnitude 8.0 or greater and 20 earthquakes in the magnitude 7.0 to 7.9 range. Release of seismic energy in the form of earthquakes has occurred throughout geologic time, and recorded history contains many references to strong earthquakes. More than 3,000 years of seismicity is documented in China, and Strabo's *Geography* mentions an earthquake in 373 B.C. in Greece. Thus humans have probably always been subject to earthquakes. Noteworthy examples from the 16th century to the present may be found in ■ Table 4.1.

This section presents case histories of four recent "world-class" earthquakes, those of Northridge, Kobe, Mexico City, and Loma Prieta. These earthquakes have some things in common. They all occurred in large cities with strict building codes; each city, in its own way, was considered a civic role model for seismic safety. The earthquakes were very costly and resulted in fatalities. Fatalities were heavy in two of the four; 10,000 people died in Mexico City, and about 5,500 in Kobe. The Northridge earthquake held the record as the world's most expensive earthquake ($13–$20 billion) for a short time, until the $100 billion Kobe event unenviably entered the record books as the most costly natural disaster ever!

Northridge, California, 1994

The largest earthquake in Los Angeles's short recorded history occurred at 4:30 A.M. Monday, 17 January 1994, on a hidden fault below the San Fernando Valley. The $M_w = 6.7$ earthquake started at a depth of 18 kilometers (11 mi) and propagated upward in a matter of seconds to a depth of 5–8 kilometers (3–5 mi, ● Figure 4.31). There were thousands of aftershocks, and their clustered pattern indicated the causative fault to be a reverse fault with a shallow dip of 35° to the south (Case Study 4.3 and Appendix 3). Such a low-angle reverse fault is called a *thrust fault,* and, because the thrust did not rupture the ground surface, it is described as "blind" (● Figure 4.32, page 95). Blind thrusts were first recognized after the 1987 Whittier earthquake 40 kilometers (25 mi) southwest of Northridge, but they were not fully appreciated until after the Northridge event.

Blind thrust faults are a newly identified seismic hazard, and there is evidence that a belt of them underlies the northern Los Angeles basin. They are especially dangerous because they cannot be detected by traditional techniques such as trenching and field mapping (thus they do not fall within the Earthquake Fault Zones Act as it is written) and yet, as we know now, they can generate significant earthquakes. Just as the residents of earthquake country were

■ TABLE 4.1 Selected Significant* Worldwide Earthquakes, in Chronological Order

LOCATION	YEAR	MAGNITUDE	IMPACT
Northern China	1556	—	800,000 killed
Lisbon, Portugal	1755	—	60,000 killed, fire
San Francisco	1906	8.3	700 killed, $7 million damage, fire
Messina (Sicily)	1908	7.5	160,000 killed
Tokyo, Japan	1923	8.3	140,000+ killed, fire
Assam, India	1950	8.4	30,000 killed
Chile	1960	8.3	5,700 killed, 58,000 homes destroyed, tsunami
Alaska	1964	8.5**	131 killed, tsunami
Guatemala	1976	7.5	23,000 killed
T'ang-shan, China	1976	7.9	655,000 killed, 779,000 injured
Mexico City	1985	8.1	10,000+ reported killed
Armenia	1988	6.8	55,000 killed
Loma Prieta (California)	1989	7.1	67 killed, $6+ billion damage
Northern Iran	1990	7.7	40,000 killed
Maharashtra, India	1993	M_w6.4	30,000+ killed, widespread collapse
Northridge, California	1994	M_w6.7	60 killed, $13–$20 billion damage
Neftegorsk, Russia	1995	M_w7.5	2,000+ killed, widespread collapse
Kobe, Japan	1995	M_w6.9	5,378 killed, $100 billion damage

* A "significant" earthquake is defined as one that registers a moment magnitude (M_w) of at least 6.5 or a lesser one that causes considerable damage or loss of life. The world averages 60 significant earthquakes per year.
** Moment magnitude 9.2

SOURCES: U.S. Geological Survey: *Earthquakes and Volcanoes;* California Division of Mines and Geology; *Academic American Encyclopedia* (1990, Grolier Electronic Publishing Company).

beginning to feel confident that geologists knew the location and behavior of most active faults, blind thrusts made their presence known.

Field inspection of the epicentral region was a depressing experience, because it revealed widespread damage. Thirteen thousand buildings were found to be severely damaged; 21,000 dwelling units had to be ordered evacuated; 240 mobile homes had been destroyed by fire; 11 major freeway overpasses were damaged at 8 locations (Case Study 4.4). The reason for the extensive damage was the high horizontal and vertical accelerations generated by the earthquake (◉ Figure 4.33, page 98). Accelerations of more than 0.30 *g* are considered dangerous, and the vertical acceleration of 1.8 *g* measured by an instrument bolted to bedrock in nearby Tarzana is probably a world record. The high ground acceleration explains why many people were literally thrown out of their beds and objects as heavy as television sets were projected

◉ **FIGURE 4.31** (right) In the 1994 Northridge earthquake, the fault rupture progressed up the fault plane from the focus at the lower right of the figure to the upper left in 8 seconds. Rather than rupturing smoothly, like a zipper opening, it moved in jerks along the fault plane, as shown by the pink patches. Total displacement was about 4 meters (12 ft).

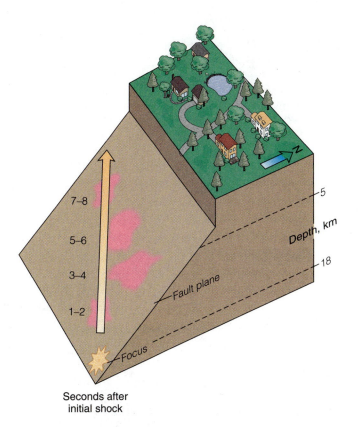

Seconds after
initial shock

Predictable "Future Shocks"

A large earthquake is normally followed by thousands of smaller-magnitude earthquakes known as **aftershocks** (⊙ Figure 1). If the main shock is small, say in the magnitude-4.0–5.0 range, aftershocks are small and nonintimidating. Following Northridge-size and bigger earthquakes, however, the strongest aftershocks ($M = 5.0$–6.0 at Northridge) can cause buildings damaged by the main earthquake to collapse and, even worse, can greatly increase anxiety in the already damaged psyche of the local citizenry. Aftershocks are caused by small adjustments (slips) on the causative fault or on other faults close to the causative one. For example, try this: push the eraser on the end of a lead pencil across a desktop. You'll find that it does not slide smoothly; it moves in jerky jumps and starts. This is called "stick–slip" and it is what happens along faults that are adjusting after a big earthquake. If aftershocks did not occur, seismologists would be very concerned about where all the remaining, unexpended energy from the main shock was going. Aftershocks relieve these still-stored stresses.

Los Angeles experienced 2,500 aftershocks in the week following the Northridge earthquake. Three strong ones, magnitude 5 or greater, occurred the first day. The largest ($M = 5.6$) occurred eleven hours after the main shock, causing concern among rescuers digging for victims beneath the rubble and the already traumatized citizens. Aftershocks follow statistically predictable patterns, as exemplified by the Northridge sequence. On

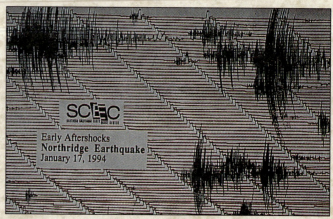

⊙ **FIGURE 1** Seismogram of early aftershocks following the Northridge earthquake, one of them a magnitude-5.6.

the first day there were 188 aftershocks of magnitude 3 or greater, and on the second day only 56 were recorded (⊙ Figure 2). By fitting an equation to the distribution of aftershocks during the first few weeks, seismologists were able to estimate the number of shocks to be expected in the future. Statistically, there was a 25 percent chance of another magnitude-5 or greater aftershock occurring within the following year, but it did not happen.

⊙ **FIGURE 2** Daily record of aftershocks of magnitude 3.0 to 5.9 during the three weeks following the main shock at Northridge. Note the sharp drop in aftershock frequency in the first four days.

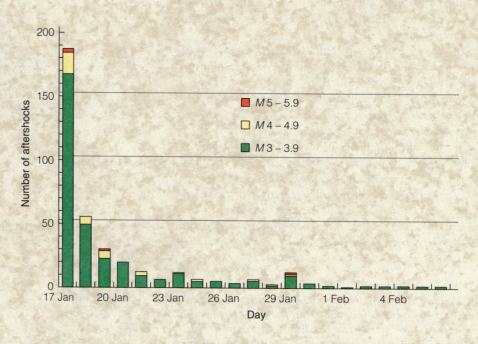

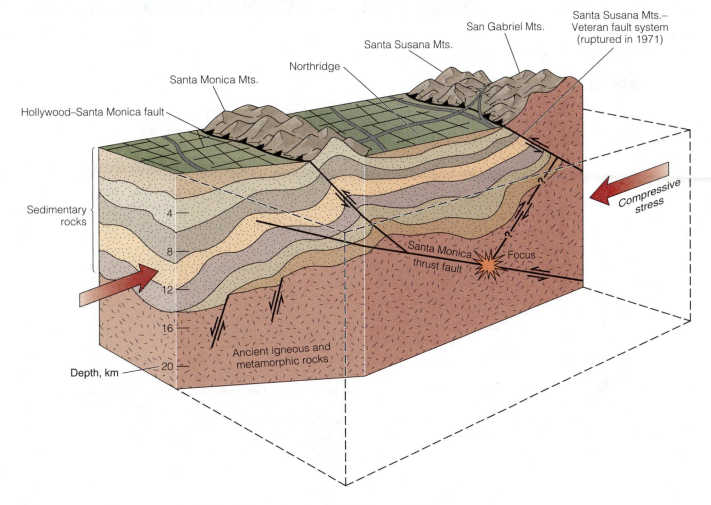

Santa Monica Mts.

Northridge

Santa Susana Mts.

San Gabriel Mts.

Santa Susana Mts.–
Veteran fault system
(ruptured in 1971)

Hollywood–Santa Monica fault

Sedimentary
rocks

Compressive
stress

Santa Monica
thrust fault

Focus

4

8

12

16

Depth, km — 20

Ancient igneous and
metamorphic rocks

◉ FIGURE 4.32 Interpretation of the fault system in the San Fernando Valley and surrounding area of Southern California. Compressive stresses built up over a long period, causing the subsurface blind thrust to rupture. The ensuing Northridge earthquake impacted the entire valley, and extended from the Santa Susana Mountains on the north to the city of Santa Monica on the south.

several meters from their stands. Although the damage was not as visually spectacular as that at Mexico City in 1985, it was heartbreaking to many residents—brick chimneys, both reinforced with steel bars and unreinforced, came down; houses moved off their foundations; concrete-block walls crumbled; gaps broke open in plaster walls; the number of shattered storefront windows was beyond counting (◉ Figure 4.34, page 99). The horizontal ground motion was directional; that is, it was strongest in the north-south direction. With few exceptions, block walls oriented east–west tipped over or fell apart, whereas those oriented north–south remained standing. Many two- and three-story apartment buildings built over open first-floor garages collapsed onto the residents' cars. These open parking areas' lack of shear resistance led to failure of the vertical supports, and everything above came down.

California State University at Northridge sustained almost "textbook" earthquake damage. In its library is "Leviathan II," recognized as one of the most advanced automated book-withdrawal systems in the country. On this day it recorded a record withdrawal of about 500,000 books—all of them onto the floor of the library. Bottled chemicals fell off their storage shelves (a major concern at any university) and caused a large fire in the Chemistry Building. In addition, one wall of a large open parking structure collapsed inward to produce the stunning art deco architecture seen in Figure 4.34, part a.

Kobe, Japan, 1995

Disaster Prevention Day is observed yearly in Japan on September 1, the anniversary of the disastrous 1923 earthquake that took 143,000 lives. Schoolchildren perform safety drills in earthquake simulators, room-size cubicles furnished with traditional furniture, gas cooking stoves, and kerosene heaters. When the rooms are shaken, the children practice

Rx for Failed Expressways

Extensive damage to freeway bridges and overpasses typically accompanies earthquakes in large urban areas. Such damage has occurred in recent years in Alaska, California, and Japan. Overpass damage commonly is due to failure of the shorter columns, which lack the flexibility of longer ones. During an earthquake the tall columns supporting a bridge or overpass system bend and sway with the horizontal forces of the quake. Because the parts of the overpass system are tied together, the stresses are transferred through the structure to the short columns, which are designed to bend only a few centimeters (● Figure 1). Failure causes the short columns to bulge just above ground level, which breaks and pops off the exterior concrete, exposing the warped, "birdcaged" steel in the interior. In addition, high vertical accelerations can cause some columns to actually punch holes through the platform deck. With excessive horizontal motion some deck spans may slip off their column caps at one end and fall to the ground like tipped dominoes.

The California Department of Transportation (CalTrans) performed an extensive engineering study after the 1971 San Fernando earthquake and decided to retrofit 122 overpasses to alleviate these problems. One aspect of the retrofitting was jacketing the short columns with steel or a composite substance and filling the space between the jacket and the original column with concrete (● Figure 2, part a). This allows the columns to bend 12.5 centimeters (5 in) instead of 2.5 centimeters (1 in) without shattering. Another solution to elevated roadway failures is to increase columns' horizontal strength by wrapping heavy steel rods around their clusters of vertical support bars, particularly on short columns (Figure 2, part b). This allows the columns to bend but prevents "birdcaging" or permanent bending. To prevent the decks from slipping off their supports and dropping to the ground, steel straps or cables are installed at the joints (Figure 2, part c). This limits the motion between the deck and supporting columns. Of the 122 overpasses that CalTrans retrofitted, not one collapsed in 1995. Ten of the eleven that collapsed were slated for future retrofitting.

The experience at Kobe was much the same, but column failures there were mostly due to inadequate steel reinforcement, particularly the wrapping of steel rods around the vertical supports (Figure 2, part b). In 1994 Japan's engineers had told the public that expressway failures such as those in California could never occur in Japan, and that their roadways would withstand even stronger earthquakes than the one at Northridge. That belief was shattered in Kobe the following January. Some of the problem there was unstable local geology (soft river deposits and bay fill), as was also the case in the I-880 failure in San Francisco

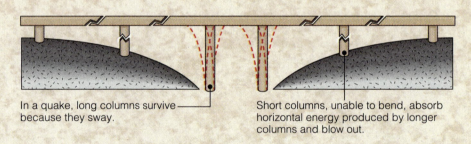

In a quake, long columns survive because they sway.

Short columns, unable to bend, absorb horizontal energy produced by longer columns and blow out.

● FIGURE 1 The difference in long and short columns' flexibility results in failure of the short ones.

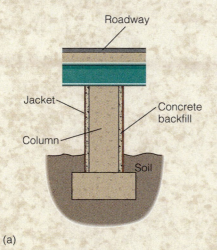

Roadway

Jacket

Column

Concrete backfill

Soil

(a)

during the 1989 earthquake (water-saturated bay muds). Most of Japan's elevated expressways are supported by short columns, and nearly every column along the elevated Hanshin Expressway through Kobe was damaged. The most spectacular failure was a 600-meter (2,000-ft) elevated section that simply rolled over.

Postmortem studies of expressway failures demonstrate that, like geologists, engineers also learn things from disasters that no theory, mathematical formula, or computer model could have predicted.

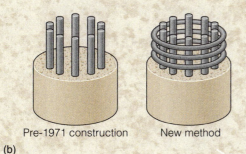

Pre-1971 construction   New method

(b)

◉ FIGURE 2 Earthquake-resistant design for bridges and overpasses. *(a)* Short columns are retrofitted with steel or composite jackets that allow them to bend five times as much. *(b)* Comparison of old and new methods of constructing highway support columns. Wrapping the vertical steel supports with steel rods allows the column to bend without "birdcaging." *(c)* Steel straps are used to hold deck sections together, preventing them from falling off their support columns.

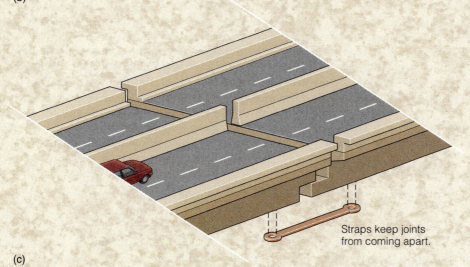

Straps keep joints from coming apart.

(c)

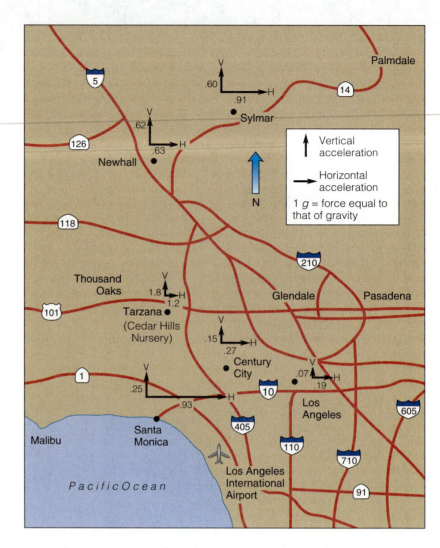

● FIGURE 4.33 Ground accelerations during the Northridge earthquake at selected sites in the Southern California area. An acceleration of 0.3 g is considered dangerous, and the 1.8 g vertical acceleration measured at Tarzana probably set a world record.

earthquake survival by ducking under tables, turning off gas and kerosene appliances, and remaining calm. Ironically, 90 percent of the deaths attributed to the 1995 earthquake in Kobe and its suburbs occurred in structures almost identical to the simulators.

Kobe, a city of 1.5 million on the island of Honshu, was struck at 5:46 A.M. a year to the day after the Northridge earthquake. The M_w = 6.9 earthquake killed 5,378 people, damaged or destroyed 152,000 buildings, and incinerated 6,913 buildings in an area equivalent of 70 U.S. city blocks. Officially known as the Hanshin–Awaji earthquake, its epicenter was off the northeast tip of Awaji Island in Osaka Bay southwest of Kobe. The islands comprising Japan are at a convergent plate boundary, and they have many active faults paralleling the boundary (● Figure 4.35; see also Figure 3.12). The January 17 earthquake occurred on the northeast-trending Nojima fault. A 50-kilometer (30-mi) rupture on it passed directly under the city of Kobe. Ground rupture with a right-lateral displacement of 1.7 meters (5½ ft) was found on Awaji Island.

The city of Kobe rests on soft river sediments and 27 square kilometers (10½ mi²) of reclaimed land in Osaka Bay. When the sand and gravel fill was dumped into the bay no real attempt was made to compact or drain it. Measured ground accelerations during the earthquake were as high as 0.5 g at ten sites, and the strong shaking lasted 10–15 seconds. Not surprisingly, the artificial fill liquified—about 17 square kilometers (6½ mi²) of it—and in so doing it expelled huge amounts of water and sand and settled as much as a meter in places. Splatter marks on walls show that water spouts from the ground reached heights up to 2 meters and typically reached 0.5 meter.

Nearly 90 percent of the fatalities occurred in crowded older neighborhoods when wood-frame houses of traditional *shinkabe* and *okabe* construction collapsed (● Figure 4.36). These are post-and-beam construction methods; unbraced or lightly braced vertical posts are spanned at the top with a thick beam and crowned with a heavy tile roof. "It's like putting a heavy book up on top of a frame of pencils," according to one engineer. There are no interior shear walls; lateral and diagonal bracing are minimal; wood pieces

(a)

(c)

(b)

◉ FIGURE 4.34 *(a, top left)* The partial collapse of this $11.5 million, 2,500-car parking structure at California State University–Northridge, was among the most dramatic failures in the Northridge area. *(b, at right)* The left side of this apartment building collapsed into a lower-level garage. The stairway led nowhere; the former second floor dropped several meters. *(c, bottom left)* Partially collapsed first-floor garage of an apartment building in Sherman Oaks. Temporary shoring was placed by the owners until the building could be jacked up and the cars extricated.

are not nailed, but fitted together by tongue and groove. The collapse of these structures was so extensive that witnesses said nothing remained to suggest that they had ever been anything but a pile of rubble (◉ Figure 4.37). In spite of the survivors' misery, their watchword was the stoic *"Shoganai"* ("It can't be helped").

Some of the most spectacular damage in Kobe was to the transportation system. Two elevated structures serving

◉ FIGURE 4.35 (right) Mapping of seismicity in part of Japan from 1961 to 1994 and the epicenter of the 1995 Hanshin–Awaji earthquake. Also shown are surface projections of the subduction-zone earthquakes that shook Kobe in 1944 and 1946.

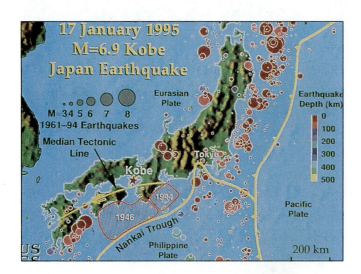

(a)

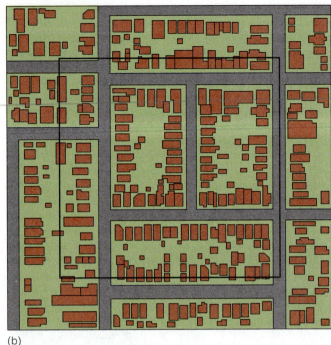

(b)

⦿ FIGURE 4.36 Plans of *(a)* a typical Japanese residential area (Osaka) and *(b)* a typical U.S. urban residential area (Oakland, California) drawn to the same scale.

(a)

(b)

⦿ FIGURE 4.37 *(a)* This traditional Japanese wood-frame house collapsed so completely that it is hardly recognizable as a former home. About 90 percent of the fatalities occurred in houses such as this. *(b)* A 0.6-kilometer-long (⅓ mi) section of the Hanshin Expressway in eastern Kobe was overturned and railway transportation was totally disrupted by the earthquake. *(c)* Comparison of Oakland, California, and Kobe, Japan. Both are on bays, both have areas of extensive fill, and both are underlain by active faults. An earthquake of about magnitude 7 has a 25 percent probability of occurring on the Hayward fault underlying Oakland between 1990 and 2020, according to the U.S. Geological Survey Working Group.

(c)

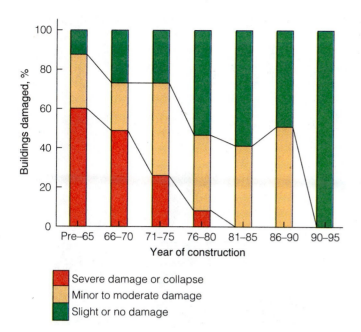

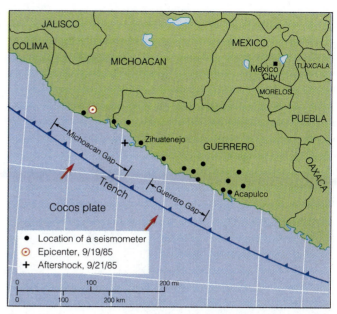

⊙ **FIGURE 4.38** Correlation between damage to reinforced concrete buildings in downtown Kobe and their year of construction. For each group, the percentage of buildings in each damage-level category is shown.

⊙ **FIGURE 4.39** Generalized map of the Middle America trench subduction zone, seismic gaps, and the epicenter of the 1985 Mexico City earthquake.

Kobe, the Hanshin and Harbor expressways, were both badly damaged. A 600-meter (⅓-mi) section of the Hanshin expressway was overturned. Supported by single short, rigid columns without sufficient reinforcing bars to contain the concrete under the dynamic earthquake forces, the elevated highways ruptured (see Case Study 4.4).

In Kobe's downtown area 1,558 reinforced concrete buildings were damaged, and 80 collapsed. Some high-rise structures collapsed at mid-height due to design problems that led to a lessening in building rigidity in the higher stories. Not surprisingly, most of the severe damage and collapse occurred in pre-1980 buildings, as there was a major revision of the building code in 1981. ⊙ Figure 4.38 makes this very clear. Even though engineering design and construction are different in Japan than in the United States and other parts of the world, the Kobe experience served to demonstrate the effectiveness of state-of-the-art seismic codes. This lesson should not be lost on other urban areas close to tectonic plate boundaries.

Mexico City, 1985

On 19 September 1985 a magnitude-8.1 earthquake shook southwestern Mexico. The earthquake waves traveled 350 kilometers (220 mi) from the epicenter near the Pacific coast and hit Mexico City like a World War II bombardment (⊙ Figure 4.39). The city's high-rise buildings began to shake, and then to sway in increasingly larger arcs. Within a few minutes' time 3,000 structures had collapsed or toppled. More than 10,000 people died.

To North American scientists, this earthquake was not a surprise. Historically, the Cocos plate (Figure 4.39) is the most active in the Western Hemisphere, having generated 42 earthquakes greater than magnitude 7 in this century. Seismologists had wired the subduction zone with 29 seismographs, concentrating them at **seismic gaps** along the plate boundary. Seismic gaps are segments on active faults where there has been little activity and the fault is "stuck," thereby accumulating strain energy (see Figure 4.39). Interestingly, before the seismic array was in place, the Oaxaca (pronounced "wha-hä´-ka") gap experienced a magnitude-7.8 earthquake (1979), which "closed" the gap, in a manner of speaking. Interest was then concentrated on the so-called Michoacán-Guerrero gap, precisely where the earthquake occurred.

Why was the damage so intense 350 kilometers from the epicenter but only slight or moderate at the epicenter? Mexico City has three geologic zones: (1) a hilly zone underlain by hard rock, (2) a transition zone, and (3) a zone underlain by soft, water-saturated lake sediments. All serious damage was concentrated in downtown areas underlain by lake beds. The lake sediments reacted as a bowl of gelatin does when it is shaken, and the earthquake caused rolling ground motions of long wave periods (⊙ Figure 4.40). Even though high-rise buildings in Mexico City had been built to withstand strong seismic forces, an earthquake of such long duration (several minutes on the lake beds) and long wave period (about 2 seconds) had not been anticipated. The city's building codes specified a natural period of about 0.1 second per story, so that a 20-story building would have a natural oscillation period of 2 seconds. Each time buildings in the 15–20 story range swayed (2-second

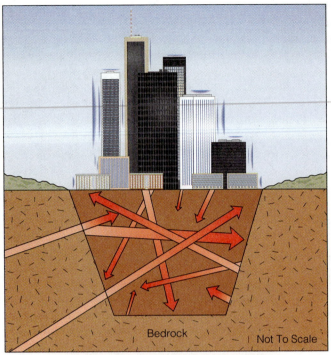

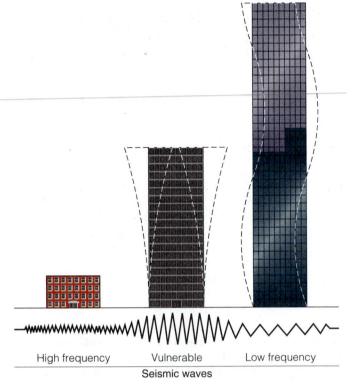

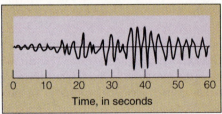

FIGURE 4.40 An illustration of seismic waves in the saturated and poorly consolidated lake beds on which much of Mexico City rests. The seismogram shows that the wave amplitude increased until about 40 seconds after the first waves arrived, and that the strong motion lasted more than a minute.

(b)

FIGURE 4.41 The effect of amplification of motion on structures of varying heights during the Mexico City earthquake. (a) Low buildings (with natural high frequency) and very tall buildings (with natural low frequency) were less impacted by the quake. (b) Collapsed high-rise.

period), they would get an extra push by the next earthquake wave of a similar period (a resonant condition). Eventually, extreme sway exceeded the design limits, and buildings either collapsed vertically (pancaked) or completely tipped over. A seismograph in the city recorded ground shaking four times the limits of building design. It should be noted that both low and very tall buildings were little affected by long waves with a two-second period (Figure 4.41).

On 15 September 1995, 10 years almost to the day after the 1985 event in Mexico City, an earthquake occurred at a focal depth of 12 kilometers (7½ mi) in the Middle America subduction zone just south of Acapulco (see Figure 4.39). A Richter magnitude-7.2 event, its damage and deaths (about 20) were confined to rural areas in the state of Guerrero near the epicenter. Nonetheless, memories of the violent earthquake a decade earlier gave rise to widespread panic in Mexico City 400 kilometers (250 mi) to the north. A combination of geological conditions at the epicenter and the shallow focus are believed to have saved the 20 million people of Mexico City from a repeat disaster. Mexico City is built on a dry lake underlain by water-saturated, soft sediments that enhance earthquake waves. Even though the city's geologic conditions may be unique, the 1985 event demonstrated that earthquake waves that travel long distances can still do considerable damage if geologic conditions serve to amplify long-period (2-second) waves. Countries and cities near plate boundaries, such as the Pacific Rim countries, should factor this concern into their seismic-safety planning.

Loma Prieta, 1989

The Loma Prieta earthquake of 17 October 1989, a magnitude-7.1 event, was certainly the most thoroughly televised earthquake to date. It occurred at 5:04 P.M. (PDT), just prior to the third game of the World Series in San Francisco, canceling the game and leaving a hundred million television fans bewildered when their screens went blank. Strong motion lasted only 15 seconds, but caused 67 deaths and more than $6 billion in property damage in the heavily populated San Francisco Bay area.

The earthquake occurred when a 25-mile-long segment of the San Andreas fault ruptured beneath the Santa Cruz Mountains, moving 6 feet (2 m) horizontally and 4 feet (1.3 m) vertically (● Figure 4.42). Prophetically, this segment had been designated by the U.S. Geological Survey in 1988 as the segment of the northern San Andreas fault most likely to rupture within the next 30 years. The basis of the prediction was the existence of a seismic gap along the San Andreas fault in the southern Santa Cruz Mountains. The Loma Prieta earthquake filled the gap (● Figure 4.43), and seismologists predict that the next event will occur farther north on the fault. There was no primary surface rupture from faulting; rather, a wide zone of large ground cracks occurred parallel to the fault, causing damage to houses and roads.

Most of the damage and loss of life were caused by structural failures due to local amplification of ground shaking or liquefaction in soft sediments or artificial fill surrounding San Francisco Bay. The 30-year-old pillars supporting the upper deck of Oakland's Cypress Street viaduct, part of Interstate 880, splayed outward, collapsing 2 kilometers (1¼ mi) of the top deck onto the vehicles below (● Figure 4.44). Forty-two rush-hour commuters were killed, and 108 were injured in this most deadly disaster of the quake. Although the double-deck structure contained 9,000 tons of steel reinforcing bars, it could not resist the highly amplified 2-meter horizontal and 1-meter vertical motion, and the columns simply collapsed.

The area of most serious damage was probably the Marina district of San Francisco, which is underlain by natural and artificial deposits of sand and mud that liquefied during strong shaking (● Figure 4.45). This liquefaction resulted in lateral spreading and opening of ground cracks, which in turn caused tilting and collapse of structures. Many of the failures were in the same general locations as

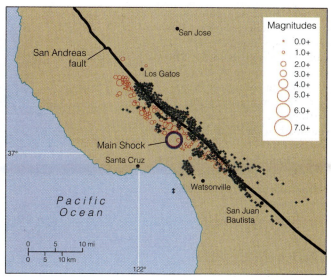
(a)

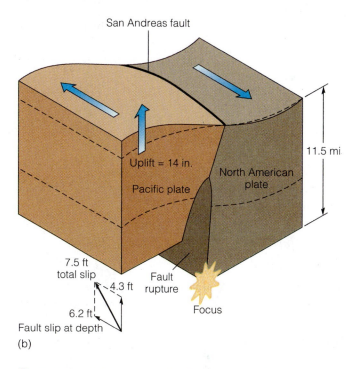
(b)

● FIGURE 4.42 *(a)* Mainshock and aftershocks of the Loma Prieta earthquake in relation to the surface trace of the San Andreas fault. Size of the symbols is relative to the magnitudes of the aftershocks. *(b)* Inferred slip along the San Andreas fault during the Loma Prieta earthquake. The Pacific plate moved upward and northward relative to the North American plate.

those of the great quake of 1906. These failures in deposits around the edge of San Francisco Bay illustrate that bay muds and poorly engineered landfill, such as those in the Marina district and the Kobe waterfront, can fail at much lower levels of shaking than firmer geologic foundation material (● Figure 4.46).

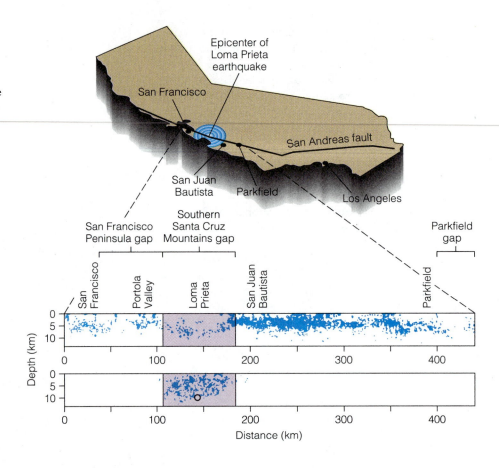

● FIGURE 4.43 Seismicity south of San Francisco, showing three seismic gaps along the San Andreas fault. The first is between San Francisco and the Portola Valley, the second is in the Loma Prieta Mountain area prior to the 1989 earthquake, and the third is southeast of Parkfield. The bottom section shows the southern Santa Cruz Mountains gap after it was filled by the 1989 Loma Prieta earthquake (open circle) and its aftershocks.

The magnitude and impact of several selected California earthquakes can be seen in ■ Table 4.2.

Does Earthquake Country Include Idaho, Missouri, and New York?

Although the vast majority of the world's earthquakes occur at plate boundaries, areas hundreds and even thousands of

● FIGURE 4.44 Failure of columns supporting the double-decked Cypress Street viaduct (I-880) caused the upper deck to fall onto the lower one.

kilometers away are not free of seismic activity. In the United States the five most seismically active states between 1980 and 1991 were

STATE	RECORDED EARTHQUAKES 1980–1991	LARGEST M_w
Alaska	10,253	7.9
California	6,732	7.2
Washington	615	5.5
Idaho	536	7.3
Nevada	398	5.6

Although Alaska and California lead the United States in the number of shakers, the earthquakes that have been felt over the largest area occurred in Missouri in 1811 and 1812, and significant seismic hazards are recognized in 39 states. No state is earthquake proof, as the seismic-risk map of the United States shows (● Figure 4.47). It appears that U.S. residents who want to be seismically on the "safe" side should move to Texas, Florida, or Alabama.

Why do such strong earthquakes occur *intraplate*—that is, far from a plate boundary (■ Table 4.3)? Intraplate earthquakes have several characteristics in common:

1. The faults causing them are deeply buried and have not broken the ground surface.

● FIGURE 4.45 Life goes on amid damage due to ground failure in the Marina district of San Francisco, 17 October 1989.

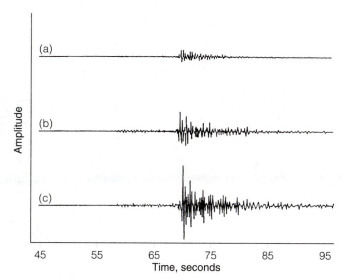

● FIGURE 4.46 Seismograms for a magnitude-4.1 aftershock on *(a)* firm bedrock, *(b)* alluvium (stream-deposited sediment), and *(c)* areas of fill and bay muds, where the most damage occurred in the San Francisco earthquakes of 1906 and 1989.

2. Because the rocks of the continental interior are stronger than those at plate boundaries, which are laced with faults, they transmit seismic waves better, causing ground motion over a huge area. In the United States this is usually many states (● Figure 4.48).
3. They do not appear to be random events.

The Charleston, South Carolina, earthquake of 1886 was larger than the 1989 San Francisco shaker. It killed scores of people, ruined the city, and slowed the South's recovery from the Civil War. The New Madrid (pronounced "mad´rid") earthquakes of 1811 and 1812 caused extensive topographic changes, locally reversed the course of the Mississippi River, and may have been felt over a

larger area than any other earthquake in recorded history. Ground motion was felt as far away as Washington, D.C., where it caused church bells to ring and scaffoldings on the Capitol to collapse.

Recent studies suggest that intraplate earthquakes are concentrated in areas where normally stable crust has been stretched and weakened by faults. Such weakened zones form where continents have been split apart *(rifted),* forming two continents, as occurs at divergent plate boundaries (see Chapter 3). When North America was separated from Africa at the Mid-Atlantic Ridge about 180–200 million

■ TABLE 4.2 Selected Significant California Earthquakes, in Order of Magnitude

LOCATION	YEAR	MAGNITUDE	IMPACT
San Francisco	1906	8.3	700 killed, fire, $7 million damage
Tehachapi–Arvin	1952	7.8	no deaths, $48 million damage
Landers	1992	7.5	1 death, $millions damage
Loma Prieta	1989	7.1	67 deaths, $6 billion damage
Northridge	1994	6.7	60 killed, $13–$20 billion damage
Imperial Valley	1990	6.7	9 deaths, $6 million damage
Mammoth Lakes	1980	6.0–6.6	no deaths
Coalinga	1983	6.5	extensive damage
San Fernando	1971	6.4	65 deaths, $500 million damage
Long Beach	1933	6.3	120 deaths, $50 million damage

SOURCES: U.S. Geological Survey: *Earthquakes and Volcanoes;* California Division of Mines and Geology; *Academic American Encyclopedia* (1990, Grolier Electronic Publishing Company).

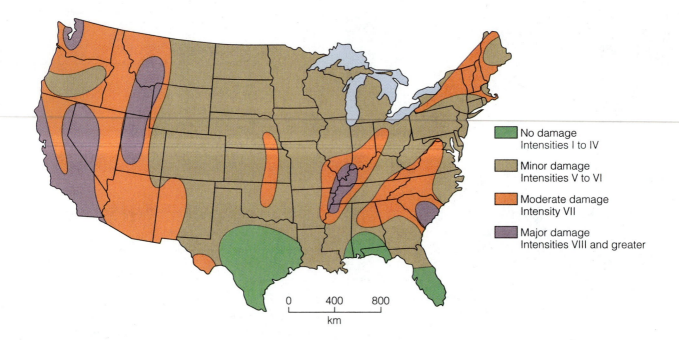

No damage
Intensities I to IV

Minor damage
Intensities V to VI

Moderate damage
Intensity VII

Major damage
Intensities VIII and greater

0   400   800
km

● **FIGURE 4.47** Seismic-risk map of the 48 contiguous states based upon historical records and intensities collected by the U.S. Coast and Geodetic Survey. The Coast and Geodetic Survey gathers intensity data from questionnaires after earthquakes.

years ago, the continental crust being rifted at the ridge was thinned, stretched, and faulted. The North American continent was transported westward with the plate, and the weakened part of the crust along the East Coast was hidden by a thick cover of younger sedimentary rocks. Buried rifted crust under the Eastern seaboard states forms the intraplate earthquake zone from the Carolinas to Canada.

The Midwest earthquake zone through Arkansas, Missouri, and Illinois is believed to be where an ancient (pre-Pangaea) divergent boundary started to form but "failed"

for some reason, leaving behind a significant buried fault zone. Known as the *Reelfoot Rift,* the buried block of crust is down-dropped between faults (the hachured lines on the map of ●Figure 4.49). The rift is 60 kilometers wide and 300 kilometers long (roughly 40 mi × 190 mi) and formed at least 500 million years ago. The linear trend of earthquakes from Marked Tree, Arkansas, northeastward to Caruthersville, Missouri, reveals upwarped sedimentary rocks along the rift axis. Detailed analysis of the geology across the Reelfoot fault scarp by trenching revealed evidence of three large earthquakes within the past 2,000

■ **TABLE 4.3** Selected North American Intraplate Earthquakes

LOCATION	YEAR	MOMENT MAGNITUDE*	IMPACT
New Madrid, Missouri	1811	8.2	Reelfoot Lake formed in NW Tennessee.
New Madrid	1812	8.3	Elevation changes caused Mississippi River to reverse its course locally.
New Madrid	1812	8.1	
Charleston, S. Carolina	1886	7.6	Felt from New York to Chicago; 60 killed.
Grand Banks, Newfoundland	1929	7.4	Submarine landslides broke trans-Atlantic cable, disrupting communications.

* Moment magnitude is used in reconstructing the strength of "preinstrument" earthquakes, because we know something of fault length, rupture length, and area felt.

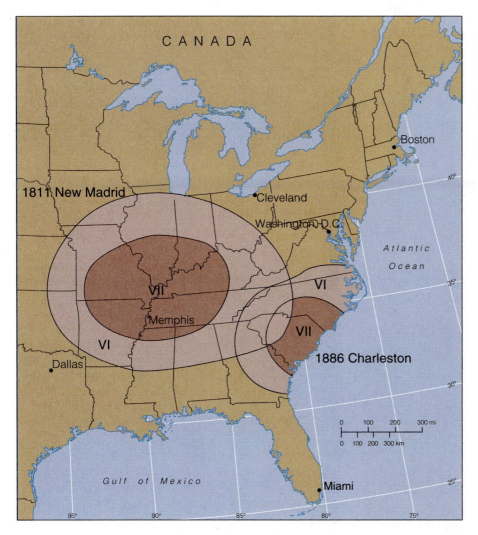

FIGURE 4.48 Isoseismals for intensity ≥VI for the New Madrid, Missouri (1811), and Charleston, South Carolina (1886), earthquakes. These earthquakes were "felt" over a much larger area of the United States than the area within isoseismals. *(b)* Damage from the 1886 earthquake along East Bay Street in Charleston, South Carolina.

(b)

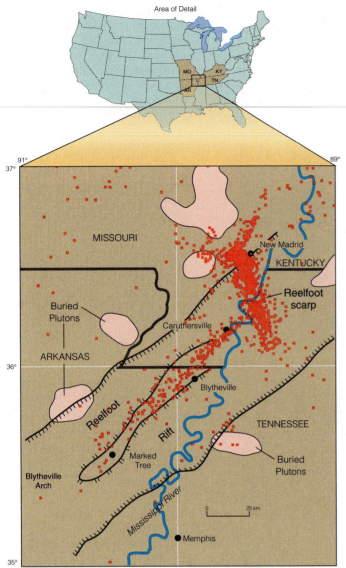

Area of Detail

91° 37°   89°

MISSOURI

New Madrid

KENTUCKY

Reelfoot
scarp

Buried
Plutons

Caruthersville

ARKANSAS

36°

Blytheville

Reelfoot

Rift

TENNESSEE

Marked
Tree

Buried
Plutons

Blytheville
Arch

Mississippi River

0   25 km

35°

Memphis

● **FIGURE 4.49 The New Madrid seismic zone covers parts of four states and consists of an ancient, fault-bounded depression (rift) buried below 1.5 kilometers (almost a mile) of sedimentary rocks. An area of major seismic hazard, it is studied intensively.**

years, which yields a recurrence interval estimate of 600–900 years for the fault.

Nobody knows how many people died in the 1811–1812 earthquakes. An amateur observer in Louisville, Kentucky, recorded almost 2,000 shocks with a homemade pendulum. Liquefaction was widespread along the Mississippi River, and the town of New Madrid, Missouri, was completely destroyed when it settled from 8 meters (25 ft) above sea level to only 4 meters (12 ft). Subsidence caused swamps to drain and others to become lakes; an example is Reelfoot Lake in northwest Tennessee, which is more than 50 feet deep. The New Madrid seis-

mic zone is a major geologic hazard in the United States, and efforts are being made to reduce the impact of a large earthquake there.

Iben Browning, a scientist with a Ph.D. in physiology but who is best-known for his work on climate, predicted there would be a repeat of the 1811 New Madrid earthquake on 3 December 1990. His prediction was based upon alignment of the planets and consequent gravitational pull, which he believed would be sufficient to trigger an earthquake on that date. Although the prediction was discounted by scientists, it created considerable anxiety in Arkansas, Tennessee, Alabama, and Missouri. Months before "the day," earthquake insurance sales boomed, moving companies were booked up, and bottled-water sales rose dramatically—as did sales of quake-related souvenirs. One church sold "Eternity Preparedness Kits," and "Survival Revivals" were held. On the day of the scientifically discredited prediction nothing earthshaking occurred.

Many earthquakes with epicenters in weakened intraplate continental crust have shaken parts of the Midwest, eastern Canada, New England, New York, and the East Coast of the United States. New York does not have a major fault such as California's, but it has many faults. In 1984 the state experienced two earthquakes, one in Westchester ($M = 4.0$) and the other off Coney Island ($M = 5.0$). The Westchester quake toppled chimneys, shattered windows, and caused some panic. As a result, the state implemented new seismic codes that add 2–5 percent to the cost of building, and the response has been favorable. By the mid 1990s New York and Massachusetts were the only Northeastern states that had earthquake building codes.

The Pacific Northwest

In years past earthquake hazards were considered minor in Oregon and Washington. During the 1980s research changed the perception, however; it revealed geologic evidence that "major" (≥ 7.0 but <8.0) or "great" (≥ 8.0 but <9.0) subduction-zone earthquakes have occurred in the past and that they can occur in the future. The Cascadia subduction zone (see Figure 5.4), extending 1,200 kilometers (740 mi) from northern California to Vancouver Island in Canada, has destructive earthquake potential. Geological evidence, consisting of carbon-dated tsunami deposits and drowned red-cedar forests, suggests that great earthquakes strike the Pacific Northwest roughly every 500 years, the last one in 1700 A.D., approximately 300 years ago. Minor earthquakes occur daily, however, and the cities of Portland, Seattle, and Vancouver, British Columbia, are in earthquake country (● Figure 4.50). The states now are working to minimize damage should the "big one" occur on the subduction zone or on the Seattle fault, which runs through downtown Seattle.

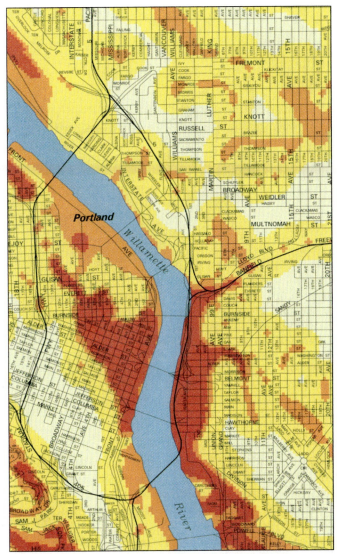

● FIGURE 4.50 Seismic-hazard map of downtown Portland, Oregon. The hazard zones are based upon liquefaction potential, landslide potential, and the probable degree of ground shaking. Red is greatest hazard; yellow indicates lowest hazard.

⊚ EARTHQUAKE PREDICTION

A guy oughta be careful about making predictions— particularly about the future.

YOGI BERRA

Earthquake prediction has great potential for saving lives and reducing property damage. The criteria for a good earthquake prediction are that it gives the location, the approximate time frame, and the magnitude of the earthquake with a high probability of correctness. Although it is a very difficult task because of the generally unpredictable behavior of faults, there are identifiable phenomena that tend to precede an earthquake—called **precursors**—that

can be measured. Nonetheless, since the 1970s only three earthquakes have been predicted.★ The methods currently used to predict earthquakes fall into three categories: statistical, geophysical, and geological.

Statistical Methods

By compiling the evidence of past earthquakes in a region, we acquire basic data for calculating the statistical probability for future events of given magnitudes. These calculations may be done on a worldwide scale or on a local scale, such as the example in ● Figure 4.51. Analysis of the graph indicates that for the particular area in Southern California, the statistical **recurrence interval**—that is, the length of time that can be expected between events of a given magnitude— is 1,000 years for a magnitude-8 earthquake (0.1/100 years), about 100 years for a magnitude-7 earthquake (1/100 years), and about 10 years for a magnitude-6 earthquake (almost 10/100 years). The probability of a magnitude-7 occurring in any one year is thus 1 percent, and of a magnitude-6, 10 percent. On an annual basis worldwide, we can expect at least two magnitude-8 earthquakes, 20 magnitude-7 earthquakes, and no less than 100 earthquakes of magnitude 6. Thus for seismically active regions, historical seismicity data can be used to calculate the *probability* of damaging earthquakes. Although these numbers are not really of predictive value as we defined it, they can be used by planners for making zoning recommendations, by architects and engineers for designing earthquake-resistant structures, and by others for formulating other life- and property-saving measures. A new approach is to evaluate the probability of a large earthquake occurring on an active fault during a given time period. This should be more properly called earthquake *forecasting*. ● Figure 4.52 shows the probabilities established in 1988 for the San Andreas fault for the 30-year period 1988–2018. One can see in the figure that along the Parkfield segment of the San Andreas there is almost a 100 percent probability for the period (the bases for this are explained later in this section) and that a very high probability was forecast for the southern Santa Cruz Mountains section. This latter section is the location of the epicenter of the 1989 Loma Prieta earthquake. Notice the high probability (near 50 percent) of a large earthquake on the southern section of the fault in the Imperial Valley. This was partially fulfilled by the $M = 7.5$ event at Landers in June 1992.

Geophysical Methods

Geophysical methods of prediction rely on instrumental measurements of changes in the physical state of the earth that precede earthquakes. Such precursors have potential

---

★Haicheng in 1975; Songpan–Pingwu in 1976; and Longling in 1976, all in China. In each case evacuation orders were issued that saved thousands of lives.

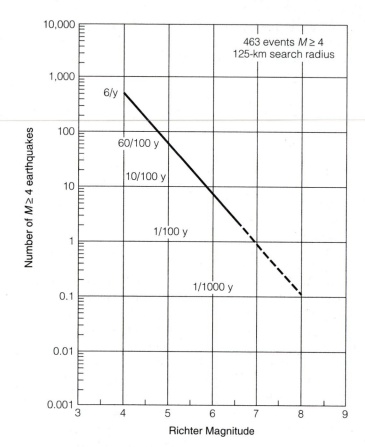

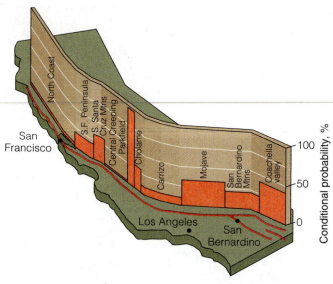

● FIGURE 4.52 Conditional probabilities of major earthquakes ($M \geq 7$) along segments of the San Andreas fault (1988–2018) established by the United States Geological Survey Working Group on California Earthquake Probabilities in 1988. Note the very high probability assigned to the Parkfield segment and the generally higher probabilities at the south end of the fault.

● FIGURE 4.51 A graphing of 463 earthquakes of magnitude 4 or greater in a small area near a nuclear reactor in Southern California over a period of 44 years. Statistically, the graph shows that 600 magnitude-4 earthquakes can be expected in 100 years, or 6 per year on average. The probability of a magnitude-8 earthquake in 100 years is 0.1, which would be 1 in 1,000 years. Such curves can be constructed for a region or for the world. The graphs are all similar in shape; only the numbers differ.

for providing significant insight. For example, it has long been known that before an earthquake, there are anomalously rapid changes in the ground tilt, earth magnetism, and electrical conductivity of rocks near the causative fault. Soviet scientists' reports at an international meeting in 1971 of a previously unrecognized precursor gave impetus to a new line of prediction research. Seismologists had observed a period of decreased velocity of P-waves of about 10 percent relative to S-waves before an earthquake, with the V_P/V_S ratio returning to a normal value just prior to the event. The duration of the period of decreased V_P/V_S prior to an earthquake appeared to be a function of the magnitude of the earthquake, ranging from a few days for small ones up to 40 years or more for a great earthquake. Excited U.S. scientists began looking for such precursors in North America. Such a reduction in V_P/V_S had indeed been reported in the Blue Mountain area of New York

before an earthquake, and a review of the 1971 San Fernando earthquake data indicated a similar anomaly. Although the reduction in this ratio has not proven to be a reliable predictor, the model of why it happens explains other observed changes, and it has initiated new areas of research.

A mechanism known as the **dilatancy–diffusion model** proposes that as rocks are stressed, microcracks form, which results in an increase in volume (dilatancy). Water then flows into the fractured rocks (diffusion), causing V_P/V_S to drop. Rock dilatancy can explain many precursory phenomena such as decreased electrical resistivity, increased water levels in wells, and increased radon emanation into ground water. For example, radon gas forms from the decay of radioactive elements at depth; as cracks develop, this gas rises into the local ground water. Because the amount of radon gas dissolved in ground water should thus increase before an event, well water is now being monitored for radon content in California. How these potential precursors are thought to change with time before and after an event are shown in ● Figure 4.53.

Seismic gaps are stretches along known active fault zones within which no significant earthquakes have been recorded. It is not always clear whether these fault sections are "locked" and thus building up strain energy, or if motion (creep) is taking place there that is relieving strain. Such gaps existed and were "filled" so to speak during the Mexico City (1985) and Loma Prieta (1989) earthquakes. Seismic gaps serve as alerts or warnings of possible future events

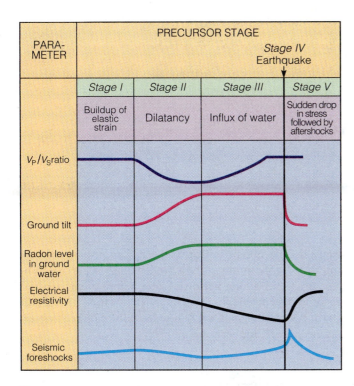

PARA-METER	PRECURSOR STAGE			
				Stage IV Earthquake
	Stage I	Stage II	Stage III	Stage V
	Buildup of elastic strain	Dilatancy	Influx of water	Sudden drop in stress followed by aftershocks
V_P/V_S ratio				
Ground tilt				
Radon level in ground water				
Electrical resistivity				
Seismic foreshocks				

⊙ FIGURE 4.53 Expected changes in five physical parameters before, during, and after a hypothetical earthquake.

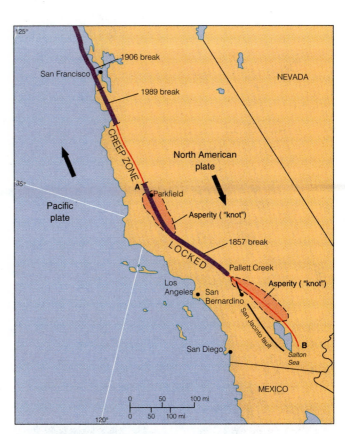

⊙ FIGURE 4.54 The San Andreas fault in Southern California. Segment A–B is the part of the fault where the greatest amount of plate motion has occurred. The purple lines represent the rupture zones of the 1906, 1857, and 1989 great earthquakes.

and can be used as forecasting tools. However, it is not possible with present knowledge to base earthquake predictions upon the presence or absence of a seismic gap.

Geological Methods

Research suggests that active faults and segments of long, active faults tend to have recurring earthquakes of characteristic magnitude, rupture length, and displacement. For example, six earthquakes have occurred along the San Andreas fault at Parkfield since 1857. They had Richter magnitudes of around 5.6, rupture lengths of 13–19 kilometers, and displacements averaging 0.5 meter. These earthquakes have recurred about every 22 years since the earliest recorded earthquake. The 95 percent probability assigned to this segment for 1988–1993 (see Figure 4.52) was established on these bases. This Parkfield prediction was the first one in the United States to be endorsed by scientists and subsequently issued by the federal government. The prediction had not been fulfilled by 1996, which led to the misconception that the scientists had failed. Parkfield remains the best-identified area on the San Andreas fault to "trap" an earthquake.

Geologists now believe that stable rough spots along fault planes, called **asperities,** determine earthquake characteristics along fault segments. Where asperities are large in size and far from other asperities, earthquakes tend to be large and infrequent; that is, to have long recurrence intervals.

Where asperities are close together and relatively small, shorter recurrence intervals are the rule. Where there are no significant rough spots, creep dominates, and no earthquakes occur. Somewhat like knots in a rope, these rough spots "lock" the fault and serve as stopping points of fault rupture. The larger the "knot" (asperity), the longer the time interval during which strain accumulates and the larger the earthquake that is produced. Smaller, unstable asperities are associated with small earthquakes and shorter time intervals between earthquakes. Using asperities as working models for characteristic earthquakes gives us a basis for studying fault behavior. Two large asperities that appear to govern the behavior of the San Andreas fault in Southern California are shown in ⊙ Figure 4.54.

Since recorded history in North America is short, geologists need other means of collecting frequency data of large prehistoric earthquakes. One method is to dig trenches into marsh or river sediments that have been disrupted by faulting in an effort to decipher a region's **paleoseismicity,** its rock record of past earthquake events. Kerry Sieh of the California Institute of Technology has done this across the San Andreas in Southern California (⊙ Figure 4.55). Sieh found an intriguing history of seismicity

● FIGURE 4.55 Disrupted marsh and lake deposits along the San Andreas fault at Pallett Creek near Palmdale, California. Sediments range in age from about A.D. 200 at the lower left to A.D. 1910 at the ground surface. Several large earthquakes are represented here by broken layers and buried fault scarps.

recorded in disrupted marsh deposits at Pallett Creek and liquefaction effects extending from the seventh century to a great earthquake in 1857. Ten large events, extending from A.D. 650 to 1857, were dated using ^{14}C. The average recurrence interval for these ancient earthquakes is 132 years, but they are clustered in four groups. Within each cluster, the recurrence interval is less than 100 years, and the intervals between the clusters are two to three centuries in length. The last big one, also the last one of a cluster, was in 1857. Thus it appears that this section of the San Andreas may remain dormant well into the next century or beyond.

An old saying in earthquake country is that "the longer it's been since the last one, the closer we are to the next one." Unfortunately, this is about the status of our present predictive ability. We remain uncertain that an earthquake will indeed follow well-identified percursory phenomena, such as anomalous animal behavior (see Case Study 4.5). It

seems that no one precursor is a perfectly reliable predictor, and that a large number of physical changes will have to be measured over a long period of time before prediction will become dependable. In China they say, "Though an earthquake does not occur for 1,000 years, do not let your observations lag for one second."

 ## MITIGATION

Reducing earthquake risks for the general public is an admirable goal of lawmakers and scientists. Lawmakers set policies, and scientists and engineers develop the technical expertise to recommend these policies and to carry them out. It has been shown that strict building codes in seismic regions, such as the Pacific Rim, reduce damage and loss of life (see Figure 4.38). Laws passed after the 1933 and 1971 Southern California earthquakes have proven the effectiveness of strict earthquake-resistant design.

The primary consideration in earthquake design is to incorporate resistance to horizontal ground acceleration, or "base shear." Strong horizontal motion tends to topple poorly built structures and to deform more flexible ones. In California, high-rise structures are built to withstand about 40 percent of the acceleration of gravity in the horizontal direction (0.4 g), and single-family dwellings are built to withstand about 15 percent (0.15 g). *Base isolation* is now a popular design option. The structure, low- or high-rise, is placed upon Teflon plates or even springs, which allows the ground to move but minimizes building vibration and sway.

Survival Tips

Knowing what to do before, during, and after an earthquake is of utmost importance to you and your family.

BEFORE AN EARTHQUAKE: ● Figure 4.56 provides the Federal Emergency Management Agency's suggestions for minimizing the possibility of damage and injuries in the home.

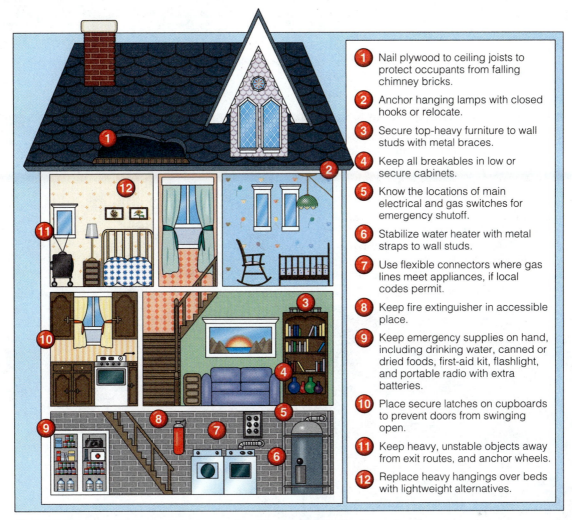

1. Nail plywood to ceiling joists to protect occupants from falling chimney bricks.
2. Anchor hanging lamps with closed hooks or relocate.
3. Secure top-heavy furniture to wall studs with metal braces.
4. Keep all breakables in low or secure cabinets.
5. Know the locations of main electrical and gas switches for emergency shutoff.
6. Stabilize water heater with metal straps to wall studs.
7. Use flexible connectors where gas lines meet appliances, if local codes permit.
8. Keep fire extinguisher in accessible place.
9. Keep emergency supplies on hand, including drinking water, canned or dried foods, first-aid kit, flashlight, and portable radio with extra batteries.
10. Place secure latches on cupboards to prevent doors from swinging open.
11. Keep heavy, unstable objects away from exit routes, and anchor wheels.
12. Replace heavy hangings over beds with lightweight alternatives.

◉ FIGURE 4.56 How to minimize earthquake damage in the home in advance.

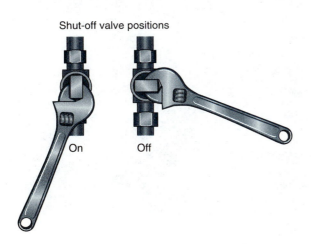

Shut-off valve positions

On   Off

◉ FIGURE 4.57 Keep a small crescent wrench at the gas meter. Turn off the gas by turning the valve end 90°.

DURING AN EARTHQUAKE:

- Remain calm and consider the consequences of your actions.
- If you are indoors, stay indoors and get under a desk, bed, or a strong doorway.
- If you are outside, stay away from buildings, walls, power poles, and other objects that could fall. If driving, stop your car in an open area.
- Do not use elevators, and if you are in a crowded area, do not rush for a door.

AFTER AN EARTHQUAKE:

- Turn off the gas at the meter (◉ Figure 4.57).
- Use portable radios for information.
- Check water supplies, remembering that there is water in water heaters, melted ice, and toilet tanks. Do not drink waterbed or pool water.
- Check your home for damage.
- Do not drive.

Seismic Oddities

S ometimes the pain of tragedy and loss can be softened by humor or humorous situations. This collage of photos presents examples of humorous, puzzling, and even downright comical results of the trembling earth. Please view them in this light (⊚ Figures 4.58 and 4.59).

⊚ FIGURE 4.58 (below) Statue of Louis Agassiz, the great geologist, naturalist, and educator, at Stanford University after the 1906 earthquake. Strong horizontal motion knocked the statue off its pedestal, prompting a noted scientist of the day to remark, "I have read Agassiz in the abstract. Now I see him in the concrete."

(a)

(b)

(c)

⊚ FIGURE 4.59 (at right) San Fernando earthquake, 9 February 1971, 6:01 A.M. (a) Compression of sidewalk panels makes for rough skateboarding. (b) The approach to an Interstate-210 overpass settled during the earthquake, providing an upramp that launched an unknown motorist and his auto 46 feet through the air as indicated by the skidmarks. The flight ended with a safe landing. (c) The sign *said* NO PARKING!

Depressed Tigers, Restless Turtles, and Earthquakes!

Anomalous animal behavior preceding earthquakes is well-documented. Domesticated animals such as barnyard fowl, horses, cats, and dogs have been known to behave so peculiarly before big events that they attracted the notice of people not knowledgeable in what is normal or abnormal animal behavior (● Figure 1). It should be remembered that although anomalous animal behavior precedes many earthquakes, it does not appear to precede every earthquake, and abnormal animal behavior is not always followed by an earthquake. Other natural phenomenon, such as atmospheric disturbances, also can cause animals to behave strangely. Here are some examples of unusual animal behavior noted before earthquakes:

- Tientsin Zoo, China, 1969: 2 hours before the magnitude-7.4 earthquake the tiger appeared depressed, pandas screamed, turtles were restless, and the yak would not eat.
- Haicheng, China, 1975: 1½ months before the magnitude-7.3 earthquake snakes came out of hibernation; 1–2 days before, pigs would not eat and they climbed walls; 20 minutes before, turtles jumped out of the water and cried.
- Tokyo, Japan, 1855: 1 day before the magnitude-6.9 earthquake wild cats cried, and rats disappeared.
- Concepción, Chile, 1835: 1 hour 40 minutes before the earthquake flocks of sea birds flew inland, and dogs left the city.

● FIGURE 1 Anomalous animal behavior. An earthquake may be in the offing if your dog dons a hard hat.

- San Francisco, 1906: dogs barked all night before the magnitude-8.3 earthquake.
- Friuli, Italy, 1976: 2–3 hours before the magnitude-6.7 earthquake cats left their houses and the village, mice and rats left their hiding places, and fowl refused to roost.

CONSIDER THIS...

Something to think about! Imagine that a scientific breakthrough allows us to predict earthquakes accurately. Let's say a $M_w = 6.7$ is predicted at a location that directly affects you. How would you receive the information and know that it is credible? What would you do if you had several days in which to act? How about only 4 hours' advance notice?

 ## SUMMARY

Earthquakes

CAUSE Movements on fractures in the crust known as *faults* that result in three types of wave motion: P- and S-waves, which are generated at the focus of the earthquake and travel through the earth, and L-waves, which are surface waves.

DISTRIBUTION Most (but not all) large earthquakes occur near plate boundaries, such as the San Andreas fault, and represent the release of stored elastic strain energy as plates slip past, over, or under each other. Intraplate earthquakes can occur at locations far from plate boundaries where deep crust has been weakened, probably at "failed" continental margins.

MEASUREMENT SCALES Some of the scales used to measure earthquakes are the modified Mercalli intensity scale (based on damage), the Richter magnitude scale (based on energy released as measured by maximum wave amplitude on a seismograph), and the moment magnitude (based on the total seismic energy released as measured by the rigidity of the faulted rock, the area of rupture on the fault plane, and displacement).

Earthquake-Related Hazards and Mitigation

GROUND SHAKING Damaging motion caused by shear and surface waves.

Ways to reduce effects—seismic zoning; building codes; construction techniques such as shear walls, seismic joints, and bolting frames to foundations.

LANDSLIDES Hundreds of landslides may be triggered by an earthquake in a slide-prone area.

Ways to reduce effects—proper zoning in high-risk areas.

GROUND FAILURE (SPONTANEOUS LIQUEFACTION) Horizontal (lateral) movements caused by loss of strength of water-saturated sandy soils during shaking and by liquefaction of quick clays.

Ways to reduce effects—building codes that require deep-drilling to locate liquefiable soils or layers.

GROUND RUPTURE/CHANGES IN GROUND LEVEL Fault rupture and uplift or subsidence of land as a result of fault displacement.

Ways to reduce effects—geologic mapping to locate fault zones, trenching across fault zones, implementation of effective seismic zonation like the Earthquake Fault Zones Act in California.

FIRE In some large earthquakes fire has been the biggest source of damage.

Ways to reduce effects—public education on what to do after a quake, such as shutting off gas and other utilities.

TSUNAMIS Multidirectional sea waves generated by disruption of the underlying sea floor.

Ways to reduce effects—early warning systems, building design that lessens impact of wave runup in prone areas.

Earthquake Prediction

STATISTICAL METHODS Historical information on earthquakes in a region is used to calculate recurrence intervals and probabilities of damaging earthquakes.

Upside/downside—good for planning purposes, but not for short term warnings.

GEOPHYSICAL METHODS Measurements of physical changes in the earth are used to identify seismic precursors.

Upside/downside—vigorous and promising research is active on many potential precursors; limited results so far.

GEOLOGICAL METHODS Active faults are studied to determine the characteristic earthquake magnitudes and recurrence intervals of particular fault segments; sediments exposed in trenches may disclose historic large fault displacements (earthquakes) and, if they contain datable C-14 material, their recurrence intervals.

Upside/downside—useful for long-range forecasting along fault segments and for identifying seismic gaps; not useful for short-term warnings.

KEY TERMS

aftershock	fault creep
asperity	focus (pl. *foci*)
body wave	intensity scale
dilatancy–diffusion model	isoseismal
elastic rebound theory	liquefaction
epicenter	longitudinal wave
fault	L-wave

magnitude	seismic gap
modified Mercalli scale (MM)	seismograph
	seismogram
moment magnitude (M_w or M)	spontaneous liquefaction
	stress
paleoseismicity	strain
precursor	S-wave
P-wave	transverse (shear) wave
recurrence interval	tsunami (pl. *tsunamis*)
resonance	wave period *(T)*
Richter magnitude scale	

STUDY QUESTIONS

1. What is "elastic rebound," and how does it relate to earthquake motion?
2. Distinguish among earthquake intensity, Richter magnitude, and moment magnitude. Which magnitude scale is most favored by seismologists today? Why?
3. Why should one be more concerned about the likelihood of an earthquake in Alaska than of one in Texas?
4. In light of plate tectonic theory, explain why devastating shallow-focus earthquakes occur in some areas and only moderate shallow-focus activity takes place in other areas.
5. What are some earthquake precursors, and how useful is each as a predictor?
6. What should people who live in earthquake country do before, during, and after an earthquake (the minimum)?
7. Describe the motion of the three types of earthquake waves discussed in the chapter and their effects on structures.
8. Why do wood-frame structures suffer less damage than unreinforced brick buildings in an earthquake?
9. What geologic conditions were responsible for the extreme damage and loss of life in the 1985 Mexico City earthquake?
10. Explain why structures close to the epicenter of the 1989 Loma Prieta earthquake suffered less damage than buildings tens of miles away in the San Francisco and Oakland Bay areas.

FURTHER INFORMATION

BOOKS AND PERIODICALS

Bolt, Bruce. 1988. *Earthquakes:* New York: W. H. Freeman.

Bolt, B. A.; W. L. Horn; G. McDonald; and R. F. Scott. 1975. *Geological hazards.* New York: Springer-Verlag.

Davis, G. A. 1994. Have you heard? Geological Society of America, *Structural Geology and Tectonics Division Newsletter* 13, no. 1.

Dolan, James F.; Kerry Sieh; Thomas Rockwell; Robert Yeats; John Shaw; John Suppe; Gary Juftile; and Eldon Gath. 1995.

Prospects for larger or more frequent earthquakes in the Los Angeles metropolitan region. *Science* 267:199–205.

Earthquake Engineering Research Institute (EERI). 1995. The Hyogo–Ken Nambu earthquake, January 17, 1995; Preliminary reconnaissance report. Oakland, Calif.: EERI, 116 pp.

Gere, James. 1984. *Terra non firma.* New York: W. H. Freeman.

Gore, Rick. 1995. Living with California's faults. *National Geographic,* April:2–34.

Hodgson, J. H. 1964. *Earthquakes and earth structures.* Englewood Cliffs, N.J.: Prentice-Hall, Inc.

Holzer, Thomas L. 1995. The 1995 Hanshin–Awaji (Kobe), Japan, earthquake. *GSA Today* (Geological Society of America), August.

Iacopi, Robert. 1971. *Earthquake country.* Menlo Park, Calif.: Lane Books.

Kockelman, William J. 1984. *Reducing losses from earthquakes through personal preparedness.* U.S. Geological Survey open file report 84-765.

Kovachs, Robert. 1995. Earth's fury: An introduction to natural hazards and disasters. Englewood Cliffs, N.J.: Prentice-Hall, Inc.

Mori, James J. 1994. Overview: The Northridge earthquake: Damage to an urban environment. *Earthquakes and volcanoes* 25, no. 1 (special issue).

Nance, John. 1989. *On shaky ground: America's earthquake alert.* New York: Avon Books.

Reid, T. R. 1995. Kobe wakes to a nightmare. *National Geographic,* July: 112–136.

Richter, C. 1958. *Elementary seismology.* New York: W. H. Freeman.

U.S. Geological Survey. 1994. Northridge, California, earthquake of January 17, 1994. *Earthquakes and volcanoes, 25, no. 1 (special issue).*

————. *Earthquakes and volcanoes.* Bimonthly publication. Washington: Government Printing Office.

————. 1990. *Probabilities of large earthquakes in the San Francisco Bay region, California.* USGS circular 1053 (free). Washington: Government Printing Office.

Wuethrich, Bernice. 1995. Cascadia countdown. *Earth: The science of our planet,* October:24–31.

VIDEO

WGBH, Boston. Predictable disaster. *Nova.* 60 minutes.

Earth exists by geological consent, subject to change without notice.

WILL DURANT, HISTORIAN (1885–1981)

Although people commonly regard volcanoes and volcanic activity as frightening, environmentally harmful geological phenomena, volcanoes are not all bad. They are attractive in a variety of ways: many of them are quite beautiful, they offer some of the best skiing and hiking in the world, and volcanic soils are highly productive. So why worry? Certainly the wealthy Romans vacationing on the Bay of Naples during the hot summer of A.D. 79 weren't concerned. Not until August 24, that is, when Mount Vesuvius suddenly erupted and buried the cities of Pompeii and Herculaneum. Vesuvius had been quiet for hundreds of years and was thought to be extinct. Vegetation covered its slopes, and its central crater showed no signs of recent activity. In two days Pompeii was buried so completely that its ruins were not uncovered

OPENING PHOTO

Mount Ruapein, New Zealand, during a spectacular eruption beneath the summit crater lake, September 1995. Steam, lake water, rocks, and volcanic bombs were ejected into the snow-covered slopes of Whakapapa Skifield that attract as many as 10,000 skiers per day.

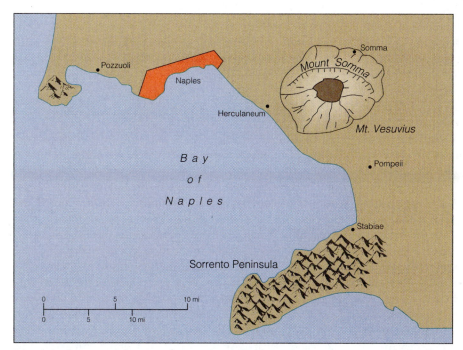

(a)

● FIGURE 5.1 *(a)* Map showing the locations of Mount Vesuvius, Herculaneum, and Pompeii on the Bay of Naples. *(b)* Air view of exhumed Pompeii as it looks today. *(c)* Naples with sleeping Vesuvius in the background. Note the remnants of an older, higher cone to the left of Vesuvius. Known as Mount Somma, it was obliterated during the A.D. 79 eruption.

(b)

(c)

for almost 17 centuries. Pompeii was rediscovered in the sixteenth century during excavation for a water line, but systematic work on the buried ruins did not begin until 1738. This date marks the beginning of the discipline we know as archaeology (● Figure 5.1). To date about half of Pompeii's 160 acres (65 hectares) have been exhumed. Herculaneum, settled by the Greeks and named after Hercules, is less well exposed because of its thick volcanic covering and because Portici, a new city, rests on top of its grave.

The heavy fall of ash at Pompeii suffocated more than 3,000 inhabitants and eventually buried the city under 6 meters (20 ft) of **pyroclastic** ("fire-broken") material. The story is preserved as molds in stone. Some people died attempting to flee the heavy fallout, whereas others, such as gladiators and slaves who were behind locked doors, had no chance to escape. An aristocratic woman was found in the gladiators' barracks; another died while attempting to retrieve her jewels. Imagine the desolation in late August

A.D. 79 caused by the sudden blanketing of hundreds of square kilometers with volcanic ash and cinders, covering everything but an occasional spire or other tall structure.

Herculaneum, to the north and away from the wind-driven plume of ash that destroyed Pompeii, suffered a different fate. Whereas Pompeii suffered for several days before being buried by a rain of volcanic ash, Herculaneum was overcome in a matter of minutes by a hot, fast-moving mixture of cinders and pumice known as a *pyroclastic flow*. These deposits and later mudflows buried the town to a depth of 20 meters (65 ft). Until the 1980s it was thought that the residents had been able to evacuate the city before being overwhelmed, as only 30 bodies had been discovered by then. Knowledge of Herculaneum was limited because excavation in the area is difficult and only the equivalent of eight city blocks had been uncovered. The excavation problems combined with the presence of a village on top of the deposits make it doubtful that Herculaneum will ever be explored to any great extent.

Recent excavations have revealed a whole different ending for some of the people of Herculaneum. Excavations along the waterfront uncovered a poignant scene. Buildings there were supported by a wall with arched chambers open to the sea. Fishermen used the chambers for storing their gear, boats, and tools of their trade. Hundreds of skeletons were found huddled together in these cavelike openings, some in what appear to be family groups. Apparently the victims ran to the beach seeking refuge and were overcome by hot pyroclastic surges there. More about Herculaneum's final hours will be revealed by the archaeologists' shovels, and this sad story of humans' attempts to survive the frightening force of a volcanic eruption will continue to unfold.

The A.D. 79 eruption of Vesuvius illustrates the nature of the volcano problem. Volcanoes may be intermittently active for a million years, and some continental ones are known to have erupted sporadically over ten-million-year periods. Thus, if dormant periods are sufficiently long, settlements may grow on or near a volcanic cone that later erupts. However, knowing that a potential volcanic problem exists allows policy makers, scientists, and others to plan for various eruption—and evacuation—scenarios, thus reducing casualties should an eruption occur. If successful, this can enable people to live in harmony with a sleeping giant.

WHO SHOULD WORRY

Most geologic activity that is dangerous to humans occurs along plate boundaries. This is where we find earthquakes and active volcanoes. Explosive volcanic activity, which presents the greatest challenge to life and property, occurs mostly at convergent plate boundaries. Inspection of the Pacific Ocean basin shows that it is surrounded by trenches, which form at convergent plate boundaries. Associated with these subduction zones is the so-called Ring of Fire, the location of two thirds of the world's active violent volcanoes. A recent catalog of the 1,350 volcanoes that have been active within the last 10,000 years (*Holocene time* to the geologist) shows 900 of them around the Pacific Rim. The largest number are found in New Zealand, Japan, Alaska, Mexico, Central America, and

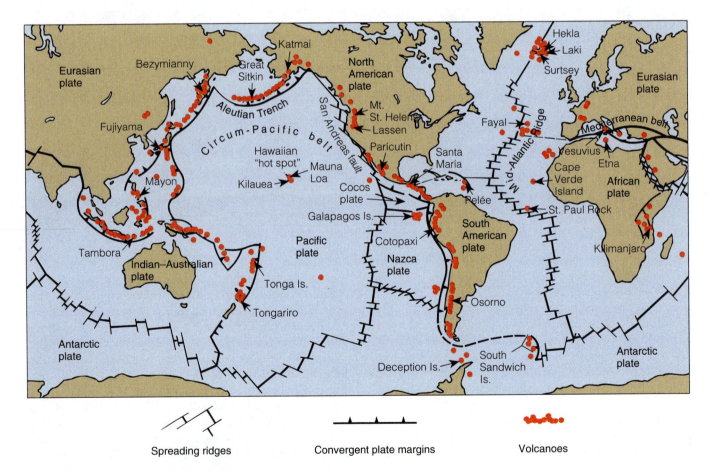

⦿ FIGURE 5.2 Distribution of earth's active volcanoes at plate boundaries and hot spots. Note the prominent "Ring of Fire" around the Pacific Ocean, which contains 900 (66 percent) of the world's active volcanoes. The remaining 450 are in the Mediterranean belt (subduction zones) and at mid-ocean-ridge spreading centers (divergent boundaries). A few important volcanic centers are related to hot spots, such as those in the Hawaiian and Galápagos Islands.

Chile (● Figure 5.2). Mount Erebus in Antarctica is the southernmost active volcano in this belt. About 250 active volcanoes are found in the Mediterranean, including the famous volcanoes Vesuvius, Etna, and Stromboli. Stromboli has been called the "lighthouse of the Mediterranean" because of the nearly continuous activity in its crater for more than a century.

The most productive volcanic centers, which exhibit less spectacular, comparatively mild activity, occur along divergent plate boundaries at mid-ocean ridges. Hot-spot volcanism may be explosive or mild and may occur on continents or in ocean basins (see also Chapter 3).

There are twice as many volcanoes north of the equator as there are south of it. This is particularly interesting to meteorologists, because large amounts of volcanic ash in the stratosphere result in a net heat loss for the earth, which affects climate. (High-flying jets are also affected; see Case Study 5.1.) Any concentration of active volcanoes at a particular latitude must be considered as a potential agent of change in global climate. Although most climatic variability is related to causes other than volcanism, historical examples of volcanic eruptions affecting weather are those of Tambora (Indonesia, 1815), El Chichón (Mexico, 1982), and most recently, Mount Pinatubo (the Philippines, 1991). Because two thirds of the world's active volcanoes and land area are north of the equator, the Northern Hemisphere is more vulnerable to climatic impact by volcanic eruptions than is the southern half of the world.

Mapping of the potential volcanic risk in the United States, such as ● Figure 5.3, reveals that the risk is seen as nil east of New Mexico. This is understandable in light of plate tectonics. The odd-shaped projections from the high-risk area on the map is due to the estimated transport of ash on the prevailing winds. The northeast–southwest trending high-risk zone through Idaho and into the corner of Wyoming is the trace of a southwestward transport of North America over a stationary hot-spot. A line of calderas mark the path, with Yellowstone being the currently active center over a fixed hot-

spot plume. This sequence is similar geologically to that of the Hawaiian chain of volcanoes.

The Cascadia subduction zone is responsible for the many active volcanoes and volcanic centers in Washington (6), Oregon (21), and California (16), (see ● Figure 5.4, page 124). The zone of active volcanoes extends 1,100 kilometers (700 mi) from Mount Lassen in California northward into British Columbia. Alaska ranks as the second-most active volcanic region in the world, and Hawaii is not far behind. Thus the study of volcanoes is important to citizens of the United States.

THE NATURE OF THE PROBLEM

How a volcano erupts determines its impact on humankind. A cataclysmic eruption, such as that of Mount St. Helens in 1980, has serious results. Simple outpourings of lava such as at Kilauea in Hawaii, on the other hand, can be good for tourism and business (except agriculture). Unfortunately, there are no measurement scales for describing the "bigness" of volcanic eruptions as there are for earthquakes. ● Figure 5.5, page 125, shows some of the criteria commonly used to describe an eruption's "explosivity." **Volcanic Explosivity Index (VEI)** values range from 0 to 8 according to the volume of material ejected, the height to which the material rises, and the duration of the eruption. Its "Classification" scale associates the particular eruption with a well-known volcano that exhibited the same kind of activity. The "Description" scale employs adjectives such as those used in newspaper headlines to describe the eruption. The 1980 eruption of Mount St. Helens, for example, could be indexed as a *4* and appropriately described as an *explosive-to-cataclysmic* eruption.

A volcano's potential explosivity has been found to increase with time since the start of its previous eruption (● Figure 5.6, page 125). This fact allows a region's risk of volcanic eruptions to be assessed when the date of the volcano's last eruption is known.

Volcanic Ash and the Friendly Skies

One night in 1982 a British Airways Boeing 747 was on a routine flight from Kuala Lumpur, Malaysia, to Australia, cruising at an altitude of 12,000 meters (37,000 ft). Just before midnight, sleeping passengers were awakened by a pungent odor filling the cockpit. Looking out the windows, they saw that the huge plane's wings were lit by an eerie blue glow. Suddenly the

In December 1989, a KLM Boeing 747 encountered airborne ash from Redoubt Volcano at about 8,500 meters (28,000 ft) during a descent to land at Anchorage, Alaska (⊙Figure 1, opposite). All four engines flamed-out, causing the large aircraft to suddenly become a glider. Thousands of feet lower, the engines were restarted, and the aircraft continued on to make

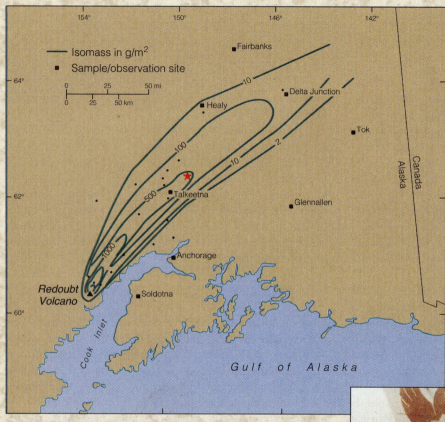

(a)

⊙ **FIGURE 1** *(a)* Deposits of tephra from Mount Redoubt's eruption of 15 December 1989. (The star indicates the KLM 747's position en route from Fairbanks to Anchorage when it encountered the ash cloud and flamed out.) Isomass contours enclose areas of equal tephra weight. Note that the weight of tephra near the vent was greater than a kilogram per square meter. *(b)* A worker removes ash from a Boeing 747 at Moses Lake, Washington, following the 1980 eruption of Mount St. Helens.

number-4 engine flamed out. Practically immediately the other three did as well, and the plane glided silently for an agonizing 13 minutes. At 4,500 meters (14,500 ft), the number-4 engine was restarted, followed by numbers 2, 1, and 3. Nonetheless, an emergency was declared, and the plane landed in Jakarta, Indonesia, with only three engines operating.

This near-death experience gained the attention of the world's airline passengers and pilots. Before this, such a failure had seemed virtually impossible in modern aircraft, which have redundant (backup) systems for almost every contingency. Unfortunately, flying air-gulping jet engines through clouds of volcanic ash is not one of these.

(b)

what was described as an "uneventful" landing. All four engines required replacement, as did the windshield and the leading edges of the wings, flaps, and vertical stabilizer, all of which had been "sandblasted." One may wonder what is required to make a landing "eventful." The interior of the plane was so filled with ash that the seats and avionic equipment had to be removed and cleaned. The total cost of returning the aircraft to service was $80 million.

Volcanic eruptions constitute a significant hazard to civil air transportation (● Figure 2). Numerous encounters with ash clouds have been reported by aircraft since the late 1970s, most notably from the Mount Pinatubo eruptions of June 1991. No fewer than 11 commercial aircraft declared emergencies in the air during the Mount Pinatubo eruptions and required extensive maintenance after landing (● Figure 3). Similar incidents have been reported over Central America, Chile, and Malaysia, going back as far as World War II, when Allied bombers encountered ash in the skies over Italy.

The U.S. Weather Service has established procedures for issuing in-flight Aviation Weather Advisories Services (AWAS) regarding conditions associated with volcanic eruptions. Such advisories state that an eruption has occurred and give information regarding the extent and movement of the airborne volcanic material. These advisories are issued as Significant Meteorological Advisories (SIGMETS), which are the most urgent weather warnings the service issues. Some business executives flying to the Far East in 1994 received a lesson in volcanology when Mount Kliuchevskoi on Russia's Kamchatka Peninsula erupted in full glory and spewed a plume of

● **FIGURE 3** Heavy ashfall from Mount Pinatubo made this World Airways DC-10 very tail heavy; Cubi Point Naval Air Station, Philippines.

ash that disrupted North Pacific air traffic for three days (● Figure 4). The lesson was shared by the astronauts on board the *Endeavor,* who photographed the 4,750-meter (15,580-ft) volcano in action.

● **FIGURE 4** Pacific air traffic to Asia was disrupted for nearly three days in September 1994 when Mount Kliuchevskoi (4,750 m, 15,580 ft) on Russia's Kamchatka peninsula erupted into airlanes at great altitude.

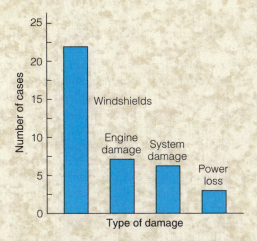

● **FIGURE 2** Reported damage to aircraft (worldwide) due to encounters with volcanic ash plumes, 1976–1986.

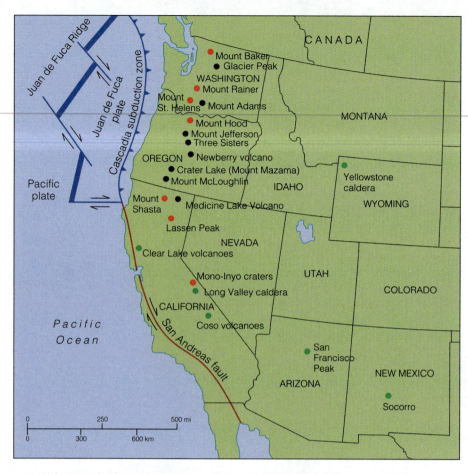

● **FIGURE 5.4** Active volcanoes and volcanic centers in the continental United States. The volcanoes of the Cascades are geologically related to the Cascadia subduction zone.

● Volcanoes that have short-term eruption periodicities (100–200 years or less) or that have erupted in the past 200–300 years, or both.

● Volcanoes that appear to have eruption periodicities of 1,000 years or greater and that last erupted 1,000 years or more ago.

● Volcanic centers that are greater than 10,000 years old, but beneath which exist large, shallow bodies of magma that are capable of producing exceedingly destructive eruptions.

A lava's *viscosity*—its resistance to flow—is another determinant of a volcano's explosivity, and viscosity is a function of lava's temperature and composition. Viscosity varies inversely with temperature for most fluids, including lava: the higher the temperature, the less viscous; the lower the temperature, the more viscous, other factors being equal. A lava's composition, most importantly its silica (SiO_2) content, also influences its viscosity. The silicon–oxygen bond is very strong; considerable heat energy is required to break it. The fluidity of siliceous (containing silica) lavas depends on the continual breaking and remaking of this bond. Thus, the higher the SiO_2 content, the more viscous is the lava—again, other factors being equal.

Viscous lavas with a high gas content can build up such high gas pressures as to be explosive, whereas gases escape readily from fluid lavas, and hence, they are much less dangerous. This fact gives us another method of categorizing large volcanoes and their activity: the nature of the magmas

they tap. Magmas form at depths of 50 to 250 kilometers (30–150 mi), where temperatures are sufficiently high to melt rocks completely or partially. At mid–ocean ridges and hot spots, volcanoes draw from magmas in the upper mantle that are high in iron and magnesium and low in silica (50 percent or less). These are **mafic** magmas (*ma-* for magnesium and *-f-* for the Latin word for iron, *ferrum,* plus *-ic*), which yield fluid (low-viscosity) lavas that retain little gas and hence do not erupt violently. Volcanoes that are adjacent to subduction zones, on the other hand, tap magmas that are a mixture of oceanic crust and sediment, upper-mantle material, and melted continental rocks. These are **felsic** magmas (*fel-* for feldspar and *-s-* for silica, plus *-ic*), which yield thick, pasty lavas with SiO_2 contents up to 70 percent. Even though hot lavas have lower viscosities and flow more readily than do cool ones, just as with honey and various other common fluids, the major determinant of lava fluidity is silica content.

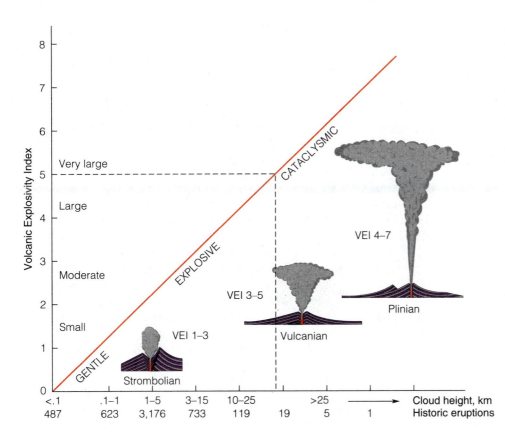

FIGURE 5.5 Graphic representation of the Volcanic Explosivity Index (VEI). As an example, a VEI 5 would be a very large eruption, described as cataclysmic, Vulcanian in its type of eruption, and have an ash plume up to 25 kilometers (15 mi) high. There have been 19 historic VEI-5 eruptions.

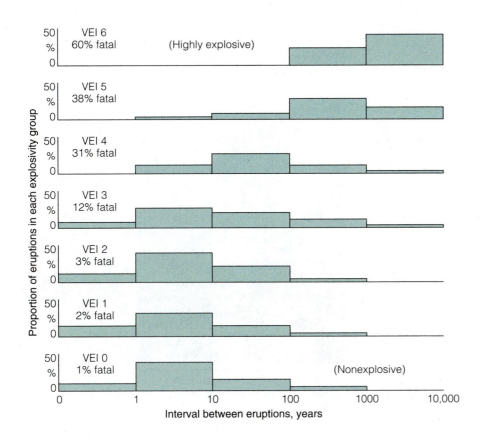

FIGURE 5.6 The longer the time interval since an eruption, the greater is the next eruption's potential explosivity. These data for 4,320 historic eruptions relate known intervals between eruptions to the Volcanic Explosivity Index (Figure 5.5). Also shown is the percentage of eruptions in each VEI that have caused fatalities.

High-silica, high-viscosity lavas retain gases, which leads to violent, explosive-type eruptions. The geologically important boundary between mild oceanic eruptions and the more explosive continental ones of the Pacific basin is called the **andesite line** (after the rock andesite from the Andes Mountains). The line is generally drawn southward from Alaska to east of New Zealand by way of Japan, and along the west coasts of North and South America. The andesite line is also a petrologic boundary between mafic magmas, which yield basalt, and felsic magmas, which yield andesite or some other viscous, high-silica rock. Thus, explosive volcanoes are always found on the continental side of the andesite line, and gentle volcanoes are found within the Pacific Ocean basin.

TYPES OF ERUPTIONS AND VOLCANIC CONES

The type of volcanic eruption determines the shape of the structure or cone that is built, and by appearance alone, one can get an idea of a volcano's hazard potential (◉ Figure 5.7).

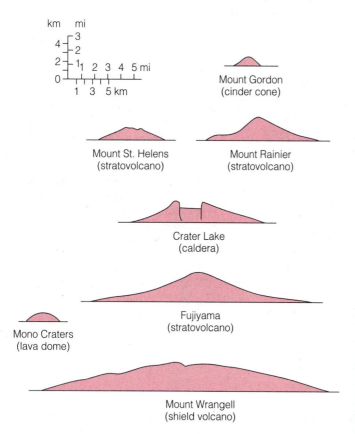

◉ **FIGURE 5.7** Comparative profiles and relative sizes of some Pacific Rim volcanoes and the Mono Craters in the western interior of the United States. Mount Gordon and Mount Wrangell (similar in shape to Hawaiian volcanoes) are in Alaska, the others are well-known stratovolcanoes.

Quiet Eruptions

SHIELD VOLCANOES. Shield volcanoes are built by gentle outpourings of fluid lavas from a central vent or conduit, and the lavas cool to form basalt, the most common volcanic rock. The name is due to the fact that a shield volcano's profile is gently convex upward like that of a shield laid on the ground. Mostly oceanic in origin, shield volcanoes are found in Iceland (where the *shield* name was first applied), the Galápagos Islands, and the Hawaiian Islands (◉ Figure 5.8). They are built up from the sea floor, layer upon layer, to elevations thousands of meters above sea level. Mauna Loa ("long mountain") and Mauna Kea ("white mountain") volcanoes on the Island of Hawaii are shield volcanoes that project 4.5 kilometers (2.8 mi) above sea level, and their bases are in water about 5 kilometers (3 mi) deep. With a total height of 9.5 kilometers (31,000 ft), these are the highest mountains on earth, exceeding Mount Everest by about 650 meters (2,150 ft). The Island of Hawaii is composed of five separate volcanoes (◉ Figure 5.9), the most active of which is Kilauea ("much spreading"), the most easterly of the group. This is exactly what plate movement would predict as the sea floor moves northwesterly over the Hawaiian "hot spot" (see Chapter 3). Loihi, a new volcano, is forming on the sea floor southeast of Kilauea, providing further evidence that the Pacific plate is moving northwest over the plume.

It is not uncommon for a shield volcano to erupt from a *fissure,* or crack, on its flanks, rather than from a central vent. This is typical of Kilauea's east fissure zone, where the countryside has been flooded under a sea of lava. Pu'u O'o ("hill of the o'o bird") is a large crater built upon a vent in the east fissure zone. Pu'u O'o's crater has reached impressive proportions as a result of scores of eruptions since 1983.

◉ **FIGURE 5.8** Shield volcano Fernandina in the Galápagos Islands. This active volcano is related to the Galápagos "hot spot." Its convex profile is similar to those of the volcanoes of Hawaii.

⦿ FIGURE 5.9 Satellite photograph of the Island of Hawaii. North is at the top of the photo. The two most active volcanoes visible in the photograph are the older, snowcapped Mauna Kea in the north and Mauna Loa, conspicuous for the abundant, fresh-looking basalt flows on its flanks in the south-central part of the island. The east fissure zone and Kilauea, the youngest active volcano, are to the southeast of Mauna Loa. The Pacific plate is moving northwesterly.

(a)

(b)

During Pu'u O'o's eruptions, Kilauea's summit deflates slightly, rising again or reinflating between eruptive episodes. This indicates that the fissure zone and Kilauea's vent plumbing are connected with the main lava reservoir. Lava flows associated with Pu'u O'o caused significant damage in the Royal Gardens subdivision near Kalapana about 8 kilometers (5 mi) from the vent. Flows destroyed at least 75 homes and covered 10 kilometers (6 mi) of residential streets (⦿ Figure 5.10). In addition, lava has built out hundreds of meters into the sea, creating new lands.

Spectacular lava fountains 400 meters (1,300 ft) high have occurred with fissure eruptions at Pu'u O'o (⦿ Figure 5.11). Some of the more fluid droplets set into glassy, tear-shaped blobs known as "Pele's tears," named after the

'(c)

⦿ FIGURE 5.10 (right) (a) A lava flow ignores a stop sign in the Royal Gardens subdivision below the east fissure zone. (b) A flow set this house afire at Kalapana near the Royal Gardens. (c) The historic church at Kalapana was moved just before lava engulfed the area.

● FIGURE 5.11 Lava fountaining at Pu'u-O'o, east fissure zone, Kilauea Volcano, Island of Hawaii.

Hawaiian goddess of volcanoes. Trailing behind these droplets may be tiny threads of lava, some more than a meter in length, that detach and fall to earth as "Pele's hair" (● Figure 5.12). Some of the detached fibers are carried many kilometers by the wind. Birds have been known to make nests from some of the more pliable threads of volcanic glass. Volcanic glass, which has a disordered or random atomic structure, forms when rapid cooling or quenching prevents the formation of orderly crystals.

FISSURE ERUPTIONS. In some shield volcano areas, lava flows *only* from long fissures, forming broad **lava**

● FIGURE 5.12 "Pele's hair," thin filaments of glassy lava that have been carried by wind.

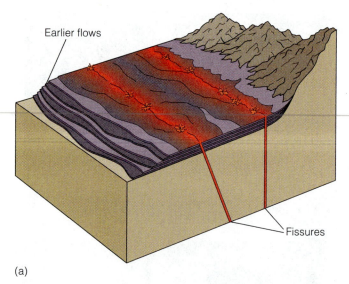

(a)

(b)

● FIGURE 5.13 *(a)* Fissure eruptions forming a lava plateau. *(b)* Antoja caves, Buddhist shrines excavated into the basalts of the Deccan lava plateau, India.

plateaus (● Figure 5.13, part a). Lava plateaus are found in Iceland, India, and the Columbia River plateau in Washington, Oregon, and Idaho in the United States. The Deccan lava plateau of India (Figure 5.13, part b) is enormous. Its massive outflow with degassing of toxic sulfurous fumes during Cretaceous time has been linked by some paleontologists to the extinction of the dinosaurs. Also, many submarine eruptions at oceanic spreading centers are fissure eruptions.

Explosive Eruptions

STRATOVOLCANOES. Explosive volcanic activity builds **stratovolcanoes,** sometimes referred to as *composite cones.* Typically thousands of feet high and 10–20 kilometers (6–12 mi), across at the base, they have a concave upward profile and a central vent. They are stratified, thus the name *strato*volcano, consisting of alternating layers of ash, cinders, and lava.

(a)

(b)

⊙ FIGURE 5.14 Two beautiful examples of stratovolcanoes from widely separated areas. *(a)* Mount Rainier, Washington, with active glaciers on its flanks. *(b)* Agung Volcano, Bali.

(a)

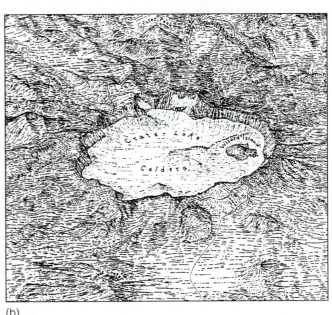

(b)

⊙ FIGURE 5.15 Crater Lake caldera, Klamath County, Oregon. *(a)* Aerial view and *(b)* oblique map view.

The upper, steep slopes are formed mostly of pyroclastic volcanic ejecta, and the less steep, lower slopes are composed of alternating layers of lava and pyroclastics. These cones are some of the most beautiful tourist attractions in the world. Noteworthy examples are Mount Vesuvius in Italy, Mount Fujiyama in Japan, Mount Hood in Oregon, Mount Rainier in Washington, and Agung in Bali (⊙ Figure 5.14). They occur on the landward side of subduction zones, where melting of oceanic crust and mantle forms magmas that rise because they are expanded and thus more buoyant than the surrounding lithosphere. As bodies of magma rise in the lithosphere, they mix with continental rocks and form thick, viscous, gas-charged lavas. These lavas do not flow easily; rather, they congeal in the volcano's central conduit (vent), which permits gas pressures to build to explosive proportions. For this reason, strato-

volcanoes present the most immediate threat to humans. Some eruptions have been so great that large cones have simply disappeared in the explosion. Crater Lake in Oregon represents the stump of the former Mount Mazama, a very large stratovolcano; it was probably the size of Mount Rainier. About 6,900 years ago, 70 cubic kilometers (17 mi^3) of Mount Mazama disappeared, due in part to explosive activity, but mostly due to the collapse of the remaining cone into the deflated magma chamber beneath it. The "crater" of Crater Lake is a **caldera**, defined as a volcanic crater that is many times larger than the vent that feeds it (⊙ Figure 5.15). Calderas may form by explosive disintegration of the top of a volcano, by collapse into the magma chamber, or by both mechanisms. Yellowstone National Park occupies several huge calderas, as does Long Valley in the eastern Sierra Nevada, both of them lying above shal-

● FIGURE 5.16 Lava domes; Mono Craters, east-central California. The Sierra Nevada range is in the background.

(a)

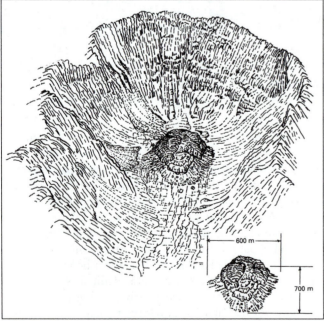

(b)

● FIGURE 5.17 *(a)* Steaming lava dome in the crater of Mount St. Helens, October 1981. *(b)* Oblique map of the same lava dome, 8 October 1981.

low bodies of magma that are capable of producing destructive eruptions.

LAVA DOMES. Lava domes are formed when bulbous masses of lava pile up around the vent because the lava is too thick and viscous to flow any significant distance from its source. Sometimes called *volcanic domes,* they usually grow by expansion from within. As the outer surface cools, the brittle crust breaks and tumbles down the sides, as at the Mono Craters on the eastern Sierra Nevada of California (● Figure 5.16). Relatively small domes may form in the crater of a larger composite cone, such as the one that formed in the crater of Mount St. Helens after the 1980 eruption (● Figure 5.17). The source vents of large domes may be substantially plugged, which offers the potential for explosive eruptions, particularly where the lavas have access to underground water or seawater. Mount Pelée on Martinique in the West Indies and Mount Lassen, Mono Craters, and Mammoth Mountain in California are all dormant lava domes. They extrude lavas such as rhyolite with silica (SiO_2) contents of 65–75 percent, and they also extrude glassy rocks that cool quickly, such as obsidian and pumice. Although pumice is largely SiO_2, it does not look like glass as does obsidian. This is because it is derived from gas-charged magmas and is thus "spongy" and full of gas holes, in contrast to smooth, glasslike obsidian. It should be noted that not all volcanic glasses are silica-rich; basalts that are chilled by extrusion underwater or by flowing into the sea sometimes form a glass.

CINDER CONES. Cinder cones are the smallest and most numerous of volcanic cones (● Figure 5.18). They are built of pyroclastic material of all sizes—from blocks and bombs to the finest ash (■ Table 5.1). Pyroclastic material of all shapes and sizes is collectively known as **tephra.** *Bombs* are blobs of still–molten lava that assume an aerodynamically induced spindle shape and solidify before striking the ground (● Figure 5.19). They can be found in great numbers around certain cinder cones and are prized by collectors and geologists. Cinder cone activity is local, within a few kilometers of the source vent, and is usually short–lived. The only new volcano to form in historic time in North America erupted in a farm field in 1943 near the village of Parícutin in the state of Michoacán, Mexico (Figure 5.18). Eruptions caused by frothing gases hurled blobs of lava into the air that became cinders and then fell back around the vent, eventually building a cone nearly 400 meters (1,300 ft) high. An observatory was established on a nearby hill, and the volcano's every burp and belch was recorded for nine years. After the cone-building stage had passed, basaltic lava flowed from the crater, inundating the nearby pueblo of San Juan and removed about 160 square kilometers (100 mi²) from agriculture. Fortunately, no lives were lost. Cinder cones are generally thought to be one-shot events; that is, once the eruption sequence ends, it is rarely reactivated. Nonetheless, exceptions are known, and one usually finds many cinder cones within a volcanic area.

● FIGURE 5.18 Parícutin Volcano, Michoacán, Mexico, in 1943. The only totally new volcano in North American history, it began as a fuming crack in a farmer's field. In a short time it was 400 meters (1,312 ft) high, and lava from it covered two nearby villages, leaving only a church steeple protruding through the lava. Volcanic activity decreased rapidly, and the volcano was inactive by 1952.

VOLCANIC PRODUCTS

Pyroclastic Materials

Volcanic activity provides products that the earth's peoples utilize in many ways. The final polish your teeth receive when you get them cleaned is an example. Dental pumice is refined and flavored volcanic pumice with a hardness just slightly less than tooth enamel. "Lava" brand hand soap, a rough, abrasive bar soap, contains the same powdered rock. Light-weight bricks, cinder blocks, and many road-foundation and decorative stone products originated deep within the earth. Where volcanic cinders are mined for road base, they may also be mixed with oil to form asphalt pavement. The reddish pavement of state highways in Nevada and Arizona contains basalt cinders that are rich in oxidized iron, which gives it the red color. Pieces of rock pumice are sold in drugstores, supermarkets, and hardware stores for use as a mild abrasive for removing skin calluses and unsightly mineral deposits from sinks and toi-

		CONDITION
NAME	SIZE	WHEN EJECTED
Blocks	>32 mm	Cold, solid
Bombs	>32 mm	Hot, plastic
Lapilli (cinders)	4–32 mm	Molten or solid
Ash	¼–4 mm	Molten or solid
Dust	<¼ mm	Molten or solid

■ TABLE 5.1 Classification of Pyroclastic Ejecta (Tephra)

lets. Powdered, it is used in abrasive cleaners and in furniture finishing.

Glassy volcanic rock, such as obsidian, is easy to chip and form. Hence, Native Americans and stone-age people in many parts of the world used it to make tools and arrowheads. Today obsidian and many similar volcanic rocks are the raw material of "rock hounds" and artisans for producing polished pieces and decorative art.

Geothermal Energy

By far the most important and beneficial volcanic product is also the most hazardous; it is heat. The same heat energy that causes eruptions also drives geysers and hot springs, and when controlled, it can be converted to other uses. Not surprisingly, the prospects for geothermal energy are best at or near plate boundaries where active volcanoes and high heat flow are found. The Pacific Rim (Ring of Fire), Iceland on the Mid-Atlantic Ridge, and the Mediterranean belt offer the most promise (● Figure 5.20). Energy from earth heat is discussed in Chapter 13.

● FIGURE 5.19 A volcanic bomb; Galápagos Islands, Ecuador. Note the spindle shape that the bomb acquired as it was flung through the air.

● FIGURE 5.20 Wairaki, one of several large volcanic centers on the North Island of New Zealand. It produces significant amounts of geothermal energy in a country that is not energy-rich.

Recreation

Volcanoes provide opportunities for recreational activities, ranging from mountain climbing and skiing to more passive activities such as photography and birdwatching. World-class ski resorts such as Mount Hood and Mount Bachelor in Oregon provide varied terrain and abundant snow for downhill and cross-country skiing and snowboarding. A big factor in their attractiveness is their relatively easy accessibility, as they do not require prolonged driving through mountainous terrain. Most of these resorts' ski lifts operate year round, serving snow sports in the winter and mountain biking and sightseeing in the summer. Of course, skiing on an active volcano poses some measure of risk, as skiers at Mount Ruapehu in New Zealand found out in 1995.

 ## VOLCANIC HAZARDS

Sicily's Mount Etna (from the Greek *aitho,* "I burn"), in the Mediterranean volcanic belt, exhibits almost continuous activity in its crater. In 1992 eruptions and lava flows were threatening several villages. Empedocles (circa 490–430 B.C.), Greek philosopher and statesman, is perhaps most notable for throwing himself into the crater of Mount Etna to convince his followers of his divinity. His dramatic suicide inspired Matthew Arnold's epic poem "Empedocles on Etna," which bears no resemblance to the little rhyme attributed to Bertrand Russell:

> Empedocles that ardent soul—
> Fell into Etna and was roasted whole!

Volcanoes comprise the third-most dangerous natural hazard in terms of loss of life, after coastal flooding (hurricanes and typhoons) and earthquakes. Following the Mount St. Helens eruption of 1980, U.S. civil defense agencies published suggestions for what to do when a nearby volcano erupts. The Federal Emergency Manage-

ment Agency's suggestions appear in the next section. For now, however, let us examine the various types of hazards associated with eruptions, several of which might occur during a single eruptive phase.

Lava Flows

Most hazards, natural and human-made, decrease in severity with distance from the point of origin. This is true of earthquakes generally, tornadoes, and falls of volcanic ash, for example. Lava flows may be the exception to this, because they generally burn or bury everything in their path, even to their farthest limit. Holocene lava flows in Queensland, Australia, traveled 100 kilometers (62 mi) from their vents down riverbeds with very low gradients. Knowledge of the paths that flows might take from given vents makes it possible to delineate low- and high-risk volcanic-hazard areas. Basaltic lavas, such as those from Mount Etna in 1992, may emerge from a vent or fissure at temperatures in excess of 1,100° C and flow rapidly downslope like a river because of their low viscosity. As a flow cools, its viscosity increases, and then it may only "chug" along slowly. In either case, the end product is a layer of rock of varying thickness that covers everything in its path.

Two kinds of basalt flows are recognized. **Pahoehoe** ("pa-ho′-e-ho′e") forms when a skin develops on a flow and buckles up as the flow moves, making a smooth or ropy-looking surface (● Figure 5.21, part a). Other basalt flows develop upper and lower surfaces that are rough and blocky. This rough, angular lava is called **aa** ("ah′-ah") and it is characterized by a very slow advance in which the top, cold rubble rolls over the front of the flow—moving much like the track of a Caterpillar tractor (see Figure 5.21, part b; also see Figures 5.10, part a, and 5.18).

A volcanic eruption cannot be prevented, but in some cases, it is possible to divert or chill a flow so as to keep it from encroaching onto croplands or structures. Flows have been diverted by constructing earthen dikes in Hawaii and Italy. Icelandic fire fighters have successfully chilled flow fronts by spraying them with large volumes of seawater from fire hoses. During World War II, the U.S. Army Air Corps tried, unsuccessfully, to divert lava flows by bombing them.

Ashfalls and Ash Flows

Roman naturalist and historian Pliny the Elder (born A.D. 23) died during the eruption of Vesuvius in A.D. 79 from complications brought on by inhaling noxious fumes and overexertion. He is revered as geology's first martyr, having been quoted by his nephew, Pliny the Younger, as saying, "Fortune favors the brave." Bravery was to be his undoing, as corpulent Pliny died in an area of heavy fallout from the eruption. Fortunately, however, Pliny had enjoyed life: he is credited with coining the phrase *In vino veritas* ("In wine there is truth"). Before Gaius Plinius (Pliny the Elder) died, he may have seen a great vertical plume of ash with a mushroom- or anvil-shaped head rising from Vesuvius's

(a)

(b)

◉ FIGURE 5.21 *(a)* Smooth, ropy pahoeloe lava; Galápagos Islands, Ecuador. *(b)* Rough, blocky aa flow in Hawaii. The trees engulfed by the flow burn, but their impressions remain as molds in the lava.

crater. This high-velocity column of steam, gas, and fragmented lava is called, for Pliny, a **Plinian eruption.** Plinian eruptions have been seen at Mount Lassen, California, in 1915, at Mount St. Helens in 1980, and at Mount Pinatubo in 1992 (◉ Figure 5.22 and Case Study 5.2). Such an ash cloud, consisting of tephra (jagged pumice, rock fragments, and bits of glass), may be carried away from the crater by prevailing winds and deposited as an **ashfall.** Ashfalls, like the ones at Pompeii and Mount St. Helens, bury objects close to their source. At Mount St. Helens, westerly winds carried ash eastward, causing difficulties in Yakima and Spokane where several inches of ash were deposited. At Butte, Montana, about 750 kilometers (465

◉ FIGURE 5.23 (right) Satellite photo of an ash cloud from Mount St. Helens two days after the eruption superimposed on an outline map. Note that the cloud covered western Washington and parts of Idaho and Montana. Ash fell as far away as New York State.

◉ FIGURE 5.22 Plinian ash cloud over Mount Pinatubo.

mi) east, the fine ash and dust in the atmosphere forced the closure of the airport. People in Yakima wore masks to filter out the choking dust. Bankers politely requested that all patrons remove their masks before entering their banks because of potential identification problems. ◉ Figure 5.23

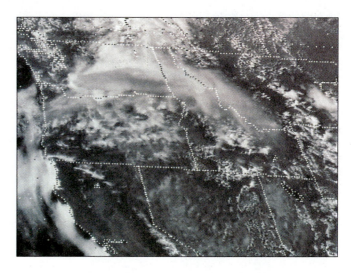

Mount Pinatubo—Biggest Eruption of the Century?

Mount Pinatubo, with a summit elevation of 1,745 meters (5,670 ft), is a volcano 100 kilometers northwest of Manila on Luzon Island, the Philippines. Before 2 April 1991, its dome had been dormant for 600 years. Throughout April minor eruptions and low-level seismicity were observed and recorded. These were followed by stronger and more frequent tremors and by heavy SO_2 emissions (as much as 500 metric tons per day) in late May and early June. Finally, things culminated in spectacular cataclysmic eruptions on June 15–16 (see Figure 5.22). This "finale" lasted 15 hours and sent tephra more than 30 kilometers (18 mi) into the atmosphere. Ash flows 200 meters (650 ft) thick and 10 kilometers (6 mi) long filled all low spots in the topography, and the volcano lost 400 meters (1,300 ft) of summit elevation, leaving a crater 3 cubic kilometers (0.73 mi³) in volume. Following the pyroclastic flows, heavy monsoon rains and winds triggered lahars that traveled down major drainage systems, adding to the Filipinos' misery by destroying more homes and causing more fatalities. It is estimated that there were no survivors within 10 kilometers (6 mi) of the volcano, and only a few within 15 kilometers (9 mi). Estimates of deaths directly attributable to the eruption, volcanic mudflows, and disease in evacuation camps range from 323 to 435.

Precursors to the main event—harmonic tremors and SO_2 and ash emissions—increased up to the time of the main eruptions, giving notice of the impending disaster. The 2½-month warning time allowed areas considered to be in the path of ash flows or lahars to be evacuated, which saved many lives. When Clark Air Force Base, a major U.S. installation 15 kilometers (9 mi) east of the volcano, was ordered evacuated on June 10, 15,000 military personnel and their families were moved to Subic Bay Naval Base in a matter of hours. An estimated 250,000 villagers were evacuated, but others remained in their homes to face the imminent danger of falling ash and mudflows (see ◉ Figure 1). Tragically, mudflows continued well into July due to Typhoon Amy. By July 26, 100,000 homes had been destroyed or severely damaged, and 90,000 people remained in evacuation camps.

The cloud of ash and SO_2 from Pinatubo expanded rapidly to the west and southwest, reaching Bangkok, 2,000 kilometers (1,240 mi) away, by June 17. Pumice the size of apricots fell at Clark Air Force Base (◉ Figure 2), and the size of marbles at Olongapo, south of Pinatubo, where the ash thickness reached 30

◉ **FIGURE 1** Map showing Mount Pinatubo's relationship to cities, the Manila Trench subduction zone, and other volcanoes.

is a satellite photograph of the ash cloud being carried eastward across several states.

Ash flows are turbulent mixtures of hot gases and pyroclastic material that travel across the landscape with great velocity. They are generated quickly and with such force that obstructions in their path may be blown down or carried away. The term **nuée ardente,** French for "glowing cloud," is commonly applied to this kind of eruption because of the intense heat in the flow's interior. A nuée ardente caused most of the fatalities at Mount St. Helens and devastated an area in excess of 160 square kilometers (62 mi²). ◉ Figure 5.24 is a graphic representation of the sequence of events at Mount St. Helens that culminated in the devastating ash-flow eruption of May 1980. This area was made into a national monument in 1982.

No other ash-flow eruption is as famous as that at Mount Pelée on the Island of Martinique in the West Indies on 8 May 1902. It caused the deaths of at least 30,000 people in Saint-Pierre at the foot of the volcano and initiated the study of this type of eruption and the unique volcanic deposits it forms. Like Mount St. Helens, Mount Pelée rumbled for some time before its main eruption. Because of

⦿ FIGURE 2 Ash-filled sky and parking lot at Clark Air Force Base.

centimeters (1.1 ft). Satellite data (Nimbus-7) showed very high concentrations of SO_2 over a large area, with a mass twice that of the sulfurous eruption of El Chichón in Mexico in 1982. By mid July the ash cloud had encircled the earth (⦿ Figure 3), and by September, atmospheric ash was creating spectacular sunsets off the west coast of North America.

The Pinatubo eruptive sequence was a major volcanic event. It may have put 15–20 million tons of sulfur dioxide into the atmosphere, which could make it the largest eruption of the twentieth century. The stratospheric dust and aerosols (SO_2) exerted a cooling influence of about 1°C on the earth during 1992–93, partly offsetting the greenhouse warming of the atmosphere. Also, gases from the volcano may have damaged the stratospheric ozone layer, which serves to shield the earth from ultraviolet radiation (see Chapter 12). Great Britain's Faraday Research Station in Antarctica recorded the lowest amount of

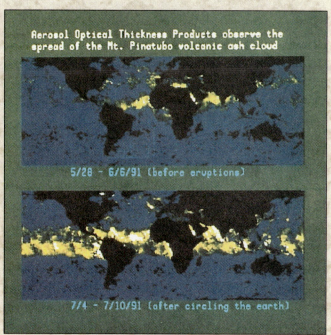

⦿ FIGURE 3 Satellite images showing the thickness of aerosols, including SO_2, before (May 28–June 6) and after (July 4–10) Mount Pinatubo's eruption. By about 20 days after the major eruptions, the aerosol cloud completely encircled the earth.

stratospheric ozone ever in early 1993. The eruption was 150 times larger than that of Mount St. Helens, but it was not as large as the eruption of Krakatoa in 1883. (The subsection "Weather and Climate" in this chapter provides some historic data about the effect of volcanic eruptions on climate.)

this precursor activity, the population of Saint-Pierre swelled from 19,700 to at least 30,000 as refugees fled there from outlying villages. The concentration of sulfurous gases in the air at Saint-Pierre became so great that horses, suffocating from the fumes, dropped in their tracks. Citizens were forced to cover their faces with wet cloths as ash fell in the avenues and streets. Many inhabitants wanted to leave the city, but officials encouraged them to stay, assuring them that there was little to fear from Mount Pelée. On the fateful day there was a tremendous explosion and a lateral blast from Pelée that boiled and rolled down the slopes of

the mountain at a velocity estimated at 100 to 150 kilometers per hour (60–90 mph). The searing cloud engulfed the town in seconds and burned or suffocated all its inhabitants except for two survivors, one of them a condemned prisoner who had been in a dungeon (⦿ Figure 5.25).

Observations at Pelée in the 1930s by F. A. Perret of the Carnegie Institution led to the first accurate description of an ash flow:

First of all it should be realized that the horizontal movement, three kilometers in three minutes, is due to an avalanche of an

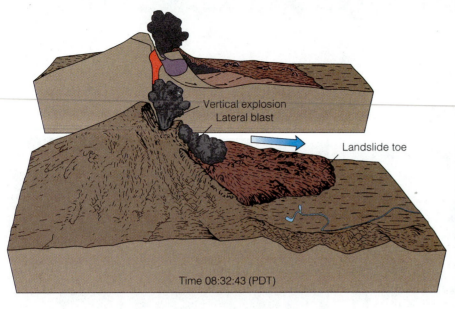

(a)

Vertical explosion
Lateral blast

Landslide toe

Time 08:32:43 (PDT)

(b)

(c)

Lateral blast
Landslide toe

Time 08:32:56 (PDT)

(d)

(e)

⊙ FIGURE 5.24 The sequence of events at Mount St. Helens on Sunday morning, 18 May 1980. (a and b) A landslide opens the vent, and a skyward eruption begins. The cross section shows the slide plane. (c and d) Lateral blast begins 13 seconds later. The toe of the landslide is still exposed. (Both a and c are drawn as viewed toward the west, with no vertical exaggeration.) The eruption culminated in a powerful ash flow (nuée ardente). (e) The force of the ash flow is evidenced by the downed trees in the so-called blow-down area. This is the area where most of the fatalities occurred.

● FIGURE 5.25 Saint-Pierre, Martinique, after the 1902 ash-flow eruption of Mount Pelée in which 30,000 to 40,000 people were killed.

● FIGURE 5.26 Lahar-filled north fork of the Toutle River at Mount St. Helens, formerly a pristine V-shaped stream valley. This photograph was taken in July 1980, well after the main eruption on May 18, and the lahar was still steaming at the time.

exceedingly dense mass of hot, highly gas-charged, and constantly gas-emitting fragmental lava, much of it finely divided, extraordinarily mobile, and practically frictionless, because each particle is separated from its neighbors by a cushion of compressed gas. For this reason, too, its onward rush is almost noiseless.

Suffocating ash flows are a major volcanic hazard and they have been observed in historic times in Alaska, Washington, and California. Prehistoric but geologically young ash-flow deposits are found in all the Western states. They are known as **welded tuffs,** because the minerals and glass composing them were fused together by intense heat (Case Study 5.3).

Debris Flows

Water from heavy precipitation, melting snow, a lake, or a river may mobilize debris on the flanks of a volcano and cause it to move a great distance downslope as a thick mush of rock, ash, and cinders. **Lahar** is the Indonesian word for such a fast-moving volcanic debris flow. A lahar accompanied the eruption of Mount St. Helens and filled the north fork of the Toutle River, a pristine mountain valley and stream before the eruption (● Figure 5.26). Lahars are basically identical to nonvolcanic mudflows except for their volcanic source (see Chapter 7). Lahar-generating volcanoes may show little activity when such flows occur, and there may be little warning.

Another example of a destructive lahar is the one caused by the mild eruption of Colombia's Nevado del Ruiz, 5,432 meters (17,800 ft) high, in November 1985. Although, as one observer put it, the volcano simply "ran a fever and cleared its throat," a lahar generated by melting snow and ice cover devastated the town of Armero 50 kilometers (30 mi) to the east. A wall of mud 40 meters high (130 ft) careened down the narrow canyon of the Lagunilla River and overwhelmed Armero at its mouth, leaving more than twenty thousand victims entombed in mud. Because the disaster struck at 11 o'clock at night, many of the victims were sleeping and unable to run to higher ground.

The potential for destructive debris flows is predictable. The map of the Mount Rainier area in ● Figure 5.27, page 141, specifies the various identified potential volcanic hazards from this stratovolcano. Because Mount Rainier is covered with ice and has some areas of escaping steam and hot water, it is considered capable of generating lahars. Even a small eruption would subject the area at its base to mudflow inundation. Similar maps have been developed for most of the active volcanoes near urban areas along the west coast of the United States.

Tsunamis

Destructive tsunamis associated with volcanic activity are rare and they occur mainly in the western Pacific Ocean and

New Zealand's Blast in the Past, a Skifield, and Fluidization

The two islands of New Zealand are clearly distinguishable geologically. Whereas North Island exhibits explosive volcanism and some of the most spectacular volcanic landforms in the world, South Island lacks volcanism and is more alpine in nature. North Island's volcanic zone includes Lake Taupo, the Tongariro–Taupo area as sacred. They believed that their gods and demons had brought volcanic fire with them from Hawaiki, the Maori's Pacific homeland.

The aftermath of this eruption was a huge explosion caldera, which is now filled with water and known as Lake Taupo. The

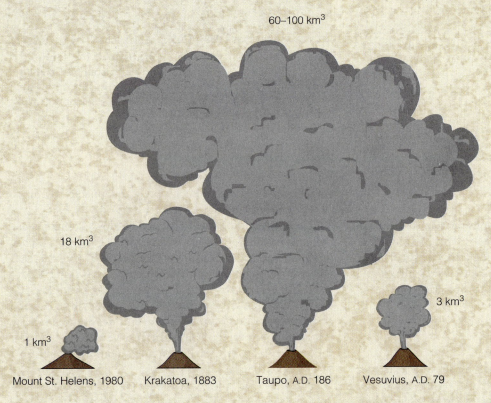

● FIGURE 1 *(a)* The relative sizes of four large eruptions of the past 2,000 years based on estimated volumes of ejected tephra. *(b)* The very large Taupo eruption left a huge caldera that is now Lake Taupo. The cones in the foreground and in the distance suggest the lake's volcanic origin.

(a)

site of a cataclysmic volcanic eruption in A.D. 186 that was arguably the largest eruption on earth in the past 5,000 years (●Figure 1). A volcanic source, now beneath Lake Taupo, vented a pyroclastic column more than 50 kilometers (30 mi) high, which spread downwind, forming light-yellow ash-fall deposits more than 5 meters (8 ft) thick in some places. At 100 kilometers (60 mi) distant the deposits are 25 centimeters (10 in) thick (●Figure 2). Toward the end of the eruption sequence the hot column became so heavy that it collapsed and flowed outward in all directions as a pyroclastic flow. Flow velocity is estimated to have been 600 kilometers per hour (375 mph), and the material traveled more than 80 kilometers (50 mi) from its source, overwhelming all the topography and vegetation in its path. Fortunately, the native New Zealanders, the Maori, had not yet settled the island when the eruption occurred. When they did come to the island about A.D. 1000, they regarded the

(b)

area is still hot, and north of the lake are the Rotorua thermal springs and geysers, which attract many tourists, and the Wairaki geothermal energy complex (see Chapter 13).

How can a pyroclastic material move at such a phenomenal speed and travel such a great distance, seemingly defying friction and gravity? Some pyroclastic flows are **fluidized,** a process by which solids are transported that is well known in chemical engineering. No, a "fluidized bed" is not a water bed; rather, it is a body of fine-grained solids through which a stream of high-pressure gas flows so that the grains separate and move as fluid. Many chemical engineering professors demonstrate this by forcing air through a powder on the bottom of a see-through box. As the air pressure is increased, the powder begins to quiver; then particles start dancing around, and eventually the whole mass is in motion. When the professor has it just right, he or she places a plastic duck (professor's choice) into the container, and it literally floats on the air-solid mixture. If "ducky" is pushed down, it immediately pops up, proving that it is floating on a denser medium. Part of the weight of the solids is taken by the air, reducing the interparticle friction to such a degree that the par-

ticles will not stand in a heap. The particles are not *entrained* (carried away bodily) like snowflakes in a blizzard; rather, the mass behaves as a fluid substance would. The point of this is that a well-understood process demonstrates how pyroclastic flows can fluidize and travel far from their source, destroying everything in their path. Geologic evidence of this is seen at Mount St. Helens, Mount Pelée in the French West Indies, and many other places.

At the south end of the Taupo Volcanic Zone is a cluster of stratovolcanoes in the Tongariro National Park (Figure 2). Two of the volcanoes are active and are popular playgrounds for New Zealanders, primarily for skiing and tramping (hiking). Mount Ruapehu, with a summit elevation of 2,797 meters (9,100 ft), is the highest point on North Island and has a lake in its summit crater. It is also the location of the Whakapapa Skifield, a very popular, well-equipped ski area. The terrane at Whakapapa (Maori, pronounced "whack-a-papa") is bare aa lava and requires a lot of snow to be skiable (● Figure 3). Throughout the skifield (or lava field, depending on the time of year) signs alert skiers and trampers to stay on high ground and out of the canyons should Ruapehu erupt, which it did in 1969, 1971, 1982, and sensa-

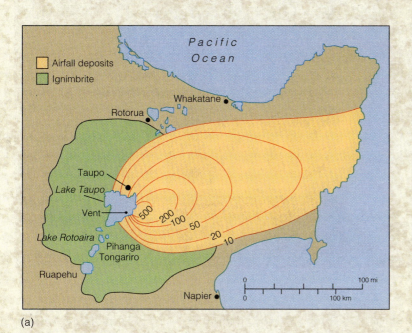

(a)

(b)

● FIGURE 2 *(a)* Map of east-central North Island showing ash-fall thickness contours (in centimeters) from Taupo's eruption of A.D. 186. The area covered by welded tuff (ignimbrite) formed from pyroclastic flows is also shown. *(b)* Tectonic setting of the Taupo Volcanic Zone, North Island, New Zealand.

● FIGURE 3 *(a)* Whakapapa Skifield is on volcanic Mount Ruapehu, the highest point on New Zealand's North Island and the site of a beautiful summit crater lake. Signs on the slopes warn skiers to stay out of canyons and low spots should a phreatic eruption create a dangerous lahar. *(b)* On 25 September 1995 a combination of hot magma and cold water produced a violent steam eruption that hurled boulders over the upper slopes of the Whakapapa Skifield. The eruption did little damage, but projectiles as big as cars were thrown over the rim. The Maori name for Ruapehu means "exploding pit".

(a)

(a)

tionally in 1995. The eruptions are *phreatic;* that is, superheated water at the bottom of the lake flashes to steam and overflows the crater. The lake's surface-water temperature which ranges from 20°C to 60°C (68°–140° F), is monitored as a potential eruption predictor. When the water heats up, some kind of activity can be expected. For example, a nighttime steam eruption in 1975 caused the lake level to drop 8 meters (25 ft) and melted the snow on the slopes, leading to lahars that destroyed several structures.

Had this occurred during the day, many skiers' lives would have been threatened. A lahar warning system at Whakapapa is based on detection of earthquake swarms, which precede phreatic and magmatic eruptions. Tongariro (Maori *tonga-,* "south wind," and *-riro,* "carried away") National Park is another case of humans putting nature to good use. However, those who use the park should keep in mind the Maori caveat *Whakangarongaro he tangata; Toitu he whenua,* "Man passes, but the land endures."

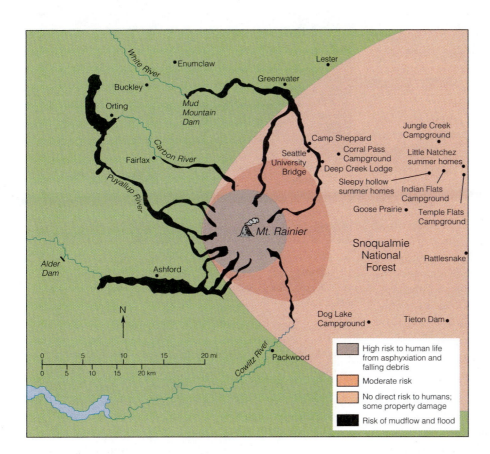

around Indonesia. Of the 405 tsunamis recorded since 1900, 12 were caused by submarine volcanic eruptions, and only two of those resulted in significant damage. The most famous tsunami of all time, however, was caused by Krakatoa's eruption in 1883. (Krakatoa Volcano is located in the Sunda Strait between Java and Sumatra.) The eruption was so enormous that the explosion was heard 3,000 miles away and it spewed a dust cloud 80 kilometers (50 mi) into the atmosphere. About 30 minutes after the eruption, a huge tsunami swept the west coasts of these two large islands, rising to a height of 35 meters (115 ft) above sea level and killing an estimated 36,000 people. The wave traveled across the Pacific, still a meter high when it reached South America, and was noticed as far away as the English Channel.

Weather and Climate

Benjamin Franklin is regarded as the first person to recognize the connection between cold weather and volcanic eruptions while he was serving as U.S. ambassador to France in 1783–84. Laki Fissure in Iceland erupted in June 1783, putting sufficient ash into the atmosphere that the following winter was extremely cold in Europe. The summer of 1783 was bleak in Iceland, resulting in famine and the loss of a fifth of its population and more than half its sheep, cattle, and horses. The eruption of Tambora Volcano in Indonesia in April 1815 hurled so much fine ash into the upper atmosphere that the following year became known as

the "year without a summer." This led to a famine that killed no less than 80,000 humans, produced the coldest three years in British history, and brought snow to New England in July. Crop failures caused the price of flour to double in Great Britain during the two crop years of cold weather.

The reason for these drastic climate changes is that the tiny particles blown into the upper atmosphere by a volcano are very effective in reducing incoming solar radiation. The average length of time volcanic dust particles (0.0001–0.005 mm) remain in the upper atmosphere, the *residence time,* has been found to be 1 to 2 years. Sulfur dioxide gas emitted by volcanic action produces white coatings on the particles, which makes them superreflectors of solar energy. This loading of the upper atmosphere with volcanic dust reduces the amount of incoming radiation relative to outgoing radiation, resulting in a net heat loss for the earth. An attempt has been made to quantify these so-called dust veils into a Dust Veil Index (DVI) that can be used to relate a specific eruption to its impact on climate. Indexes for selected volcanoes are shown in ■ Table 5.2. As is apparent in the table, Tambora and Krakatoa have had potentially the most significant effects on climate in recent history, whereas Mount St. Helens, with a DVI of 1, had little effect. Research suggests that weather deteriorates and stormy conditions prevail after large eruptions. The destructive coastal storm of March 1983 following the eruption of El Chichón on Mexico's Yucatan Peninsula is cited as evidence. The storm

TABLE 5.2 Dust Veil Indexes of Selected Volcanoes	
VOLCANO	**DVI**
Tambora, Indonesia, 1815	1,500
Krakatoa, Indonesia, 1883	1,000
Mount Pelée, Martinique, 1902	500
Mount St. Helens, Washington, 1980	1

SOURCE: Lamb, H. H. Volcanic Dust in the Atmosphere (Phil. Trans. Royal Society of London, Ser. A, 266, 425–533).

◉ FIGURE 5.28 These pine trees died due to a buildup of volcanogenic CO_2 in the soil, which resulted in oxygen starvation; Horseshoe Lake campground, Mammoth Mountain, California.

coincided with high tides and caused extensive property damage along the California coast from San Francisco to San Diego. Dust veils are obviously sporadic and transient, and many volcanoes would have to erupt in concert to produce significant long-term effects on climate (see Case Study 5.2).

Gases

Denver, Mexico City, and Los Angeles have their smog; the Island of Hawaii, the "big island," has its *vog*—short for "*volcanic gas.*" Since 1986 Kilaeua Volcano near the island's southeast coast has been working almost nonstop producing lava and, as a biproduct, 1,000 tons of sulfur dioxide (SO_2) per day. This sulfurous gas quickly combines with water to form sulfuric acid, which is hazardous to health and corrosive to metallic components of structures and machinery. There are reports of the nails and hinges of houses being eaten away to such an extent that the structures have almost collapsed and of damage to plants and crops by the acidic gases. During most the year the trade winds blow the vog toward the west side of the island, where it swirls behind massive Mauna Loa and gets trapped over the island's main tourist area, the city of Kailua Kona and the Kona coast. (The satellite photo in Figure 5.9 gives a general idea of the island's geography.) Health officials and citizens are concerned, and there is much debate about vog as a health hazard. However, the volcano itself will end the debate when this activity cycle ends.

Mammoth Mountain is a large rhyolitic volcano on the east side of the Sierra Nevada in east-central California. Its long history of volcanism extends back about 200,000 years, and it produced a steam eruption as recently as 300–700 years ago. One of the largest ski areas in the United States, it experienced two very strong earthquakes (*M* 5.6 and *M* 6.0) within days of the Mount St. Helens' eruption in 1980. In 1989 the area experienced earthquake swarms, which sometimes indicate a pending eruption. As the area's only highway access is located in the direction of the prevailing wind over the mountain, an "escape road" was built in the opposite direction to provide for safe evacuation. Then in 1990, tree-killing emissions of CO_2 were

discovered, indicating magma activity at shallow depth. (Similar emissions are known at Mount Etna and Mount Vulcano in Italy.) About 30 hectares (75 acres) of Mammoth Mountain's pine and fir trees had suffocated by 1995 and very high concentrations of CO_2 were found around their root systems—up to 90 percent of the total soil gas (◉ Figure 5.28). Where CO_2 concentrations exceeded 30 percent most of the trees were dead. One campground was closed when a park ranger reported symptoms of asphyxia in the area and high CO_2 levels were found in cabins and restrooms. Again, in early 1996, there were more earthquakes. Some people believe the volcano is waving a flag to residents, scientists, and skiers to signal future activity. The intimate association of lethal carbon dioxide gas and volcanism yielded tragic results in 1986. A cloud of carbon dioxide gas emitted from a crater lake in Cameroon, West Africa, killed more than 1,700 people (see Case Study 5.4).

Selected historic volcanic eruptions and prehistoric eruptions indicated by thick, widespread tephra deposits of Holocene age are listed in ■ Table 5.3. Although other significant eruptions are known, those in the table were chosen to illustrate the volcanic hazards and impacts discussed in this chapter.

CONSIDER THIS . . .

You are taking your dream vacation at the Kailua Kona coast on the Island of Hawaii. On the second day of your stay some guests in your hotel complain of burning sensations in their eyes and throats. Should they take their complaint to the hotel management?

■ TABLE 5.3 Selected Notable Worldwide Volcanic Eruptions★

YEAR	VOLCANO NAME AND LOCATION	VEI	COMMENTS
4895 B.C.±	Crater Lake, Oregon	7	Posteruption collapse formed caldera
1390 B.C.±	Santorini (Thera), Greece	6	Late Minoan civilization devastated
79	Vesuvius, Italy	5	Pompeii and Herculaneum buried; at least 3,000 killed
186	Taupo, New Zealand	7	16,000 km² devastated; largest eruption in last 5,000 years
1631	Vesuvius, Italy	4	Modern Vesuvius eruptive cycle begins
1783	Laki, Iceland	4	Largest historic lava flows; 9,350 killed
1792	Unzen, Japan	2	Debris avalanche and tsunamis killed 14,500
1815	Tambora, Indonesia	7	Most explosive eruption in history; 92,000 killed; weather changed
1883	Krakatoa, Indonesia	6	Caldera collapse; 36,000 killed, mostly by tsunamis
1902	Mount Pelée, Martinique, West Indies	4	Saint-Pierre destroyed; 30,000–40,000 killed; spine extruded from lava dome
1912	Katmai, Alaska	6	Perhaps largest 20th-century eruption; 33 km³ of tephra ejected
1914–17	Lassen Peak, California	3	California's last historic eruption
1943	Parícutin, Mexico	3	New cone formed; event observed and documented from first eruption
1959	Kilauea, Hawaii	2	Lava lake formed that is still cooling
1963	Agung, Bali	4	1,100 killed; climatic effects
1968	Fernandina, Galápagos	4	Caldera floor drops 350 meters
1980	Mount St. Helens, Washington	4–5	Ash flow, 600 km² devastated
1982	El Chichón, Mexico	4	Ash flows kill 1,877; climatic effects
1991	Mount Pinatubo, Philippines	5–6★★	Probably the second largest eruption of the 20th century; huge volume of SO_2 emitted
1991	Unzen, Japan	4★★	Pyroclastic flows killed 41 people including 3 volcanologists; lava dome
1991–1993	Etna, Italy	1–2★★	Longest continuous activity (473 days) in 300 years ended; 300 million m³ lava extruded
1983–1996	Kilauea, Hawaii	1–2★★	Longest continuing eruption, with more than 50 eruption events

★Historic lava or tephra volume >2 km³, Holocene >100 km³, fatalities >1,500.
★★Author's estimate.
SOURCE: Data adapted from Smithsonian Institution, Global Volcanism Program.

MITIGATION AND PREDICTION

Diversion

As mentioned earlier in the chapter, various tactics have been used in attempts to prevent lava flows from overwhelming the land. At various times, people have dammed them, diverted them, and even bombed them. Diversion barriers, high earth and rock embankments piled up by bulldozers, have been constructed on Mauna Loa to protect the city of Hilo, and also to divert potential flows away from the National Oceanographic and Atmospheric Administration (NOAA) observatory at Mauna Loa (● Figure 5.29, page 146). Practically, diversion can be considered only if the topography allows the diverted lava to flow onto unim-

proved lands whose owners do not object to it. Flows have been slowed and even stopped by spraying them with seawater at Heimaey Island, Iceland (● Figure 5.30, page 146). This was done in 1973 at the fishing village of Vestmannaeyjar. Both damming and diversion were used in Sicily in 1992 when eruptions of Mount Etna generated extensive lava flows.

Volcano Hazard Zones and Risk

The purpose of a hazard-zone map of any type is to give accurate information on the frequency and severity of the particular hazard. In the 1950s, the area below Kilauea's east rift zone on the Island of Hawaii was developed. The vol-

Volcanic CO$_2$ and Fountaining Lakes

Lake Nyos is contained within a volcanic crater, one of 40 such craters that are roughly aligned southwest to northeast in Cameroon, West Africa (⊙ Figure 1). Specifically, the lake is in a **maar,** a low-relief crater produced by explosive interaction of magma with underground water. Such a high-pressure steam explosion, called a **phreatic eruption,** does not produce much new magmatic material, which is why the resulting crater is relatively shallow. Lake Nyos and several nearby crater lakes and springs are known to be highly charged with carbon dioxide, CO$_2$, just as is a bottle of soda water.

On the evening of 21 August 1986, almost noiselessly and without warning, the lake erupted, discharging 80 million cubic meters (100 million yd^3) of CO$_2$ dissolved in its waters into the atmosphere. So much gas escaped from the lake that its level dropped 1 meter. Because carbon dioxide weighs about one and a half times as much as air, the suffocating gas descended downslope, following stream valleys and drainages for 16 kilometers (10 mi). The CO$_2$ cloud was 50 meters (160 ft) thick and it killed all animal life in its path, including 1,200 residents of Nyos village and 500 persons in nearby communities. The entire event took less than an hour (⊙ Figure 2).

Below its surface waters, Lake Nyos contains so much dissolved gas that a liter of water is charged with 1 to 5 liters of gas, 99 percent of which is carbon dioxide. (At a depth of 208 meters—680 ft—the lake's depth, water can hold 15 times its own volume in CO$_2$ at saturation.) The source of the CO$_2$ is believed to be volcanic, although none of the gases usually associated with volcanic activity—H$_2$S, CO, SO$_2$—is found in the waters. Carbon-14 study of the CO$_2$ indicates that it is more than 35,000 years old. This rules out a biogenic origin for the gas, as the lake is less than 1,000 years old. However, the isotopic ratios of He^3/He^4 and C^{13}/C^{12} in the gas are consistent with a magmatic source. Therefore, since evidence to the contrary has not been found, the CO$_2$ of Lake Nyos is believed to be escaping from magma at depth and sweeping upward through a shattered volcanic vent into the bottom of the lake.

What happened the evening of the tragedy that caused the gas in the lake to explode catastrophically? Was it a small volcanic disturbance, or did some other physical phenomenon upset the lake? It is known that lakes periodically *overturn;* that is, their surface water sinks to the bottom and is replaced by bottom water rising to the surface. This usually occurs in temperate climates in the fall, because as the surface water cools, it becomes denser than the bottom water and displaces it. Overturn usually requires a trigger, such as strong winds, a landslide, or a drastic change in atmospheric pressure. Lake Nyos is only seven degrees north of the equator, however, so seasonal temperature change alone would not be sufficient to cause overturn. However, Cameroon lakes are known to be least stable during the rainy season of August and September, when cool fresh water is added to the lake surface. The best hypothesis appears to be that winds associated with the rainy season blew the lake's surface water to one side of the crater. This reduced the hydrostatic (confining) pressure on the bottom

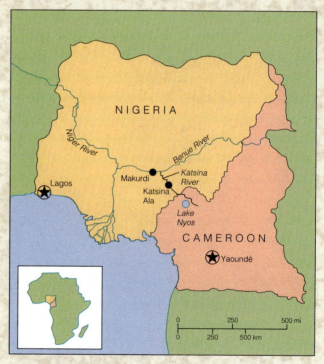

⊙ **FIGURE 1 The location of Lake Nyos in Cameroon, West Africa.**

cano had been dormant for 17 years and it had not erupted on the lower east rift zone since 1840. Kilauea became active in 1952, however, and by 1996 there had been 13 separate rift-zone eruptions, the last of which has been almost continuous since 1983 (⊙ Figure 5.31). In 1990 lava-flow hazard maps were prepared for the island, delineating zones ranging from low risk (9) for Kohala at the north end of the island to high risk (1 and 2) within and below the east rift zone (⊙ Figure 5.32). The Royal Gardens residential subdivision, a development project of the 1950s, has been almost totally buried under lava (see Figure 5.10), and soon after the hazard-zone maps were published, the town of Kalapana was covered by aa flows. Had the developers known that 90 percent of the land surface of Kilauea has been covered by lava

(a)

(b)

⦿ FIGURE 2 *(a)* Deadly carbon dioxide belched from the bottom of volcanic Lake Nyos and stood muddy and full of debris for several days. The gas settled into the valleys below, tragically suffocating 1,700 people. *(b)* Cattle suffocated by carbon dioxide at Lake Nyos.

waters, which allowed the highly charged gas to come out of solution and form bubbles. The formation of bubbles further reduced the density of the water, and the gas in solution frothed and rose with dramatic speed. That this sequence of events indeed happened is supported by stripped vegetation along the lake shore. The leaves and other parts of plants were most likely removed by a surge of water associated with the explosion—just as a fountain of foam is associated with popping the top of a warm can of soda. Evidence to support the instability hypothesis, rather than a volcanic cause, for the Nyos tragedy is a similar explosion on 16 August 1984 at Lake Manoun, 95 kilometers (60 mi) south, with a loss of 37 lives. The occurrence of these two events during the season of minimum stability for Cameroon lakes suggests that future events may occur during this season.

Bantu folklore is filled with stories of exploding lakes and "rains" of dead fish. According to legend, there have been at least three earlier episodes of catastrophic lake explosions. Dead fish found floating on the water surface during minor degassing events are believed by the local people to be gifts from their ancestors who live in the lake, and they believe that they cannot drown there without prior approval of their ancestors.

Since 1986 the gas content of the lake has increased by additions from subterranean springs and magmatic sources. Engineering studies indicate that Lake Nyos and other dangerous crater lakes in the region could be defused by installing pipes in them to bring charged water from the lake bottom to the surface as fountains. Once started, the fountains would be driven entirely by the lifting force of the released gas.

The danger is far from over at Lake Nyos. In addition to its high concentration of carbon dioxide, the lake has a very fragile dam of soft volcanic ash on its perimeter. Piping through this thin ash dam could cause failure and the release of a flood calculated at twice the flow of the 1899 Johnstown, Pennsylvania, flood (see Chapter 9). There are plans to lower the lake level and to improve the natural dam in order to minimize the flood threat, but the plans cannot be carried out before there is a better understanding of the potential impact of a lower lake level on the dissolved gas.

since the arrival of the first Hawaiians 1,500 years ago, they may have reconsidered building there. May have, that is!

Prediction

The ability to predict when and where a volcanic eruption will occur has attained a high degree of refinement at the

U.S. Geological Survey's Hawaiian Volcano Observatory. It is located at Kilaeau crater in Hawaii Volcanoes National Park on the east slope of Mauna Loa, Island of Hawaii. Kilauea and Mauna Loa are two of the most active volcanoes in the world. These volcanoes and the magma reservoir that feeds them can be viewed as balloons buried below thin layers of sand and clay. As magma enters the reservoir

FIGURE 5.29 Lava diversion dike; Mauna Loa, Hawaii.

● FIGURE 5.30 Lava flows on Heimaey Island off the south coast of Iceland were successfully chilled by spraying them with seawater in 1973.

prior to an eruption, the volcano's surface layers are pushed upward and outward—just as if gas were being blown into the balloon. This expansion is measurable with tiltmeters. As magma works its way to the volcano surface, earthquakes are so numerous that they are referred to as *swarms*. Earthquake swarms have been found to be quite reliable precursors of an impending eruption. ● Figure 5.33 illustrates the three stages of an eruption and the accompanying changes in tilt and seismicity. Increasing tilt indicates that something is about to happen, and the locations of earthquake epicenters and foci indicate where the outbreak is likely to occur.

Observations of similar phenomena at Mount St. Helens led geologists to expect an eruption—though not such a cataclysmic one. The swelling there produced a noticeable change in the shape of the cone, and the harmonic tremors (caused by magma moving continuously within the reservoir) caused such surface motion on the cone that many persons were uncomfortable. Unlike earthquakes, the erup-

● FIGURE 5.31 Lava flows on Kilauea Volcano's east rift zone. By late 1992 more than 83 square kilometers (33 mi²) had been covered by the 1983 to 1992 eruptive sequence shown here, and activity has continued into 1996.

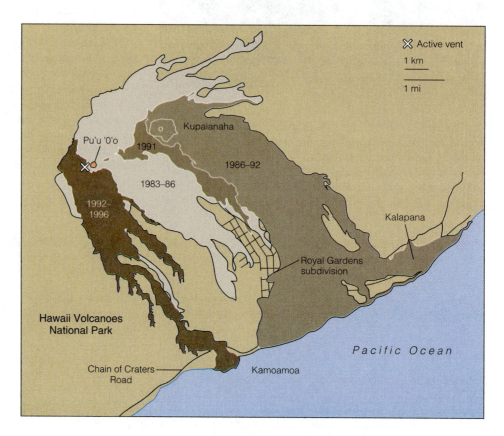

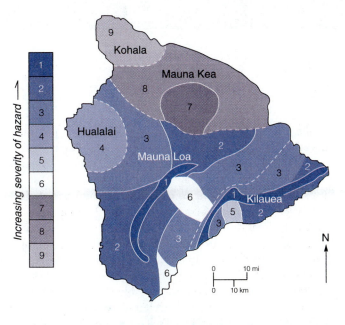

FIGURE 5.32 (at left) Lava-flow hazard zones for the Island of Hawaii. The zones range from low risk (9) to high risk (1). Volcano boundaries are denoted by dashed lines.

FIGURE 5.33 (below) Hawaiian volcanoes' typical eruption sequence. *(a–c)* Inflation and deflation of the cone are accompanied by *(d)* measurable changes in slope and seismicity. Three stages are apparent.

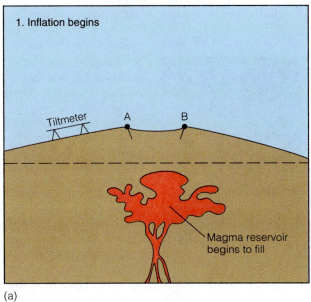

1. Inflation begins

Tiltmeter

A   B

Magma reservoir begins to fill

(a)

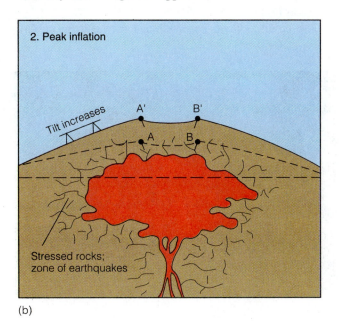

2. Peak inflation

Tilt increases

A′   B′

A   B

Stressed rocks; zone of earthquakes

(b)

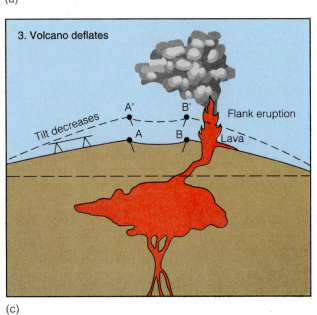

3. Volcano deflates

Tilt decreases

A′   B′

A   B

Flank eruption

Lava

(c)

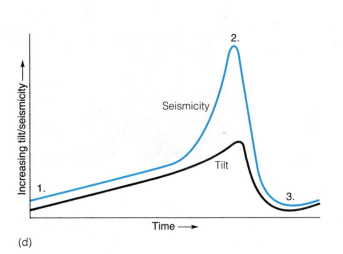

(d)

MITIGATION AND PREDICTION   **147**

Volcanic Wonders

Volcanism creates landscapes that are attractive to humans. Yellowstone National Park in Wyoming, Oregon's Crater Lake, and California's Devil's Postpile National Monument are but a few examples. One of the most eye-catching volcanic features is *columnar jointing,* which forms as cooling volcanic rock contracts around many equally spaced centers, causing the lava to split into polygonal columns. Six-sided columns are most common, but four-, five-, seven-, and eight-sided columns also occur. Excellent examples are seen at the Devil's Postpile (◉ Figure 5.34, part a), the Giant's Causeway in Northern Ireland (Figure 5.34, part b), and at the Devil's Tower, Wyoming, a shallow intrusion (◉ Figure 5.35).

The columns form perpendicular to the main cooling surfaces of the flows, which is why they are sometimes curved. The 264-meter-tall (865 ft) Devil's Tower, called "Grizzly Bear Lodge" and "Mato Tepee" by early Native Americans, is surrounded by colorful Cheyenne and Lakota Sioux legends. According to one legend, a number of children were being chased by a grizzly bear. The Great Spirit saw their plight and raised the ground they were on. The bear scratched deep grooves into the rocks in his attempts to reach the frightened children. It is not difficult to visualize the tribal elders passing on this legend to wide-eyed children. It is a much more memorable explanation for the origin of the columns than the scientific one of contraction cracks. The columns at Devil's Tower are a challenge to rock climbers, but they have been scaled many times.

(a)

(b)

◉ **FIGURE 5.34**
(a) Beautiful columnar jointing in basalt at Devil's Postpile National Monument, Mammoth Lakes area, east-central California. *(b)* Closeup of Giant's Causeway columnar jointing in basalt on the north coast of Northern Ireland. The Tertiary basalts rest on the Cretaceous chalk that forms the famous white cliffs on the English Channel. Columns are about 0.5 meters (1.5 ft). across.

◉ **FIGURE 5.35** Devil's Tower National Monument in northeastern Wyoming is an exposed volcanic neck with spectacular columnar jointing. Also known as "Grizzly Bear Lodge" and "Mato Tepee," it rises 865 feet above the surrounding landscape.

CONSIDER THIS . . .

Although a few areas on the Island of Hawaii are regarded as relatively free from geologic hazards, you might not want to live in one of those "safe" places. Select an area on the island that appeals to you (for its view, beach access, culture, or some other amenity) and determine what geologic hazards you could encounter there.

tions of active volcanoes can be predicted with fair accuracy as to time and place; the magnitude of the pending eruption, however, is more difficult to assess. Predicting the behavior of dormant volcanoes is much more difficult; much remains to be learned about predicting their future activity.

What to Do When an Eruption Occurs

Based upon the experience of people at Mount St. Helens, the Federal Emergency Management Agency (FEMA) made the following suggestions to ease the trauma of a volcanic eruption.

AT HOME.

- Stay calm and get children and pets indoors.
- If outside, keep eyes closed or use a mask and seek shelter.
- If indoors, stay there until heavy ash has settled.
- Close doors and windows, and place damp towels at door thresholds and other draft openings.
- Do not run exhaust fans or use clothes dryers.
- Remove ash accumulations from low-pitched roofs and gutters.

IN YOUR CAR.

- Drive slowly because of limited visibility.
- Change oil and air filters every 50–100 miles (80–160 km) in heavy dust, or every 500–1,000 miles (800–1,600 km) in light dust (visibility up to 200 feet, or 60 meters).

CONSIDER THIS . . .

Volcanic activity is perhaps the most spectacular of all geologic hazards, and the destructive potential of many volcanoes is immense. Fortunately, hazards associated with volcanoes are not spread equally around the earth's surface. What areas in the United States are most likely to be affected by future volcanic eruptions, and how have they been identified?

- Use both windshield washer and wipers.
- Volcanic ash is abrasive and can ruin engines and paint.

PETS.

- Keep pets indoors.
- Brush or vacuum them if they are covered with ash.
- Do not let them get wet, and do not try to bathe them.

 ## SUMMARY

Volcanoes

DEFINED A vent or series of vents that issue lava and pyroclastic material.

DISTRIBUTION Adjacent to convergent plate boundaries around the Pacific Ocean (the Ring of Fire); in a west-east belt from the Mediterranean region to Asia; along the mid-ocean ridges; and in the interior of tectonic plates above hot spots in the mantle. In the U.S., a 1,100-kilometer (680-mi) belt extends northward from northern California through Oregon and Washington.

MEASUREMENT The Volcanic Explosivity Index assigns values of 0 to 8 to eruptions of varying size.

EXPLOSIVITY Depends upon the lava's viscosity and gas content. Viscosity, a function of temperature and composition, determines the type of eruption: felsic (high-SiO_2) lavas are viscous and potentially explosive; mafic (low in SiO_2, high in magnesium and iron) lavas are more fluid and less explosive.

Types of Volcanoes and Volcanic Landforms

SHIELD (HAWAIIAN OR "QUIET" TYPE) Gentle outpourings of lava produce a convex-upward edifice resembling a shield.

FISSURE ERUPTION Lava erupts from long cracks, building up broad lava plateaus such as those found in the Columbia River Plateau (Oregon–Washington–Idaho), Iceland, and India.

STRATOVOLCANOES (EXPLOSIVE TYPE) Characterized by concave-upward cone thousands of meters high. Vesuvius (Italy), Fujiyama (Japan), and Mount Hood (Oregon) are examples. Crater Lake (Oregon) is the stump of a stratovolcano that exploded and collapsed inward, leaving a wide crater known as a *caldera*.

LAVA DOMES Bulbous masses of high-silica, glassy lavas, such as found at Mono Craters (California). Domes may form in the crater of a composite cone after an eruption, such as at Mount St. Helens (Washington).

CINDER CONES The smallest and most numerous of volcanic cones, composed almost entirely of tephra.

Volcanic Products

BUILDING MATERIALS Cinder blocks, road-base materials, pumice, light-weight concrete, decorative stone.

GEOTHERMAL ENERGY Cooling magma at depth heats water that can be converted to steam to drive electrical generators.

Hazards

LAVA FLOWS Destroy or burn everything in their path, can travel a distance of 100 kilometers or more.

ASHFALLS AND ASH FLOWS Heavy falls of volcanic ash from Mount Vesuvius buried Pompeii in A.D. 79. There were heavy ashfalls from Mount Pinatubo, the Philippines, 1991. Vertical plumes of ash may rise many kilometers from a vent, which are then carried by the wind and blanket the terrain. Such plumes rising from a vent are known as *Plinian eruptions*. Ash flows are hot, fluid masses of steam and pyroclastic material that travel down the flanks of volcanoes at high speeds, blowing down or suffocating everything in their path. Such flows have killed thousands of people.

DEBRIS FLOWS (LAHARS) Catastrophic mud flows down the flanks of a volcano. They cause the most volcanic fatalities, more than ash flows.

TSUNAMIS Submarine volcanic eruptions (such as that of Krakatoa in 1883) create enormous sea waves that may travel thousands of miles and still do damage along a shoreline.

WEATHER Sulfur dioxide coatings on ash and dust particles increase their reflectivity and cause cooling of weather, such as occurred after the eruption of Mount Pinatubo, 1991–1993. Large eruptions have been related to stormy conditions and El Niño. The effects of ash and dust in the stratosphere are short-term—a few years.

GASES Corrosive gases emitted from a volcano can be injurious to health, structures, and crops. Such gasses are called *vog (volcanic gas)* on the island of Hawaii. Carbon dioxide emitted from the bottom of volcanic Lake Nyos in Cameroon formed a cloud that sank to the ground and suffocated 1,700 people in 1986.

Mitigation and Prediction

DIVERSION AND CHILLING Flows have been diverted in Hawaii and elsewhere and have been chilled with seawater in Iceland.

PREDICTION Prediction in Hawaii, based upon seismic activity (earthquake swarms and harmonic tremors) and the tilt of the cone as magma works its way upward, has become quite accurate. Similar phenomena were observed at Mount St. Helens and Mount Pinatubo.

KEY TERMS

aa	cinder cone
andesite line	felsic
ashfall (air fall)	fluidization
ash flow *(nuée ardente)*	lahar
caldera	lava dome (volcanic dome)
lava plateau	shield volcano
maar	stratovolcano
mafic	(composite cone)
pahoehoe	tephra
phreatic eruption	Volcanic Explosivity Index
Plinian eruption	(VEI)
pyroclastic	welded tuff

STUDY QUESTIONS

1. What are the four kinds of volcanoes, and what type of eruption can be expected from each kind? What are fissure eruptions, and what have they produced to shape the earth's landscape?

2. How would you expect the eruption of a volcano that taps mafic magma to differ from the eruption of one that taps felsic magma? What kind of rocks will result from each eruption?

3. Explain the distribution of the world's 1,350 active volcanoes in terms of plate tectonic theory.

4. List the five geologic hazards connected with volcanic activity in decreasing order of their threat to human life and limb.

5. Where in the United States are the most dangerous volcanic hazards encountered? Which geologic hazard related to volcanic activity presents the greatest danger to humans, and how can this hazard be mitigated, or at least minimized?

6. Why can eruptions on the Island of Hawaii be predicted more accurately than volcanic eruptions at subduction zones?

7. How may calderas be formed? What geologic evidence is found at Crater Lake, Oregon, that indicates it was formed by a special set of circumstances?

8. In what ways are volcanic activity and its products useful to humankind?

9. Name five volcanic areas of great scenic wonder, and indicate the geologic feature found there; for example, Crater Lake, caldera.

10. How do volcanic eruptions affect climate and weather? What impact does this have on populations, particularly Third World peoples?

11. In what ways do gases, including water vapor, emitted by a volcano impact the nearby area and people living there?

12. Define or sketch these volcanic features: aa, pahoehoe, lahar, Dust Veil Index, lapilli, tephra, columnar jointing, welded tuff.

FURTHER INFORMATION

BOOKS AND PERIODICALS

Bullard, Fred M. 1984. *Volcanoes of the Earth,* 2d ed. Austin: University of Texas Press.

Casadevall, T. J.; ed. 1991. *Volcanic ash and aviation safety: First international symposium,* U.S. Geological Survey circular 1065. Washington, D.C.: Government Printing Office.

Crenson, Matt. 1994. Where North America is going: *Earth: The science of our planet,* November.

Decker, Robert, and Barbara Decker. 1989. *Volcanoes,* 2d ed. New York: W. H. Freeman.

Farrar, C. D., and others. 1995. Forest-killing diffuse CO_2 emissions at Mammoth Mountain as a sign of magmatic unrest. *Nature* 376 (no. 6542): 675–678.

Francis, Peter. 1976. *Volcanoes.* New York: Penguin Books.

———. 1993. *Volcanoes: A planetary perspective.* New York: Oxford University Press.

Hazlett, Richard W. 1987. *Kilauea volcano: Geological field guide.* Hilo, Hawaii: Hawaiian Natural History Association.

Kling, George W., and others. 1987. The 1986 Lake Nyos gas disaster in Cameroon, West Africa. *Science* 236: 169–174.

McClelland, Lindsay; D. Leschinsky; and K. Kivimaki. 1991. Global volcanism network. *Bulletin of the Smithsonian Institution,* March–July.

McDonald, G. A.; A. T. Abbott; and F. L. Peterson. 1983. *Volcanoes in the sea: The geology of Hawaii,* 2d ed. Honolulu: University of Hawaii Press.

McPhee, John. 1989. *The control of nature.* New York: Farrar, Straus, and Giroux.

Schuster, Robert L., and J. P. Lockwood. 1991. Geologic hazards at Lake Nyos, Cameroon, West Africa. *Association of Engineering Geologists News,* April.

Sharpton, V. L., and P. E. Ward. 1991. *Global catastrophes in earth history: An interdisciplinary conference on impacts, volcanism, and mass mortality.* Geological Society of America special paper 247.

Simpkin, Tom, and Lee Siebert. 1994. *Volcanoes of the world,* 2d ed. Tuscon, Ariz.: Geoscience Press.

Smith, A. L., and M. J. Roobol. 1991. *Mt. Peleé, Martinique: An active island-arc volcano.* Geological Society of America memoir 175.

Stager, Curt. 1987. Killer lake. *National Geographic,* September: 404–420.

Tilling, Robert I. 1984. *Monitoring active volcanoes.* Denver, Colo.: U.S. Geological Survey.

Williams, Karen. 1990. *Volcanoes of the south wind,* Wellington, New Zealand: Tongariro Natural History Society.

Wright, Thomas L., and T. C. Pierson. 1992. *Living with volcanoes.* U.S. Geological Survey circular 1073. Denver, Colo.: U.S. Geological Survey.

VIDEO

National Geographic Society with WQED, Pittsburgh. 1983. *Born of fire.* 60 minutes.

WEATHERING AND SOILS

A few inches between humanity and starvation.

ANONYMOUS

S oil is the solid earth's most fundamental resource. Through it we produce much of our food, and it supports forest growth, which gives us essential products. Soil material, organisms within the soil, and vegetation constitute an ecological system in which the four ingredients needed for plant growth are recycled. These ingredients are water, air, humus (organic matter), and mineral matter (● Figure 6.1).

The multiple uses of **soil** are reflected in the many definitions and classification schemes that have been developed for soils. To the engineer, a soil is the loose material at the earth's surface; that is, material that can be moved about without first being dynamited and upon which structures can be built. The geologist and the soil scientist see soil as weathered rock and mineral grains that are capable of supporting plant life. A farmer, on the other hand, is mostly interested in which crops a soil can grow

OPENING PHOTO
Bryce Canyon National Park, Utah. The park was created in 1928 to protect an area of 14,500 hectares (36,000 acres) of colorful, eroded sandstones and limestones. Its geologic story is closely related to those of Grand Canyon and Zion, except that the rock formations of the Bryce Canyon are much younger than those of the adjacent parks. The canyon was named for Ebenezer Bryce, an early settler in the area.

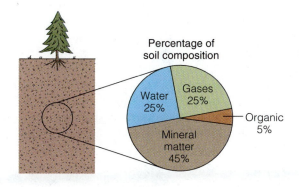

● FIGURE 6.1 Composition of a typical soil. The organic component includes humus (decomposed plant and animal material), partially decomposed plant and animal matter, and bacteria.

and in whether the soil is rich or depleted with respect to humus and minerals. Soils form by the weathering of **regolith**, the fragmental rock material at the earth's surface. By studying soils we can make inferences about their parent materials and the climate under which they formed and can determine their approximate age.

The carrying capacity of our planet—that is, the number of people the earth can sustain—depends on the availability and productivity of soil. This is why reducing soil erosion is so important. For all practical purposes, once productive topsoil is removed, it is lost to human use forever.

Five environmental factors determine the development of soils: (1) climate, (2) organic activity, (3) relief of the land, (4) parent material, and (5) the length of time that soil-forming processes have been acting. These factors can be entered into the "CLORPT" equation," a memory aid for these five factors:

$$\text{Soil} = f\,(Cl,\ O,\ R,\ P,\ T)$$

where f is read "a function of" and Cl, O, R, P, and T represent climate, organic activity, relief, parent material, and time, respectively. The rate at which a soil weathers is a function of climate. In order to understand how soils form, change with time, and are distributed on the surface of the earth, we must first understand how rocks, minerals, and even the structures we place on the earth are broken down by weathering—their interaction with water and air.

WEATHERING

Weathering is a destructive process by which rocks and minerals are broken down by exposure to atmospheric agents. Specifically, **weathering** is the physical disintegration and chemical decomposition of earth materials at or near the earth's surface. **Erosion,** on the other hand, is the removal *and transportation* of weathered or unweathered materials by wind, running water, waves, glaciers, underground water, and even gravity.

Physical Weathering

Physical weathering makes little rocks out of big rocks by processes such as wedging. Rocks are wedged apart along planes of weakness by thermal expansion and contraction and by the activities of organisms. **Frost wedging** occurs when water freezes in joints or other rock openings and expands. This expansion, which amounts to almost ten percent in volume, can exert tremendous pressures in irregular openings and joints (● Figure 6.2). As one would suspect, this process operates only in temperate or cold regions that have seasonal or daily freeze and thaw cycles. Roots of even the smallest plants can act as wedges in rock in any climate. The most dramatic example of this is probably the damage to paving and structures caused by tree roots. Although this process is independent of climate, it operates in tandem with the effects of climate.

Thermal expansion and contraction may loosen mineral grains and contribute to rock disintegration in areas of temperature extremes. In some desert regions temperature variations of 40° C may occur in one day, for example. Although the quantitative importance of temperature variation on rock disintegration is difficult to evaluate, laboratory experiments indicate that it is probably not great.

Animals, on the other hand, contribute significantly to weathering and soil formation by aerating the ground and mixing loose material. Insects, worms, and burrowing mammals, such as ground squirrels and gophers, move and mix soil material and carry organic litter below. Some tropical ants' nests are tunnels hundreds of meters long, for example, and common earthworms can thoroughly mix soils to depths of 1.3 meters (4¼ ft). Earthworms are found in all soils where there is enough moisture and organic matter to sustain them, and it is estimated that they eat and pass their own weight in food and soil minerals each day, which amounts to about ten tons per acre per year on average. An Australian earthworm is known that grows to a length of 3.3 meters (11 ft) and is a real earth mover!

Chemical Weathering

Chemical reactions also work on rock and mineral debris at the earth's surface. These reactions are complex and they involve many steps, but the fundamental processes are reactions between earth materials and atmospheric constituents such as water, oxygen, and carbon dioxide. The products of chemical weathering are new minerals and/or dissolved minerals. The specific reactions are *solution,* the dissolving of minerals; *oxidation,* the "rusting" of minerals; *hydration,* the combining of minerals with water; and *hydrolysis,* the complex reaction that forms clay minerals and the economically important aluminum oxides.

SOLUTION. Carbon dioxide released from decaying organic matter and from the atmosphere combines with water to form weak carbonic acid:

● FIGURE 6.2 Frost wedging has fragmented this outcrop of jointed rock occupied by an Adelie penguin; Antarctic Peninsula, Antarctica.

$$CO_2 + H_2O \rightarrow H_2CO_3 \rightarrow H^+ + (HCO_3)^-$$

An excess of hydrogen ions (protons) creates an acid; and the more excess hydrogen ions there are, the more acidic the solution is. Since carbonic acid produces few hydrogen ions, it is a weak acid. Yet this natural weak acid attacks solid limestone (the mineral calcite), dissolving it and yielding an aqueous solution, *(aq)*, of calcium and bicarbonate ions:

$$CaCO_3 + H_2CO_3 \, (aq) \rightarrow Ca^{++} \, (aq) + 2(HCO_3)^- \, (aq)$$
limestone   carbonic acid   calcium ions   bicarbonate ions

Solution of a rock produces no new minerals, only dissolved chemicals; hence, solid calcite disappears. Solution in limestone terrane creates caverns underground, such as the Mammoth Caves in Kentucky and Carlsbad Caverns in New Mexico. (Chapter 8 explains this.) Acid rain, which forms when sulfur dioxide or nitrogen oxides combine with water droplets in the atmosphere to form sulfuric and nitric acids, takes its toll on rock monuments (●Figure 6.3).

OXIDATION AND HYDRATION. Oxidation produces iron oxide minerals (hematite and limonite) in well-aerated soils, usually in the presence of water. It is the "rusting" commonly seen on metal objects that are left outdoors. Iron in minerals combines with oxygen and water to form hydrated iron oxides as follows:

$$4Fe^{++} + 3O_2 + 6H_2O \rightarrow 2(Fe_2O_3 \cdot 3H_2O)$$
iron-rich minerals   oxygen   water   limonite (yellow)

Pyroxene, amphibole, magnetite, pyrite, and olivine are most susceptible to oxidation because they have a high iron content. It is the oxidation of these and other iron-rich minerals that provides the bright red and yellow colors seen in many of the rocks of the Grand Canyon and the Colorado Plateau.

HYDROLYSIS. The most complex weathering reaction, hydrolysis, is responsible for the formation of clays, the most important minerals in soil. A typical hydrolytic reaction occurs when orthoclase feldspar, a common mineral in granites and some sedimentary rocks, reacts with slightly acidic carbonated water to form clay minerals, potassium ions, and silica in solution:

$$2KAlSi_3O_8 + 2H^+ + 9H_2O \rightarrow$$
orthoclase feldspar   acid   water

$$Al_2Si_2O_5(OH)_4 + 2K^+ + 4H_4SiO_4 (aq)$$
clay mineral   potassium ion   soluble silica

The ions released from silicate minerals in the weathering process are salts of sodium, potassium, calcium, iron, and magnesium, which become important soil nutrients. Clay minerals are important in soils, because their extremely small grain size—less than 4 microns (0.004 mm)—gives them a large surface area per unit of weight. Soil clays can adsorb significant amounts of water on their surfaces, where it stays within reach of plant roots. Also, because the clay surfaces have a slight negative electrical charge, they attract positively charged ions, such as those of potassium and magnesium, making these important nutrients available to plants. Other soil components being equal, the amounts of clay and humus in a soil are what determine its suitability for sustained agriculture. Sandy soils, for example, have much less water-holding capacity, less humus, and fewer available

(a)   (b)

⊙ FIGURE 6.3 Comparative weathering of granite obelisks, both believed to have been shaped 3,500 years ago (1500 B.C.). *(a)* Obelisk amid the ruins of Karnak near Luxor, Egypt. The hieroglyphs are sharp, and the polished surface is still visible. *(b)* The red-granite obelisk known as "Cleopatra's Needle" that stands in New York's Central Park was a gift from the government of Egypt in the late 19th century. New York City's humidity and other environmental conditions have caused chemical weathering of the incised hieroglyphs that had endured several thousand years in Egypt's arid climate.

CONSIDER THIS...

Tennis is a sport that is played throughout the world on a variety of court surfaces. For example, the five most prestigious professional tennis tournaments are played on grass (in England), clay (in France and Italy), and concrete (in the United States and Australia). Why do you suppose clay-surface courts are more common in the eastern United States, and in tropical and humid countries, than in the southwestern United States, where concrete is the preferred surface?

nutrients than do clayey soils. But clay-rich soils drain poorly and are hard to work. The best agricultural soils are a compromise between these extremes called **loam**.

The Rate of Weathering

The rate of chemical weathering is controlled by the surface environment, grain size (that is, the surface area), and climate. The stability of the individual minerals in the parent material is determined by the pressure and temperature conditions under which they formed. Quartz is the last mineral to crystallize from a silicate magma, and the temperature at which it does so is lower than the temperatures at which olivine, pyroxene, and the feldspars crystallize. For this reason, quartz has greater chemical stability under the physical conditions at the earth's surface. The relative resistances to chemical weathering of minerals in igneous rocks thus reflect the opposite order of their crystallization sequence in a magma. Most stable to least stable, the sequence is

quartz (most stable; crystallizes at lowest temp.)

  orthoclase feldspar

    amphibole

      pyroxene

        olivine (least stable; crystallizes at highest temp.)

Micas and plagioclase feldspars form over a range of temperatures and thus vary in weathering stability. Note that ferromagnesian (iron and magnesium) minerals—olivine, pyroxene, and amphibole—are least stable and most subject to chemical weathering. Igneous rocks made up of iron-rich minerals weather readily in humid climates.

A monolithic mass of rock is less susceptible to weathering than an equal mass of jointed and fractured rock. This is because the monolith has less surface area on which atmospheric processes can act. A hypothetical cubic meter of solid rock would have a surface area of 6 square meters. If we were to cut each face in half, however, we would have 8 cubes with a total surface area of 12 square meters (⊙ Figure 6.4). Thus smaller fragments have more area upon which chemical and physical weathering can act. For example, whereas soils have developed on volcanic ash in the West Indies after only 30 years, some nearly 200-year-old solid basalt flows in Hawaii have yet to grow a radish. This reflects differences in particle size and surface area.

The effect of climate has already been discussed. It is important to remember that chemical weathering dominates in humid areas, and physical weathering is most active in arid or dry alpine regions. (Figure 6.3 may help you remember this.) However, humans have accelerated chemical weathering by polluting the atmosphere with acid-forming compounds. Limestone monuments in English cemeteries are reported to require 250–500 years to weather to a depth of 2.5 centimeters and thus to obliterate inscriptions. Buildings in northern England are known to

Surface area = 6 m²

Surface area = 12 m²

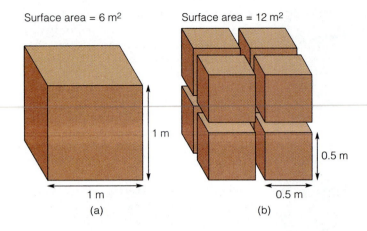

1 m

1 m

(a)

0.5 m

0.5 m

(b)

◉ **FIGURE 6.4** (left) Surface area increases with decreasing particle size. A cube that is 1 meter on each side has a surface area of 6 m². If the cube is cut in half through each face, 8 cubes result that have a total surface area of 12 m².

weather more on the side that receives sulfur-laden winds from a nearby industrial area. A sixteenth-century monastery in Mexico that was built of volcanic tuff has been ravaged by chemical weathering to the extent that some of the carved statues on its exterior are barely recognizable (◉ Figure 6.5). Salt used to remove ice from Michigan Avenue in Chicago has deeply corroded (9 mm) the granite facing of the Intercontinental Hotel. A rock wall

(a)

(b)

◉ **FIGURE 6.5** *(a)* A 16th-century monastery in Mexico shows the ravages of weathering mostly from wind and wind-driven rain. The rock is volcanic tuff. *(b)* Deep pitting (9 mm) in Mount Airy granite due to corrosion of feldspars by street salt and water; Intercontinental Hotel, Chicago. *(c)* A Lake Michigan seawall built of Indiana limestone is chemically weathered by waves overtopping the structure. Solution of the limestone has caused "honeycomb" weathering.

(c)

● **FIGURE 6.7** Spheroidal weathering in jointed granite; Alabama Hills, California. Mount Whitney, just right of the twin spires, and the Sierra Nevada are in the background.

protecting Chicago's Lakeshore Drive was built of Indiana limestone. Chemical weathering by water overtopping the wall has dissolved clearly visible pits in a pattern known as "honeycomb" weathering. You can see examples of weathering in your own community by visiting a cemetery. Compare the legibility of the inscriptions on monuments of various ages and materials. Slate and granite monuments last much longer than those of limestone or marble, as you might suspect.

Geologic Features of Weathering

Talus refers to a loose pile of angular boulders at the base of the cliff or steep slope from which the debris has been derived. Also called *scree,* talus is a common geologic feature in alpine and arid environments (● Figure 6.6). **Spheroidal weathering** is chemical weathering in which concentric shells of decayed rock are separated from a block bounded by joints or other fractures (● Figure 6.7). It occurs when a rectangular block is weathered from three sides at the corners and from two sides along its edges, while the block's plane faces are weathered uniformly. It is also called "onion-skin" weathering, since the layers of an onion provide a good analogy. **Exfoliation domes** (Latin *folium,* "leaf," and *ex-*, "off") are the most spectacular examples of physical weathering. These domes result from the unloading of rocks that had been deeply buried. As erosion removes the overburden, it reduces pressure on the underlying rocks, which in turn begin to expand. As the rocks expand, they crack and fracture along **sheet joints** parallel to the erosion surface. Slabs of rock then begin to slip, slide, or spall (break) off the host rock, revealing a large, rounded domelike feature. Sheeting along joints occurs most commonly in plutonic rocks like granite and in massive sandstone. Half Dome in Yosemite Valley, California, and Stone Mountain in Georgia are granite exfoliation domes (● Figure 6.8).

 ## SOILS

Soil Profile

As we have noted, loose rock and mineral fragments at the surface of the earth are referred to as *regolith*. Regolith may

● **FIGURE 6.6** Talus slopes caused by frost wedging and other physical weathering processes; Banff National Park in the Canadian Rocky Mountains.

● FIGURE 6.8 Half Dome in Yosemite National Park is a product of glaciation (the flat side) and exfoliation (the dome-shaped side).

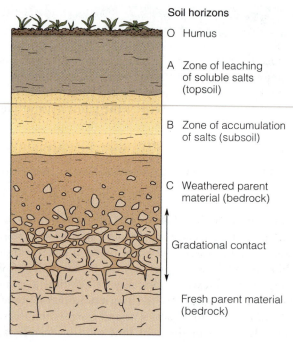

Soil horizons

O   Humus

A   Zone of leaching of soluble salts (topsoil)

B   Zone of accumulation of salts (subsoil)

C   Weathered parent material (bedrock)

Gradational contact

Fresh parent material (bedrock)

● FIGURE 6.9 An ideal soil profile. Heavily irrigated soils and very young ones exhibit variations of this profile.

consist of sediment that has been transported and deposited by rivers or wind, for example, or it may be rock that has decomposed in place. Where weathering has been sufficient, a surface layer that can support plant life—soil—has developed. Climate is the most important factor in soil development; given enough time, soils formed on different rocks under the same climate will look much the same. The influence of the parent material is most apparent in young soils such as decomposed granite, which contains quartz and feldspar grains, some of them having been altered to clay minerals.

A mature **soil profile** has several recognizable **soil horizons,** layers roughly parallel with the ground surface that are products of soil-forming processes (● Figure 6.9). The uppermost soil horizon, the "A horizon," is the *zone of leaching.* Here, mineral matter is most strongly dissolved by downward-percolating water. It may be capped by a zone of organic matter *(humus)* of variable thickness called the "O horizon," which provides CO_2 and organic compounds that make the percolating water slightly acidic. As the dissolved chemicals move downward, some of them are redeposited as new minerals and compounds in the "B horizon," called the *zone of accumulation.* This horizon is usually lighter in color and harder than the A horizon. Below this is the "C horizon," the weathered transition zone that grades downward into fresh parent material.

The thickness and development of the soil profile are profoundly influenced by climate and time. In desert regions of the Southwest, the A horizon may be thin or nonexistent, and the B horizon may be firmly cemented by white, crusty calcium carbonate known as **caliche.** Caliche forms when downward-percolating water evaporates and deposits calcium carbonate that has been leached from above. Also, if the water table is close to the ground surface, upward-moving waters may evaporate into the layers, leaving deposits of caliche. Some of these deposits are so thick

and hard that unusual measures are required to till or excavate through them.

RESIDUAL SOILS. Residual soils are soils that have developed in place on the underlying bedrock. For example, in temperate climates granite decomposes into soils that contain quartz, weathered feldspar, and clay minerals formed from the weathering of feldspar. This soil is usually light brown or gray in color and can be mistaken easily for its parent rock. Engineers and geologists call slightly weathered granite *decomposed granite,* or simply *d.g.* It is an excellent foundation material for structures and roads. Heavy clay soils may develop on shale bedrock. The expansive nature of some of these clays makes them troublesome. (This is discussed in the next section of this chapter.) Residual soils may take hundreds or even thousands of years to develop and they are always subject to erosion and removal by one of many geologic processes.

TRANSPORTED SOILS. Some soils have formed on transported regolith. These transported soils are designated by the geologic agent that is responsible for their transportation and deposition: *alluvial* soils by rivers, *eolian* soils by the wind, *glacial* soils, *volcanic* soils and so on. ● Figure 6.10 illustrates a soil that developed on volcanic ash overlying an older basalt on which little or no soil had developed. Known as the *Pahala ash,* this soil supports such exotic crops as macadamia nuts, sugar cane, and papaya on the Island of Hawaii.

As Pleistocene glaciers expanded and moved over northern North America, they scraped off Canada's soils and

● FIGURE 6.10 (at left) Volcanic ash (horizontal layers) overlying weathered basalt; Pahala, Hawaii. This volcanic soil supports rich crops of sugar cane, pineapple, and macadamia nuts.

redeposited them in a band across the United States from Montana to New Jersey. Soils formed on these widespread deposits—collectively known as *glacial drift*—are extremely variable; they may be thin or thick, bouldery or fine-grained, fertile or relatively barren. Some areas of drift are so bouldery that they are impossible to farm. This is one of the reasons that pastures for grazing dairy cattle, rather than cultivated grain crops, are the agricultural mainstay of such states as Wisconsin and New Hampshire.

Loess (approximately "luss," to rhyme with *cuss*) are wind-blown silt deposits composed of angular grains of feldspar, quartz, calcite, and mica. Covering about 20 percent of the United States and about 10 percent of the earth's land area (● Figure 6.11), loess is arguably the earth's most fertile soil. The sources of loess are glacial deposits and deserts. Strong winds blowing off Ice Age glaciers that covered most of Canada and the northern United States swept up nearby meltwater deposits and redeposited them far away as loess in valleys and on ridge tops. In the United

● FIGURE 6.11 (below) Loess-covered regions of the world. Note that the largest areas are in China, central and western Asia, and North America.

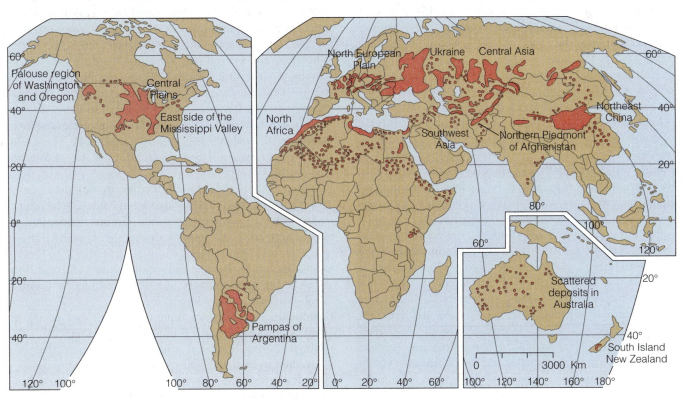

● FIGURE 6.12 Almost-vertical slopes supported by loess in People's Republic of China. The small holes are probably insect nests and are testimony to loess's easily eroded nature.

States loess has produced especially rich agricultural lands in the Palouse region of eastern Oregon and Washington, and in the central Great Plains, sometimes called the "breadbasket" of the United States. The fertility of the Ukraine, formerly part of the Soviet Union but now a republic, is due to a favorable climate and its loessal soils. So highly regarded is this soil that Adolph Hitler had large quantities of it moved to Germany during World War II.

Without doubt, the largest loess deposits are in China, where they cover 800,000 square kilometers (312,500 mi²) and are as thick as several hundred meters. These deposits were derived from the Gobi and Takla Makan deserts (see Figure 12.8) to the north and west. On windy days airborne silt from these deserts is noticeable in Beijing, hundreds of miles away. Similarly, the loess in Africa's eastern Sudan was carried there by winds from the Sahara Desert.

Loess has several noteworthy physical properties. It is tan to almost yellow in color and rather loosely packed (low density), which makes it easily excavated and eroded by running water. Its interlocking angular grains give it cohesive strength, however, and it is capable of standing in almost-vertical slopes without falling. These unusual properties served the Union and the Confederacy well in Mississippi during the Civil War. Both sides excavated tunnels, trenches, revetments, and other earthworks in the loess soil prior to the siege and battle at Vicksburg, which ultimately was won by U. S. Grant's Union forces. In northern China's Shansi Province in the 16th century elaborate caves scraped out of loess provided refuge for more than a million people (● Figure 6.12). Unfortunately for the cave residents, loess is subject to collapse under dynamic stresses such as those of an earthquake, and an estimated 800,000 people were buried in their homes during the earthquake of 1556. The load of loess carried by the Huang Ho River

as it flows through this area gave the river its other name, the *Yellow River*—which empties into the Yellow *Sea*.

Soil Classification

ZONAL CLASSIFICATION. Pedologists (Greek *pedon,* "soil," and *logos,* "knowledge of") have long known that soils that form in tropical regions are different from those that form in arid or cold climates. Where rainfall is heavy, soils are deep, acidic, and dominated by chemical weathering. Soluble salts and minerals are leached (removed) from the soil, iron and aluminum compounds accumulate in the B horizon, and the soil supports abundant plant life. In arid regions, on the other hand, soils are usually alkaline and thinner, coarse-textured or rocky, and dominated by physical weathering. Because of low precipitation there, salts are not leached from soils, and when soil moisture evaporates, it leaves additional salts behind (Case Study 6.1). These salts may form crusts and lenses of caliche ($CaCO_3$ and other salts) in the B horizon.

These understandings led to the development of a "zonal" classification of soils based upon climate, which then evolved into a system of Great Soil Groups. These groups may be simplified into four major types (● Figure 6.13):

- **pedalfers,** soils that are high in aluminum (Al) and iron (Fe) and that are characteristic of humid regions (rainfall >50 cm/year, or >20 in/y);
- **pedocals,** soils that are high in calcium and found in desert or semiarid regions (rainfall <50 cm/y);
- **laterites,** the brick-red soils of the tropics that are enriched in hydrated iron oxides (rust); and
- **tundra soils,** the soils of severe polar climates, found mostly in the Northern Hemisphere.

Salinization and Waterlogging

Saline soils contain sufficient soluble salts to interfere with plant growth. Salinization is the oldest soil problem known to humans, dating back at least to the fourth century B.C. At that time a highly civilized culture was dependent upon irrigation agriculture in the southern Tigris–Euphrates Valley of Iraq, the "Cradle of Civilization." By the second century B.C., the soils there were so saline that the area had to be abandoned. Soils in parts of California, Pakistan, the Ukraine, Australia, and Egypt are now suffering the same fate.

Salinization can be human-induced by intensive irrigation, which raises underground water levels very close to the soil. Capillary action then causes the underground water, containing dissolved salts, to rise in the soil, just as water is drawn into a paper towel or sugar cube. The soil moisture is then subject to surface evaporation, and as the water in the soil evaporates, it leaves behind salts that render the soil less productive, and eventually barren (see ◉ Figure 1). The productivity of 15 percent of all U.S. cropland is totally dependent upon irrigation, and 30 percent of U.S. crops are produced on this land.

For millenia, annual flooding of the Nile River added new layers of silt to the already rich soil and removed the salts. After completion of the Aswan Dam in 1970, however, the yearly flooding and flushing action stopped, and salts have now accumulated in the soils of the floodplain. Salinization is particularly evident in the arid Imperial Valley of California. Soils of the Colorado Desert, or any desert for the matter, can be made productive if enough water is applied to them. Because deserts have low rainfall, irrigation water must be imported, usually from a nearby river. The All-American Canal brings relatively high salinity water from the Colorado River to the Imperial Valley, where it is used to irrigate all manner of crops from cotton to lettuce. In places the water table has risen to such a degree that the cropland lies barren, its surface covered by white crusts of salt. The Impe-

◉ **FIGURE 1 Accumulation of salts at the soil surface due to evaporation of irrigation and ground water; Imperial Valley, California.**

rial Valley's Salton Sea is one of the largest saltwater lakes in the United States, and it grows saltier and larger every year by the addition of saline irrigation water. The return of degraded irrigation water in canals to the Colorado River in southern Arizona has required that a desalinization facility be established near Yuma to make the water acceptable for agricultural use in Mexico.

The good news is that soil salinization can be reversed. It can be done by lowering the water table by pumping and then applying heavy irrigation to flush the salts out of the soil. Salinization can also be mitigated, or at least delayed, by installing perforated plastic drain pipes in the soil to intercept excess irrigation water and carry it offsite. Neither remedy is an easy or inexpensive solution to this age-old problem.

Pedalfers are the rich soils of the prairies and steppes that support forest and grassland growth. Pedocals form where there is little rainfall, resulting in desert or scrub vegetation. Pedocals and pedalfers are mid-latitude, temperate-climate

soils. In contrast, laterites (Latin *later,* "brick, tile") are the residual soils of the tropics, where heavy rainfall leaches out all soluble materials, leaving behind the clay minerals and hydrated aluminum and iron oxides that impart a red color

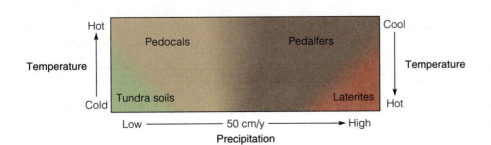

◉ **FIGURE 6.13 The Great Soil Groups and their climatic environments. Note that the temperature scale is reversed on the left and right sides of the diagram.**

● FIGURE 6.14 Rocky laterite soil in Nigeria.

(● Figure 6.14). Laterite is not uniquely identified with any particular rock type, but it does require an iron-containing parent, abundant rainfall, and well-drained terrain. Laterites are currently forming in humid tropical and subtropical regions. Leaching of these soils may be so extensive that with the right parent rock, economically valuable aluminum-rich *bauxite* deposits are the end products of the soil-forming process (● Figure 6.15). Finally, tundra soils form in arctic and subarctic regions and support only such vegetation as mosses, sedges, lichens, and dwarf shrubs, making for bleak, forestless landscapes (● Figure 6.16).

U.S. COMPREHENSIVE SOIL CLASSIFICATION SYSTEM. Because of dissatisfaction with existing classification schemes, the U.S. Soil Conservation Service instituted the first of seven trial classification systems in 1952. Each trial system was sent to a select group of international pedolgists for their comments. After incorporating the suggestions and findings of these experts, the "Seventh Approximation," as it is called, was officially adopted by the

Soil Conservation Service in 1965. The distinguishing features of this system are that:

1. Soils are classified by their physical characteristics, rather than their origin; that is, the system is nongenetic.
2. Soils that have been modified by human activities are classified along with natural soils.
3. The soils' names convey information about their physical characteristics.

The system subdivides soils into a hierarchy of six levels — orders, families, series, and so forth, with ten orders at the highest level in the system. With a practiced eye, each soil order can be easily identified. Two examples of soil orders are *oxisols,* which are the intensely leached, iron-rich, oxidized red soils of the tropics (laterites), and *aridisols,* the alkaline soils of desert regions (pedocals). Because the number of soil entries increases astronomically at the lower levels (illustrated by the fact that 14,000 soil series are recognized in the U.S.), this classification is for the professional, not the layperson. Nonetheless, one should be aware that it is the classification used by today's soil scientists.

 SOIL PROBLEMS

Soil Erosion

Although soil is continually being formed, it is for practical purposes a nonrenewable resource, because hundreds or even thousands of years may be necessary for soil to develop. Ample soil is essential for providing food for the earth's burgeoning population, but yet, throughout the world it is rapidly disappearing. Although the United States has one of the world's most advanced soil conservation programs, soil erosion remains a problem after 60 years of conservation efforts and expenditures of billions of dollars. Erosion removed an estimated two billion tons of topsoil from U.S. farmlands in 1992—down from slightly more than three bil-

● FIGURE 6.15 Bauxite ore developed on aluminum-rich igneous rock; Jamaica.

● FIGURE 6.16 Treeless tundra landscape near Kotzebue, Alaska. Most of the soil in tundra regions is frozen much of the year.

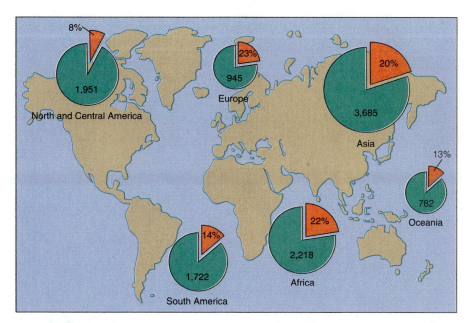

FIGURE 6.17 World soil degradation, 1945–1991. The areas degraded are in millions of hectares. (A hectare, 10,000 square meters, equals 2.47 acres.) Table 6.2 summarizes the relative contributions of overgrazing, deforestation, and bad agricultural practices to world soil degradation.

lion tons a decade earlier. With the average annual soil loss estimated as 10–12 tons per hectare (4–4.8 tons/acre) and an assumed annual soil-formation rate of 2–4 tons per hectare (0.8–1.6 tons/acre), three to five times as much soil is being lost as is being formed in the United States. Not only does soil erosion destroy fertility, but tons of soil settle in lakes and clog waterways and drainages with sediment, pesticides, and nutrients each year. Losses of only 2–3 centimeters (1 in) of topsoil represent about 200 tons per hectare. Such erosion cuts crop yields and may render the land unproductive.

On a global basis, an area the size of China and India combined has suffered *irreparable degradation* from agricultural activities and overgrazing—mostly in Asia, Africa, and Central and South America (● Figure 6.17). Lightly damaged areas can still be farmed using the modern conservation techniques explained later in this chapter, but "fixing" serious erosion problems is beyond the resources of most Third World farmers and their governments.

EROSION PROCESSES. The agents of soil erosion are wind and running water. Heavy rainfall and melting snow create running water, which removes soil by sheet, rill, and gully erosion. **Sheet erosion** is the removal of soil particles in thin layers more or less evenly from an area of gently sloping land. It goes almost unnoticed. **Rill erosion,** on the other hand, is quite visible in discrete streamlets carved into the soil (● Figure 6.18). If these rills become deeper than about 25–35 centimeters (10–14 in), they cannot be removed by plowing, and gullies form. Gullies are created in unconsolidated material by widening, deepening, and headward erosion of rills (● Figure 6.19). **Gullying** is a big problem in the easily eroded loess soils of the Palouse

● FIGURE 6.18 Rill erosion in cultivated soil.

● FIGURE 6.19 Extreme gullying in soft sediments encroaching on tilled land near Lumpkin, Georgia. The width of the area in the photo is approximately 1¾ kilometers) (1 mi).

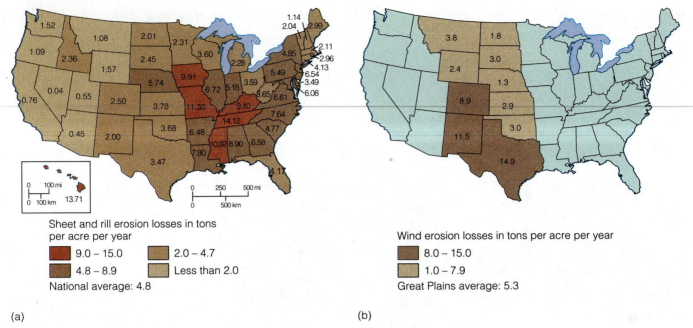

Sheet and rill erosion losses in tons per acre per year

- 9.0 – 15.0
- 4.8 – 8.9
- 2.0 – 4.7
- Less than 2.0

National average: 4.8

(a)

Wind erosion losses in tons per acre per year

- 8.0 – 15.0
- 1.0 – 7.9

Great Plains average: 5.3

(b)

◉ FIGURE 6.20 Estimated annual rates of cropland soil loss in the United States, 1980. *(a)* Sheet and rill erosion losses. *(b)* Wind erosion losses in the Great Plains.

■ TABLE 6.1 Significant U.S. Soil Erosion by State, 1980

RANK	STATE	TONS/ACRE/YEAR
Sheet and Rill Erosion		
1	Tennessee	14.12
2	Hawaii	13.71
3	Missouri	11.30
4	Mississippi	10.92
5	Iowa	9.91
Wind Erosion		
1	Texas	14.9
2	New Mexico	11.5
3	Colorado	8.9
4	Montana	3.8
5	Oklahoma	3.0
Total Soil Erosion, Water and Wind*		
1	Texas	18.40
2	Tennessee	14.12
3	Hawaii	13.71
4	New Mexico	13.50
5	Colorado	11.40

*The U.S. average annual rate of total soil erosion is estimated at 4.0–4.8 tons/acre (10–12 tons/hectare).
SOURCE: *Environmental Trends* (Washington, D.C.: Council on Environmental Quality, 1981), cited in Sandra Batie, *Soil Erosion: A Crisis in America's Cropland?* (Washington, D.C.: The Conservation Foundation, 1983).

region of eastern Washington. Snow melting there saturates the loess soil, causing it to become fluid and then to flow like wet concrete across the underlying still-frozen ground. Estimated annual U.S. cropland soil loss from sheet and rill erosion is shown by state in ◉ Figure 6.20, part a. Tennessee headed the list in 1980, followed by Hawaii, Missouri, Mississippi, and Iowa, as shown in ■ Table 6.1.

Wind erodes soils when tilled land lies fallow and dries out without vegetation and root systems to hold it together. The dry soil breaks apart, and wind removes the lighter particles. If the wind is strong, a dust storm may occur. The great "black blizzards" of the 1930s created the "dust bowl" of the southern Great Plains during a period of severe drought (Case Study 6.2). Much of the reason for the dust bowl was the widespread conversion of marginal grasslands, which receive only 25–30 centimeters (10–12 in) of rainfall per year, to cropland. Unfortunately, this is still occurring some places in the United States, and the rest of the world. Figure 6.20, part b, presents estimates of the amount of soil lost each year to wind erosion in the Great Plains states. When the 50 states' estimated wind and water erosion losses are combined (Table 6.1), the state with the greatest annual loss is Texas, followed by Tennessee, Hawaii, and New Mexico, respectively. ■ Table 6.2 summarizes the recognized causes of accelerated soil erosion worldwide and assesses their relative impact.

In addition to the soil erosion caused by overgrazing, deforestation, and bad agricultural practices, some recreational pursuits also cause soil erosion, particularly those that involve offroad vehicles (ORVs). A 1974 study indicated that recreational pressure on semiarid lands of Southern California by ORVs was "almost completely uncon-

■ **TABLE 6.2 Causes of World Soil Erosion and Deterioration**

CAUSE	ESTIMATED % OF TOTAL
Overgrazing	35
▪ reduced vegetation cover	
▪ trampling that leads to decreased waterholding capacity	
Deforestation	30
Bad agricultural practices	28
▪ salinization due to improper draining of irrigated land	
▪ fertilization that results in acidification	
▪ lack of terraces on tilled sloping land	
▪ wind erosion of fallow lands	
Other causes, natural and human	7

SOURCE: World Resources Institute, United Nations Environment and Development Program, *World Resources: A Guide to the Global Environment.* (New York: Oxford University Press, 1992).

trolled" and that those desert areas had experienced greater degradation than had any other arid region in the United States (◉ Figure 6.21). In the two decades since that study, much has been accomplished in controlling offroad traffic in California and other states, but ORVs continue to exert severe impacts upon areas established (translate *sacrificed*) for their use.

MITIGATION OF SOIL EROSION. Thomas Jefferson was among the first people in the United States to comment publicly on the seriousness of soil erosion. He called for such soil conservation measures as crop rotation, contour plowing, and planting grasses to provide soil cover during the fallow season. These practices are now common in industrialized nations, but unfortunately, only minimally employed in Third World countries. Proper choices of where to plant, what to plant, and how to plant are the components of soil conservation. The major erosion-mitigation practices are described here.

- *Terracing*—creating flat areas, terraces, on sloping ground—is one of the oldest and most efficient means of saving soil and water (◉ Figure 6.22).
- *Strip-cropping,* by which close-growing plants are alternated with widely spaced ones, efficiently traps soil that is washed from bare areas and it also supplies some wind protection. An example of this is alternating strips of corn and alfalfa.
- *Crop rotation* is the yearly alternation of soil-depleting crops with soil-enriching crops. Rotating corn and wheat with clover in Missouri, for example, has been found to reduce soil runoff from 19.7 tons/acre for corn to 2.7 tons/acre, a huge loss reduction. Groundcovers such as grasses, clover, and alfalfa tend to be soil-conserving, whereas rowcrops like corn and soybeans can leave soil vulnerable to runoff losses. Retaining crop residues, such as the stubble of corn or wheat, on the soil surface after harvest has been found to reduce erosion and to increase water retention by more than half.
- *Conservation-tillage* practices minimize plowing in the fall and encourage the contour plowing of furrows perpendicular to the field's slope so that they will catch runoff, increasing water infiltration (◉ Figure 6.23, page 168).

◉ **FIGURE 6.21** Hill-climbing motorcyclists have devastated this slope in Jawbone Canyon in the Mojave Desert, California. The bikers have compacted the soil surface, causing reductions in vegetation and soil permeability. The reduced soil moisture content has seriously harmed the local plant and animal ecology. Desert rains have eroded the tire tracts, transforming them into gullies, which increases water runoff and sediment yield (erosion).

◉ **FIGURE 6.22** Terracing steep hillsides is an ancient soil conservation practice; China.

The Wind Blew and the Soil Flew—
The Dust Bowl Years

The stock market crash of 1929 ushered in the Great Depression of the 1930s, which brought widespread unemployment without unemployment benefits or entitlements such as welfare, social security, and free health care. It was a bad time, and to make things worse, there was protracted drought in the Great Plains, an area that was totally dependent upon rainfall for crop production at the time. Thousands of acres of marginally productive land, mostly grasslands, had been tilled because the government had begun guaranteeing wheat prices. Farmers thought they couldn't lose—but they were banking on rain. The entire United States except New England dried up, but between 1932 and 1940 the southern Great Plains was hit the worst (● Figure 1).

A longstanding agricultural practice is to plow up cropland after the fall harvest and allow the land to stand fallow through the winter. Plowing turns crop stubble into the soil, gets rid of weeds, and enhances soil water and air uptake. This is a fine prac-

tice as long as it doesn't rain too much—or too little. If too much rain falls, the soil is subject to severe sheet and rill erosion. If the pulverized soil dries out and the wind blows, the result is the same: extreme soil erosion.

The wind did blow in the southern Great Plains in the mid 1930s, and huge quantities of topsoil were simply blown away. There were many dust storms during this period, but the one of April 1935 was the worst. Soil silt, clay, and organic matter were lifted 5 kilometers (16,000 ft) into the air and carried eastward to Washington, D.C., and even 500 kilometers (300 mi) beyond the east coast into the Atlantic Ocean. The source area of the dust storms became known as the "Dust Bowl." It is a zone roughly 900 kilometers long by 450 kilometers wide (540 mi x 270 mi) in Colorado, Kansas, New Mexico, Texas, and Oklahoma (● Figure 2).

The wind and dust made farming impossible, and farm families, burdened by debt for equipment, seed, and supplies, left their

● FIGURE 1 A dust storm hits a southwestern Great Plains village in October 1937. Storms like this one destroyed crops and buried pasturelands. Minutes after this photograph was taken the village was completely engulfed in choking dust.

● FIGURE 2 (right) Map of the area most seriously affected by blowing dust during the Dust Bowl years, 1935–1936, and 1938. Kansas was the state most devastated in terms of percentage of cultivated area degraded by strong winds.

farms to become migrant workers. Many traveled to California, where their lot was hardly improved, and the entire country was soon referring to this huge unfortunate group, derisively, as "Okies" (● Figure 3). This subculture of the 1930s, innocent victims of severe soil erosion, was the subject of John Steinbeck's touching novel *The Grapes of Wrath*.

Even today wind erosion exceeds water erosion in Texas, Colorado, Montana, New Mexico, Oklahoma, and North Dakota. Mismangement of plowed lands and overgrazing are making parts of these six states potential dust bowls. In fact, the worst wind erosion since the Dust Bowl years took place in the southwest Great Plains in 1996.

● FIGURE 3 (below) Dust Bowl America. In 1938 Dorothea Lange, who worked for the Farm Security Administration, took this historic photograph of a homeless migrant family of seven on a highway in Pittsburgh County, Oklahoma. By 1939, half of the families in Oklahoma were on relief.

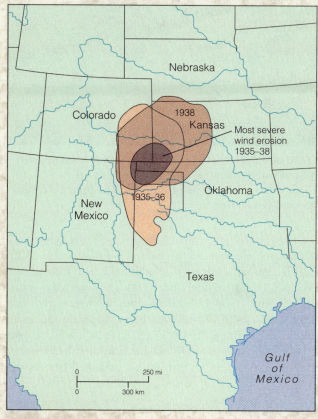

● FIGURE 6.23 Contour plowing. Runoff is intercepted in furrows that follow an elevation and that are perpendicular to the slope of the land. The practice retains water and reduces erosion.

- *No-till* and *minimum-till* practices also conserve soil. With no-till farming, seeds are planted into the soil through the previous crop's residue, and weeds are controlled solely with chemicals. Specialized no-till equipment is required (●Figure 6.24). "No-till" has been especially successful in soybean farming. In 1993 U.S. farmers left 30 percent of their crop debris on 37 million cultivated hectares (92 million acres), nearly a third of all cultivated land in the United States. Ten percent of the U.S. planted acreage was not plowed at all. No-till agriculture is most beneficial during times of drought, when plowed fields dry out and topsoil is more vulnerable to wind erosion. No-till practices saved much of the agricultural soil that would have otherwise been eroded by floodwaters of the Great Flood of 1993 in the Mississippi River valley.

Wind-caused soil losses can also be reduced by planting windbreaks of trees and shrubs near fields and by planting strip crops perpendicular to the prevailing wind direction. Wind losses are almost zero with cover crops such as grasses and clover.

Expansive Soils

Certain clay minerals have a layered structure that allows water molecules to be absorbed between the layers, which causes the soil to expand. (This process differs from the *adsorption* of water to the surface of nonexpansive soil clays.) Soils that are rich in these minerals are said to be **expansive soils**. Although this reaction is somewhat reversible, as the clays contract when they dry, soil expansion can exert extraordinary uplift pressures on foundations and concrete slabs, with resulting structural and cosmetic damage (●Figure 6.25). Damage caused by expansive soils costs about $6 billion per year in the United States, mostly in the Rocky Mountain states, the Southwest, and Texas

● FIGURE 6.24 Mechanics of the no-till drill. (1) An opener disc digs a trench. (2) A gauge wheel regulates the depth of the trench. (3) A press wheel pushes seed and fertilizer through tubes into soil. (4) A closing wheel returns soils to the trench. All this is done without greatly disturbing the soil or the stubble remaining from the previous season's crop, thus minimizing the amount of soil surface that is exposed to erosion.

● FIGURE 6.25 This well-built concrete-block wall in Colorado was uplifted and tilted by stresses exerted by the underlying expansive soil.

and the Gulf Coast states (● Figure 6.26). The swelling potential of a soil can be identified by several standard tests, usually performed in a soils engineering laboratory. An obvious test is to compact a soil in a cylindrical container, soak it with water, and measure how much it swells against a certain load. Clays that expand more than 6 percent are considered highly expansive; those that expand 10 percent are considered critical. Treatments for expansive soils include (1) removing them, (2) mixing them with nonexpansive material or with chemicals that change the way the clay reacts with water, (3) keeping the soil moisture constant, and (4) using reinforced foundations that are designed to withstand soil volume changes. Identification and mitigation of damage from expansive soils is now standard practice for U.S. soil engineers.

Permafrost

The term **permafrost**, a contraction of *permanent* and *frozen,* was coined by Siemon Muller of the U.S. Geologi-

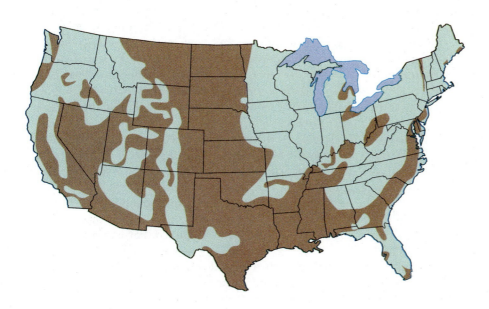

● FIGURE 6.26 Distribution of expansive soils (shown in brown) in the United States.

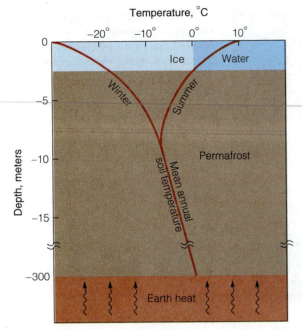

Temperature, °C

● **FIGURE 6.27** The interaction of soil temperature and depth establishes the lower and upper limits of permafrost. The upper limit occurs where and when the summer soil temperature exceeds 0° C. The lower limit may be at great depth. It is the point where earth heat raises the soil temperature above the freezing temperature of water. Note the break in the depth scale.

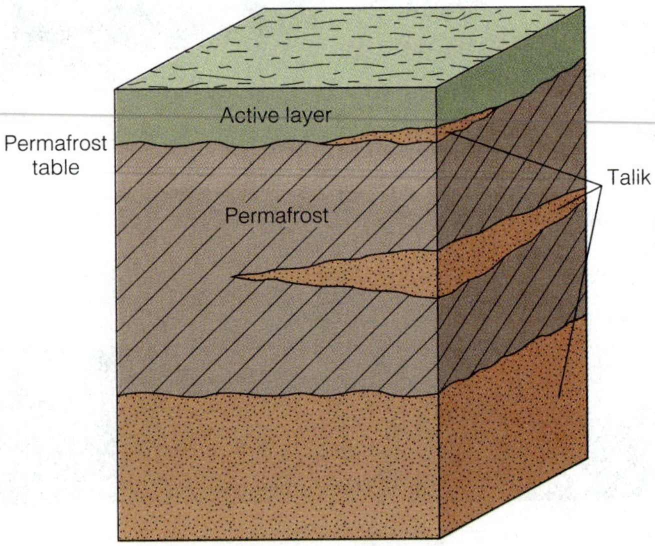

● **FIGURE 6.28** Permafrost zones. Talik, unfrozen layers, or "lenses," may occur in the active layer and in the permanently frozen layer.

cal Survey in 1943 to denote soil or other surficial deposits in which temperatures below freezing are maintained for several years. More than 20 percent of the earth's land surface is underlain by permafrost, and it is not surprising that most of our knowledge about frozen ground has been derived from construction problems encountered in Siberia, Alaska, and Canada. Permafrost becomes a problem for us humans when we change the surface environment by our activities in ways that thaw the near-surface soil ice, since this results in soil flows, landslides, subsidence, and related phenomena. Permafrost forms where the depth of freezing in the winter exceeds the depth of thawing in the summer (● Figure 6.27). If cold ground temperatures continue for many years, the frozen layer thickens until the penetration of surface cold is balanced by the flow of heat from the earth's interior. This equilibrium between cold and heat determines the thickness of the permafrost layer. Thicknesses of 1,500 meters and 600 meters (almost 5,000 and 2,000 ft) have been reported in Siberia and Alaska, respectively. Such thick layers are very old; they must have formed during the Pleistocene Epoch of geologic time, the time known as the Great Ice Age.

The top of the permanently frozen layer is called the **permafrost table** (● Figure 6.28). Above this is the **active layer,** which is subject to seasonal freezing and thawing and is always unstable during summer. When winter freezing does not penetrate entirely to the permafrost table,

water may be trapped between the frozen active layer and the permafrost table. These unfrozen layers and lenses, called *talik* (a Russian word), are of environmental concern, because the unfrozen ground water in the talik may be under pressure. Occasionally, pressurized talik water bursts through to the surface, forming a domical mound 30–50 meters (100–160 ft) high called a *pingo.* The presence of ice-rich permafrost becomes evident when insulating vegetation, usually tundra, is stripped away for a road, runway, or building. Thawing of the underlying soil ice results in subsidence, soil flows, and other gravity-induced mass movements (● Figure 6.29). Even the casual crossing of tundra on foot or in a jeep can upset the thermal balance and cause thawing. Once melting starts, it is impossible to control, and a permanent scar is left on the landscape (● Figure 6.30).

Structures that are built directly on or into permafrost settle as the frozen layers thaw (● Figure 6.31). Both active and passive construction practices are used to control the problem. Whereas active methods involve complete removal or thawing of the frozen ground before construction, passive methods build in such manners that the existing thermal regime is not disturbed. For example,

- a house can be built with an open space beneath the floor so that the ground can remain frozen,
- a building can be constructed on foundation piers whose

FIGURE 6.29 Utility poles extruded from permafrost ground; Alaska. One solution is to place the poles in tripods on the ground surface.

FIGURE 6.30 Scars made by vehicles on tundra soil remain for many years; Alaska.

temperature is held constant by heat dissipators or coolant, and

■ roads are constructed on top of packed coarse-gravel fill to allow cold air to penetrate beneath the roadway surface and keep the ground frozen (⊙ Figure 6.32).

The most ambitious construction project on permafrost terrain to date is the 800-mile oil pipeline from Prudhoe Bay on Alaska's North Slope to Valdez on the Gulf of Alaska. About half the route is across frozen ground, and because oil is hot when it is pumped from depth and also during transport, it was decided that the pipe should be above ground so as to avoid the thawing of permafrost soil (⊙ Figure 6.33). The four-foot-diameter pipeline also crosses several active

faults, three major mountain ranges, and a large river. The pipeline's above-ground placement allows for some fault displacement and for easier maintenance if it should be damaged by faulting. This placement has the added advantage of accommodating the longstanding migratory paths of herding animals such as caribou and elk.

Settlement

Settlement occurs when a structure is placed upon a soil, rock, or other material that lacks sufficient strength to support it. Settlement is an engineering problem, not a geological one. It is treated here, however, because settlement and poor soil seem inextricably associated. All structures

(a)

(b)

FIGURE 6.31 Two casualties of improper construction for permafrost conditions in Alaska, *(a)* a modern home in Fairbanks and *(b)* Bert and Mary's Roadhouse on the Richardson Highway. The front center portion of the log-construction roadhouse was the area of greatest subsidence, because that was the location of the furnace. The 14-year-old structure had to be removed in 1965 because of distress due to continued sinking.

Soils, Health, and Locoweed

Humans need trace elements in their systems in order to function properly. **Trace elements** occur in small amounts in the body, usually a few parts per million, as opposed to the bulk elements, which are found in large quantities. The body obtains these elements naturally by this sequence: rocks to soils to plants and finally to animals. Humans get trace elements by eating plants or by eating animals that ate plants that grew on soils containing them or by using dietary supplements of them. Some water-soluble trace elements, such as fluoride, are sometimes added to municipal drinking-water supplies.

The best-known relationship between soils and human health is the one between soil iodine deficiency and the enlargement of the thyroid gland known as *goiter*. Besides being cosmetically demoralizing, goiter is debilitating in other ways. Iodine deficiency during pregnancy can lead to "cretinism" in newborns, a severe mental and physical disorder. Endemic goiter has been firmly established in the northern United States in the so-called goiter belt (see ◉ Figure 1). Because of the wide availability of iodized table salt, this disease has practically disappeared except among the poorly educated. Additionally, little of the U.S. diet is "home grown" anymore. This is another reason goiter is no longer an endemic problem. Soil-iodine thyroid diseases are also known in England, Thailand, Mexico, the Netherlands, and Switzerland.

Several trace elements—zinc, copper, iron, cobalt, manganese, and molybdenum—are associated with and stimulate enzymes, the catalysts of biochemical reactions. These enter the body mostly in plant and animal foods. Their importance might be better appreciated when we realize that silver, mercury, and lead are toxic because they are enzyme *inhibitors*. Some of the beneficial effects of trace elements are that zinc aids in healing wounds and burns, iron transports oxygen in the blood, and cobalt is a

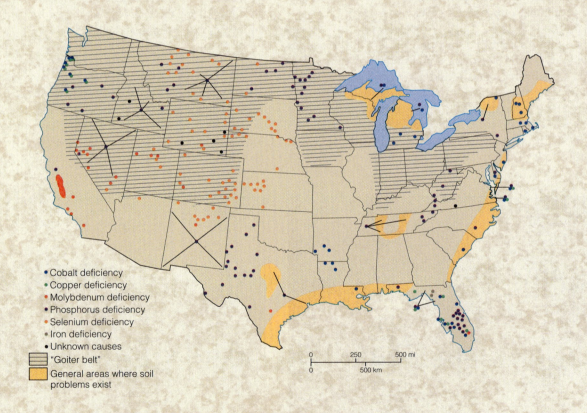

- Cobalt deficiency
- Copper deficiency
- Molybdenum deficiency
- Phosphorus deficiency
- Selenium deficiency
- Iron deficiency
- Unknown causes
- "Goiter belt"
- General areas where soil problems exist

◉ **FIGURE 1** Distribution of mineral-nutritional diseases of animals in the United States. Dots show specific locations where problems have been reported.

vital component of vitamin B_{12}. Deficiencies of these may result in anemia, slowed growth, vitamin-B_{12} deficiency, and other ailments.

A most interesting trace element is selenium. It is poisonous at high dosages, but deficiencies cause abnormalities in plants and animals. It is abundant in soils of the high plains and western United States, and certain plants there concentrate it in their tissue. Eating these plants, known as *locoweeds,* causes cattle to get the "blind staggers" and to appear drugged. Figure 1 shows areas where mineral-nutritional diseases are known to occur in animals. These are all localities where soils have deficiencies or excesses of vital or toxic trace elements.

Although molybdenum is probably best known as a metal that strengthens steel, it is also essential to plant and possibly animal health. Many plant enzymes contain molybdenum, two of which are important in metabolizing nitrogen, a major plant nutrient. In large concentrations, molybdenum interferes with the absorption of copper, an essential element, and may be detrimental to certain livestock. *Molybdenosis* can cause anemia, depressed growth, poor reproduction, bone malformation, and loss of normal hair color (⊙ Figure 2). Black Angus cattle with molybdenosis turn a mouse-gray color, for example. If treated early enough, the symptoms of the disease can be reversed by corrective nutrition.

Lignite coal (see Chapter 13) is known to be rich in molybdenum. Ranchers in South Dakota reported problems resembling molybdenosis in cattle pastured near an abandoned open-pit lignite mine. Chemical analysis revealed that grass growing near the exposed lignite contained high levels of molybdenum. This led to the discovery that molybdenum was being leached from the coal and transported by surface-water runoff to nearby pastures, where it was taken up by range grass and thereby entering the food chain.

Molybdenum-related disease has not been found in humans in such countries as the United States where food-production and distribution are widespread. In a region in southern India, however, some individuals developed *genu valgun* (knock-knee), a crippling bone deformity, with a devastating impact on the local population. The outbreak was traced to a high level of molybdenum in sorghum, a staple in the residents' diet. High levels of the metal were then found in the ground water. Ultimately it was discovered that construction of a large dam and reservoir nearby had raised the water table, making the high-molybdenum ground water accessible to the sorghum crop (see Chapter 10).

(a)

(b)

⊙ FIGURE 2 Molybdenosis results from an imbalance of copper to molybdenum in the diet of cattle. The visible symptoms are hair discoloration and loss and weight loss. *(a)* A healthy Hereford steer. *(b)* A steer that is suffering from molybdenosis.

(a)

● FIGURE 6.32 Passive construction methods for mitigating damage due to permafrost. *(a)* Raised WW II–era Quonset hut; Nome, Alaska. *(b)* Heat-radiating foundation piers on a modern structure; Kotzebue, Alaska.

(b)

settle, but if settlement is uniform, say two or three centimeters over the entire foundation, there is usually no problem. When differential settlement occurs, however, cracks appear or the structure tilts.

The Leaning Tower of Pisa, Italy, is probably the best-known, most-loved example of poor foundation design. It may have scientific importance as well, as Galileo is said to have conducted gravity experiments at the 54.5-meter (163 ft) structure. Begun in 1174, the tower started to tilt when the third of its eight stories was completed, sinking into a two-meter-thick layer of soft clay just below the ground surface. Initially the tower leaned north, but after the initial

(a)

● FIGURE 6.33 The 800-mile Alaskan oil pipeline was constructed to accommodate permafrost and wildlife. *(a)* The meandering pattern allows for expansion and contraction of the pipe and for potential ground movement along the Denali fault, which it crosses. The pipeline supports feature heat-dissipating radiators. *(b)* Moose and caribou are able to pass beneath the elevated pipeline.

(b)

settlement and throughout the rest of its history it has leaned south. To compensate for the lean, the engineer in charge had the fourth through eighth stories made taller on the leaning side. The added weight caused the structure to sink even further. It now leans about 5.2 meters (17 ft) from the vertical, about 5°, which gives one an ominous feeling when standing on the south side of the tower. The tower even has a slight bend like that of a banana, because the builders continued to construct vertically after tilting began (●Figure 6.34).

Careful measurements made between 1911 and 1990 showed that the top of the tower was moving south by an average of 1.2 millimeters per year. This was in spite of an attempt to stabilize the foundation in 1935 by pumping cement grout (a cement–water mixture) into the soft clays through 36 half-inch drill holes on the tower's south side. In 1989, a similarly constructed leaning bell tower at the cathedral in Pavia, Italy, collapsed, causing officials to close the tower at Pisa to visitors. In 1990 the Italian government established a special commission of structural engineers, soils engineers, and restoration experts to determine new ways to save the monument. The commission's goal is to reverse the tilt 10–20 centimeters (4–8 in), which they believe will add 100 years to the lifetime of the 800-year-old structure. More than 750 metric tons (825 of our 2,000-pound, "short," tons) of lead were placed on the foundation that rings the tower on the north side in 1993. This brought the top of the tower northward 2.5 centimeters (an inch) within nine months. Because the south side of the tower is overloaded due to the lean, steel bands were placed around the second story, which was in danger of collapsing. In 1995, another band was installed and anchored by steel cables to a sand layer 50 meters (162 ft) below the ground surface on the north side. This will ultimately replace the lead weight and straighten the tower even more.

It should be noted that tilting towers are not exceptional in Italy. Few of the 170 campaniles (bell towers) still standing in Venice are vertical, for example, and many other towers throughout Italy and Europe tilt. These graceful towers are certainly testimony to past architects' ability and talent, but also to their failure to understand that a structure can be truly beautiful only if local geology permits.

Other Soil Problems

Laterite soils, the red soils of the tropics, are soft when they are moist but become brick-hard and durable when they dry out (●Figure 6.35). The Buddhist temple complex at

● FIGURE 6.34 Inadequate foundation investigation and design; Pisa, Italy.

● FIGURE 6.35 Dried and hardened laterite; Brazil. This soil is so hard that soil scientists call it "ironstone."

Rx for Contaminated Soils

Although spills of hazardous materials frequently contaminate soils, the Environmental Protection Agency (EPA) did not include a soil component in their system for ranking Superfund cleanup sites until 1990 (see Case Study 1.3, Table 1). Soils become contaminated when they are forced to absorb a wide variety of contaminants, the most common being oil-field brines (salty water), heavy metals, pesticides, cleaning agents, and petroleum products. Fortunately, this contamination by absorption is reversible, and soils can be "cleaned," *remediated,* by a variety of methods. A large soil remediation industry has sprung up in response to EPA requirements.

Various soil remediation techniques are used, depending upon the thickness of the soil to be cleaned and the chemical nature of the contaminant. *Dilution,* mixing contaminated soil with clean soil, is one technique. It works, but it is expensive and it contaminates an even larger volume of soil. The old saying "the solution to pollution is dilution" is relevant to many pollution problems, but it is not practical when large volumes of soil are contaminated.

Bioremediation, a natural process, uses microorganisms to degrade and transform organic contaminants. Bacteria and fungi flourish naturally in soils where oil, sewage sludge, or food-production waste has been discarded. Biodegradation can be stimulated by increasing the amount of air in the soil through cultivation, vapor-extraction pumping (see below), or stockpiling the soil and forcing air into it. Adding easily decomposed organic matter such as hay, compost, or cattle manure also speeds up degradation of organic contaminants. Bioremediation is relatively inexpensive, and some contaminated soils simply require turning (disking or plowing) to speed up oxidation. A successful bioremediation project at Mobil Oil Company's natural-gas refinery at Liberal, Kansas, is shown in ⦿ Figure 1. The soil contained up to 26,000 parts per million (2.6%) of hydrocarbons. It was decided that the contaminated soil should be spread in a layer less than 20 centimeters (8 in) thick and that a nutrient and microbial amendment (cow manure) should be applied to it at a rate of about 60 tons per hectare (15 tons/acre) (Figure 1, part b). The soil was periodically disked to promote biological activity and was subsequently reclaimed and planted (Figure 1, part c). Under careful management even heavy metals such as selenium and arsenic can be converted to gaseous forms and released to the atmosphere. The three-year project was closely coordinated with the Kansas Department of Health and Environment.

Vapor extraction is used for removing volatile (easily evaporated) organic compounds such as cleaning solvents from soils. Air is drawn through the soil by means of dry wells connected to vacuum pumps. The air passing through the soil evaporates the compounds and exhausts them to the atmosphere. This technique is especially desirable for cleaning soils beneath buildings or roads, because no excavation is required except for the drilling of shallow wells. The down side is that it may take 6 to 12 months to sanitize the soil adequately.

Phytoremediation, the use of plants to remove contaminants, is used mainly for extracting salts. This has been used in the Netherlands, where periodic seawater flooding contaminates soils. Plant scientists are developing *hyperaccumulator* plants that clear soils of heavy metals such as cobalt, nickel, and zinc by concentrating them in their tissues. The cost of phytoremediation materials, equipment, and labor is relatively low, but up to a decade can be required to remediate soil to the point that it can be used again.

The justification for soil remediation is obvious. It allows vegetation to be established; it decreases soil erosion, ground-water contamination, and air pollution; and it can restore soils' agricultural productivity. The soil remediation industry is in its infancy, and a considerable amount of research is underway for developing faster and more economical means of "cleaning dirt."

CONSIDER THIS . . .

It is jokingly said that the twelfth-century architect of the Leaning Tower of Pisa saved money by not having a foundation investigation. Had engineers and geologists with today's knowledge and tools existed at that time, what routine tests would they have performed on bore-hole soil samples that would show settlement was inevitable? What recommendations would they have made to the client?

Angkor Wat in Cambodia is partly constructed of laterite bricks that have endured since the thirteenth century. Deforestation of areas underlain by laterites is an invitation to this hardening, as U.S. forces discovered in Vietnam when they attempted to grade roads and runways on lands that had been defoliated for military reasons. Deforested laterites make for swampy conditions during the rainy season, because they lack permeability; there is little infiltration of the water. This can raise havoc with wheeled vehicles and restrict mobility in such regions.

Although laterite soils support some of the densest rain forests in the world, they are relatively infertile because of

(a)

(b)

(c)

● FIGURE 1 A successful bioremediation project. *(a)* Oil-contaminated soil (2.6% hydrocarbons) at a Kansas refinery is removed for bioremediation by a bulldozer. Note the patchy distribution of the oil contamination. *(b)* The soil is spread in a layer about 20 centimeters (8 in) thick. Nutrient amendment (cow manure) is added to stimulate biological degradation of the hydrocarbons. *(c)* Remediated soil (< .01% hydrocarbons) is spread and planted with grass. Note the clods of manure and the white chunks of caliche in the redistributed soil. The structure in the background is a grain silo in which wheat is stored.

extensive leaching of minerals. The nutrients that support the rain forest are in the vegetation, and they are recycled when the plants die and decompose into the soil. Deforested laterites used for agricultural purposes must be fertilized intensively after a few harvests in order to support productive crops.

In some areas of extensive irrigation, agricultural soils form impervious clayey layers called **hardpans** in the B horizon. Hardpans inhibit waters from percolating and thus reduce crop yields. They must be broken up by deep plowing using large, heavy tractors pulling long, single-tooth

rippers that penetrate the subsoil. Hardpans and claypans are the scourge of farmers, as they can render their farmland unproductive.

The presence or absence of beneficial elements in soils can result in various dietary benefits or deficiencies in humans. This is because the soil is the actual beginning of the food chain. Soil elements are taken up by plants (vegetables, fruits, and grains), which are then eaten by animals. Some examples of soil elements and health are discussed in Case Study 6.3. The presence of pollutants in the soil now requires expensive cleanup (Case Study 6.4).

GALLERY

A Few Inches between Humanity and Starvation

Our planet's carrying capacity depends upon soil, and it is being lost faster than weathering processes can produce it. Poor agricultural practices and attempting to make marginal soils productive are major problems (◉ Figure 6.36). Some soils defy conversion to agricultural use because of their origin (◉ Figure 6.37), and some productive soils are damaged by soil inhabitants (◉ Figure 6.38). Other soils are "mined" for clays to make pottery or bricks (◉ Figure 6.39).

◉ FIGURE 6.36 (right) The "soil" of the Sahel of Africa is little more than dry, wind-blown sand. A region of severe drought and famine in recent years, the Sahel encompasses an east–west belt that separates the arid Sahara Desert from tropical Africa. It includes parts of Chad, Niger, Mali, Mauritania, Senegal, and Gambia.

◉ FIGURE 6.37 Remnants of a boulder fence in an area that was once cleared for agriculture; Attleboro, Massachusetts. Rocky glacial soils are difficult to farm, because freezing and thawing cycles continually lift boulders up into the soil root zone. This produces a new "crop" of rocks each spring in some areas. Boulder fences and walls are seen in fields and forests throughout the northeastern United States.

● FIGURE 6.38 (above) A soil-covered slope that was literally honeycombed by burrowing animals; England. Such burrows lead to heavy soil erosion and sediment loading of nearby streams.

● FIGURE 6.39 (left) Children in some developing nations do not have time to make mud pies or sand castles, because they are employed in brickyards.

 ## SUMMARY

Soil

DEFINED   Many definitions depending upon use (engineering, agriculture, geology, etc.). For our purposes, however, soil is defined as weathered rock and mineral grains that can support plant life.

ENVIRONMENTAL FACTORS OF SOIL FORMATION   Climate, organic activity, relief (topography), parent material, and time. Soils form from the weathering of regolith.

WEATHERING   Physical and chemical breakdown of bedrock and regolith.

1. *Physical weathering* is caused by temperature differences and disruption by plants and animals.
2. *Chemical weathering* processes include solution, oxidation, hydration, and hydrolysis. It is important in the formation of clays, the most important mineral group in soils. Clays are important because they hold moisture and exchange nutrients with plants.

RATE OF WEATHERING   Depends upon the rock or regolith's mineral composition and surface area. Smaller particles weather faster than large masses, and ferromagnesian minerals weather faster than quartz or feldspar.

FEATURES AND LANDFORMS   Talus, spheroidal weathering, and exfoliation domes are common features.

Soil Profile

DEFINED   Distinguishable layers, or horizons, in mature soils composed of the A horizon, or zone of leaching (where minerals are dissolved), and a lower B horizon known as the zone of accumulation, where salts and clays accumulate. Below this is the C horizon of weathered parent material. The thickness of the horizons and the development rate of soil are profoundly influenced by climate.

RESIDUAL SOIL   Developed in place on underlying bedrock. Laterite, for example, is the brick-red soil of humid (subtropical and tropical) regions that represents intensive weathering of parent rock.

TRANSPORTED SOIL   Developed on regolith transported and deposited by wind (eolian), glaciers, rivers, or volcanic action. Loess, for example, is an eolian soil associated with ice-age continental glaciation. It is highly productive and is found throughout the central U.S. "breadbasket."

Soil Classification

ZONAL   A climatic classification based upon rainfall and temperature. The four main zonal soils are pedalfers (rich brown soils of temperate regions), pedocals (alkaline soils of arid regions), laterites (tropical and subtropical soils), and tundra soils of polar regions.

U.S. COMPREHENSIVE SOIL CLASSIFICATION SYSTEM   Known as the *Seventh Approximation,* a classification of soils by their physical characteristics. Soil names convey information about the soils' characteristics; e.g., *aridisols* are alkaline soils of desert regions.

Soil Problems

EROSION   A major problem, because soil determines the earth's carrying capacity. Globally, irreparable soil degradation has impacted an area the size of China and India combined. U.S. soil losses vary greatly with area.

PROCESSES   Sheet erosion by running water (thin layer is removed) can become rill erosion (defined streamlets are carved in the soil), which leads to gullying (the most destructive). Sheet and rill erosion can be halted, but once gullying starts, it is extremely difficult and costly to mitigate. Wind erodes soil on marginal lands that were once cultivated and then dried out. Without plant root systems to hold the soil, wind erosion can lead to "dust bowl" conditions. Off-road vehicles also seriously degrade soils.

MITIGATION   Good agricultural practices such as terracing, no-till and minimum tillage, strip cropping, and crop rotation. Wind losses can be minimized by planting windbreaks, strip crops, and cover crops such as clover.

Other Soil Problems

EXPANSIVE SOILS   Certain clay minerals absorb water and expand, exerting high uplift pressures that can crack or even destroy structures. Mitigated by removal of soil, mixing, or keeping the moisture content constant.

PERMAFROST   Permanently frozen ground (soil or rock) underly more than 20 percent of the earth's surface. Structures built directly on soil permafrost will settle, so mitigation involves building with an open space below the structure so as not to disturb the thermal regime in the soil and melt the permafrost. Removal of permafrost zones is sometimes possible.

SETTLEMENT   Occurs when the applied load is greater than the bearing strength of the soil; e.g., Leaning Tower of Pisa. An engineering problem that can be mitigated by proper design or by strengthening of the underlying soil.

MISCELLANEOUS SOIL PROBLEMS   Salinization and waterlogging (high water table), hardening of laterites due to deforestation and desiccation, hardpans (cemented B horizon), and presence or absence of trace elements essential to human health.

CONTAMINATION   Brines, petroleum products, crude oil, and chemicals of all sorts may contaminate soils. Common methods of cleansing the soil are bioremediation (bacteria), phytoremediation (plants), and vapor extraction (oxidation).

 ## KEY TERMS

active layer	caliche
bioremediation	erosion

exfoliation dome
expansive soils
frost wedging
gullying (erosion)
hardpan
laterite
loam
loess
pedalfer
pedocal
pedologist
permafrost
permafrost table
phytoremediation

regolith
residual soil
rill erosion
sheet erosion
sheet joint
soil
soil horizon
soil profile
spheroidal weathering
talus
trace elements
tundra soils
vapor extraction
weathering

STUDY QUESTIONS

1. Sketch a soil profile and describe each soil horizon.
2. Distinguish chemical weathering, physical weathering, and erosion.
3. What kinds of soils would you expect to develop on basalt in Hawaii, New Jersey, and Arizona? Explain the differences in the light of the CLORPT equation.
4. What is permafrost, and where is it found?
5. What soil problem does permafrost present to engineering works, and what techniques are used to prevent damage from this problem?
6. What is expansive soil, and what problems and challenges does it pose to soils engineers? What can be done to prevent distress due to this condition?
7. Soil erosion and degradation are two of humankind's oldest problems. How has human-induced soil loss come about? What agricultural practices have led to salinization and water-logging of soils?
8. An inscribed rock obelisk dating from 1500 B.C. was resurrected from the ruins at Karnak, Egypt. Its inscriptions were crystal clear when it was brought to New York's Central Park as a gift of the Egyptian government in 1880, but today they are totally illegible. Explain the reason for this.
9. Laterite is one of the most interesting of the zonal soil groups. What special problems have laterites presented in deforested areas of Brazil and wartorn Vietnam?
10. Compare the rates of soil erosion and soil formation in the United States.

FURTHER INFORMATION

BOOKS AND PERIODICALS

Batie, Sandra S. 1983. *Soil erosion: A crisis in America's croplands?* Washington, D.C.: The Conservation Foundation.

Birkeland, Peter. 1984. *Soils and geomorphology.* New York: Oxford University Press.

Brady, Nyle C. 1990. *The nature and properties of soils.* New York: Macmillan Publishing Co.

Brown, K. W. 1994. New horizons in soil remediation. *Geotimes.* American Geological Institute, September.

Ferrians, O. J., and others. 1969. *Permafrost and related engineering problems,* U.S. Geological Survey professional paper 678.

Heiniger, Paolo. 1995. The leaning tower of Pisa. *Scientific American* 273, no. 6: 62–67.

Hunt, Charles B. 1972. *Geology of soils.* San Francisco: W. H. Freeman.

Rahn, Perry H. 1987. *Engineering geology: An environmental approach.* New York: Elsevier Publishing Co.

Tank, Ronald W. 1983. *Environmental geology: Text and readings.* New York: Oxford University Press.

World Resources Institute and United Nations Environment and Development Programs. 1992 *World resources, 1992–1993: A guide to the global environment.* New York: Oxford University Press.

LANDSLIDES AND MASS WASTING

Nature to be commanded, must be obeyed.

FRANCIS BACON, PHILOSOPHER (1561–1626)

Mass wasting is the general term that denotes any downslope movement of soil and rock under the direct influence of gravity. Mass-wasting processes, which include landslides, rapidly moving debris flows, slow-moving soil creep, and rockfalls of all kinds, are a significant geologic hazard throughout North America. As a hazard to humans, mass wasting is more likely to occur than are volcanic eruptions, earthquakes, and floods, but it is much less spectacular and thus less newsworthy. Annual damage from landsliding alone in the United States is estimated at between one and two billion dollars. If we include other ground failures, such as subsidence, expansive soils, and construction-induced slides and flows, total losses are many times greater than the annual combined losses from earthquakes, volcanic eruptions, floods, hurricanes, and tornadoes.

OPENING PHOTO

The February 1993 La Ventana (Spanish, "the window") landslide near San Clemente, California. The ocean bluff collapsed during heavy rains, destroying four homes. When the landslide began, the owners of the white house on the right sawed off their ocean-view rooms in an attempt to save part of their home.

Mass wasting is also a global environmental problem. Earthquake-triggered landslides in Kansu Province in China killed an estimated 200,000 people in 1920, and debris flows left 600 dead and destroyed 100,000 homes near Kobe, Japan, in 1938. The largest loss of life from a single landslide in U.S. history, 129 fatalities, occurred at Mameyes, Puerto Rico, in 1985.

The term **landslide** encompasses all moderately rapid falls, slides, and flows that have well-defined boundaries and that move downward and outward from a natural or artificial slope. They occur in all the states and are an economically significant factor in more than 25 of them (◉ Figure 7.1). Wherever unstable slopes exist, some form of mass wasting usually occurs. Areas with the greatest topographic relief are at highest risk, because gravity acts to smooth out topography by reducing the high areas through mass wasting and filling in the low areas with slide debris. For example, more than two million mappable slides exist in the Appalachian Mountains from New England to Alabama. Almost ten percent of the land area of Colorado is landslide terrane, and landslides and debris flows in Utah caused $300 million worth of damage in 1983–84. In Southern California, eight wet years between 1950 and 1993 *averaged* $500 million in landslide damage. Slope stability problems have such impact that a national landslide-loss-reduction program was implemented by the U.S. Geological Survey in the mid-1980s. A high priority of the program is to identify and map landslide-prone lands.

The classification of slope movements used in this book is similar to the one widely used by engineering geologists and soil engineers (◉ Figure 7.2). The bases of the classification are

- the type of material involved, such as rock or soil;
- how the material moves, such as by sliding, flowing, or falling;

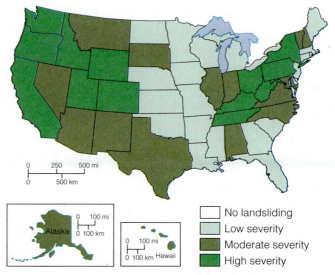

◉ **FIGURE 7.1 Severity of landsliding in the United States. Note that the most serious problems are found in the Appalachian Mountains, the Rocky Mountains, and the coastal mountain ranges of the Pacific Rim.**

No landsliding
Low severity
Moderate severity
High severity

- the moisture content of the material; and
- how fast the material moves, its velocity.

For instance, soil creep occurs at imperceptibly slow rates and is relatively dry; debris flows are water-saturated and move swiftly; and slides are coherent masses of rock or soil that move along one or more discrete failure surfaces, or **slide planes.** Failure occurs when the force that is pulling the slope downward (gravity) exceeds the strength of the earth materials that compose the slope. Because gravity acting on a given slope is essentially constant, either the earth materials' resistance to sliding must decrease or the gradient of the slope must increase in order for sliding to occur. In

◉ **FIGURE 7.2 Classification of landslides by mechanism, material, and velocity.**

MECHANISM		MATERIAL			Velocity
		Rock	Fine-grained Soil	Coarse-grained Soil	
SLIDE		Slump	Earth slump	Debris slump	Slow
		Block glide	Earth slide	Debris slide	Rapid
FLOW		Rock avalanche	Mudflow, avalanche	Debris flow, avalanche	Very Rapid
		Creep	Creep	Creep	Extremely slow
FALL		Rockfall	Earthfall	Debrisfall	Extremely rapid

this chapter we restrict our discussion to the mechanics of slides and flows, because rockfalls are discussed in Chapters 6 and 11.

FLOWS

Types of Flows

Creep is the slow (a few millimeters per year), essentially continuous downslope movement of soil and rock on steep slopes. It involves either freezing and thawing or alternate wetting and drying of a hill slope, which causes upward expansion of the ground surface perpendicular to the face of the slope. As the slope dries out or thaws, the soil surface drops vertically, resulting in a net downslope movement of the soil (◉ Figure 7.3). Burrowing animals and other biological processes that produce openings in the soil also contribute to soil creep. Bent trees, leaning fence posts and telephone poles, and bending of tilted rock layers

down-slope are all evidence of soil creep. Homes built with conventional foundations 18–24 inches deep may develop cracks due to soil creep. This is seldom catastrophic; typically this is a cosmetic and maintenance problem. The influence of soil creep can be overcome by placing foundations through the creeping soil into bedrock (Figure 7.3).

Debris flows are dense, fluid mixtures of rock, sand, mud, and water. They may be generated quickly during heavy rainfall or snowmelt where there is an abundant supply of loose soil and rock. Moving with the consistency of wet concrete, they are very destructive, with velocities up to many meters per second. Because of their high density, commonly 1.5–2.0 times the density of water, debris flows are capable of transporting large boulders, automobiles, and even houses in their mass. The house shown in ◉ Figure 7.4 is in Shields Canyon in the San Gabriel Mountains of Southern California. Thirteen cars were packed around the house, and several others were deposited in the backyard

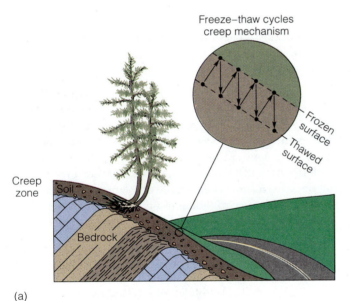

(a)

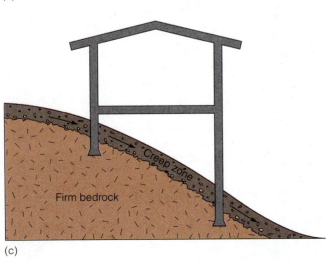

(c)

(b)

◉ **FIGURE 7.3** *(a)* Diagram and *(b)* photograph of soil and rock creep; east slope of the Sierra Nevada, California. *(c)* Securing a structure on pilings that extend through the creep zone into firm bedrock can prevent damage due to soil motion.

● FIGURE 7.4 Debris flow in 1977 Shields Canyon, San Gabriel Mountains, California 1977. Note that both the car and the man are on the roof of a house.

swimming pool. The family was rescued from inside the house, floating on mud that reached nearly to the ceiling.

The areas most subject to debris flows are characterized either by sparse vegetation and intense seasonal rainfall, or are in regions that are subject to drenching rains associated with hurricanes. Debris flows are relatively common in Canada, the Andes (particularly Peru), and alpine and desert environments worldwide. Debris flows associated with volcanic eruptions (lahars) are potential hazards in all volcanic zones of the world. They pose a particular danger in populated areas near volcanoes in Washington, Oregon, and California. The eruption of Mount St. Helens in Washington displaced water in Spirit Lake at its base and produced a lahar that filled the north fork of the Toutle River with the largest (2.8 km³; 0.67 mi³) and most destructive debris flow of modern times (look back to Figure 5.26). These types of debris flows are the most significant threat to life of all the landslide hazards.

A **debris avalanche** is simply a fast-moving (>4 m/sec) debris flow (● Figure 7.5). A structure in the path of one of these flows can be severely damaged or completely flattened with attendant injury or loss of life. During the winter of 1982, debris flows and avalanches in the San Francisco Bay area killed 25 people and resulted in $66 million worth of damage. In Mill Valley, a small town north of San Francisco, a debris flow ripped a house from its foundations and deposited it 45 meters (150 ft) downslope—without serious injury to the residents (passengers). The house effectively dammed the canyon and prevented debris-flow damage farther downslope.

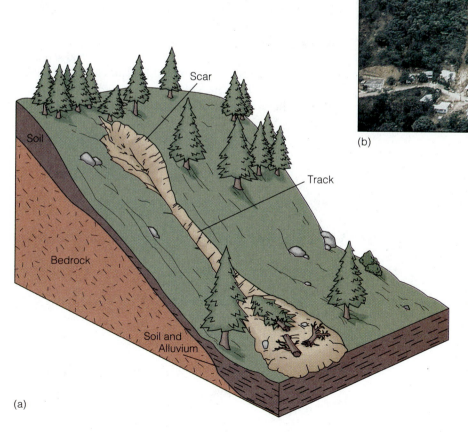

(a)

(b)

● FIGURE 7.5 *(a, left)* Debris avalanche track and zone of deposition. A debris avalanche may travel thousands of feet beyond the base of the slope. *(b, above)* Debris avalanche scars at Panuelos, Puerto Rico, following heavy rains of October 1985. One of these avalanches flattened the local Lions Club building, fortunately unoccupied at the time.

Causes of Debris Flows and Debris Avalanches

The combination of heavy rainfall and loose soil on steep slopes promotes debris flows and avalanches. Studies of debris flows indicate that two rainfall conditions are necessary to cause them: (1) an initial period of rainfall, known as *antecedent rainfall,* that saturates the soil and (2) a subsequent period of intense rainfall that puts even more water into the soil and initiates flowage. Thus the onset of debris flows may be estimated during the rainy season if we have data on the amount of rainfall needed to initiate flowage. Such data have been compiled for flows and slides in the San Francisco Bay area of northern California (⊙ Figure 7.6). They indicate that a short period of intense rainfall or less-intense rainfall over a longer period can produce debris flows. Specifically, the curves of Figure 7.6 show that for a rainfall intensity of 0.5 inch per hour, the threshold time for the onset of debris flows is 8 hours in Marin County and 14 hours in Contra Costa County. The differing rainfall thresholds in these relatively close areas are due to the variability of geologic materials and topography. The lower curve of Figure 7.6 shows that less-intense rainfall will produce debris flows in semiarid and arid areas of California. This is because these dry areas have little vegetation with root systems that retain soils and, therefore, have abundant loose surface debris. Although these curves are only preliminary, they will serve to alert residents in critical areas once the threshold conditions for flows have been attained, and they will be refined as more data become available.

Mountain slopes burned by range and forest fires are also susceptible to debris flows during the wet season. The loss of active root systems to bind soil particles can result in an extremely dangerous condition. In addition, debris flows from burned slopes have been found to have longer runout distances into foothill areas than those from vegetated slopes.

Not all debris flows and avalanches are triggered by intense rainfall. Slide Mountain, about halfway between Reno and Carson City, Nevada (and between Washoe Lake and Lake Tahoe), is not named for the ski area on its slopes; it is the site of many flows and slides. On 30 May 1983, a debris flow was generated on Slide Mountain that killed one person, injured many more, and destroyed a number of homes and vehicles, all in less than 15 minutes (⊙ Figure 7.7). About 720,000 cubic meters of weathered granite gave way from the mountain's steep flank and slid into Upper Price Lake. This mass displaced the water in the lake, causing it to overflow into a lower lake, which in turn overflowed into Ophir Creek gorge as a water flood. Picking up sediment as it went, it became a debris flow that emerged from the gorge, spread out, destroyed homes, and covered a major highway. Because this is a popular recreational area and debris flows can be sudden and hazardous to one's health, geologic hazard warning signs are posted on trails throughout the area. There is a history of debris flows in this part of Nevada, as documented by the 8 July 1890 edition of the *Carson Appeal:* ". . . On Sunday afternoon about quarter of 5 o'clock, Price's Reservoir at the foot of Slide Mountain burst, and the water, rushing down the canyon, submerged the V&T Railroad track at Franktown. . . ." The 1983 Slide Mountain debris flow and the one associated with an earthquake at Yungay, Peru, that killed 18,000 people (discussed in Chapter 4) were both initiated by landslides into water masses.

◉ LANDSLIDES

Types of Landslides

Two kinds of slides are recognized according to the shape of the slide surface. They are **slumps,** or **rotational slides,** which move on curved, concave-upward slide surfaces and are self-stabilizing, and **block glides,** or **translational slides,** which move on inclined slide planes (see Figure 7.2). A block glide moves like a glacier until it meets an obstacle or until the slope of the slide plane changes.

⊙ **FIGURE 7.6** Rainfall thresholds for debris avalanches and landslides for two central California counties and California semiarid regions in general. For example, a rainfall intensity of 0.5 inch/hour for 8 hours is sufficient to initiate flows in Marin County. The same intensity for 14 hours is the threshold for Contra Costa County.

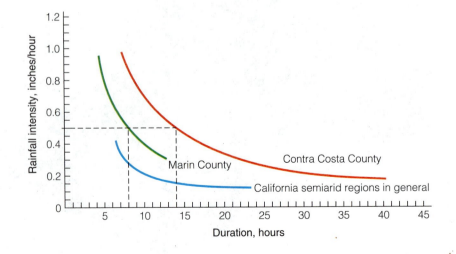

(a)

(b)

(c)

(d)

● **FIGURE 7.7 Slide Mountain in western Nevada, site of the 1983 disaster.** *(a)* Looking up Ophir Creek and the runout area of the debris flow. The scar of the debris-avalanche source area is the treeless area on the mountain. *(b)* Debris flow and a damaged home in lower Ophir Creek. *(c)* Abraded bark on tree serves to record the thickness of the flow. *(d)* A warning to be taken seriously.

SLUMPS. Slumps are the most common kind of landslide and range in size from small features a few meters wide to huge failures that can damage structures and transportation systems. They are spoon-shaped, having a slide surface that is curved concave-upward and exhibiting a backward rotation (● Figure 7.8). They occur in geologic material that is fairly homogeneous, such as soil or badly weathered or fractured bedrock, and the slide surface cuts across geologic boundaries. Slumps move about a center of rotation; that is, as the toe of the slide rotates upward, its mass eventually counterbalances the downward force, causing the slide to stop (Figure 7.8, part b). Slumping produces repeated uniform depressions and flat areas on otherwise sloping ground surfaces. Slumps grow headward (upslope). This is because as one slump mass forms, it removes support from the slope above it. Thus a stair-step surface is common in landslide terrain (Figure 7.8, part c). Slump-type landslides are the scourge of highway builders, and repairing or removing

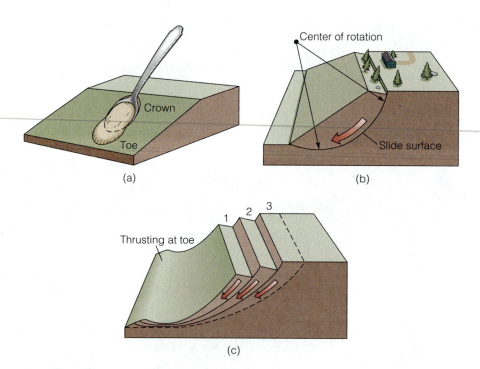

● FIGURE 7.8 A rotational landslide. *(a)* Stylized oblique view of its spoon shape, crown, and toe. *(b)* Cross section, showing the curved slide plane, the center of rotation, and backward-tilting trees at the top of the slide. *(c)* Headward growth of slump blocks results in stair-step topography at the crown.

Center of rotation

Crown

Toe

(a)

Slide surface

(b)

Thrusting at toe

1 2 3

(c)

them costs millions of dollars every year (Case Study 7.1). The Mussel Rock landslide on the San Francisco peninsula is a rotational slump block consisting of 8 million cubic meters (10 million yd³) of debris from the soft and highly fractured Merced Formation of Tertiary age (● Figure 7.9). The San Andreas fault runs through the slide mass and undoubtedly contributes to the instability here (Figure 7.9, part b). Several homes in Daly City have been lost due to headward growth of the scarp into the subdivision at the crown of the slide. The toe of the slide was a landfill from the 1960s until 1979 and is currently protected by a seawall.

BLOCK GLIDES. Block glides are coherent masses of rock or soil that move along relatively planar sliding surfaces (failure planes), which may be sedimentary bedding planes, metamorphic foliation planes, faults, or fracture surfaces.

(a)

● FIGURE 7.9 The Mussel Rock landslide at Daly City, California (looking eastward). A landfill was operated at the toe of the landslide until 1979. *(b)* Location of the landslide in relation to the San Andreas fault. The fault extends offshore here and reappears on land just north of the Golden Gate Bridge about 18 miles to the north.

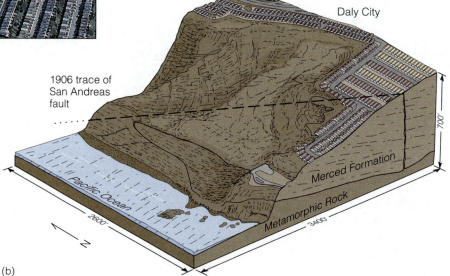

Daly City

1906 trace of San Andreas fault

Pacific Ocean

Merced Formation

Metamorphic Rock

700'

2600'

3400'

N

(b)

(a)

Potential slide planes

(b)

Potential slide (slip) plane

(c)

(d)

◉ **FIGURE 7.10 Cutting a natural slope for homesites leads to potential block glides on bedding or foliation planes.** *(a)* **Natural slope and proposed cut.** *(b)* **After cutting, the slope exhibits unstable conditions.** *(c)* **A slide plane in clay shale shortly after the overlying slide mass began moving; Santa Monica Mountains, California.** *(d)* **Striations gouged in wet clay by the landslide mass shown in** *(c)*.

For a block glide to occur, it is necessary that the failure plane be inclined *less steeply* than the inclination of the natural or manufactured hill slope. Slopes may be stable with respect to block glides until they are steepened by excavating for building subdivisions or roads, thus leading to landsliding (◉ Figure 7.10).

A classic block glide of about ten acres is seen along a sea cliff undercut by wave action at Point Fermin in San Pedro, California. Movement was first detected there in January 1929, and by 1930 the landslide had moved two meters seaward. It has been intermittently active ever since (◉ Figure 7.11, page 193). The rock of the slide mass is coarse sandstone (not the type of earth material usually involved in block glides), but a thin layer of **bentonite** dipping 15° seaward forms the slide plane. Bentonite is volcanic ash that has chemically weathered to clay minerals, which become

plastic and slippery when wet. Bentonite is very commonly involved in slope failures; addition of water is all that is needed to initiate a landslide where dips are as slight as 5°.

Some of these translational slides are large and move with devastating speed and tragic results. In 1985, a tropical storm dumped a near-record 24-hour rainfall that averaged almost 470 millimeters (18.5 in) on a mountainous region near the city of Ponce on the south coast of Puerto Rico. Rainfall intensities peaked in the early morning hours of October 7, reaching 70 millimeters (2.8 in) per hour in some places. At 3:30 that Monday morning, much of the Mameyes residential district of the city was destroyed by a rock block glide initiated during the most intense period of rainfall. This resulted in the worst loss of life from a landslide in U.S. history—129 deaths. The landslide was in a sandstone whose stratification parallels the natural slope of

Plate Tectonics and Highway Maintenance— Whose Fault Is It Now?

Highways along the west coast of North America, some of the most beautiful drives in the world, are plagued with slope stability problems. There are literally thousands of landslides along these roads from Baja California northward through California, Oregon, Washington, and to some extent, into the Canadian province of British Columbia. A major reason for this is that the coast is near the North American–Pacific plate boundary, and the high frequency of landslides results from a tectonically active coast with steep slopes and sea cliffs largely supported by faulted, shattered, and weathered rocks. The condition is compounded by the large storm waves that periodically batter away at the base of the cliffs. Now and then the coastal communities of Pacifica and Montara on the San Francisco peninsula, for example, become cul-de-sacs on scenic California Highway 1 when the Devil's Slide landslide activates (●Figure 1). In the southern part of the state, Malibu and other beach colonies are cut off almost annually by landslides onto the coast highway.

Devil's Slide is about 15 kilometers (9 mi) south of downtown San Francisco on the east side of Highway 1 (●Figure 2). Extending 730 meters (2,400 ft) parallel to the coast, it rises 260

meters (860 ft) above sea level and involves several slide planes. Many planar landslides controlled by joints in the sedimentary rocks (●Figure 3) have occurred within the larger, arcuate slide since the late 1800s, and a massive landslide occurred during construction of the highway in 1937. The roadway has dropped 14 meters (45 ft) below its original alignment in places. In the 1950s an inland bypass project was proposed, and the right of way for it was purchased. Meanwhile, the National Environmental Policy Act (NEPA) and the California Environmental Quality Act (CEQA) were enacted, which require extensive environmental impact studies, analyses of alternative mitigation measures, and public input on such projects. Litigation ensued, and eventually the bypass project was dropped for lack of financing. Lengthy road closures in the 1980s resulted in $50 million federal storm-damage funding, however, and the bypass route studies were resumed. In 1986 a two-lane 2.7-kilometer (1.67-mi) bypass was approved, but by then McNee State Park had been created on either side of the bypass alignment on Montara Mountain. The Sierra Club and other opponents initiated lawsuits to prohibit construction of the bypass through the park.

● FIGURE 1 The Devil's Slide on scenic but high-maintenance California Highway 1 periodically reactivates, stopping north- and south-bound traffic. The landslide occupies most of the left-central part of the photograph.

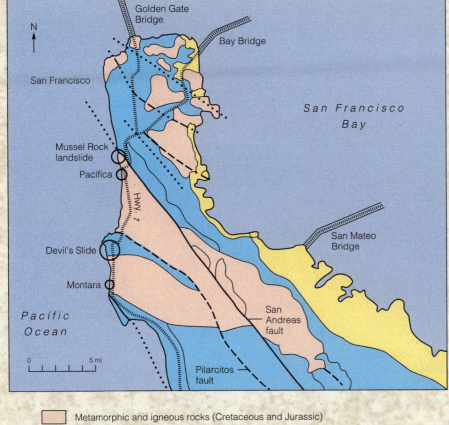

● FIGURE 2 Geologic map of the San Francisco peninsula showing the locations of the Devil's Slide and the nearby Mussel Rock landslide (see Figure 7.9). Note the close proximity of the San Andreas fault system (a transform plate boundary), the landslides, and the city of San Francisco.

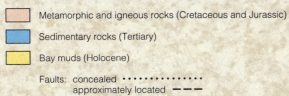

Metamorphic and igneous rocks (Cretaceous and Jurassic)

Sedimentary rocks (Tertiary)

Bay muds (Holocene)

Faults: concealed ················
         approximately located – – –
         accurately located ———

Spirited public debate continued into 1996. The proposed remediation plans included:

- in-place stabilization involving a minimum of regrading,
- reconstruction of the slide involving a large amount of regrading,
- a 4-kilometer (2½-mi) bypass—favored by the state, and
- a 1.6-kilometer (1-mi) tunnel—favored by environmentalist groups.

Geotechnical reconstruction, as proposed by plans 1 and 2, was determined to be infeasible because of the nature of the rock and the steepness of the slopes. This left rerouting as the only alternative. Although the closures of Highway 1 have negatively impacted the economies of nearby Pacifica, Montara, and Half Moon Bay, those communities fear the growth and other societal problems that the bypass might bring.

In early 1995 the spectacular La Conchita (Spanish, "the little shell") landslide occurred north of Ventura on the heavily used coast highway between Los Angeles and Santa Barbara. Well documented in the news media, it is best described as a very large slump with flowage at the toe. No lives were lost, but the landslide destroyed nine homes (● Figure 4). A water main broke a few days later, and a quagmire of sand and mud spread across the highway and forced its closure. "It moved like soft Jell-O," said one observer. The slide came after weeks of renewed

continued on next page

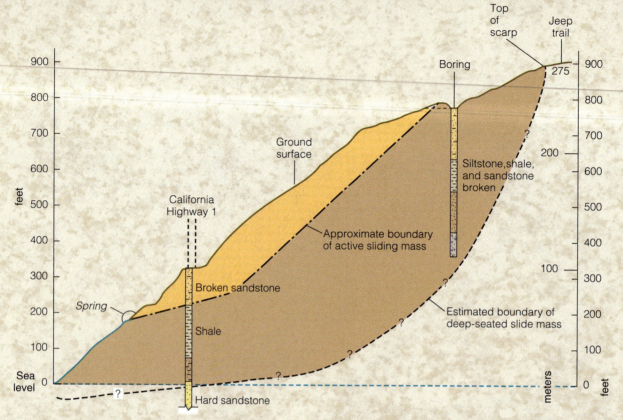

FIGURE 3 Cross section of the Devil's Slide. Shown are the postulated location of the main, deep-seated slide surface and the more surficial planar slide surfaces along joint planes. Jeep trail is one of the bypass options.

concern over widening gashes at the top of the slope, gashes that were first observed in 1988 and were the subject of Ventura County–authorized geological studies in the early 1990s. The slide occurred exactly where it had been predicted by the geologists who conducted the studies. The final scene will undoubtedly occur in some courtroom.

(a)

(b)

FIGURE 4 (a) La Conchita landslide on the Pacific Coast Highway (U.S. 101 and California 1) north of Ventura, California. The main part of the landslide is a slump, but the toe has mobilized and started to flow. The scarp at the top left of the slope indicates another area of instability. (b) Closeup of the slide toe. Note the damaged homes and the hazardous positions of several other homes.

● FIGURE 7.11 The Point Fermin landslide at San Pedro, California, is a rock block glide that has moved intermittently since 1929. Damaged homes have been removed from the slide area in the foreground. The slide plane emerges just in front of the houses in the background, which are on relatively stable ground.

the slide mass, a condition called a *dip slope*. It moved at least 50 meters (165 ft), probably on a clay layer in the sandstone, before breaking up into the large blocks that destroyed 100 homes (● Figure 7.12). The scarp at the top of the slide is 10 meters high (32 ft), and the maximum thickness observed at the toe of the landslide is 15 meters (49 ft).

The importance of water in initiating landslides and debris flows cannot be overemphasized. South-central Puerto Rico is the "dry" side of the island, where the average annual rainfall is about 1,000 millimeters (39 in). Ponce received almost half of that amount in the 24 hours preceding the landslide (● Figures 7.13 and ● 7.14). Note in Figure 7.14 that the cumulative rainfall at Ponce was more than 400 millimeters (15.6 in) in the 20-hour period from 8:00 A.M. October 6 until 4:00 A.M. the following day. Investiga-

tion by the U.S. Geological Survey showed that most of the flows and landslides occurred in the area enclosed by the 400-millimeter *isohyet* in Figure 7.13. Isohyets are map contours enclosing areas of equal rainfall during a set period of time—an hour, a day, or a year. The year is the most commonly used time period for engineering purposes, and size requirements for sewers, storm drains, and the like are typically determined by annual precipitation.

In addition to the heavy rainfall and adverse geologic conditions, two other factors probably contributed to the disaster: (1) Mameyes was a densely populated district with no sewer system; sewage was discharged directly into the ground. (2) A water main at the top of the slide was reported to have been leaking for some time. If this is true, the subsurface rocks there were already saturated with

(a)

(b)

● FIGURE 7.12 The 1985 Mameyes landslide near Ponce, Puerto Rico. *(a)* Disrupted surface and destroyed homes. *(b)* Two phases of movement are visible from the left side of the photograph to the right side.

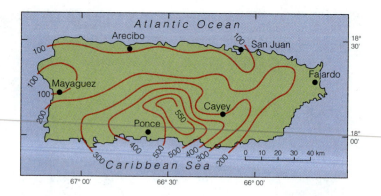

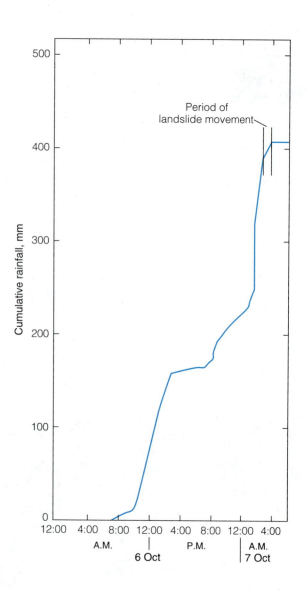

• FIGURE 7.14 Graph of cumulative rainfall during 28 hours 6–7 October 1985 and the landslide at Ponce. The steeper the curve, the more intense was the rainfall. About 400 millimeters (15.6 in) fell in the 18 hours preceding the landslide.

water, and the heavy rains of Tropical Storm Isabel were all that was needed to trigger the slide. The slide mass closely followed the boundaries of the Mameyes residential district, testimony to the impact of urban development on the landscape.

A similar landslide occurred in the Alps of northeastern Italy in 1963 at the site of Vaiont Dam, the highest thin-arch dam in the world at the time (275 m; 900 ft). The geology at the reservoir consists of a sedimentary-rock structure that is bowed downward into a concave-upward **syncline** with its axis parallel to the Vaiont River canyon. Limestones containing clay layers dip toward the river and reservoir from both sides of the canyon due to this down-folded structure (● Figure 7.15). Slope movements and slippage along clayey bedding planes above the reservoir had been observed before the dam was constructed. This condition gave engineers and geologists sufficient concern that they placed survey monuments on the slope above the dam for monitoring such movement. Heavy rains fell for two weeks before the disaster, and slope movements as large as 80 centimeters (31 in) per day were recorded. On the night of October 9, without warning, a huge mass of limestone slid into the reservoir so fast it generated a wave 100 meters (330 ft) high. The wave burst over the top of the dam and flowed into the Piave River valley, destroying villages in its path and leaving 2,500 people dead. The landslide velocity

CONSIDER THIS...

You must choose between two hillside homesites. One site is underlain by schist, the other by gneiss, both with foliation dipping out of the hill ("daylighting"). Which site would you choose, and why? (Hint: Examine Figure 7.10, parts c and d, and refer to the discussion of sedimentary and metamorphic rocks in Chapter 2).

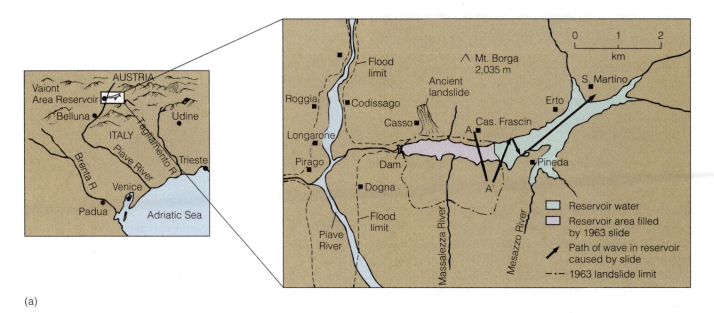

(a)

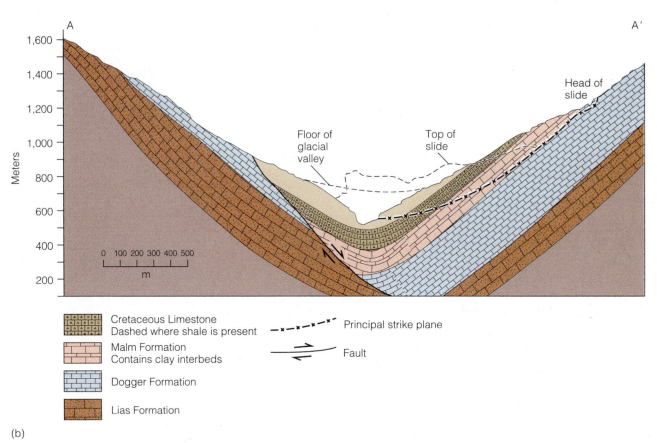

Cretaceous Limestone
Dashed where shale is present

Malm Formation
Contains clay interbeds

Dogger Formation

Lias Formation

Principal strike plane

Fault

(b)

⊙ FIGURE 7.15 The 1963 Vaiont Dam landslide. *(a)* Map of the slide mass that hurtled into the reservoir and the area that was impacted by the giant wave and flooding. Longarone and several other villages along the Piave River below the dam were devastated. *(b)* Cross-section through the Vaiont River valley showing a syncline with the sedimentary layers dipping toward the valley axis. The principal slide plane and resultant slide mass are indicated. The cross-section line is shown in *(a)*.

The geologists' report on a building site that you are considering buying states that the sedimentary strata—shale and sandstone—have a "bad attitude." Does this mean they are not nice? Is there some geological interpretation of this term?

into the reservoir was so speedy that the slide mass almost emptied the lake. The dam did not fail and it is still standing today—a monument to excellent engineering but poor site selection.

Slump and block glides are the most common types of landslides, but there are also **complex landslides,** combinations of the two types. These landslides have elements of both rotational and translational failure, and there are many examples, most of them large landslides such as the Slumgullion landslide in Colorado (Case Study 7.2, page 198).

Selected significant historic landslides are listed chronologically in ▪ Table 7.1. All of these slides were considered major disasters, and some of them were triggered by earth-

quakes, which allowed little if any advance warning. Most of them moved with high velocity, and many were accompanied by shock waves or by walls of water from displaced lakes or rivers.

◉ Figure 7.16 portrays the kinds of failures that might be experienced in regions of abundant rainfall, steep slopes, and weak rock or soil. The diagram illustrates the principal slide and flow types: slump, block glide, rock avalanche, soil creep, and rockfall.

The Mechanics of Slides

Landslides do not just happen. They are explainable as a change in the balance between **driving forces,** the gravity forces that pull a slope downward, and **resisting forces,** the cohesive and frictional forces that hold a slope in place. A change in the balance can result from water seepage into slope material, an oversteepening of the slope by natural erosion or artificial cutting, or the addition of weight at the top of a slope. If any of these things happen, the driving forces may eventually exceed the resisting forces, which will cause a landslide to occur.

This can be demonstrated using a simple model of a sliding block on an inclined plane (◉ Figure 7.17). The driving force, d, is the component of gravity acting parallel to the inclined plane at an angle α with the horizontal. The coef-

▪ TABLE 7.1 Significant Historic Landslides

YEAR	LOCATION	TYPE	DEATHS
1512	Biasca, Switzerland	Landslide dam broke	>600
1556	Hsian, China	Quake-triggered landslides	~1,000,000
1806	Goldau, Switzerland	Rock glide	457
1843	Mt. Ida, Troy, N.Y.	Slump and flow	15
1881	Elm, Switzerland	Rockfall	115
1903	Frank, Alberta, Canada	Rock glide	70
1920	Kansu Province, China	Quake-triggered landslides, caves collapse	~200,000
1938	Kobe, Japan	Debris flows	600
1959	Hebgen Dam, Montana	Quake-triggered landslide	~26
1962	Mt. Huascarán, Peru	Ice avalanche, debris flow	~4,000
1963	Vaiont Dam, Italy	Landslide into reservoir, flood	~2,000
1964	Anchorage, Alaska	Quake-triggered quick-clay landslide	114★
1966	Rio de Janeiro, Brazil	Landslides	279
1966	Aberfans, Wales	Mine-dump collapse, debris flow	144
1970	Mt. Huascarán, Peru	Avalanche, debris flow after quake	25,000
1972	Buffalo Creek, West Virginia	Mine-dump collapse, debris flow	400
1982	San Francisco Bay area	Debris avalanches and flows	25
1985	Mameyes, Puerto Rico★★	Hurricane-triggered debris flows	129
1992	Mt. Pinatubo, Philippines	Typhoon-triggered lahars	~350

★Combined toll from quake and slide
★★Largest in U.S. history
SOURCES: Various.

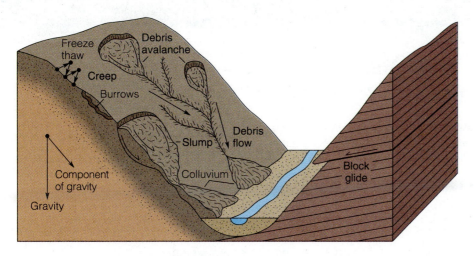

ficient of friction between the block and the plane is *f*. Static friction between the block and slide plane increases as the normal force, *n,* acting across the slide plane increases. Friction between the block and the plane resists sliding. Thus,

resisting forces = friction coefficient × normal force

or

$$r = f \times n$$

A delicate balance exists, and sliding is imminent when

$$d(\text{driving force}) = r(\text{resisting force})$$

that is, when

$$r/d = 1.0$$

The Factor of Safety (F. S.) relates resisting force to driving force. When the two forces are equal, the F. S. equals 1.0. When resisting forces are greater than driving forces, the F. S. is greater than 1.0; and when resisting forces are less than driving forces, the F. S. is less than 1.0. Most modern building codes require that manufactured or natural slopes have a Factor of Safety of 1.5 or greater. In practice, one simply sums up all the driving forces and resisting forces acting along a potential slide plane. From this it can be calculated whether the F. S. of the particular slope is greater than 1.0 (relatively stable) or less than 1.0 (relatively unstable).

If water pressure builds up in the pore spaces between sediment grains, it pushes them apart and reduces the soil's effective grain-to-grain friction—referred to as its **effective**

stress. This is analogous to injecting water under pressure beneath the block on an inclined plane. The water pressure will support a portion of the block's weight, thereby reducing the normal force *n* and the sliding resistance of the block. Water in soil or sediment pore space supports part of the grain-to-grain pressure, thereby reducing the frictional resistance across the grain contacts. The effective stress increases as the normal force increases and decreases as the pore-water pressure increases. A "quick" condition exists when the pore pressure is equal to intergranular friction; in other words, when the net effective stress is reduced to zero. At this point, sand becomes a dense sand-water mixture known as *quicksand*. It is often said that water "lubricates" a failure plane, making it easier to slip. It has been demonstrated, however, that a certain amount of moisture actually increases the strength of some materials. Moist sand, for example, builds a better castle and stands on a steeper slope than does dry sand. Water itself has no lubricating qualities. Rather, it is pore-water pressure that reduces grain-to-grain friction and therefore soil strength.

Lateral Spreading

Horizontal movement of a mass of soil overlying a liquefied or plastic layer characterizes a form of mass wasting called **lateral spreading.** Such slides are complex, as they involve elements of translation, rotation, and flow. Typically triggered by earthquakes, lateral spreading may result in the spontaneous liquefaction of water-saturated sand layers or in the collapse of **sensitive clays**—also known as **quick clays** (see Chapter 4). Spreading failures in the United States occur mostly in sand layers; however, areas underlain by glacial

Friction plane

α = slope angle
f = friction
n = normal force
d = driving force
W = weight of block = mass · gravity

● FIGURE 7.17 Resolution of driving forces *(d)* and resisting forces *(f × n)* acting on a slide plane inclined at an angle α.

Colorado's Slumgullion Landslide—A Moving Story 300-Years Old

The Rocky Mountain states, and particularly Colorado, have some of the nation's highest levels of landslide hazard. Each year landslides in the Rockies take several lives and cause millions of dollars in damage to forests, roads, pipe and electrical-transmission lines, and buildings. What may be the largest active landslide in the United States is between Gunnison and Durango, Colorado, and bears the unusual name *Slumgullion*, a name that was probably bestowed upon it by early prospectors (⊙ Figure 1).

> **Slumgullion** (slum-gul´yən), *n*. **1** *Slang.* a meat stew with vegetables, as potatoes and onion. **2.** *Mining.* a muddy red residue in the sluice. *(Webster's New International Dictionary)*

Slumgullion, the stew, was a standard of mining-camp cooks, and slumgullion, the red residue, was what miners usually found instead of gold in the sluice box.

The Slumgullion landslide originated 700 years ago when highly weathered and altered Tertiary volcanic rocks gave way on a ridge above the Lake Fork tributary of the Gunnison River. It is a few kilometers upstream from Lake City, a historic mining town in the San Juan Mountains. The landslide dammed Lake Fork River, forming the largest natural lake in Colorado, Lake San Cristobal. The landslide is huge—6 kilometers (3.6 mi) long, 1 kilometer (0.6 mi) wide, and an average of 40 meters

(132 ft) thick. As ⊙ Figure 2 shows, the Slumgullion is a landslide within a landslide; its 300-year-old active portion is flowing within the larger, 700-year-old mass. The total drop from headscarp to toe is about 762 meters (2,500 ft), and the slide moves at about 6 meters (20 ft) per year. The Slumgullion exhibits three distinct regions of deformation (⊙ Figure 3):

- The top of the slide is in tension (extension). Normal faulting dominates this region.
- The middle region flows as a rigid block, called *plug flow*. Nearly vertical strike-slip faults are found along plug's lateral edges.
- The toe is a region of compression. It exhibits thrust (reverse) faults.

Within the active landslide the terrane is jumbled, and trees are either highly tilted or dead.

The landslide poses several geologic hazards. If it overruns Colorado Highway 149, it will isolate recreational development upstream. Downstream flooding will result if the lake level rises and overtops the natural dam. For these reasons, the U.S. Geological Survey is conducting multidisciplinary studies of the area. In addition to 30 years' worth of observation data, the USGS is evaluating:

- precise measurements of the landslide's rate of movement using conventional survey techniques as well as Global Positioning System (GPS) technology.
- old photographs of the area. These enable the researchers to estimate the slide's rate of headward

⊙ **FIGURE 1** The Slumgullion landslide in southwestern Colorado has a tremendous length and a sinewy path. The head scarp is more than 500 meters (1,600 ft) high.

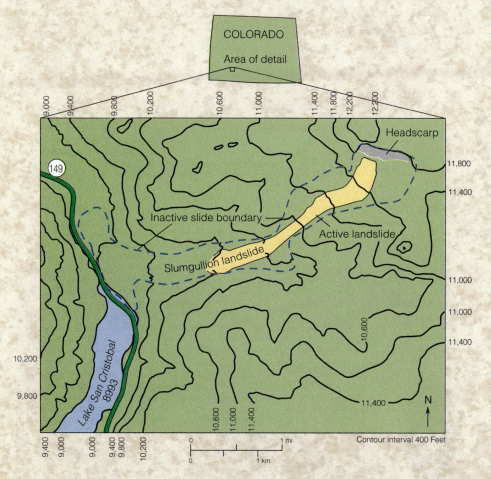

COLORADO

Area of detail

Headscarp

11,800

11,400

Inactive slide boundary

Active landslide

Slumgullion landslide

11,000

11,000

11,400

N

Contour interval 400 Feet

Lake San Cristobal 8993

149

0 1 mi
0 1 km

● FIGURE 2 Generalized contour map showing the 300-year-old active landslide within an inactive older slide mass. Lake San Cristobal formed when the landslide dammed the Lake Fork River.

growth. Because a slide's retrogression feeds rock debris and mass to the head of the slide, it serves to stimulate movement and growth of the slide.

■ detailed studies of Lake San Cristobal and its landslide dam. These will provide information that is applicable to other landslide dams in Colorado and the Rocky Mountains.

■ three-dimensional models of the active and inactive portions of the slide based upon the accumulated data. These facilitate prediction of the Slumgullion's future behavior.

It is hoped that the results of this study will be applicable to large landslides in other states.

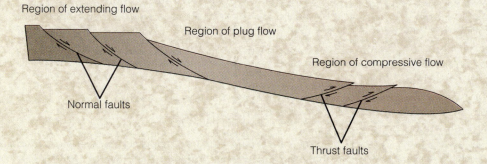

Region of extending flow

Region of plug flow

Region of compressive flow

Normal faults

Thrust faults

● FIGURE 3 Flow regions within the landslide. Extension (pull apart) at the top of the landslide leads to slumping there. The coherent plug flow in the middle of the landslide moves with lateral slip at the sides of the plug. Thrusting takes place at the toe of the landslide as the coherent plug attempts to override the shattered but stable toe.

● FIGURE 7.18 The Turnagain Heights landslide following the 1964 magnitude-8.6 Alaska earthquake. *(a)* Simplified geology of the slide. The earthquake caused failure of the sensitive (quick) Bootlegger Cove Clay and liquefaction of the sand and silt layers. *(b)* Turnagain Heights after the landslide.

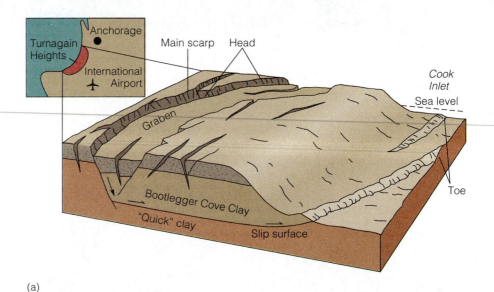

(a)

(b)

sediments may spread because of quick clay layers within the deposits.

Much of the damage in the Marina district of San Francisco in the 1906 and 1989 earthquakes was due to lateral spreading caused by liquefaction (see Figure 4.22). Most of the damage in Anchorage, Alaska, during the 1964 quake was due to quick-clay-induced lateral spreads in the Turnagain Heights residential neighborhood and the downtown area. The spreading was so extensive that two houses that had been more than 200 meters (more than 2 football fields) apart collided within the slide mass. The Turnagain Heights area is now a tourist attraction known as "Earthquake Park" (● Figure 7.18).

Probably no lateral spread in history has received more attention in geology and soils textbooks than the one at Nicolet, Quebec, Canada, in 1958. Here, water-saturated quick clays collapsed because of vibrations generated by artillery fire at a nearby military installation. The result was that a large section of the town and a historic monastery slid into the Nicolet River (● Figure 7.19). It was discovered later that the old monastery's wooden sewer system had leaked water into the subsurface glacial-marine clays, causing them to liquefy.

FACTORS THAT LEAD TO LANDSLIDES

Seldom can a landslide be attributed to a single cause; rather, landslides result from series of events that lead to failure. Nonetheless, the weakening of slope materials due

to the addition of water is the most important causative factor of all slides and flows. Thus heavy rainfall, rapid snowmelt, leaking water mains, private sewage-disposal (cesspool) inflow, and poor building-pad drainage can all lead to landsliding. Excess water causes a buildup of pore-water pressure, which weakens the materials supporting the slope. This is why mass wasting is closely correlated with a series of heavy-rain years in semiarid Southern California and with torrential flooding in the eastern United States.

Recall that when driving forces exceed resisting forces, a failure is imminent. Factors that increase driving force are

- an increase in the slope angle,
- removal of lateral support at the toe of a slope, and
- added weight at the top of a slope.

The effect of increasing slope angle is best illustrated by the natural **angle of repose** of granular material such as dry sand or gravel. The angle of repose of such material is the maximum slope angle at which it can remain stable. For example, no matter how steeply you try to pile dry sand, it will always form a slope of about 32°–34°, its angle of repose (◉ Figure 7.20). Moist sand has a greater angle of repose because of the temporary cohesion (added strength) imparted to it by moisture. This is why a castle built of moist sand collapses when the sand dries and loses its temporary cohesion. The angle of repose represents the point at which the frictional resistance between sand grains (shear strength) and the downslope component of gravity are in balance.

Theoretically, mass movements will not occur as long as the angle of repose of the particular slope material is not exceeded. Because undercutting the toe of a slope is equivalent to increasing the slope angle, it causes instability. Adding weight to the slope, as with fill material or an unusually heavy structure, increases the pore pressure and

can thus induce failure. Factors that reduce a slope's strength and, therefore, its resistance to sliding are

- infiltration of water underground,
- weathering and breakdown of minerals, especially when clays are formed, and
- burrowing by animals.

In addition, it should be noted that sliding is also facilitated when a slope is cut in such a way that its bedding planes are exposed or inclined (dipping) out of the slope face.

A slope with a Factor of Safety less than 1 is ready to slide, but it might not do so unless it is "triggered" by an earthquake, heavy traffic, sonic boom, detonation of explosives, or another source of energy. Such triggers are split-second stresses that overcome *static friction,* the force resisting sliding between two surfaces at rest. Static friction is always greater than *sliding (kinetic) friction,* the friction existing between two surfaces in relative motion. This explains why landslides are very hard to stop once they start.

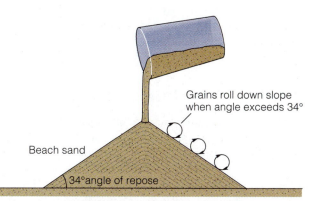

Grains roll down slope when angle exceeds 34°

Beach sand

34° angle of repose

◉ FIGURE 7.20 The angle of repose for dry sand is 34°. Coarser material, such as that found on a talus slope, will stand slightly steeper, but it also has a specific angle of repose.

REDUCING LOSSES FROM MASS WASTING

Landslide Hazard Zonation

The first step in reducing losses due to mass-wasting over the long term is to make a careful inventory or map of known landslide areas. Ancient landslides have a distinct topography characterized by hummocky (bumpy) terrain with knobs and closed depressions formed by backward-rotated slump blocks. On topographic maps, slide terrain can be evidenced by curved, closely spaced, amphitheater-shaped contour lines indicating scarps at the heads of landslides. In the field, one may see scarps, trees, or structures that are tilted uphill; unexpected anomalous flat areas; and water-loving vegetation such as cattails where water issues at the toe of a landslide.

With an adequate data base provided by maps and field observations, it is possible to define landslide hazard zones. Maps can be generated that show the *landslide risk* in an area, and then these areas can be avoided or work can be done to stabilize them (●Figure 7.21). Japan has been a leader in landslide mapping, and Russia, France, Sweden, and the Czech and Slovak Republics all have extensive mapping agendas. Mapping programs do not prevent slides, but they offer a means for minimizing slides' impact on humans. The human tragedy of landslides in urban areas is that they cause the loss of land as well as of homes. In contrast, after an earthquake, for example, a damaged structure can be repaired or replaced on the same site.

Building Codes and Regulations

Building codes dictate which site investigations must be performed by geologists and engineers and the way structures must be built. Chapter 70 of the *Uniform Building Code,* which deals with slopes and alteration of the landscape, has been widely adopted by cities and counties in the United States. It specifies compaction and surface-drainage requirements and the relationships between such planar elements as bedding, foliation, faults, and slope orientation. Code requirements for manufactured slopes in Los Angeles resemble the specifications found in codes elsewhere (●Figure 7.22). The maxi-

mum slope angle allowed by code in landslide-prone country is 2:1; that is, 2 feet horizontally for each foot vertically (27°). The rationale for the 2:1 (27°) slope standard is that granular earth materials have a natural angle of repose of about 34°; 2:1 slopes allow for a factor of safety. It is possible, however, to obtain a variance that allows a steeper slope to be manufactured in coherent, unfractured rocks such as sandstone or granite if engineering calculations justify such a slope.

Losses have been cut dramatically by enforcement of grading codes and they will continue to decrease as codes become more restrictive in hillside areas. For instance, before grading codes were established in the City of Los Angeles in 1952, 1,040 building sites were damaged or lost by slope failures out of each 10,000 constructed—a loss rate of 10.4 percent. With a new code in effect between 1952 and 1962 that required minimal geologic and soil-engineering investigation, losses were reduced to 1.3 percent for new construction. In 1963, the city enacted a revised code requiring extensive geological and soils investigations. Losses were

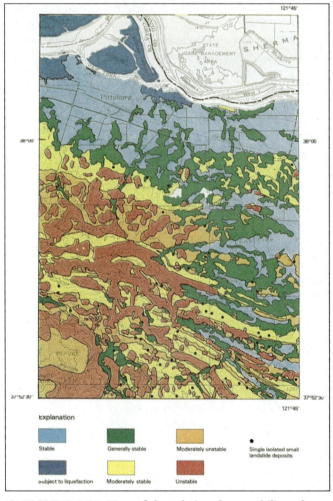

● FIGURE 7.21 Map of the relative slope stability of parts of Contra Costa County and adjacent counties derived by combining a slope map, a landslide-deposit map, and a map of susceptible geologic units. The area shown here is 15 kilometers (9 mi) wide.

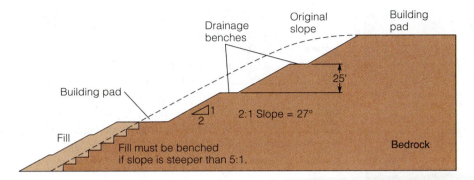

Drainage benches

Original slope

Building pad

Building pad

25'

2:1 Slope = 27°

Fill

1
2

Fill must be benched
if slope is steeper than 5:1.

Bedrock

● FIGURE 7.22 Part of a typical grading ordinance (code) that shows the limitation of cut-slope steepness, the required benches on slopes for collecting rainfall and reducing slope-face erosion, and an approved method of constructing a fill slope with bedrock benches for increased stability.

reduced to 0.15 percent of new construction in the following six years, which included 1969, the largest rain year in recent history.

 ## CONTROL AND STABILIZATION

The basic principles behind all methods of preventing or correcting landslides relate to strengthening the earth materials (increasing the resisting forces) and reducing the stresses within the system (decreasing the driving forces). For a given slide mass, this requires halting or reversing the factors that promote instability, which will probably be accomplished by one or more of the following: draining water from the slide area, excavating and redistributing the slide mass, and installing retaining devices. Debris flows require special devices that divert the path of the flow away from structures.

Water Drainage and Control

Water is the major culprit in land instability, and surface water must be prevented from infiltrating the potential slide mass. Water within the slide mass can be removed by drilling horizontal drains, called *hydrauger* (water bore) holes, and lining them with perforated plastic pipe (● Figure 7.23). This is commonly done on hillside terrain where underground water presents a problem. In some areas the land surface of a building site above a suspect slope is sealed with compacted fill to help keep surface water from percolating underground. Also, plastic sheeting is widely used to cover cracks and prevent water infiltration and erosion of the slide mass (● Figure 7.24). The efficacy of drainage control and planting is shown in ● Figure 7.25. The unimproved slope is raveling badly and will undoubtedly be an expensive maintenance headache for the owner. The terraced and planted slope, protected from the elements, is performing well.

Excavation and Redistribution

Recontouring is a textbook method of stabilizing a slide mass. By this method, material is removed from the top of the landslide and placed at the toe. Compacted-earth struc-

● FIGURE 7.23 Hydrauger drains drilled into serpentine. Perforated plastic drain pipes are connected to a large "manifold" that directs water drained from the cut slope into a natural drainage; San Luis Obispo, California.

● FIGURE 7.24 Plastic sheeting placed over the head scarp of a landslide prevents rainwater infiltration. Although this is common practice in landslide-prone areas, plastic sheeting is virtually unavailable in "home" stores during the rainy season in many communities.

● FIGURE 7.25 Erosion prevention measures on a graded slope; San Clemente, California. Whereas slope planting and terracing have maintained the middle slope in good condition, the unimproved slope on the left has raveled badly and will need maintenance in the future.

tures called *buttress fills* are often designed to retain large active landslides and known inactive ones believed to be vulnerable. They are constructed by removing the toe of a slide and replacing it, layer by layer, with compacted soil. Such fills can "buttress" huge slide masses, thus reclaiming otherwise unusable land. The finished product is required to have concrete terraces for intercepting rainwater and preventing erosion and downdrains for conveying the intercepted water

off the site (● Figure 7.26). Buttressing unstable land has aesthetic and economic advantages, in that it can be landscaped and built upon. The cost of the additional work needed to stabilize unstable land is usually offset by a relatively low purchase price, which can make it economically "buildable." Of course, the ultimate mitigating measure is total removal of the slide mass and reshaping of the land to buildable contours. When this is done, little evidence of the former instability remains. It is cost-effective only for small landslides, however.

Retaining Devices

Many slopes are oversteepened by cutting back at the toe, usually for the purpose of obtaining more flat building area. The vertical cut can be supported by constructing steel-reinforced concrete-block retaining walls with drain (weep) holes for alleviating water-pressure buildup behind the structure (● Figure 7.27). Retaining devices are constructed of a variety of materials, including rock, timber, metal, wire-mesh fencing, and a sprayed concrete known as *shotcrete* (● Figure 7.28).

Steep or vertical rock slopes that are jointed or very seamy are often strengthened by inserting long rock bolts into holes drilled perpendicular to planes of weakness. This binds the planes together much as a beam is bound. Rock bolts are used extensively to support tunnel and mine openings. They add considerably to the safety of these operations by preventing sudden rock "popouts." They are also installed on steep roadcuts to prevent rockfalls onto highways (● Figure 7.29). Retaining devices also are effective for mitigating rockfalls and debrisfalls.

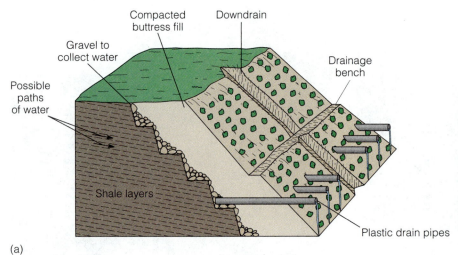

(a)

(b)

● FIGURE 7.26 *(a)* Buttress fill with horizontal drains for removing water and surface drains for preventing erosion. The fill acts as a retaining wall to hold up unstable slopes such as those that result from excavation. *(b)* A slope with surface drains built according to accepted standards.

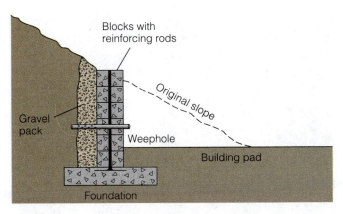

● FIGURE 7.27 Typical retaining wall with drain (weep) holes to prevent water retention and the buildup of hydrostatic pressure behind the wall. The wall is constructed of concrete blocks with steel reinforcing rods and concrete in their hollow centers.

● FIGURE 7.28 "Shotcrete" stabilization of a roadcut in fractured igneous rock; Kobe, Japan.

Diversion Techniques

Debris flows present a different challenge, because they can originate some distance away from the site of interest. It is simply not good practice to build at the bottom or the mouth of a steep ravine or gully, especially if loose soils are present on the higher slopes. This is particularly true in areas of high mean annual rainfall, areas that are subject to sudden cloudbursts, and areas that have been burned, such

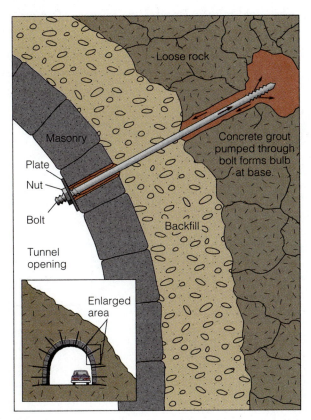

(a)

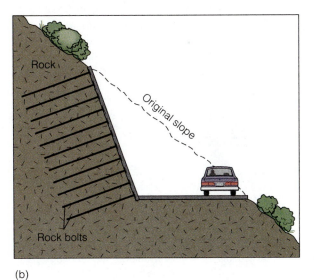

(b)

● FIGURE 7.29 (a) Typical rock bolt installation in a tunnel or mine opening. (b) Rock bolts support jointed rocks above a highway.

(a)

(b)

⊙ FIGURE 7.30 *(a)* The outlet of a debris-flow diversion structure; San Fernando Valley, California. *(b)* One effect of the May 1968 Heath Canyon mudflow caused by rapid snowmelt; Wrightwood, California. A diversion wall or dike might have saved this mountain cabin in the San Gabriel Mountains. Heath Canyon is now leveed to contain debris flows within the river channel.

as in the tragic October 1993 firestorms in Southern California. As long as people continue to build houses in the "barrels" of canyons, as is done in the foothills of Southern California, debris flows will continue to take their toll. Debris flow insurance is expensive—about $1,500 per $100,000 of house value per year in flood-prone areas in 1993—but the cost can be less if *deflecting walls* are designed into the house plans (⊙ Figure 7.30). These walls have proven to be effective in protecting dwellings, but they could have adverse consequences if they divert debris onto neighbors' structures, in which case neither an architect nor a geologist would be as helpful as a lawyer. ■ Table 7.2 summarizes mitigating measures that can slow or stop landsliding and other mass-wasting processes.

The Portuguese Bend Remediation Project

Portuguese Bend is an active complex landslide on the Palos Verdes Peninsula 25 miles south of Los Angeles (⊙ Figure 7.31). The slide is a block of mostly sedimentary

■ TABLE 7.2 Summary of Mass-Wasting Processes and Their Mitigation

LANDSLIDES

Cause	Effect	Mitigation
Excess water	Decreased friction (effective stress)	Horizontal drains, surface sealing
Added weight at top	Increased driving force	Buttress fill, retaining walls, decrease slope angle
Undercut toe of slope	Increased slope angle and driving force	Retaining walls, buttress fill
"Daylighted" bedding	Exposure of unsupported bedding in cut or natural slope	Buttress fill, retaining walls, decrease slope angle

OTHER MASS-WASTING PROCESSES

Process	Mitigation
Rockfall	Rock bolts and wire mesh on the slope, concrete or wooden cribbing at bottom of the slope, cover with "shotcrete"
Debris flow	Diversion walls or fences
Soil creep	Deep foundations
Lateral spreading	Dewater, buttress, retain (difficult to mitigate)

● FIGURE 7.31 The Portuguese Bend landslide with the Los Angeles basin in the background. The two points of land are resistant basalt intrusions, which have remained fixed as the landslide has moved around and between them.

rocks moving on a layer of bentonite that emerges at the base of a 30-meter-high (100-ft) sea cliff (● Figure 7.32). Bentonite is capable of absorbing huge amounts of water, whereupon it loses strength and becomes plastic. Several hundred homes were built in the area in the 1940s and 50s, all with septic tanks (cesspools), which released a huge volume of water into the underlying rocks. In addition, several thousand tons of rock fill were deposited on the head of the ancient slide for a thoroughfare (Crenshaw Blvd.) extension, thereby increasing the driving forces. The landslide reactivated in 1956. It involved 65 hectares (250 acres) and eventually destroyed 160 homes. The cause of this slide reactivation is labeled "what not to do" in every introductory geology course; that is, do not introduce water to a landslide mass or add weight at the top of it. Many of the "textbook fixes" discussed in this section have been tried at Portuguese Bend, beginning in 1956 with the placing of 20-foot-long, 4-foot-diameter concrete "pins" across the bentonite slide plane. The pins slowed the movement for a few months, but eventually they overturned (they did not break) and were carried in the slide mass to be ejected 30

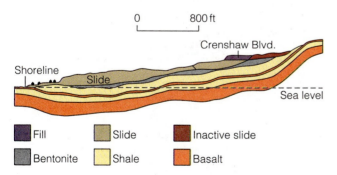

● FIGURE 7.32 Cross section through the long axis of the Portuguese Bend landslide. Miocene volcanic and sedimentary rocks rest on a volcanic tuff (bentonite) slide plane.

● FIGURE 7.33 Six-meter-long (20-ft) reinforced concrete columns were installed in large drill holes to "pin" the sliding rock mass above to the stable rock below. The columns tilted to a near-horizontal position and were carried with the sliding mass, arriving at the base of the sea cliff (background) 30 years later. They had traveled a distance of 700 feet.

years later at the sea cliff (● Figure 7.33). Regrading—that is, removing material from the head (top) of the slide (the driving force) and redistributing it at the toe and mid-slide (the resisting force)—reduced the movement rate by almost half (● Figure 7.34). Dewatering with drilled wells also reduced slide movement (● Figure 7.35). In addition, rock-filled wire-mesh baskets known as *gabions* were placed at the base of the sea cliff to prevent wave erosion from further removing the resisting mass there. The success of this project has been such that by 1993 some parts of the landslide had been stabilized, and by 1996 discussions about converting the stabilized areas to a golf course or other active-recreation facility were underway.

◉ SNOW AVALANCHES

The mechanics of snow avalanches are similar to those of landsliding, the differences being in the material and the velocity. Although snow avalanches have been around as long as there have been mountains and snow, the relatively recent emergence of the recreational skiing industry as big business has transformed avalanche control from an art to a science. Unfortunately, it's too late for Hannibal. His crossing of the Alps in 218 B.C. was plagued by avalanches. Purportedly 18,000 of his troops and who knows how many of his elephants were killed by them. It's also too late for others. In World War I 6,000 troops were buried in one day by avalanches in the Dolomite Mountains of northern Italy. Avalanches in high mountain regions have wiped out villages, disturbed railroad track alignments, blocked roads, and otherwise made life difficult if not impossible for millenia.

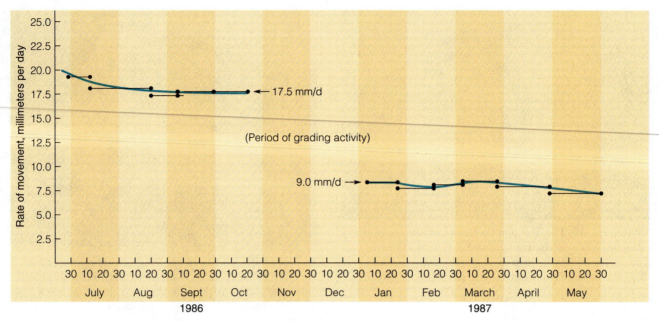

⊙ FIGURE 7.34 Regrading the landslide by moving weight from the top to the toe or mid-slide greatly reduced landslide movement from 17.5 millimeters per day to 9.0 millimeters per day.

Avalanches are caused by an addition of heavy new snowfall or a weakening of older snow. A **slab avalanche** is a coherent mass of snow and ice that hurtles down a slope of more than 30° as an entity—much like a magazine sliding off a tilted coffee table. It may be a thick, wet snow accumulation sliding on ice, or a coherent slab of snow that detached along a *depth hoar* layer. Depth hoar begins to form at the surface when snow particles evaporate in very cold, calm weather and form a loose, open array of snow crystals. As the layer of crystals is buried, evaporation continues, resulting in

a very weak, porous layer at depth. This hoar layer may collapse when it is disturbed, forming a compressed-air layer upon which the overlying snow can move with great velocity. Slab or "wet" avalanches associated with depth hoar are common on the east side of the Rocky Mountains because of the cold temperatures and relatively light snowfall there. These avalanches account for the majority of avalanche deaths because of their sheer weight and their tendency to solidify when they stop. About 25 avalanche fatalities occur each year in North America, mostly in backcountry areas,

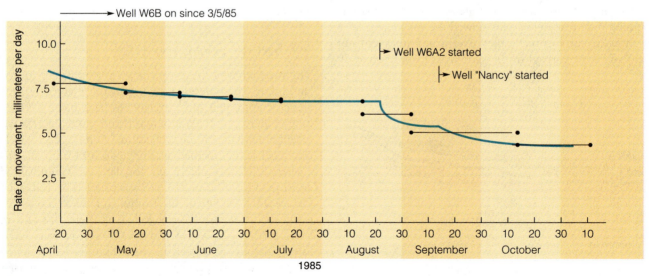

⊙ FIGURE 7.35 Record of landslide movement at one measuring station in 1985. Dewatering wells in the slide slowed movement about 30 percent in seven months at this station.

● FIGURE 7.36 The toe of an avalanche reveals a tangled mass of aspen and pine trees ripped out by their roots. The power and danger of this winter phenomenon is obvious. The photo was taken in June 1995, in the San Juan Mountains near Telluride, Colorado.

and the victims tend to be cross-country skiers, snowmobilers, climbers and hikers.

Powder-snow avalanches are more common in the Sierra Nevada and Coast Ranges, where depth hoar is rare and abundant moisture results in deep snow packs. These "dry" avalanches are dangerous because they entrain a great amount of air and behave like a fluid. Furthermore, the strong shock wave that precedes them can uproot trees and flatten structures (● Figure 7.36). They are most common in areas where heavy new snow has fallen on top of older snow with weak layers at depth. Because they usually occur during storms, they are difficult to predict. Victims of dry avalanches tend to be buried in light snow close to the surface, which makes their chances of being rescued alive better than those of wet-avalanche victims. Regardless of whether an avalanche is wet or dry, it results from a weak boundary layer between potentially unstable snow above and a stable mass below.

Avalanches can be triggered by skiers, rockfalls, loud noises, or explosives. Avalanche control is thus achieved by triggering an avalanche with explosives or cannon fire when dangerous snow conditions exist in an area frequented by humans. The explosions collapse the depth hoar or detach powder snow, causing a "controlled" avalanche.

Various means have been developed for mitigating the destructive might of avalanches. Avalanching snow leaves scars on hillsides, which serve to indicate areas where building should be avoided (● Figure 7.37). Where structures already exist, fences can be built to divert avalanches away from them. In some areas in the Alps, the uphill sides of chalets are shaped similar to a ship's prow so that they will divert avalanching snow. Railroads and highways can be protected by building snowsheds along stretches adjacent to steep slopes (● Figure 7.38). In the event of an avalanche, the snow slides *over* the road or track, rather than

● FIGURE 7.37 Avalanche scars on steep wooded slopes near Aspen, Colorado. Trees have been cleared from the slopes by past avalanches; it would not be wise to build at or near the base of these avalanche "chutes."

onto it. Much of the railroad across the Donner Pass in Placer and Nevada Counties, California, is protected by snowsheds.

● FIGURE 7.38 The historic White Pass and Yukon Railroad, near Skagway, Alaska, is protected from avalanches by snowsheds.

GALLERY

Slip and Slide

Undeniably, the results of mass-wasting processes can be tragic and devastating, as explained throughout this chapter. In some cases, however, the results are whimsical. For example, a block glide occurred along Highway 89 in Wyoming when the toe of a dormant ancient slide mass was removed to make room for the road (● Figure 7.39). This caused the landslide to reactivate, endangering a telephone line across the moving mass. Over time the solution to this problem became increasingly obvious: the telephone pole supporting the line should be cut off near its base. The pole now hangs from the telephone wires, and the base of the pole has moved 70 feet downslope. Clay-rich shale inclined toward the river in the background of the photograph formed the slide plane. A beautifully developed landslide on the planet Mars shows that gravity and water, even though the water is mostly ice, operate on this rocky planet (● Figure 7.40). Explanations for a few other "light" landslide examples (● Figures 7.41–7.43) are given in their captions.

● FIGURE 7.39 An active block glide on U.S. 89 in Montana. The telephone pole, hanging from the lines, was cut from its base, which had moved 23 meters (70 ft) downslope in the three years before this photograph was taken. The person standing downslope is near the "stump" of the pole.

● FIGURE 7.40 My favorite Martian landslide—oblique view from the south toward the Olympus Mons landslide complex, Mars. The arcuate landslide crown scarp in the background is about 25 km (15 mi) wide and 4 km (2.4 mi) high. The landslide debris flowed up to 65 km (39 mi) from the scarp toward the observer. The volume of the slide is roughly a thousand times larger than the May 18, 1980, debris avalanche from Mount St. Helens. Mass wasting is widespread on this rocky planet.

● FIGURE 7.41 Very appropriate use for a landslide. A mobile-home park is tucked onto an inactive landslide near Malibu, California. Should the landslide reactivate, residents can quickly remove their homes.

● FIGURE 7.42 Watch that centerline! Offset along the Portuguese Bend landslide looking toward stable ground. The slide moves continuously, and the road requires periodic repair and centerline repainting.

● FIGURE 7.43 "Pedestal pavement" formed by landsliding and erosion; Point Fermin, San Pedro, California. This former street was lined with houses, all destroyed by the landslide.

Mass Wasting

DEFINED Downslope movement of rock and soil under the direct influence of gravity.

CLASSIFICATION BY
1. Type of material—rock or earth.
2. Type of movement—flow, slide, or fall.
3. Moisture content.
4. Velocity.

Flows

DEFINED Mass movement of unconsolidated material in the plastic or semifluid state.

TYPES
1. Creep—slow, imperceptible downslope movement of rock and soil particles.
2. Debris flows—dense fluid mixtures of rock, sand, mud, and water that are fast-moving and destructive.
3. Debris avalanche—fast-moving (15 km/h) debris flow.

CAUSES Heavy rainfall on steep slopes with loose soil and sparse vegetation. Most common in arid, semiarid, and alpine climates and where dry-season forest fires expose bare soil.

Slides

DEFINED Movement of rock or soil (landslide) on a discrete failure surface or slide plane.

MECHANICS Driving force (gravity) exceeds resisting forces (friction and force normal to the slide surface).

TYPES
1. Rotational slide or slump on a curved, concave-upward slide surface.
2. Translational slide or block glide on an inclined plane surface.
3. Complex landslide, a combination of (1) and (2).
4. Lateral spreads of water-saturated ground (liquefaction of sand and quick clays).

CAUSES
1. Weakening of the slope material by saturation with water, weathering of rock or soil minerals, and burrowing animals.
2. Steepening of the slope by artificial or natural undercutting at toe.
3. Added weight at the top of the slope such as by a massive earth fill.
4. A trigger such as an earthquake, vibrations from explosives, heavy traffic, or sonic booms.

Reduction of Losses

GEOLOGIC MAPPING To identify active and ancient landslide areas.

BUILDING CODES To limit the steepness of manufactured slopes and specify minimum soil and fill conditions and surface-water drainage from the site.

Control and Stabilization

LATERALLY INSERTED PIPES (HYDRAUGERS) To drain subsurface water.

EXCAVATION To redistribute soils and rock from the head of the potential slide to the toe. This decreases driving forces and increases resisting force.

BUTTRESS FILLS AND RETAINING DEVICES To retain active slides and potentially unstable slopes.

ROCK BOLTS To stabilize slopes or jointed rocks.

DEFLECTION DEVICES To divert debris flows around existing structures.

Snow Avalanches

TYPES Slab avalanche is old snow and ice that moves as a coherent mass. Powder snow avalanche is composed of new snow that moves like a fluid at high velocities.

CAUSES Weak boundary layer between unstable layer above and stable mass below.

MITIGATION Controlled avalanching by shooting with light cannons, diversion structures, and snow sheds.

■ KEY TERMS

angle of repose	mass wasting
bentonite	powder-snow avalanche
block glide	resisting force
complex landslide	rotational landslide
creep	quick clay
debris avalanche	sensitive clay
debris flow	slab avalanche
driving force	slide plane
effective stress	slump
landslide	syncline
lateral spreading	translational landslide

■ STUDY QUESTIONS

1. What grading and land-use practices can reduce the likelihood of landslides?
2. What factors caused reactivation of the long-dormant Portuguese Bend landslide?
3. How can vegetation, topography, and sediment characteristics aid in identifying old or inactive slumps, slides, and debris flows?
4. Name the processes that move earth materials downslope (gravity is the *cause*, not the *process*). On what factors does each depend?
5. Modern building codes require a Factor of Safety of 1.5 or greater in order to build on or at the top of a cut slope. What is the Factor of Safety, and why should it be 1.5?
6. What are some common, cost-effective methods of landslide control and prevention?
7. What caused the disaster at Vaiont Dam?

▣ FURTHER INFORMATION

BOOKS AND PERIODICALS

Alger, C., and Earl Brabb. 1985. *Bibliography of U.S. landslide maps and reports.* U.S. Geological Survey open file report 85–585.

Coates, Donald, ed. 1977. *Landslides.* Reviews in engineering geology, no. 8. Boulder, Colo.: Geological Society of America.

Dolan, R., and H. Grant Goodell. 1976. Sinking cities. *American scientist,* no. 1.

Fleming, R. W., and T. A. Taylor. 1980. *Estimating the costs of landslides in the United States.* U.S. Geological Survey circular 832.

Highland, Lynn M. 1993. Slumgullion: Colorado's natural landslide laboratory. *Earthquakes and volcanoes* 24, no. 5: 208.

Krohn, J. P., and James Slosson. 1976. Landslide potential in the United States. *California geology,* October.

McPhee, John. 1989. Los Angeles against the mountains. *The control of nature,* ch. 3. New York: Farrar, Straus & Giroux.

National Academy of Sciences. 1978. *Landslides: Analysis and control.* Transportation Research Board special report 176. Washington, D.C.: National Academy Press.

Nilsen, Tor H., and others. 1979. *Relative slope stability and land-use planning in the San Francisco Bay region, California.* U.S. Geological Survey professional paper 944. Washington, D.C.: U.S. Government Printing Office.

Pearson, Eugene. 1995. Environmental and engineering geology of central California: Salinian block to Sierra foothills. National Association of Geology Teachers, Far Western Section, Fall Conference, University of the Pacific, Stockton, California.

Schultz, Arthur, and Randall W. Jibson (eds.). 1989. *Landslide processes of the eastern United States and Puerto Rico.* Geological Society of America special paper 236, 102 pp.

Smelser, M. G. 1987. Geology of the Mussel Rock landslide, San Mateo County. *California geology* 40, no. 3: 56.

Tremper, Bruce. 1993. Life and death in snow country. *Earth: The science of our planet,* no. 2.

U.S. Geological Survey. 1982. *Goals and tasks of the landslide part of a ground-failure hazards reduction program.* U.S. Geological Survey circular 880. Washington, D.C.: U.S. Government Printing Office.

Varnes, David. 1984. *Landslide hazard zonation: A review of principles and practice.* Paris: UNESCO.

SUBSIDENCE AND COLLAPSE

I love to think of nature as an unlimited broadcasting station, through which God speaks to us every hour, if we only will tune in.

GEORGE WASHINGTON CARVER, AGRONOMIST AND NATURALIST (1860–1943)

Two million people live below the high-tide level in Tokyo. The canals of Venice overflow periodically and flood its beloved tourist attractions. The Houston suburb of Baytown has subsided almost three meters since 1900, and 80 square kilometers (31 mi²) of this coastal region is permanently under water. Earthen dikes prevent the sea from flooding the land in some places in California, and Mexico City's famous Palace of Fine Arts rests in a large depression in the middle of the city (■Table 8.1). These areas suffer from natural and human-induced **subsidence,** a sinking or downward settling of the earth's surface. Loss of life due to subsidence is rare, and it is not usually catastrophic, but land

OPENING PHOTO
High tides and storm surges frequently flood the beautiful city of Venice, Italy. The frequency of these high waters, known as the *aqua alta* to Venetians, has increased dramatically since 1960, because Venice is sinking into the Adriatic Sea. This subsidence is due to poor geologic conditions and withdrawal of underground water.

TABLE 8.1 Subsiding Cities, 1986

WORLD		UNITED STATES	
City	Maximum Subsidence, cm*	City	Maximum Subsidence, cm
Mexico City	850	Long Beach	900
Tokyo	450	(San Joaquin Valley)	880
Osaka	300	San Jose	390
Shanghai	263	Houston	270
Niigata, Japan	250	Las Vegas	66
Bangkok	100	Denver	30
Taipei, Taiwan	190	New Orleans	22
London	30	Savannah	20

* Since measurement began
SOURCE: R. Dolan and H. Grant Goodell, "Sinking Cities," *American Scientist* 74 (1986), no. 1.

subsidence is currently observed in 45 states and is estimated to cost $125 million annually. In the United States the area of human-induced subsidence is estimated as at least 44,000 square kilometers (17,000 mi²), about the size of Maryland and New Jersey combined, but the actual area is probably even greater. Planners and decision makers need to be aware of the causes and impacts of subsidence in order to assess the risks and reduce material losses.

HUMAN-INDUCED SUBSIDENCE

Human-induced subsidence occurs when humans extract underground water, engage in mining or oil and gas production, and when they cause loose sediments at the ground surface to consolidate or compress. The effects may be local or regional in scale. In the United States and Mexico alone, an area the size of Vermont has slowly subsided 30 centimeters (1 ft) due to withdrawal of underground water. Sinking of the land changes drainage paths, and it is particularly damaging to coastal areas and lands adjacent to rivers because it increases flood potential.

Along the Texas Gulf Coast, for example, annual losses from subsidence due to fluid withdrawal were estimated at $109 million from 1943 to 1973. In the eastern United States, the lowering of the water table in several regions underlain by cavernous limestone has caused the ground surface to collapse into caverns below. Economic losses from these collapses amount to many millions of dollars per year. Each year collapse of the land surface into abandoned coal mines costs the United States $30 million. About a quarter of the seven million acres that have been mined for coal has subsided, and some of this area underlies cities. The U.S. Geological Survey estimated in 1983 that the cost of backfilling all mines beneath urban areas would be $12 billion.

NATURAL SUBSIDENCE

Natural subsidence is caused by earthquakes, volcanic activity, and the solution of limestone, dolomite, and gypsum. Earthquake-related subsidence occurs rapidly, and although it is best known in Alaska and California, it also occurs in a number of other states. Displacement along large faults (see Chapter 4) can raise or lower the land surface over a large area. For instance, subsidence of one meter over 180,000 square kilometers (70,000 mi²) took place during the Alaskan earthquake of 1964. Much of the subsidence was along the coast, and the subsided area is now flooded at high tide (● Figure 8.1). In fact, the land actually tilted, and a large area of the Gulf of Alaska was uplifted several meters.

Severe ground shaking that leads to liquefaction may also lower the ground surface. This happened along the valley of the Mississippi River during the New Madrid earthquakes of 1811–1812 and in the Marina district of San Francisco in 1989 (see Chapter 4). A very small-scale subsidence problem is the collapse of roofs of shallow lava tunnels and tubes in volcanic areas. Volcanic activity that empties magma chambers can cause collapse and subsidence over much larger areas when it forms calderas (see Chapter 5). Subsidence caused by tectonic (mountain-building) processes occurs so very slowly that it is not considered an environmental hazard.

When the roofs of limestone caves or caves in other soluble rocks collapse, they form circular depressions known as **sinkholes**. Although this is a natural phenomenon, it has been greatly accelerated by human activity in the southeastern United States. The location and distribution of natural volcanic and solution features in the United States are

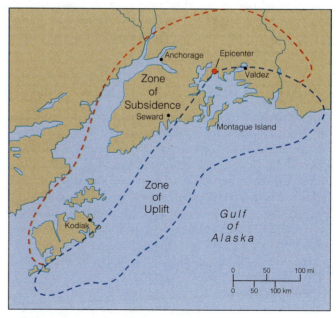

● FIGURE 8.1 Area of subsidence and uplift (tilt) that resulted from the Alaska earthquake, Good Friday 1964. As much as a meter of subsidence occurred.

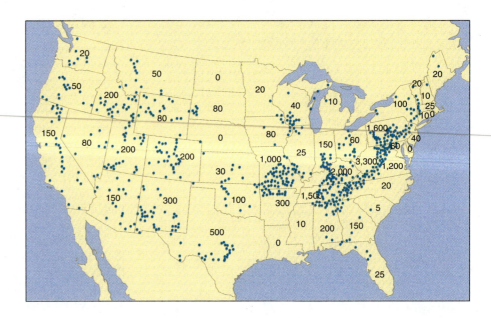

● **FIGURE 8.2** Locations of (karst) caves formed by the solution of limestone, dolomite, and gypsum in the United States. The estimated number of caves in each state is shown. Also shown are lava tunnels, mostly in Oregon, Idaho, and California. The areas where solution caves are numerous are also areas with high potential for sinkhole collapse.

shown in ● Figure 8.2. These areas are also the most likely places for ground collapse to occur in the future.

To put this geologic hazard into perspective, most subsidence that adversely affects humans is caused by human activity and is irreversible. Once the land surface sinks by any significant amount, it cannot be "jacked" back up to its original level. Furthermore, its impact upon drainage devices, sewers, canals, and other human enterprises is often impossible to remedy. For this reason, it is desirable to anticipate subsidence and to take whatever mitigating measures are needed and possible. As noted, most subsidence occurs so slowly that it poses little danger to people. There are some exceptions, however, such as sudden collapses into underground openings and subsidence-related dam failures (see Case Study 8.1, page 218). In this chapter we emphasize subsidence that has well-identified connections with human activities.

 A CLASSIFICATION OF SUBSIDENCE

One method of classifying land subsidence is according to the depth at which the subsidence is initiated. *Deep subsidence* is initiated at considerable depth below the surface when water, oil, or gas is removed from there. *Shallow subsidence,* on the other hand, takes place nearer the ground surface when underground water or solid material is removed by natural processes or by humans. In addition, many poorly consolidated shallow deposits such as peat are subject to compaction by overburden pressures, groundwater withdrawal, and in some soils, simply by being saturated with water.

Settlement differs from subsidence. It occurs when an applied load—such as that of a structure—is greater than the bearing capacity of the soil onto which it is placed. For example, the artificial fill placed for New York's La Guardia Airport settled more than two meters into the subsoil in the

airport's first 25 years of operation. Settlement is totally human-induced and it is a soils- and foundation-engineering problem. The Leaning Tower of Pisa is a classic example of foundation settlement; its interesting construction history is discussed in Chapter 6.

Deep Subsidence

Removal of fluids—water, oil, or gas—confined in the pore spaces in rock or sediment causes deep subsidence. Porewater pressure—that is, the hydrostatic pressure of water in the pores between sediment grains (see Chapter 7)—helps to support the overlying material. As pressure is reduced by extraction, the weight of the overburden gradually transfers to mineral-and-rock-grain boundaries. If the sediment was originally deposited with an open structure, the grains will reorient into a closer-packed arrangement, thus occupying less space, and subsidence will ensue (● Figure 8.3). Because clays are more compressible than are sands, most compaction takes place in clay strata.

At least 22 oil fields in California have subsidence problems, as do many fields in Texas, Louisiana, and other oil-producing states. A near world record for subsidence is held by the Wilmington Oil Field in Long Beach, California, where the ground has dropped nine meters (about 30 ft). This field is the largest producer in the state, with more than 2,000 wells tapping oil in an upward-arched geologic structure called an *anticline* (see Chapter 13). The arch of the Wilmington field's anticline gradually sagged as oil and water were removed, causing the land surface to sink below sea level (● Figure 8.4). Dikes were constructed to prevent the ocean from flooding the adjacent Port of Los Angeles facilities and the naval shipyard at Terminal Island. People who have their boats moored on Terminal Island must actually walk *up* to board them at sea level (● Figure 8.5)! When it rains it is necessary to pump water out of low spots

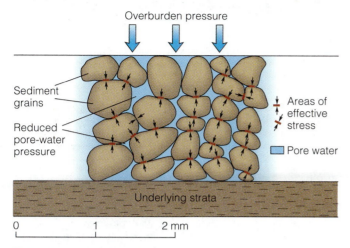

Overburden pressure

Sediment grains

Reduced pore-water pressure

Areas of effective stress

Pore water

Underlying strata

0   1   2 mm

⦿ **FIGURE 8.3** A reduction of pore pressure causes increased effective stress and rearrangement and compaction of sediment particles.

upward to the sea, and oil-well casings and underground pipes have "risen" out of the ground as the land has sunk. To remedy the situation, in 1958 the City of Long Beach initiated a program of injecting water into the ground to replace the oil being withdrawn (⦿ Figure 8.6). Subsidence

is no longer a problem, and the surface has even rebounded slightly in places. However, fluid injection cannot attain the necessary pressures to bring the surface back to its original level.

Visible effects related to ground-water withdrawals are generally restricted to water-well damage or to cracking of long structures such as canals. As water levels are lowered, however, unconsolidated sediments compact, and tension cracks and fissures can develop. In the Antelope Valley, California, home of Edwards Air Force Base, the amount of water pumped from underground storage has exceeded the amount replacing it since 1930, and water levels beneath the dry lake had been lowered by more than 60 meters (200 ft) by 1993. The dry lakebed at Edwards is the landing strip for the space shuttle and other high-performance aircraft. A ground fissure appeared in the lakebed that, initially only a few centimeters wide, became greatly enlarged by surface-water runoff and erosion (⦿ Figure 8.7). This halted operations on the lakebed until repairs could be made. Similar fissuring has occurred in Arizona, New Mexico, Texas, and Nevada. In Arizona ground-water withdrawal in some places has caused subsidence of 5 meters (more than 15 ft) and the development of cracks and earth fissures, particularly along the edges of some basins. The subsidence has changed natural drainage patterns and caused highways to settle and

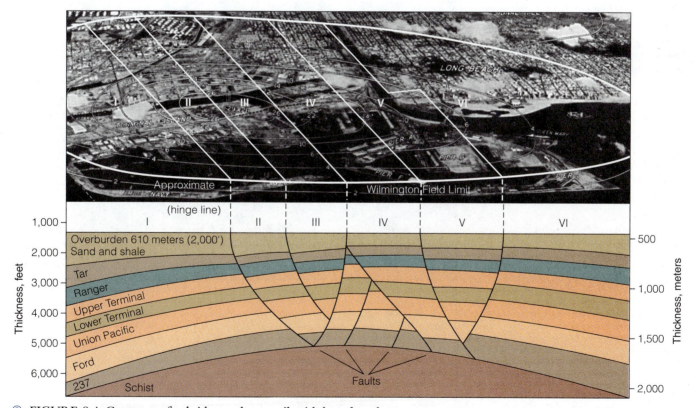

⦿ **FIGURE 8.4** Contours of subsidence due to oil withdrawal at the Wilmington Oil Field, Long Beach, California, 1928–1974. The oil is geologically trapped in the faulted anticline shown in cross section below the photograph.

Subsidence and a Dam Disaster

About 11:15 A.M. on Saturday, 14 December 1963, the care-taker on duty at the Baldwin Hills Reservoir in Los Angeles heard unusual sounds coming from a gallery of pipes beneath the reservoir. Upon inspection, he found the area to be full of mud and water and "just blowing right out and making a terrific racket." He alerted his supervisor, who immediately ordered the lowering of the water in the reservoir. Neither man realized he was witnessing a catastrophe in the making. At 3:38 that afternoon, the earth-filled dam holding back 65,000,000 gallons of water would rupture. Flood waters devastated the residential area below the dam, destroying scores of homes and causing five deaths. Only early warning and prompt action by reservoir personnel and the police prevented a greater tragedy.

The Baldwin Hills Dam and Reservoir were completed in 1950. The project was built as a backup facility for serving water needs during peak-use hours. Some redesign was required during construction, because the project geologist found that two normal faults ran through the reservoir and earthen dam (● Figure 1). Because the geologist believed that one of the faults was active, special subsurface drains were placed below the reservoir to collect any water that might percolate downward along the faults and to carry it to below the dam. Percolating underground water in sandy materials can create conduits by an erosion process called **piping,** which can seriously weaken a dam foundation. The Newport–Inglewood fault, the active fault of most concern in the greater Los Angeles area, is only a few hundred yards from the dam, and the surrounding Inglewood Oil Field is an area of known subsidence. The design of the reservoir (● Figure 2) incorporated an impervious ten-foot-thick earth seal lined with asphalt and gravel-packed tile drains below the seal. This design was intended to capture any water that might leak through the dam if one of the faults should move and cause the lining to crack.

Believing they had designed around all potential geological problems, the engineers boldly built a dam (a) near a major active

● FIGURE 1 Location of Baldwin Hills Reservoir and the oil field relative to the active Newport–Inglewood fault zone. Note the offset of the oil field on the right-lateral strike-slip Newport–Inglewood fault. Only earthquake epicenters on or near the main fault are shown.

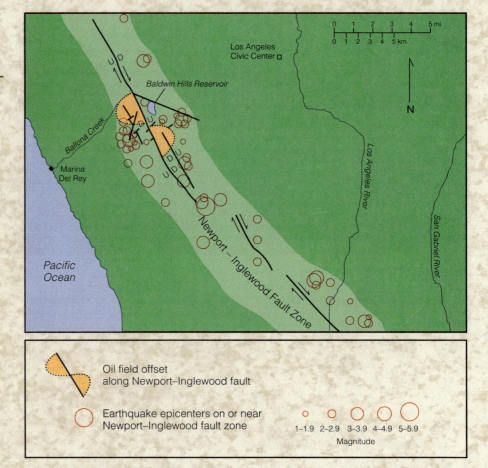

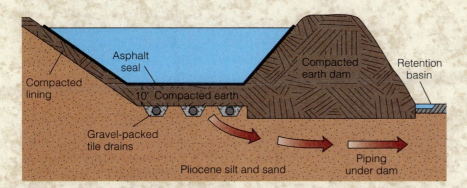

● FIGURE 2 A crack-proof lining and tile drains were designed to prevent leaking and piping under the Baldwin Hills Dam. The lining *did* crack, and the drains failed to function as they were designed.

fault, *(b)* in an area of known subsidence, and *(c)* with two inactive faults running through the site. On the day of the failure, the reservoir cracked, the tile drains were clogged, and full-reservoir water pressure piped through the foundation beneath the dam.

By 1:00 P.M. the reservoir was being drained as quickly as possible, but water began appearing on the downslope toe of the dam. Leakage was so severe by 1:30 that a decision was made to evacuate the residential area below the dam. By 2:00 the leak in the dam face was described as "a sheet of water shooting through that little crack." The streets below the dam were still dry at the time, however, and some residents were not aware of the problem. Not until 3:30—eight minutes before the dam breached—did the retention basin at the toe of the dam fill and overflow onto the home-lined street below it. The photos of ● Figure 3 were taken over a period of about 15 minutes during the final failure of the dam. Flooding was extensive, and five people who were unaware of the failure drowned or were swept away.

An investigation by the mayor's "blue-ribbon" committee of geologists and engineers determined that subsidence caused by oil extraction from the Inglewood Oil Field ultimately led to the collapse of the Baldwin Hills Dam. On the day of the failure, slippage occurred along the two inactive faults that ran through the reservoir—slippage induced by oil-field subsidence. Although the slippage was only a matter of inches, it was enough to rupture the reservoir lining, forcing water under full-reservoir pressure along the faults into the underlying sands. This percolating water eventually "piped" an opening beneath the earthen dam embankment, which led to the dam failure and flooding below.

Indeed, this dam never should have "happened." It was built in an area of known subsidence, close to a major active fault, and with inactive faults running through the reservoir. Any location with these geological problems would receive little consideration today as a potential dam site. The dam sits empty today, its floor covered by vegetation growing through cracks in the lining, mute testimony to faulty site selection. We learn much from tragedies like this one, and the education is very expensive.

(a)

(b)

● FIGURE 3 *(a)* Piping opened a hole in the face of the dam 15 minutes before the failure. *(b)* A residential street below the dam minutes after the dam was totally breached.

(a)

(b)

⊙ FIGURE 8.5 The effects of subsidence on the coastal communities of Wilmington and Long Beach, California. *(a)* The power plant *(left)* is at the point of maximum subsidence in the Wilmington Oil Field, about 9 meters (30 ft). Well below sea level, the building is protected from the sea by dikes. The yachts are moored at sea level. *(b)* This fire hydrant extruded as the ground around it sank; Long Beach.

crack. It is influencing the siting of waste-treatment and solid-waste disposal sites and generally impacting all development activity in the state. The problem is so serious that the Arizona Department of Water Resources and the Arizona Geological Survey have established the Center for Land-Subsidence and Earth-Fissure Information, with the catchy acronym *CLASEFI,* to provide information and assistance to the public.

Coincidentally, the magnitude of the subsidence due to ground-water withdrawal is similar to oil-field subsidence. In the San Joaquin Valley of California, for example, 8.9 meters (29 ft) of subsidence resulted from 50 years' water

⊙ FIGURE 8.6 Graph of oil production, land subsidence, and water injection at Wilmington Oil Field, 1937–1967. Note that the rate of subsidence decreased rapidly when water injection began.

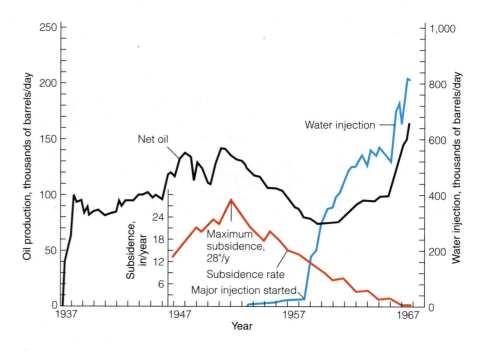

⊙ FIGURE 8.7 Giant ground fissure across Rogers Lake, the dry lakebed of Edwards Air Force Base, due to ground-water withdrawals in the Antelope Valley. This dry lake is one of the two landing strips for the space shuttle.

⊙ FIGURE 8.8 Subsidence due to excessive ground-water withdrawal for agriculture in the San Joaquin Valley, California. The numbers on the pole indicate the ground level for the years indicated. The area is still subsiding, but at a much slower rate.

extraction for agricultural irrigation (⊙ Figure 8.8). How much the land subsides depends mostly on how significantly the ground-water levels are lowered and the thicknesses of the compactable dewatered sediment layers.

The "sinking" of Venice, Italy, is of special interest because of the city's historical importance, priceless antiquities, and beauty (⊙ Figure 8.9). The central square, Piazza San Marco, is awash several times a year, and it has been raised in attempts to compensate for subsidence. Venice is sinking at a rate of about 25 centimeters (1 ft) per century due to settlement into the underlying soft sediments and to water withdrawal from 20,000 wells in the region. Sitting on a low mudbank in a lagoon sheltered by barrier islands, the city has dropped about three meters (about 10 ft) relative to sea level since it was founded. During periods of extremely high water, locally called the *aqua alta* (Italian, "high water"), the avenues and squares of Venice become flooded, and gondola traffic comes to a halt. The *aqua alta* occurs when an onshore flow of water from the Adriatic Sea, (a *storm surge*) coincides with exceptionally high tides. There were 58 high waters between 1867 and 1967, 30 of them in the last decade of that period. Although Venice has subsided only 22 centimeters (8.7 in) since 1910, the frequency of flooding has increased dramatically since 1950

(⊙ Figure 8.10). This demonstrates the sensitivity of coastal areas to subsidence. Movable barriers to prevent flooding have been under study by the Italian government since 1989. Most if not all of Venice's water wells have been shut down, and water is imported by aqueduct.

Mexico City sits in a fault valley that has accumulated nearly 2,000 meters (6,600 ft) of sediments, mostly of pyroclastic origin. It was founded by the Aztecs at the site of ancient Lake Texcoco, where loosely consolidated volcanic tuffs and other sediments were deposited (see Chapter 1 and Figure 1.2). In 1925 it was demonstrated that subsidence was occurring here due to the withdrawal of water from these sediments. Total subsidence varies throughout the city, but almost 6 meters (20 ft) of subsidence had occurred by the 1970s in the northeast part of the city, mostly in the water-bearing top 50 meters (160 ft) of sediment. In 1951

● FIGURE 8.9 When the *aqua alta* floods St. Mark's Square in Venice, raised boardwalks are provided for pedestrians.

Shallow Subsidence

COLLAPSING SOILS. Subsidence that results from heavy application of irrigation water on certain loose, dry soils is called **hydrocompaction**. The in-place densities of such "collapsible" soils are low, usually less than 1.3 grams per cubic centimeter, about half the density of solid rock. Upon initial saturation with water, such as for crop irrigation or the construction of a water canal, the open fabric between the grains collapses and the soil compacts. Subsidence of monumental proportions has occurred because of hydrocompaction. In the United States, it is most common in arid or semiarid regions of the West and Midwest where soils are dry and moisture seldom penetrates below the root zone. Any sediment that was deposited rapidly and that has an open, granular structure is subject to compaction. These are most commonly debris flows or wind-blown (eolian) deposits. Known areas of hydrocompaction are the Heart Mountain and Riverton areas of Wyoming; Denver, Colorado; the Columbia Basin in Washington; southwest and central Utah; and the San Joaquin Valley of California.

an aqueduct was completed that brought water to the city, and as a result, a number of wells were closed. The effect on subsidence rates was dramatic, and by 1970 subsidence had decreased to less than 5 centimeters per year (● Figure 8.11). Some wells tapped deeper water-bearing layers, and well casings (large pipes) extruded from the ground as the shallower layers compacted around them. Similarly, structures built on piles resting on stable layers at depth *seem* to rise out of the earth as the ground beneath these buildings sinks. Many massive older buildings have settled well below street level into the dewatered surface sediments. Underground water mains, utility lines, and surface-water drainage systems are disrupted in areas of subsidence. The famous Our Lady of Guadalupe (the "brown Madonna") Cathedral tilts, because its foundation is partly on lava flows and partly on the subsiding lake sediments (● Figure 8.12).

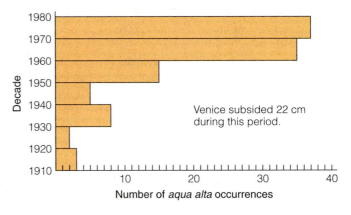

● FIGURE 8.10 Occurrences of *aqua alta* and flooding in Venice, 1910–1980. The increase in the frequency of high waters since 1950 is due to subsidence.

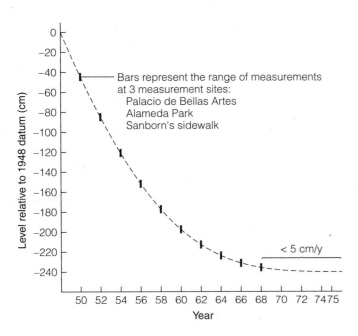

FIGURE 8.11 Cumulative subsidence at three sites in Mexico City, 1948–1975. Note the rapid decrease in subsidence (flattening of the curve) when pumping was stopped. It was almost nil by 1975. Most of the subsidence occurs at depths of less than 50 meters (160 ft).

FIGURE 8.12 Differential settlement at Our Lady of Guadalupe church, Mexico City. The foundation rests across both solid basalt and a subsiding lakebed.

Collapsing soils were a major problem for designers of the California Aqueduct in the San Joaquin Valley. In order to maintain a constant ground slope for the flow of water, it was necessary to identify those areas of loose soil that could not be avoided. Sediments were field-tested by saturating large plots and measuring subsidence at various levels beneath the surface. Subsidence was found to average 4.2 meters (close to 14 ft) over a period of 484 days, during which thousands of gallons of water had been applied to the test plots (● Figure 8.13). On the basis of the field tests, 180

FIGURE 8.13 *(a, left)* Location of the San Joaquin Valley. Collapsible soils occur almost entirely along the western border of the valley, whereas subsidence due to underground-water withdrawal is found throughout the valley. *(b, below)* Water-soaked test plot in the valley's collapsible soils. Total subsidence at this plot is on the order of 3 meters (10 ft).

(a)

(b)

kilometers (112 mi) of the aqueduct alignment in the valley was "presubsided" by soaking the ground three to six months before construction. This assured that the gradient would be maintained after construction. Twenty-two million dollars was added to the cost of the project. Water transport systems, unlike highways or utility lines, are totally determined by the slope of the land beneath them.

Sediments deposited on floodplains or on the deltas of large rivers, such as the Mississippi River delta, also are subject to compaction and subsidence. Compaction occurs as the weight of overlying sediment builds up and causes water to be expelled from the sediment pore spaces. Oxidation of organic-rich soils also can lead to subsidence of the land surface (see Case Study 8.2).

SINKHOLE COLLAPSE. The first signs of impending disaster in Winter Park, Florida, occurred one evening when Rosa Mae Owens heard a "swish" in her backyard and saw, to her amazement, that a large sycamore tree had simply disappeared. The hole into which it had fallen gradually enlarged, and by noon the next day her home had also fallen into the breach. Six Porsches in a nearby storage yard suffered the same fate, as did a city swimming pool and two residential streets. Deciding to work *with* nature, the City of Winter Park stabilized and sealed the sinkhole, and converted it to a beautiful urban lake (●Figure 8.14). Few people who read of this notorious sinkhole in May 1981

(a)

(b)

● FIGURE 8.14 Sinkhole in Winter Park, Florida. *(a)* Shortly after the initial collapse. Note the expensive sports car going down the sink. *(b)* Air view that shows its almost perfect symmetry. *(c)* Plugged and landscaped, the sinkhole area is now an urban park.

(c)

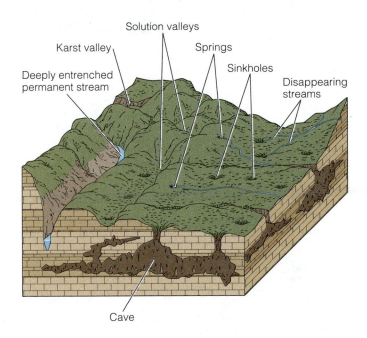

Solution valleys
Karst valley
Springs
Deeply entrenched
permanent stream
Sinkholes
Disappearing
streams
Cave

● **FIGURE 8.15 Surface and subsurface features of karst topography.**

realized the extent of the sinkhole problem in the southeastern United States.

Sinkholes are most commonly found in carbonate terranes (limestone, dolomite, and marble), but they are also known to occur in rock salt and gypsum. They form when carbonic acid (H_2CO_3) dissolves carbonate rock, forming a near-surface cavern. Continued solution leads to collapse of the cavern's roof and the formation of a sinkhole. About 20 percent of the United States, and 40 percent of the country east of the Mississippi River, is underlain by limestone or dolomite. Most of this area has the sinkholes, disappearing streams, springs, and caverns that characterize **karst terrane,** (German *Karst* for the Kars limestone plateau of northwest Yugoslavia; ●Figure 8.15). Urbanization on karst terrane typically results in problems such as increased flooding, contamination of underground water, and collapse, as in the Winter Park example. Total U.S. property damage from sinkhole collapse is estimated at several hundred million dollars. The states most impacted by surface collapse are Alabama, Florida, Georgia, Tennessee, Missouri, and Pennsylvania (●Figure 8.16).

Tens of thousands of sinkholes exist in the United States, and the formation of new ones can be accelerated by human activities. For example, excessive water withdrawal that lowers ground-water levels reduces or removes the buoyant support of shallow caverns' roofs (●Figure 8.17, page 229). In fact, the correlation between water-table lowering and sinkhole formation is so strong that "sinkhole seasons" are designated in the Southeast. These "seasons" are declared whenever ground-water levels drop naturally because of decreased rainfall and in the summer, when the demand for water is heavy. Vibrations from construction activity or explosive blasting also can trigger sinkhole collapse. Other mechanisms that have been proposed for sinkhole formation are fluctuating water tables (alternate wetting and drying of cover material reduces its strength) and

● **FIGURE 8.16 Satellite image map of central Florida showing its karst terrane with disappearing streams, hundreds of lakes, and thousands of sinkholes. Central Florida is underlain by the Tertiary Ocala Limestone, and the coastal areas by younger, detrital sand deposits. This explains the location of karst terrane along the state's north–south axis.**

Water, Water, Everywhere

New Orleans, the Crescent City, has the flattest, lowest, and youngest geology of all major U.S. cities. The city's "alps," with a maximum elevation of about 5 meters (16 ft) above sea level, are natural levees built by the Mississippi River, and its average elevation is just 0.4 meters (1.3 ft). No surface deposits in the city are older than about 3,000 years. The city was established in 1717 on the natural levees along the river. The levees' sand and silt provided a dry, firm foundation for the original city, which is now called *Vieux Carré* (French "old square"), or "the French Quarter." Farther from the river the land remained undeveloped,

⊙ FIGURE 1 *(a)* Major landforms and depositional environments in the New Orleans area before urbanization. The Pleistocene Prairie Formation dips down seaward beneath Lake Pontchartrain and New Orleans, providing a firm foundation for pilings that support the Superdome and other large buildings. The natural levees on both sides of the river also provide a firm, dry foundation. *(b)* High-altitude false-color image of New Orleans, a city almost completely surrounded by water. The light spot in the lower center is the city. Note that the Mississippi River meanders through it, Lake Pontchartrain is to the north, and smaller Lake Borgne is to the northeast. Low, swampy delta lands appear brownish to the east and south. The land width in the photo is approximately 200 kilometers (125 mi). *(c)* Cypress swamp on the outskirts of present-day New Orleans.

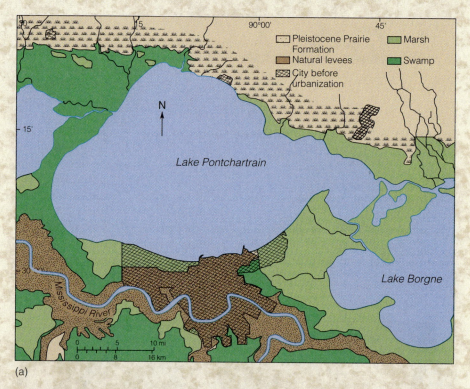

(a)

(b)

(c)

because it was mainly water-saturated cypress swamp and marsh formed between distributaries of the Mississippi River's ancient delta. These areas are underlain by as much as 5 meters (16 ft) of peat and organic muck (◉ Figure 1, opposite). When high-volume water pumps became available about 1900, drainage canals were excavated into the wetlands to the north. Swamp water was pumped upward into Lake Pontchartrain, a cutoff bay of the Gulf of Mexico. About half the present city is drained wetlands lying well below sea and river level. The lowest part of the city, about two meters below sea level, is in the lively, historic French Quarter, noted for its music, restaurants, and other tourist attractions (◉ Figure 2).

Subsidence is a natural process on the Mississippi River delta due to the great volume and the sheer weight of the sediment laid down by the river (see Chapter 9). The sediment compacts, causing the land surface to subside. The region's natural subsidence rate is estimated to have been about 12 centimeters (4 in) per century for the past 4,400 years. This is based upon C-14 dating of buried peat deposits and does not take into account any rise of sea level during the period.

Urban development added to the subsidence in the 1950s. Compaction of peaty soils in reclaimed cypress swamps coincided with the period's construction of drainage canals and planting of trees, both of which lowered the water table. Peat shrinks when it is dewatered, and also oxidation of organic matter and compaction contribute to subsidence. Most homes constructed on reclaimed swamp and marsh soils in the 1950s were built on raised-floor foundations. These homes are still standing, but they require peri-

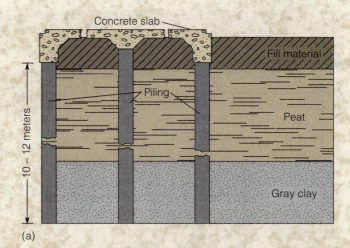

(a)

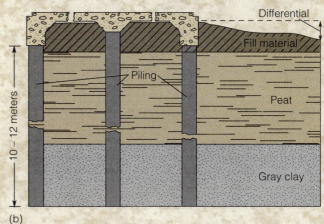

(b)

◉ **FIGURE 3 Slab-on-pilings construction** *(a)* **before and** *(b)* **after differential subsidence.**

odic leveling. Unfortunately, homes built on concrete-slab foundations, a technique that had just been introduced, sank into the muck and became unlivable. Other homes' foundations were constructed on cypress-log piles sunk to a depth of 10 meters or more (at least 30 ft). These houses have remained at their original level, but the ground around them has subsided. This *differential subsidence* requires that fill dirt be imported in order to make the sunken ground meet the house level (◉ Figure 3). It is very common to see carports and garages that have been converted to extra rooms when subsidence has cut off driveway access to them (◉ Figure 4). One also sees many houses with an inordinate number of porch steps; the owners have added steps as the ground has sunk (see Gallery, Figure 8.23). In 1979, Jefferson and Orleans Parishes (counties are called "parishes" in Louisiana) passed ordi-

continued on next page

◉ **FIGURE 2 The French Quarter as seen from atop the natural levee on the Mississippi River. Note that the levee slopes downward toward the French Quarter. The wall and gate are part of a floodwall built to hold back floodwaters should the river overtop the levee (see Chapter 9).**

● FIGURE 4 Differential subsidence around this pile-supported house has left the carport high and dry. The carport was converted to a family room, and fill was imported to bring the yard surface up to the previous level.

nances requiring 10- to 15-meter-deep wooden-pile foundations for all houses built upon former marsh and swamp land. This has been beneficial, but differential subsidence continues to damage New Orlean's sewer, water, and natural-gas lines and streets and sidewalks.

Large buildings are founded on steel or reinforced-concrete pilings driven down to a late Pleistocene erosion surface beneath the city, as this surface has the bearing strength needed to support the structures (see Figure 1). High-rise buildings of up to 50 stories may rest on as many as a thousand piles driven to depths of 15–50 meters (50–165 ft). Older 10- to 12-story buildings rest on veritable forests of cypress pilings. It is jokingly said that the city's very old, massive 3- and 4-story buildings are founded on cotton bales, but they too have cypress-cribbing foundations (wooden walls).

The Superdome arena is in a class by itself. It is supported by 2,266 concrete piles driven to a depth of 50 meters (160 ft). Each pile's design bearing weight is 175 tons; that's 350,000 pounds! Clusters of three to nine piles were driven, and the arena's main support columns were placed upon the clusters. Adjacent pile clusters are calculated to have no more than 13 millimeters (0.5 in) of differential subsidence—quite an accomplishment in an area of water-saturated, compressible soils.

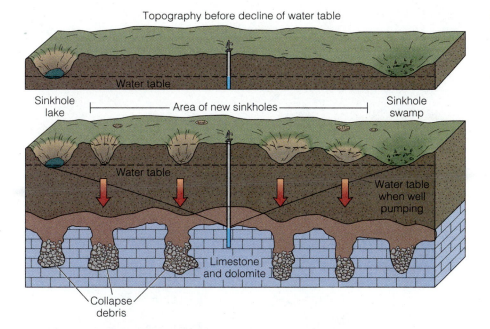

Topography before decline of water table

Sinkhole lake — Area of new sinkholes — Sinkhole swamp

Water table

Water table

Water table when well pumping

Limestone and dolomite

Collapse debris

● FIGURE 8.17 The effect of water-table decline on the formation of sinkholes. The original topography with sinkhole lake and swamp is shown for comparison.

CONSIDER THIS . . .

You could become a millionaire if you developed a relatively inexpensive method of locating caverns that may collapse and form sinkholes. Got any ideas? If not, what types of more sophisticated (and expensive) methods might be employed to locate caverns?

high water tables (which erode the roofs of underground openings).

Although the development of natural sinkholes is not predictable, it is possible to assess a site's potential for collapse resulting from human activities. This requires extensive geologic and hydrogeologic investigation prior to site development. Under the best of geologic conditions, good geophysical and subsurface (drilling) data can tell us where collapse is most likely to occur, but they cannot tell us exactly when a collapse may occur. Armed with this information, planners and civil engineers can implement zoning and building restrictions that will minimize the hazards to the public.

MINE COLLAPSE. Subsidence related to mining is caused by the removal of solids underground that leads to collapse of the overburden into the mine opening. This causes damage to buildings, alters stream courses, impacts vegetation and wildlife, and damages aquifers, pipelines, and roads. As oil and gas reserves decline, coal production will increase, most of it from subsurface mines. Lignite and bituminous (soft) coals are the largest coal reserves in the United States, and many of them occur in areas susceptible to subsidence

(● Figure 8.18). This is because they are found and exploited in near-horizontal beds at relatively shallow depths. Surface subsidence occurs above flat-lying coal beds when unsupported rock collapses into the mined-out openings (● Figure 8.19). This collapse gradually extends upward and outward, eventually resulting in surface subsidence. The extent of subsidence depends on the method of mining and the depth and thickness of the coal bed. Surface deformation may be in the form of large circular depressions, open fractures, small pits, or faulting. ● Figure 8.20 shows some of these features that have developed over a mined-out area in Wyoming.

The state of Pennsylvania has suffered severe problems due to mining. Subsidence has been documented in Pittsburgh, Scranton, Wilkes-Barre, and many other areas. In 1982 an entire parking lot disappeared into a hole above a mined-out coal seam 88 meters (290 ft) below. Underground coal-mine fires also have caused subsidence, and a variety of air-pollution problems as well. An anthracite (high-grade, hard coal) seam inclined 40° below the eastern Pennsylvania village of Centralia has been burning since 1962 in spite of efforts by the U.S. Bureau of Mines to extinguish it. The intensity and location of the fumes vary from day to day as new ground cracks appear, and since 1984 the federal government has been purchasing the homes of Centralia residents who want to leave. Underground coal fires are difficult to extinguish, and flooding is the usual means. This does not appear to be practical at Centralia, which leaves only two alternatives: physically removing the burning strata or evacuating the town. The population of Centralia dropped from 1,100 in 1962 to 63 in 1990.

Whenever coal is extracted by underground mining methods, the possibility of subsidence exists. The "room and pillar" method recovers about 50 percent of the coal,

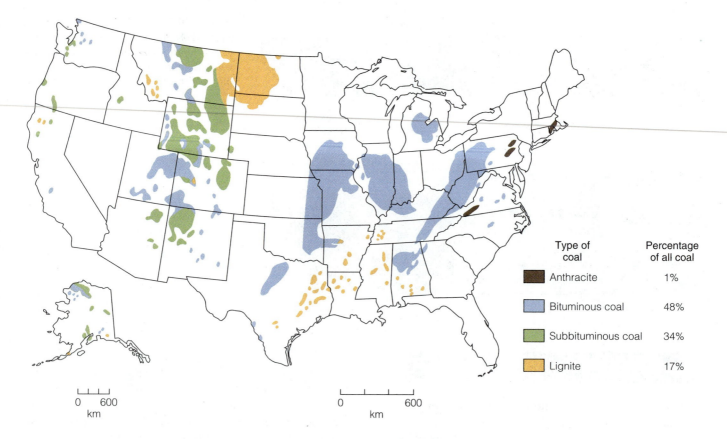

Type of coal	Percentage of all coal
Anthracite	1%
Bituminous coal	48%
Subbituminous coal	34%
Lignite	17%

0 600
km

0 600
km

● FIGURE 8.18 Major deposits of coal in the United States. Because many of these deposits are flat-lying and close to the ground surface, they pose a potential for subsidence and collapse problems if coal is extracted by underground mining techniques. Compare this map with the one in Figure 8.2.

leaving the remainder as pillars of coal to support the roof of the mine. As the mine is abandoned, the pillars are sometimes removed or reduced in size, increasing the possibility of future subsidence. Even when the pillars are left intact, failure is possible. Sometimes the weight of the overlying strata (overburden) causes the pillars to rupture or physically drives the pillars into the shale underlying the coal bed. The resulting subsidence can damage structures hundreds of feet above the mined-out area.

A newer method, "longwall mining," recovers nearly all the coal. This method utilizes movable roof supports (to protect miners and equipment), which are advanced as mining progresses. The roof in the mined-out area behind the active wall collapses. The roof collapse in underground

● FIGURE 8.19 Pattern of deformation in rocks and the ground surface over an extracted coal seam. Note that the amount of surface subsidence is a fraction of the extraction thickness and that the overburden is highly fractured above the collapse.

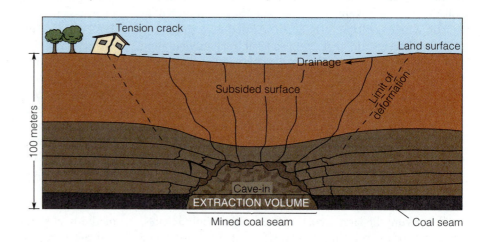

● FIGURE 8.20 Surface effects of underground mining along the Tongue River in Wyoming. The pits, troughs, depressions, and cracks have formed since the mine was abandoned in 1914.

mines can cause damage to buildings, utility lines, highways, wells, and aquifers. Persons who are purchasing real estate in a coal mining area should determine if mining has taken place there and if minable coal is present beneath or near the property under consideration, and then carefully consider their purchase and insurance options.

 ## MITIGATION

The best strategy for minimizing losses is to restrict human activities in areas that are most susceptible to subsidence and collapse. Unfortunately, most instances of land subsidence due to fluid withdrawals have not been anticipated up to now. Knowledge of the causes, mechanics, and treatment of deep subsidence has increased with each damaging occurrence, so that we can now implement measures for reducing the incidence and impact of these losses.

Some of the measures now taken to minimize subsidence are controlling regional water-table levels by monitoring and limiting water extraction, replacing fluids by water-injection, and importing surface water for domestic use in order to avoid drawing on underground supplies. The State of Texas established the Harrison-Galveston Subsidence District in 1979 to regulate the withdrawal of water and thereby control subsidence and prevent coastal flooding in the Galveston area. Underground mining operators are now required to leave pillars to support mined-out coal beds, and mine openings are being backfilled with com-

pacted mine waste to support the overburden. This latter procedure, however, may introduce chemicals that could contaminate underground water unless the backfill is first stripped of pollutants. Pennsylvania's Mine Subsidence Insurance Act of 1962 allows some property owners to buy subsidence insurance after inspection of their property. Another Pennsylvania bill allows homeowners to buy the bituminous coal beneath their property at a fair price and therefore exercise control over mining operations.

Collapsible soils along canals are now routinely precollapsed by soaking, thereby anticipating in advance any adjustments of the canal gradient. Relatively inexpensive geophysical techniques are available for locating shallow limestone caverns beneath potential building sites. Subsidence problems will continue, but we should employ every means possible to foresee the problems and minimize their impact.

Property owners' ability to control subsidence on their own land is hampered by legal problems. The legal recovery theories (tort law) that establish one's right to sue for damages due to subsidence seem to contradict laws that establish others' rights to recover resources from underground. Although it is known that pumping fluids out of the ground can cause subsidence, existing laws may not enable one to force a halt to pumping. It may be time to change the law and work toward a balance between the need to control subsidence and the need to extract underground water, coal, and petroleum products.

Wonders of Karst and Delta—That Sinking Feeling

Karst terrane exists in many parts of the United States, but it is particularly prevalent in parts of Florida and Kentucky (see Figures 8.2 and 8.16). Perhaps the most spectacular karst area in the world is along the Li River in the predominantly limestone region of Kwangsi in southern China near Canton (Guangzhou). Solution of limestone has reduced the landscape there but left enormous pinnacles of resistant limestone, called "tower karst," that present an almost extraterrestrial appearance (◉ Figure 8.21). The towers and other karst features draw tourists from all over the world.

In many places, water containing dissolved calcium carbonate percolates downward through soil and rock into caverns, where evaporation and precipitation form needle-like **stalactites**—extending down from the ceiling—and the more irregular **stalagmites**—rising up from the floor (◉ Figure 8.22). These fresh-water limestone features in caves are called **dripstone,** because they build up drip-by-drip. Carlsbad Caverns in New Mexico and Mammoth Cave in Kentucky offer spectacular examples, but literally hundreds of other limestone caves also exhibit these features and are open to the public (see Figure 8.2). **Speleology** is the study of caves, and spelunking, or "caving," is an exciting recreational pastime for many people. Cavers explore and map natural underground openings. Much of what we know about caves and caverns is attributable to these hobbyists' efforts and talents.

Subsidence on the Mississippi River delta has created some unusual sights for travelers. The house in ◉ Figure

◉ **FIGURE 8.22 Dripstone stalactites, stalagmites, and pillars in a limestone cavern.**

8.23 is an example. It was photographed over a period of 51 years during which it was occupied by one family. These photographs document the periodic, aggravating (to the owner) subsidence that has occurred in the neighborhood, a drained cypress swamp. The house originally had one very small and three normal-sized steps leading up to the porch. However, between 1941, when the house was built, and 1993, the land around the house subsided a total of 81 centimeters (32 in).

◉ **FIGURE 8.21 Residual karst towers; Kwangsi region, southern China. The former limestone plateau has been lowered by solution except for resistant-limestone pinnacles.**

(a)

(b)

(c)

(d)

 FIGURE 8.23 Differential subsidence of the underlying water-saturated peat soils in the Lakeview Subdivision of northwest New Orleans has forced the home's owner to add new steps periodically to maintain access to the front porch. Built in 1941 using a slab-on-pilings technique, the house has remained stable while the yard around it has subsided. (a) A year after construction. (b) Still four steps, but a huge first step was required. (c) 1983. (d) By 1993 eight steps were required.

THE FUTURE

Mainly because of the growing awareness of the causes of subsidence, subsidence rates have decreased in most areas of the world where they have been a serious problem. Unfortunately, the ground surface does not rebound once subsidence has been stopped, and many of the world's major cities remain at risk of flooding. If sea level rises in response to global warming, as most models predict it will (see Chapters 11 and 12), then coastal flooding will become more frequent and severe. Sea level has risen 15–20 centimeters (6–8 in) per century over the past 300 years, and recent estimates predict a sea-level rise of between one and three meters (3–10 ft) from the present level by the end of the next century. If the actual rise is even close to the upper estimate, many heavily populated areas will be inundated. Impressive bulwarks and dikes will be needed for holding back the ocean unless we choose to relocate cities, which is unlikely. Living behind such barriers will be a challenge, and it presents significant risks. It has been said that if sea level does rise significantly, construction of new dikes or the raising of the old ones will be the most expensive undertaking yet faced by humans!

 SUMMARY

Subsidence

DEFINED A sinking of the earth's surface caused by natural geologic processes or by human activity such as mining or the pumping of oil or water from underground.

TYPES

1. Deep subsidence caused by the removal of water, oil, or gas from depth.
2. Shallow subsidence of ground surface caused by underground mining or by hydrocompaction, the consolidation of rapidly deposited sediment by soaking with water.
3. Collapse of ground surface into shallow underground mines or into natural limestone caverns, forming sinkholes.
4. Settlement occurs when the load applied by a structure exceeds the bearing capacity of its geologic foundation.

Causes

DEEP SUBSIDENCE Removal of fluids confined in pore spaces at depth transfers the weight of the overburden to the grain boundaries, causing the grains to reorient into a closer-packed arrangement. Many oil fields in California have subsidence problems, and parts of the Texas Gulf Coast, San Joaquin Valley of California, Venice (Italy), and Mexico City have subsided because of withdrawal of underground water.

SHALLOW SUBSIDENCE AND COLLAPSE The ground surface may collapse into shallow mines (usually coal). Sinkhole collapse is due to a lowering of the water table, which reduces the buoyant forces of underground water. Most sinkholes result from the collapse of the soil and sediment cover into an underground limestone or dolomite cavern. Heavy application of irrigation water causes some loose, low-density sediments and soils to consolidate. This is known as hydrocompaction. Compaction due to dewatering of peat and cypress-swamp (delta) clays and oxidation of the highly organic sediments causes subsidence. In the United States this is peculiar to the Mississippi River delta.

Problems

Subsidence may cause ground cracking, deranged drainage, flooding of coastal areas, and damage to structures, pipelines, sewer systems, and canals. Whole buildings have disappeared into sinkholes.

Mitigation

Underground water can be monitored and managed. Extracted crude oil can be replaced with other fluids. Mine operators are now required to leave pillars to support the mine roof, and rock spoil can be returned to the mine to provide support. Collapsible soils along canals can be precollapsed by soaking. Foundations on organic deltaic sediments should be on piles seated on firm geologic material. This protects structures from damage if the surrounding land subsides. Relatively inexpensive geophysical techniques can be used to locate limestone caverns with roofs that might collapse. Subsidence insurance is available in Pennsylvania.

Features

Land that is underlain by limestone or dolomite and characterized by caves, caverns, lakes, and disappearing streams is called *karst* topography.

KEY TERMS

dripstone	sinkhole
hydrocompaction	speleology
karst terrane	stalactites
piping	stalagmites
settlement	subsidence

STUDY QUESTIONS

1. What are the causes of subsidence and of collapse?
2. What causes sinkholes to form, and how have humans contributed to this problem?

3. What is the difference between settlement and subsidence? What can be done in advance to prevent settlement?

4. Distinguish between subsidence and collapse that are due to natural causes and due to human causes.

5. What can be done to slow or stop subsidence that is caused by withdrawal of water or petroleum?

6. Name the common rocks that develop into karst topography. Where in the United States would you go to see karst terrane?

7. What engineering problems make constructing canals and aqueducts in areas of hydrocompaction such an insidious geologic hazard?

8. What surface manifestations of subsidence due to ground-water withdrawal make it a serious problem, particularly in California and Arizona?

9. What hazard does subsidence present in coastal areas?

10. What is speleology, and what might be the value of this study?

FURTHER INFORMATION

BOOKS AND PERIODICALS

Amandes, C. B. 1984. Controlling land surface subsidence: A proposal for a market-based regulatory scheme. *U.C.L.A. law review* 31, no. 6.

Dolan, R., and H. Grant Goodell, 1986. Sinking cities. *American scientist* 74, no. 1:38–47.

Fellows, Larry D. 1995. Land-subsidence and earth fissure information. *Arizona geology* 25, no. 2.

Halliday, William R., M.D. 1976. *Depths of the earth.* New York: Harper and Row.

Holzer, T. L., ed. 1984. Man-induced land subsidence. *Geological Society of America Reviews in engineering geology,* vol. VI.

Kolb, C. R., and R. T. Saucier. 1982. Engineering geology of New Orleans. *Geological Society of America reviews in engineering geology,* vol. V.

Menard, H. W. 1974. *Geology, resources, and society.* New York: W. H. Freeman and Co.

Saucier, Robert T., and Jesse O. Snowden. 1995. Engineering geology of the New Orleans area. *Geological Society of America annual meeting field trip guidebook,* nos. 6a and 6b.

State of California, Resources Agency. 1965. *Landslides and subsidence.* Geologic Hazards Conference, Los Angeles.

White, W. B.; D. C. Culver; J. S. Herman; T. C. Kane; and J. E. Mylroie. 1995. Karst lands. *American scientist* 83, no. 5 (September–October): 450.

WATER ON LAND

Rain added to a river that is rank
perforce will force it overflow the bank.

WILLIAM SHAKESPEARE (1564–1616)

I f extraterrestrials were to approach earth from afar, they couldn't help but be struck by our planet's heavy cloud cover and its expanse of ocean, clear indications that earth is truly "the water planet." Water is indeed what distinguishes our planet from other bodies in our solar system. Three quarters of the earth's surface is covered by water, and even the human body is 65 percent water. Water provides humans with means of transportation, and much human recreation is in water—and *on* it where it exists in solid form as ice and snow. Since it is not equally distributed on the earth's surface, we have droughts, famine, and catastrophic floods. The absence or abundance of water results in such marvelous and diverse features as deserts, rain forests, picturesque canyons, and glaciers. In this chapter we investigate the

OPENING PHOTO
The Mississippi River engulfs a village in Illinois during the great flood of 1993. The expanse of the flooded area led one observer to call this the seventh Great Lake.

LOCATION	VOLUME*		PERCENTAGE OF TOTAL
	Km^3	$Miles^3$	
Surface water			
lakes and rivers (fresh)	125,000	30,000	0.009
lakes and inland seas (salty)	104,000	25,000	0.008
Subsurface water			
shallow, fresh ground water	4,167,000	1,000,000	0.31
deep, salty ground water	4,167,000	1,000,000	0.31
Ice caps and glaciers (frozen)	29,000,000	7,000,000	2.15
Atmosphere	12,500	3,100	0.001
Oceans	1,321,000,000	317,000,000	97.2
TOTAL (rounded)	1,358,600,000	326,000,000	100

*There are 1.1 trillion gallons in a cubic mile and 242 million gallons, or 920 million liters, in a cubic kilometer.

SOURCE: J. H. Feth, "Water Facts and Figures for Planners and Managers," U.S. Geological Survey Circular 601-1, 1973.

reasons for this uneven distribution and its consequences. Water is involved in every process of human life; it is truly our most valuable resource.

 ## WATER AS A RESOURCE

Water is the only common substance that occurs as a solid, a liquid, and a gas over the temperature range found at the earth's surface. About 97 percent of the earth's water is in the oceans, and 2 percent is in ice caps and glaciers; less than 1 percent is readily available to humans as underground or surface fresh water (■ Table 9.1). Fortunately, water is a renewable resource as illustrated by the **hydrologic cycle,** the earth's most important natural cycle (◉ Figure 9.1). Water that is evaporated from the ocean or land surface goes back into the atmosphere and forms clouds. Precipita-

◉ **FIGURE 9.1 The hydrologic cycle.**

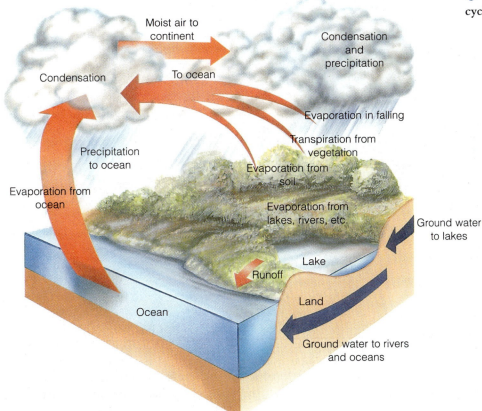

tion from these clouds that falls over land areas eventually flows in streams back to the sea, and the cycle is repeated. Remarkably, very little water is in the atmosphere at any one time, about 0.001 percent of the total. Because of this, we know that water must recycle continuously in order to supply the precipitation we measure. The average annual rainfall over the surface area of the 48 contiguous states is about 75 centimeters (30 in or 2.5 ft). **Evapotranspiration** is the term for both the direct return of surface water to the atmosphere by evaporation and its indirect return through transpiration from the leaves of plants. It accounts for the return of about 55 centimeters (22 in) of the average annual precipitation. River runoff to the sea and infiltration into the ground account for the remaining 20 centimeters, but it varies from 2.5 centimeters per year in the dry West to more than 50 centimeters per year in the East. Although runoff varies greatly with location, the net water transport from land to sea must over time balance the net transport from sea to land.

Although total runoff is many times the volume of water actually consumed in the United States, its variability really determines and limits the amount that is available in any one region for human use. In 1990, for example, total off-stream U.S. water use was about 1.5 trillion liters per day (422 billion gal/day), 80 percent of which came from surface water. This was 8 percent less than the demand in 1980, reflecting the impacts of a slowed economy and of conservation practices.

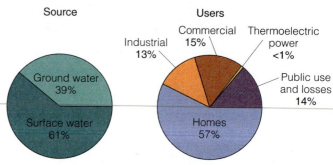

38,500 million gallons withdrawn per day

⦿ **FIGURE 9.3** The sources and users of the U.S. public water supply in 1990. This does not include fresh water supplied for agricultural use or water that is self-supplied.

Planners and forecasters recognize several categories of water use:

- *Instream use*—all uses that occur in the channel itself, such as hydroelectric power generation and navigation.
- *Offstream use*—the diversion of water from a stream to a place of use outside it, such as to your home. The amount of water available for offstream use along any stretch of a river is the total flow *minus* the amount required for instream purposes and for maintaining water quality. ⦿ Figure 9.2 is a flow diagram of public water supplies and self-supplied water. Note that the amount of water that is consumed is the difference between the amount delivered *(B)* and the amount that is returned to the stream system *(D)*.
- *Consumptive use*—water that evaporates, transpires, or infiltrates and cannot be used again immediately. Forty-four percent of all water that is withdrawn is used consumptively, and agriculture accounts for about 90 percent of that (Case Study 9.1).
- *Nonconsumptive use*—water that is returned to streams with or without treatment so that it can be used again downstream. Interestingly, domestic (household) water is used nonconsumptively, as it is returned to the cycle through sewers and storm drains.

The sources and users of public water supplies are shown in ⦿ Figure 9.3. The data do not include water used in agriculture or water that is self-supplied.

In many areas, particularly the western United States, consumption exceeds local water supplies. This necessitates importing water from other areas or mining ground water (see Chapter 10). For example, about 75 percent of Southern California's water is imported from sources more than 200 miles away (⦿ Figure 9.4). Two aqueducts import water from west and east of the Sierra Nevada, and a third aqueduct brings water from the Colorado River. Irrigation utilizes about 40 percent of the total water withdrawn in the United States, and a startling 80 percent in California. It is not surprising then that California, with the largest

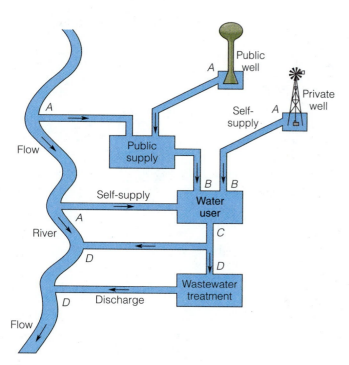

⦿ **FIGURE 9.2** Water is supplied to users from streams and from public and private wells *(A)*. Consumptive use *(C)* is the difference between the amount delivered to users *(B)* and the amount returned to the system *(D)*.

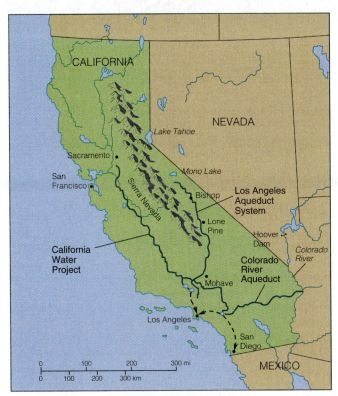

● FIGURE 9.4 Aqueducts bring water to Southern California from three source areas: the Colorado River, the east slope of the Sierra Nevada, and the west slope of the Sierra Nevada.

1970s and 80s, compounded by overtaxed water supplies, would be a disaster to those countries and their people.

RIVER SYSTEMS

Rivers are just one of several **geologic agents**—others are mass-wasting, glaciers, wind, underground water, and waves—that perform the work of erosion and deposition, which shape and modify the landscape. Rivers are the most important of the geologic agents today. Erosion (Latin *erodere,* "to gnaw away") is the carrying off of weathered materials from the earth surface (see Chapter 6). A stream's ability to erode the land and transport soil and rock is a function of the quantity and the velocity of the water in its channel. With equal **discharge**—that is, the same volume of water flowing past a point in the river channel during a certain period of time—a fast-flowing stream erodes its banks more significantly and carries more sediment than a slow-moving one. The equation of flow for a river is that its discharge *(Q) equals* its velocity *(V) times* its cross-sectional area *(A,* the product of its width and its average depth at a particular point), or

$$ Q \quad = \quad V \quad \times \quad A $$
discharge   velocity   cross-sectional area

Hydrology is the study of water and its properties, its circulation, and its distribution on and below the earth surface. English units are used in American hydrology and its literature: Q is expressed in cubic feet of water per second (cfs), V in feet per second (ft/sec), and A in square feet (ft^2). A stream with a discharge of 1,000 cfs (7,480 gal/sec) and a cross-sectional area of 100 ft^2 would have a velocity of 10 ft/sec, which is about 6.8 mph. If the river discharge increases because of rainfall or snowmelt, the stream's velocity and cross-sectional area also increase, and in a systematic way (● Figure 9.6, page 242).

Stream velocity also depends upon how much frictional resistance the bed and banks present to the flowing water. Wide, shallow streams have lower velocities than do narrow, deeper ones with the same gradient. (Gradient is explained in the next paragraph.) Thus stream velocity is more complicated than simple discharge–area relationships (Case Study 9.2). With increased velocity and discharge, there are increases in a stream's erosion potential and its sediment-

population and a huge irrigation system, is the largest U.S. consumer of water, using almost as much as the two next-largest consumers, Idaho and Texas, combined (●Figure 9.5).

Total world withdrawals of water for offstream purposes amount to about 2,400 cubic kilometers per year (1.7 trillion gal/day, *gpd*), of which 82 percent is for agricultural irrigation. This is expected to increase manyfold to 20,000 cubic kilometers (14.5 trillion gpd) by the year 2050, which is beyond the firm flow of the world's rivers. Availability of water will limit population growth in the next century, either by statute in developed countries or by some natural method, usually famine, in Third World nations. Repetition of the terrible droughts experienced in Africa in the

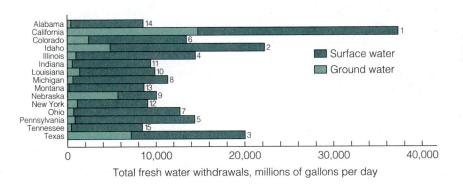

● FIGURE 9.5 The 15 largest consumers of fresh water in the United States in 1985. California, Idaho, and Texas are the biggest users by far.

The Colorado—A River in Peril

The thought grew in my mind that the canyons of this region would be a Book of Revelations in the rock-leaved Bible of geology.

JOHN WESLEY POWELL, GEOLOGIST AND EXPLORER

The Colorado River begins as a mere rivulet on the west side of the drainage boundary called the **Continental Divide**. All water that falls on the continent west of the divide flows to the west, and only two large rivers in the contiguous 48 states flow to the Pacific Ocean: the Colorado and the Columbia (Figure 1). The Colorado's flow, 15 million acre-feet per year (af/y), is puny compared to those of the Mississippi and the Columbia Rivers at 400 million and 192 million af/y, respectively. An acre-foot of water is the amount that would cover an acre of land 1-foot deep; it equals 325,000 gallons, enough to supply the water needs of a family of five for a year. However, the Colorado is one of the saltiest and siltiest rivers in the United States. The river and its two major tributaries, the Green River of Utah and Wyoming and the Gunnison River of Colorado (Figure 2), flow over sedimentary rock outcrops containing salt deposits, which contributes to the Colorado's saltiness. Its steep gradient

and predominantly sedimentary geology account for the large mass of sediment it transports, up to a half-million tons a day.

The Colorado and Green Rivers figured prominently in the exploration and settlement of the West. The Colorado's most celebrated explorer, John Wesley Powell, explored the area in 1869 and 1871–72 and noted prophetically that "all waters of . . . arid lands will eventually be taken from their natural channels." Today southern Utah's Lake Powell, impounded by the Glen Canyon Dam, bears his name. The Glen Canyon is one of ten major dams that control the Colorado's flow and meter out its precious water to more than 20 million people. Even so, the lake level rose to an all-time high in 1995–96.

The Colorado River has been called "a river drained dry." (See the Carrier entry in the Further Information list.) One of the most regulated streams in the world, legal battles over its water simmer continuously among the river states and between the United States and Mexico because the water has been over-appropriated. Problems began in 1922 when an agreement divided the river's catchment area into two water-resource basins. This agreement totally ignored Native American and Mexican claims dating back to the mid 1800s, which under the "first in time, first in right" doctrine, may establish their legal claim to sig-

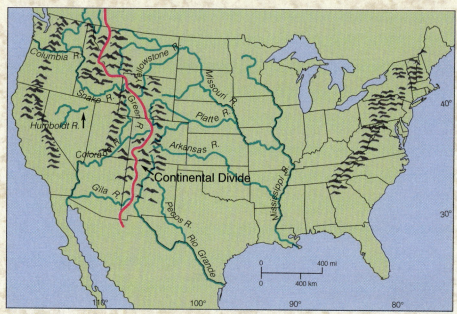

◉ **FIGURE 1** West of the Continental Divide streams flow to the Pacific Ocean. East of it, they flow to the Gulf of Mexico.

nificant amounts of river water. Under the 1922 agreement, Utah, Wyoming, Colorado, and New Mexico were to share the water of the upper basin, and Arizona, California, and Nevada were to share the lower-basin water. The users of each basin were allocated 7.5 million af/y. In addition, Mexico, by a 1944 treaty, was allocated a minimum of 1.5 million af/y.

Unfortunately, the allocations exceed the Colorado's total sustained yield, which fell to 9 million af/y during the drought of the late 1980s and early 1990s. Exacerbating the problem is the diversion of upper-tributary waters across the Continental Divide to supply cities and farms east of the Rocky Mountains. There are now 20 such diversions, and were it not for the requirement that sufficient flow be maintained to support wildlife, this drainage area would be totally diverted to Denver, Fort Collins, Aurora, and other foothill cities to the east.

So much water is taken from the Colorado River, both in the United States and Mexico, that it is a mere trickle when it flows into the Gulf of California. Much of the water is taken for agriculture by the Central Arizona Project and by California for the Imperial Valley. The water flows in unlined canals to irrigation sites, and some is returned to the river and flows to Mexico in a degraded condition. The treaty that gives Mexico its fair share of Colorado River water assumes that the quality of the water delivered will be suitable for agriculture. Returned Arizona irrigation water is so salty, however, that a purification plant had to be constructed at Yuma to bring it to agricultural standards. Nothing is simple. What started as a grand plan to give all users a fair share of water has become a legal nightmare for the users and a boon for lawyers.

In March 1996, a man-made flood on the Colorado River was accomplished by releasing 120 billion gallons of water from the Glen Canyon Dam over a 10-day period. It is hoped that the deluge, which raised the level of the river 3 to 5 meters (10–15 ft), will restore beaches and wildlife habitat in the Grand Canyon that have deteriorated since completion of the dam in 1966. This innovation in ecosystem management did not affect the level of Lake Powell.

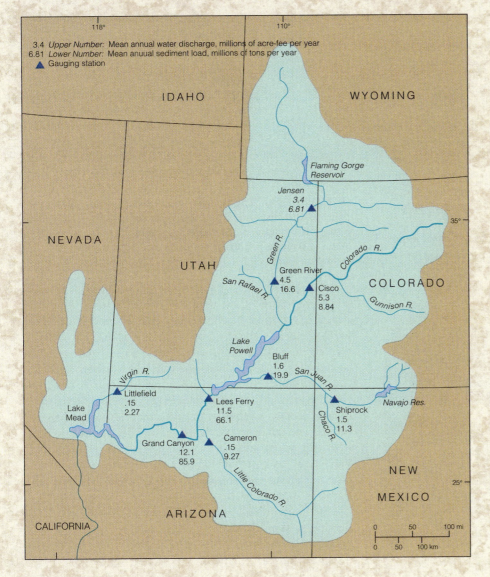

● FIGURE 2 The Colorado River drainage basin with its major tributaries. The mean annual water discharge and sediment load (1941–1957) at selected locations are shown.

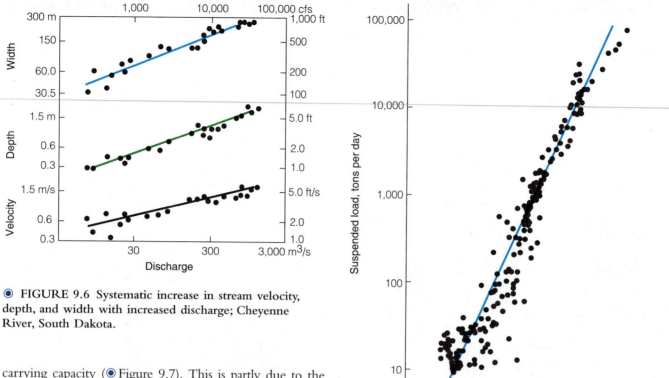

⊚ **FIGURE 9.6** Systematic increase in stream velocity, depth, and width with increased discharge; Cheyenne River, South Dakota.

⊚ **FIGURE 9.7** The relationship between water discharge (Q) and the suspended load for a typical stream. The points represent separate measurements. A tenfold increase in water discharge results in almost a 100-fold increase in sediment load.

carrying capacity (⊚ Figure 9.7). This is partly due to the fact that water flows in turbulent swirls and eddies at high velocities and in smooth stream lines called *laminar flow* at lower velocities. Turbulent flow dislodges sediment from the bed and sides of the stream channel and keeps it suspended in the flow until the stream slows, when gravity causes the particles to settle. The saddle-shaped lower boundary of the "erosion" field in ⊚ Figure 9.8 shows the velocities required to lift or suspend clay, silt, sand, gravel, and cobbles; the middle, "transportation" field shows the velocity range in which those particles remain in suspension. At flow velocities

below the lower "transportation" curve, particles settle (are deposited) to the bottom of the stream. Note that although higher velocities are required to erode clay and silt particles than to erode sand grains, once they are suspended in the stream, they can be transported at much lower velocities than can sand. This is perhaps surprising, but clay and silt particles are more cohesive than sand grains, and therefore, a higher velocity is required to pick them up (suspend them) into the stream flow. You may have experienced this phenomenon when walking on dry, fine-grained sediment on a windy day. As your foot breaks the cohesive sediment surface and creates air turbulence, the sediment becomes airborne as choking dust.

The longitudinal profile, or side view, of a stream channel has a slope known as its **gradient,** which can be expressed as the ratio of its vertical drop to the horizontal distance it travels (in meters per kilometer or feet per mile). The gradient of the Colorado River in the Grand Canyon is about 3.2 meters per kilometer, a moderately steep gradient, whereas in its lower reaches, where it flows into the Gulf of California, its gradient is only a few centimeters per kilometer. A mountain stream may have a gradient of several tens-of-meters per kilometer and an irregular profile

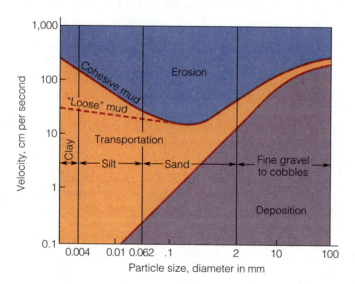

⊚ **FIGURE 9.8** So-called Hjulstrom diagram of threshold stream velocities for erosion, transportation, and deposition of varying particle sizes. Note that a higher water velocity is required to erode cohesive clay and silt than to move fine sand.

River Velocity and Friction

Hydraulic engineers have found that a stream's velocity depends on the **channel shape,** and is a function of the stream's cross-sectional area and its wetted perimeter. The **wetted perimeter** is the amount of contact the stream has with its channel in cross section—in other words, how much streambed area provides frictional resistance to the flow of water. Specifically, the relationship is stated as velocity *(V)* is proportional to the stream's cross-sectional area *(A)* divided by its wetted perimeter *(p),* or

$$V \approx A/p$$

◉ Figure 1 shows three streams of identical cross-sectional area but of significantly different wetted perimeters. The stream of shape (c) has the smallest wetted perimeter and would be the swiftest, whereas the stream of shape (a) would be the slowest.

The effect of wetted perimeter on velocity is illustrated in ◉ Figure 2, which shows three small tributaries joining to form a single stream. The cross-sectional area of the main stream is equal to the sum of the areas of the three tributaries, but the main stream's wetted perimeter is less than the sum of the wetted perimeters of the three tributaries. Thus *A/p* is greater for the main stream, and the extra water is accommodated by increased velocity.

This example assumes there is no drastic change in the roughness (frictional drag) of the streambed under consideration. All other things being equal, the smaller the wetted perimeter, the greater is the velocity. Obviously, if the slope (gradient) increases,

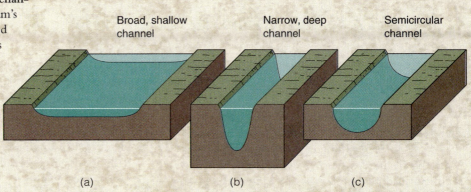

Broad, shallow channel

Narrow, deep channel

Semicircular channel

(a)　　　　(b)　　　　(c)

◉ FIGURE 1 Three stream cross sections of equal areas. Section (a) has the smallest wetted perimeter, and, assuming equal gradients, its velocity would be greatest.

the velocity also will increase. Thus, the velocity of a stream really depends on four variables:

- gradient (the slope),
- channel shape *(A/p),*
- discharge, which influences both area and wetted perimeter, and
- roughness (the frictional drag of boulders, sand, silt, clay, or whatever composes the bed).

On alluvial fans in arid regions (see Figure 9.10), just the opposite happens. The wetted perimeter increases because of changes in the channel shape and also because the fan channels divert or spread out into many *distributary channels,* rather than *coalescing* into a single main stream (see Figure 9.32). When this happens, the water slows and eventually sinks into the ground.

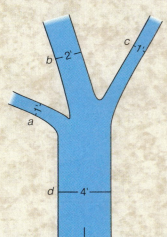

	Area, ft² *A*	Wetted perimeter, ft *p*	*A/p*
1' *a* 1' / 1'	1	3	1/3
1' *b* 1' / 2'	2	4	1/2
1' *c* 1' / 1'	1	3	1/3
1' *d* 1' / 4'	4	6	2/3

◉ FIGURE 2 Cross sections of three tributaries *(a, b,* and *c)* and the main stream *(d).* The cross-sectional area of the main stream equals the sum of the three tributaries' cross-sectional areas, but the ratio of the main stream's area to its wetted perimeter, *A/p,* is greater than the *A/p* of any tributary. Because the main stream has less bed friction (wetted perimeter), it speeds up to accommodate increased discharge.

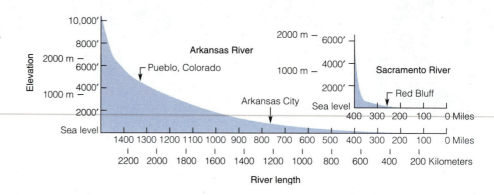

● FIGURE 9.9 Measured longitudinal profiles for the Arkansas and Sacramento Rivers. The vertical scale is exaggerated 275 times.

CONSIDER THIS...

We generally associate high sediment loads with high river discharge, but there are other variables that determine the sediment transported by a river. With this in mind, would you expect the suspended sediment load per cubic meter of water to be greater in the Columbia River, which has a high-rainfall drainage basin, or in the Colorado River, which has a low-rainfall basin?

with rapids and waterfalls. As the same stream approaches the lowest level to which it can erode, an elevation known as **base level,** its gradient may be almost flat. A stream's ultimate base level is sea level, of course, but it may have intermediate, temporary base levels such as a lake surface or a rock formation that is highly resistant to erosion. A stream whose gradient has adjusted so that an equilibrium exists between its discharge and its sediment supply is known as a **graded stream.** Longitudinal profiles of graded streams are generally concave-upward from source to end. ● Figure 9.9 shows the profiles of the Arkansas River (see Figure 9.35), which drains the wide east slope of the Rocky Mountains, and the Sacramento River, which drains the much narrower Pacific slope.

Gradients decrease in the lower reaches of a river, and sedimentation may occur as streams overflow their banks onto adjacent **floodplains.** Decreased water velocity on the floodplain results in the deposition of fine-grained silt and clay adjacent to the main stream channel. This productive soil, commonly known as *bottom land,* is much desired in agriculture.

⊚ STREAM FEATURES

All sediment deposited by running water is called **alluvium** (Latin, a flooded place or flood deposit). In the arid West, where streams lose velocity as they flow from mountainous terrain onto flat valley floors, we find conspicuous deposi-

tional features known as **alluvial fans** (● Figure 9.10). They are in fact fan-shaped and they build up as the main stream branches, or splays, outward in a series of *distributaries,* each "seeking" the steepest gradient. Exemplified by Figure 9.10, alluvial fans can provide beautiful, interesting views of surrounding terrain, but they also can be subject to flash floods and debris flows from intense seasonal storms in nearby mountains.

Deltas form where running water moves into standing water. The Greek historian Herodotus (484–420 B.C.) applied the term to the somewhat triangular shape of the Nile delta in recognition of its resemblance to the Greek letter delta (Δ). Deltas are characterized by low stream gradients, swamps, lakes, and an ever-changing channel system. The "arcuate" (arc-shaped) delta of the Nile and the "bird's foot" delta of the Mississippi also are described by their appearance on maps (● Figure 9.11). Deltas also form at the mouths of small streams, as illustrated by the arcuate delta at Malibu Creek—location of the famous Malibu Colony of the rich and famous and the early beach-boy-type surfing movies (● Figure 9.12, page 247).

Three main types of deltas are recognized according to the relative importance of streams, wave action, and tides in their formation and maintenance. The Mississippi delta exemplifies *stream-dominated* deltas. It consists of long, fingerlike distributaries that have built the delta out into the Gulf of Mexico and imparted to it the appearance of a bird's foot. The Nile delta is categorized as *wave-dominated,* because its seaward margin consists of a series of sand-barrier islands formed by currents and wave action. The tremendous sand output of the Nile has created the huge coastal

CONSIDER THIS...

Deltas are deposited where a river flows into a body of standing water, such as a lake or an ocean. Why does one of the great rivers of North America, the St. Lawrence River, have no delta, even though it flows into a landlocked estuary?

(a)

(b)

Distributaries

● FIGURE 9.10 (a) Alluvial fans in Death Valley, California. The Panamint Range is in the background. A fan forms where a mountain stream flows onto an adjacent, flat-floored valley and deposits its sediment load in a system of distributary channels. (b) A fan's symmetrical shape is the result of the active channel "migrating" back and forth across the fan surface.

dunes east of Alexandria that formed the background for the famous World War II battle at El Alamein (see Case Study 12.2, page 330). The Ganges–Brahmaputra delta of Bangladesh is *tide-dominated,* consisting of tidal flats and many low-lying sand islands offshore (Figure 9.11, part c). Because deltas are at or near sea level, they are subject to flooding by tropical storms. Coastal flooding is the most deadly of natural disasters, even more deadly than earthquakes or river floods. This type of flooding was particularly disastrous in Bangladesh in 1970, when more than 500,000 people were drowned by a surge of water onshore from a tropical cyclone. A similar tragedy occurred in 1991, when another onshore water surge left a reported 100,000 dead on the low-lying islands off the delta (see Chapter 11).

Streams and rivers with low gradients and floodplains form a number of features that both aggravate and alleviate flood damage. The courses of such rivers follow a series of S-shaped curves called **meanders** (after the ancient name of the winding Menderes River in Turkey; ● Figure 9.13). Straight rivers are the exception, rather than the rule, and it appears that running water of all kinds will meander—ocean currents, rivers on glaciers, and even trickles of water on glass. The key to these regularly spaced curves is the river's ability to erode, transport, and deposit sediments. The meandering process begins when the gradient is reduced to a slope at which a

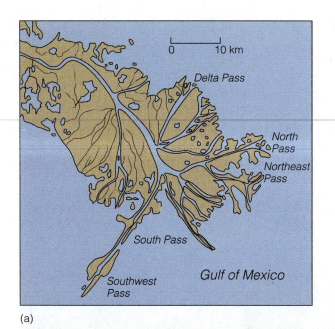

(a)

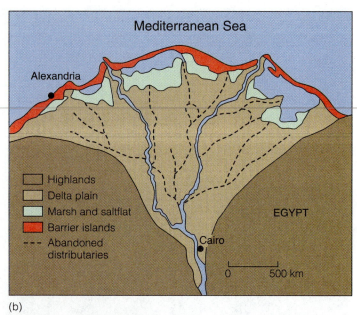

(b)

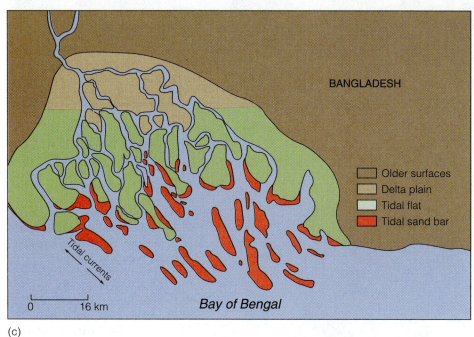

(c)

● FIGURE 9.11 *(a)* River-dominated "bird's-foot" delta of the Mississippi River. The main river channels are separated by swampy land. *(b)* Wave-dominated arcuate delta of the Nile River. *(c)* Tide-dominated Ganges–Brahmaputra delta of Bangladesh. The tidal flats and islands are subject to periodic flooding.

stream stops deepening its valley. As a stream meanders, it erodes the outside curve of each meander, the *cutbank,* and deposits sediment on each inside curve, the *slip-off slope,* producing a deposit known as a **point bar.** ● Figure 9.14 shows the velocity distribution on a stream meander, which explains the erosion–deposition pattern. Meanders also widen stream valleys by lateral cutting of valley side slopes, causing mass wasting and eventually creating an alluvium-filled floodplain adjacent to the river. It is across this surface that the river meanders, sometimes becoming so tightly curved that it shortcuts a loop to form a **cutoff** and **oxbow lake** (● Figure 9.15, page 249).

By the time a river has developed a wide floodplain with meanders, cutoffs, and oxbow lakes, a situation of poor drainage exists that increases the impact of overbank flooding. **Natural levees** build up on both banks of a river when it overflows and deposits sand and silt adjacent to the channel (● Figure 9.16). Natural levees, or manufactured ones, **artificial levees,** keep flood waters channelized so that the water level in the stream may be above the level of the adjacent floodplain. This is a tenuous situation, because a breach in the levee would cause flooding of the floodplain, which could force the river into a new course. New Orleans on the Mississippi River is an example of a city on

⊙ FIGURE 9.12 The arcuate, wave-dominated delta of Malibu Creek, Malibu, California, after a record rainfall in 1969. Note the suspended sediment in the ocean and the concentration of homes along the delta north *(left)* of the creek. Compare this shape with that of the Nile delta, Figure 9.11, part b.

⊙ FIGURE 9.13 Meanders on the Missouri River at Glasgow, Missouri. Note that the meander belt swings across the full width of the river's floodplain.

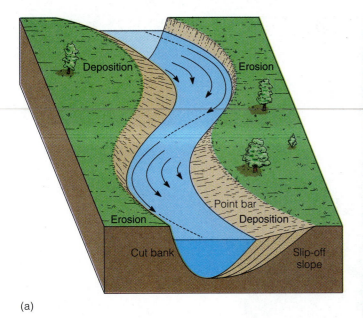

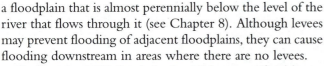

(a)

(b)

● FIGURE 9.14 Erosion and deposition patterns on a meandering stream. *(a)* Erosion of the cutbank and deposition of a point bar on the slip-off slope. Arrow length is proportional to stream velocity. *(b)* Rillito Creek rampaging through Tucson, Arizona, during the flood of 1983. Damage occurred when the cutbank of the meander migrated west *(left)* into vacant property immediately upstream from the townhouses *(center left)*. This allowed erosion to occur behind a concrete bank that was protecting the townhouse property. Note the prominent point bar that developed *(bottom center)* as the meander migrated westward.

a floodplain that is almost perennially below the level of the river that flows through it (see Chapter 8). Although levees may prevent flooding of adjacent floodplains, they can cause flooding downstream in areas where there are no levees.

During the 1993 floods in the Midwest, artificial levees saved some cities in the Mississippi River drainage basin, notably St. Louis, where the flood crested at 49.4 feet and the levee height is 52 feet. Quincy, Illinois, lacking artificial levees, was inundated, but the levees at Hannibal, Missouri, on the opposite side of the river held and spared the town. A levee on the Raccoon River at Des Moines, Iowa, failed, and floodwaters inundated a city water-purification plant. Thousands of residents were without drinking water for 19 days. Some levees were intentionally breached in order to reduce flood-crest elevation downstream. More than 800 of the 1,400 levees in the nine-state disaster area were breached or overtopped. Most of these were simple dirt berms built by local communities, which explains why so many small towns were almost completely submerged by the floodwaters. However, more than 30 levees built by the U.S. Army Corps of Engineers also failed during this record flood event (●Figure 9.17, page 250).

WHEN THERE IS TOO MUCH WATER

Excess water, whether in rivers, lakes, or along a coast, leads to flooding. The basic hydrologic unit in fluvial (river) systems is the **drainage basin,** the land area that contributes water to a particular stream or stream system. Individual drainage basins are separated by **drainage divides.** A drainage basin may have only one stream, or it may encompass a large number of streams and all their tributaries. Most of the rain that falls within a particular drainage basin after a dry period infiltrates (seeps) into the uppermost layers of soil and rock; little water runs off the land. As the surface materials become saturated and the rain continues, however, rainfall exceeds infiltration, and runoff begins. Runoff begins as a general sheet flow, then forms tiny rills and gullies, then flows into small tributary streams, and finally is consolidated into a well-defined channel of the main stream in the watershed. Whether or not a flood occurs is determined by several factors: the intensity of the rainfall, the amount of antecedent (prior) rainfall, the amount of snowmelt if any, the topography, and the vegetation.

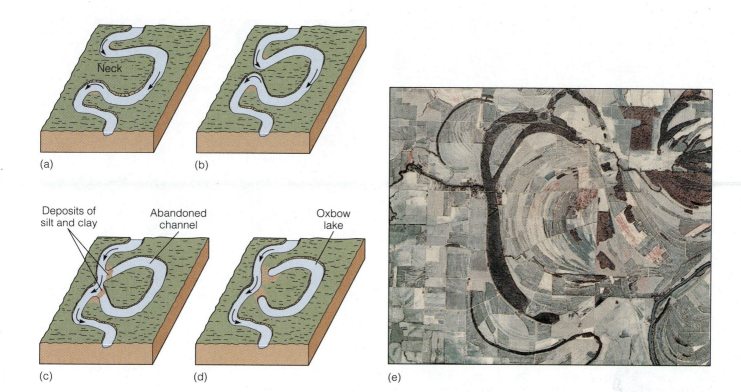

● **FIGURE 9.15** *(a–d)* Meander pattern sequence of erosion leading to a cutoff and an oxbow lake. A cutoff forms in *(c)* and completely isolates the oxbow lake in *(d)*. *(e)* Aerial photo of an oxbow lake on the Tallahatchie River in Mississippi. Note the old scars from the migration of meanders.

Upland floods occur in watershed areas of moderate or high topographic relief. They may occur with little or no warning and they are generally of short duration; thus the name *flash floods*. Such floods result in extensive damage due to the sheer energy of the flowing water. Lowland, *riverine floods,* on the other hand, occur when broad floodplains adjacent to the stream channel are inundated. The typical damage is that things get wet or silt-covered. The difference in these types of floods is due to watershed morphology (shape). Mountain streams occupy the bottoms of V-shaped valleys and cover most of the narrow valley floor; there is no adjacent floodplain. Heavy rainfall at higher elevations in the watershed can produce a wall of fast-moving water that descends through the canyon, destroying structures and endangering lives (● Figure 9.18). Lowland streams, in contrast, have low gradients, adjacent wide floodplains, and meandering paths.

A **hydrograph** is a graph of a water body's discharge, velocity, stage (height), or some other characteristic over time, and it is the basic tool of hydrologists (● Figure 9.19). The "synthetic" hydrograph in ● Figure 9.20 graphically illustrates the basic difference between upland (flash) floods of short duration and the longer-lasting lowland floods. The timing of flood crests on the main stream—that is, whether they arrive from the tributaries simultaneously or in sequence—in large part determines the severity of lowland flooding.

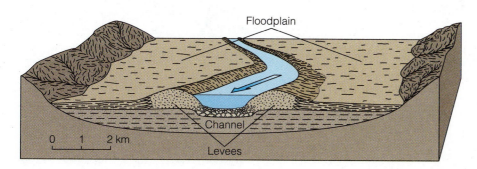

● **FIGURE 9.16** Relationship of natural levees and floodplain to a hypothetical stream. The coarsest sediment is deposited in the stream channel, where the stream's velocity is greatest, and finer silt and clay are deposited on levees and the floodplain as velocity diminishes during overflows.

● FIGURE 9.17 One of the 800 levees that failed or were overtopped along the Mississippi, Missouri, and Illinois Rivers during the 1993 floods. This levee is near Davenport, Iowa, on the Mississippi River.

● FIGURE 9.18 A wall of water from a short-lived cloudburst upstream rushed through Big Thompson Canyon, Colorado, in 1974. The flash flood destroyed many homes in the narrow canyon and caught 140 victims who could not scale the steep canyon slopes in time to avoid the water.

Floods are normal and, to a degree, predictable natural events. Generally, a stream will overflow its banks every two to three years, gently inundating a portion of the adjacent floodplain. Over long periods of time, catastrophic high waters can be expected. Depending upon their severity, these events are referred to as fifty-, one-hundred-, or five-hundred-year floods. The more infrequent the event, the more widespread and catastrophic is the inundation (● Figure 9.21). The Midwest flood of 1993 that inundated 421 counties in 9 states, with a peak flow of 720,000 cfs (5.1 million gallons per second) at Boonville, Missouri, is regarded as a 500-year flood. Flood frequency is explained in more detail later in this section.

● FIGURE 9.19 A hypothetical hydrograph that relates precipitation and discharge to time. Note that peak discharge occurs after the peak rainfall intensity. The drainage basin includes the main stream and all its tributaries.

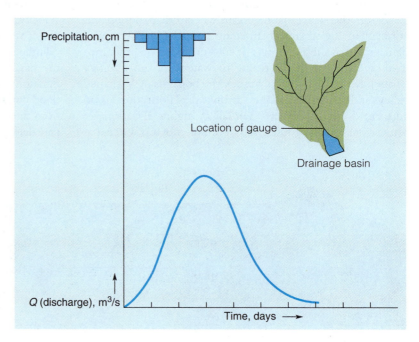

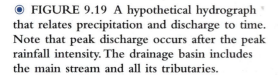

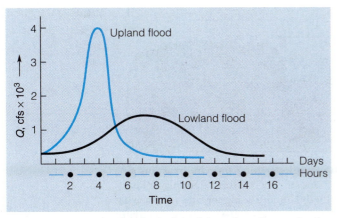

◉ **FIGURE 9.20 Hypothetical hydrographs of an upland flash flood and lowland riverine flood.**

Floods from all causes are the number–one natural disaster in the world in terms of loss of life. Twenty-three disastrous U.S. floods are listed in ◼ Table 9.2, and their locations

are shown on the map of ◉ Figure 9.22. Although by no means a complete listing of severe floods, the table and map show the general distribution of riverine floods, flash floods, and coastal floods in the United States. Coastal flooding due to hurricanes and typhoons is the greatest single killer; this is discussed in Chapter 11. The record–holding river floods are due to dam failures. The legendary 1889 Johnstown flood killed 3,000; the 1928 St. Francis Dam failure in

◼ **TABLE 9.2 Selected U.S. Floods since May 1889**

MAP NUMBER	DATE	LOCATION	LIVES LOST	ESTIMATED DAMAGE, $ MILLIONS
1	May 1889	Johnstown, Pa. (dam failure)	3,000	—
2	8 Sept. 1900	Galveston, Tex. (hurricane)	6,000	30
3	May–June 1903	Kansas and Lower Missouri Rivers	100	40
4	March 1913	Ohio River basin	467	147
5	14 Sept. 1919	South of Corpus Christi, Tex. (hurricane)	600–900	22
6	Spring 1927	Mississippi River valley	313	284
7	12–13 Mar. 1928	California (St. Francis Dam failure)	450	14
8	13 Sept. 1928	Lake Okeechobee, Fla.	1,836	26
9	Mar. 1938	Southern California	79	25
10	21 Sept. 1938	New England	600	306
11	August 1955	Northeastern U.S. (Hurricane Diane)	187	714
12	27–30 June 1957	Texas and Louisiana (Hurricane Audrey)	390	150
13	Jan.–Feb. 1969	California	60	399
14	17–18 August 1969	Mississippi, Louisiana, and Alabama (Hurricane Camille)	256	1,421
15	February 1972	Buffalo Creek, W. Va.	125	10
16	June 1972	Black Hills, S.D.	238	165
17	June 1972	Eastern U.S. (Hurricane Agnes)	105	4,020
18	June–July 1975	Red River of the North basin	<10	273
19	June 1976	Southeast Idaho (Teton Dam failure)	11	1,000
20	July 1976	Big Thompson River, Colo.	139	30
21	Sept.–Oct. 1983	Tucson, Arizona	<10	50–100
22	Summer 1993	Upper Mississippi River drainage basin	50	12,000
23	Jan.–Feb. 1996	Oregon, Washington, Idaho, Montana	1	89*

SOURCE: Various sources, including U.S. Geological Survey.
*Figure is for Oregon and Washington only.

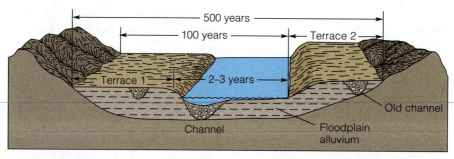

● FIGURE 9.21 The probable extent of flooding over long time intervals on a hypothetical floodplain. The 2–3-year flood inundates the existing floodplain. The 100-year flood inundates both the existing floodplain and a higher one (Terrace 1), which formed when the river stood at a higher level. The 500-year event inundates an even higher terrace (Terrace 2) and all lower terraces and floodplains.

Southern California killed 450 people (Case Study 9.3), and the 1972 collapse of an earth embankment at Buffalo Creek, West Virginia, resulted in 125 fatalities.

Flood Measurement

As we have seen, the basic unit in river studies is the drainage basin, which can be defined on any scale. A large

river's drainage basin includes the drainage basins of many tributaries, and each tributary's drainage basin includes the drainage basins of all *its* tributaries, and so on. It follows that a large drainage basin has the potential for a larger flood than does a small drainage basin.

Flood studies begin with hydrographs for a particular storm on a river within a given drainage basin. These graphs are generated from data obtained at one or more

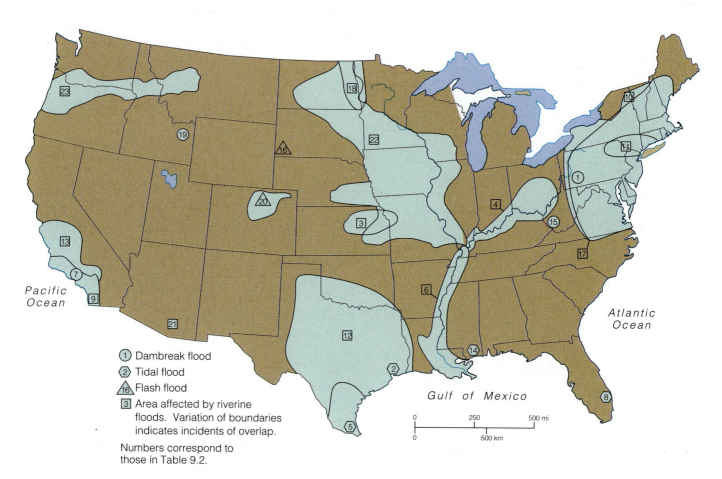

● FIGURE 9.22 Locations of 23 major floods in the conterminous 48 states, 1889–1995. The numbers on the map refer to descriptions of the floods in Table 9.2.

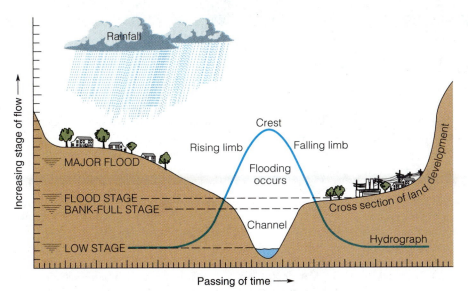

● FIGURE 9.23 A stage hydrograph superimposed on a hypothetical stream valley, which illustrates the impact of increasing stage height on the flooded area.

stream gauging stations, where stage (the water elevation above a previously established datum plane) and velocity are measured and then related to discharge. Rapid runoff in a drainage basin is accompanied by high stream discharge and high water velocity. As discharge increases, the water rises to the *bank-full stage* and then spills onto the adjacent flood-plain at *flood stage.* ● Figure 9.23 shows such a condition. It is a hydrograph superimposed on the cross section of a hypothetical stream valley. A *stage hydrograph* is most useful for flood planning purposes, because it relates stream discharge to river height, or elevation (Case Study 9.4). One can see at a glance at what point the stream will overflow its banks and at what elevations flood waters will come in contact with specific structures. Note the contrasting hydrographs in ● Figure 9.24, page 257. Whereas the flash flood in Tucson, Arizona, lasted only about five hours, the North Dakota flood lasted for almost a month. This reflects the climatic and topographic differences between an area with intense, short-duration monsoonal rains (Tucson), and an area of heavy snowfall followed by a rapid spring thaw accompanied by heavy rainfall (north-central U.S.). The Mississippi River flood of 1993 lasted for three months.

Flood Frequency

One of the most useful relationships that can be derived from long-term flood records is that of **flood frequency,** or **recurrence interval;** that is, how often on average a flood of a given magnitude can be expected in a particular location. With this kind of information we are better equipped to design dams, bridge clearances over rivers, sizes of storm drains, and the like.

The information needed for determining recurrence intervals is a long-term record of annual *peak discharge* (the largest flow) for the particular location on the river. The longer the period of record, the more reliable the statistics will be. The peak discharges are then ranked according to

their relative magnitudes—with the highest discharge being ranked as 1, the second-largest as 2, and so forth. To serve as an example, selected annual peak discharges on the Rio Grande near Lobatos, Colorado, are listed in ■ Table 9.3, with their ranks and recurrence intervals. A recurrence interval, *T,* is calculated as

$$T = (N + 1) \div M$$

where N is the number of years of record (76 years in our example), and M is the rank of the event. The analysis predicts how often floods of a given size can be expected to occur in a watershed. In this example, a flood the size of the 1906 flood ($M = 10$) would be expected to occur on average once every 7.7 years [$(76 + 1) \div 10$]. If 100 years of records are available, the largest flood to occur during that

■ **TABLE 9.3** Annual Peak Discharges on the Rio Grande River near Lobatos, Colorado, 1900–1975

YEAR	DISCHARGE, m³/sec	RANK (M)	RECURRENCE INTERVAL (T)
1900	133	23	3.35
1901	103	35	2.20
1902	16	69	1.12
1903	362	2	38.50
1904	22	66	1.17
1905	371	1	77.00
1906	234	10	7.70
1907	249	7	11.00
1908	61	45	1.71
1909	211	13	5.92
1974	22	64	1.20
1975	68	43	1.79

SOURCE: U.S. Geological Survey.

It's Another Dam Disaster

St. Francis Dam is just one of many dams that have been constructed on faults and that have later failed or presented problems because of it. This dam was to be the southern terminus of the 225-mile-long Los Angeles Aqueduct system that would bring water from the east slope of the Sierra Nevada and the Owens Valley to arid Southern California. Much has been written about the motives and the consequences of building the aqueduct; the award-winning movie *Chinatown,* for example, is a thinly disguised portrayal of the project's principals. The dam was completed in 1926, and the 38,000-acre-foot reservoir was full by 5 March 1928 (◉ Figure 1). Just one week later it failed, and 450 persons lost their lives.

The dam was built in San Francisquito Canyon, a short distance north of the now-heavily populated San Fernando Valley. The plan was to store aqueduct water here and distribute it by gravity to local storage reservoirs and eventually to users in the San Fernando Valley and the Los Angeles basin. The St. Francis Dam was a massive gravity dam; that is, it was designed to hold back the water in its reservoir by its sheer weight alone, rather than with the support of an arch to distribute stresses into the foundation rock (◉ Figure 2). The right abutment (foundation) of the dam was in mica schist, which is landslide-prone, and the left abutment was in red sandstone and conglomerate. The contact between the two rock types was an inactive fault zone of crushed rock about five feet thick. Later it was found that the fault zone and the sandstone contain considerable gypsum, which is soluble in water. Indeed, these were not desirable foundation conditions for a dam. No geologic examination was performed before the dam was built, and therefore the fault was not identified, but the rocks were tested dry and found to have adequate strength.

Shortly after the reservoir had filled, a leak was noticed at the left (sandstone) abutment. On March 10 damkeeper Tony Harnischfeger reportedly told friends who were arranging a fishing party that "if the dam's still here Wednesday, I'll come down and we'll go out [fishing]." On March 12 Harnischfeger noticed a new leak at the west (left) abutment and alerted William Mulholland and Harvey Van Norman, the men who were responsible for the dam's construction. They believed the seepage was not serious and had a photographer take their pictures on top of the dam.

Just before midnight that night, water burst through the foundation along the fault contact between the schist and sandstone, and the dam collapsed at this point. The swirling waters undermined the schist at the opposite abutment, carrying away the damkeeper's shelter and leaving only the dam's center section. A wall of water 185 feet high exploded down the canyon, wiping out powerhouses, farms, construction camps, and an Indian trading post owned by silent-movie actor Harry Carey. Huge pieces of the dam, some weighing 10,000 tons, were washed more than a half-mile downstream. Water submerged a construction-camp tent city so quickly that many of the sealed canvas tents acted as air pockets, bringing the floors, tents, and men rapidly to the surface, where they rode the wave and luckily survived the flood. A rancher shotgunned a hole in the roof of his house to provide an

◉ FIGURE 1 The St. Francis Dam in San Francisquito Canyon, March 1928.

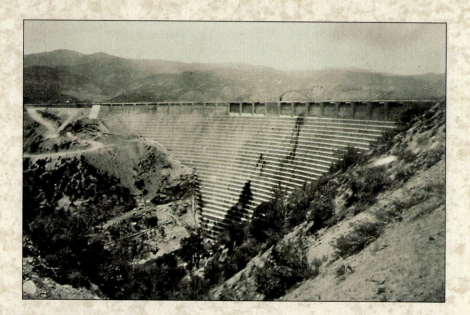

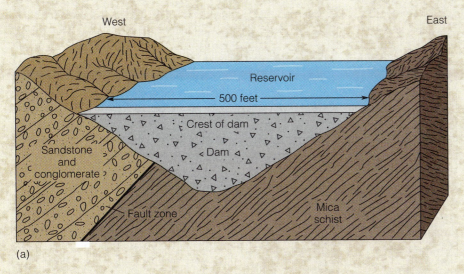

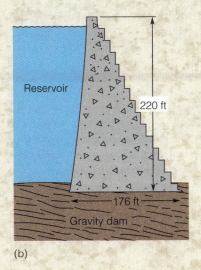

⊙ **FIGURE 2** *(a)* Geologic cross section of the St. Francis Dam through its abutments. *(b)* Cross section of a typical gravity dam. The dimensions shown are those of the St. Francis Dam (not to scale).

escape hatch for his family as the waters rose. As it turned out, this was unnecessary. The entire house was lifted and floated downstream. There it was gently deposited without harm to the house or any of the family.

Others were not so fortunate. In all, about 450 people drowned in the flood. William Mulholland died shortly afterward and was regarded as the dam's 451st victim. The dam's center section remained standing, like a monolithic gravestone (⊙ Figure 3). Eventually it was blasted into rubble so that it would cease being a reminder of the tragedy and a tourist attraction.

⊙ **FIGURE 3** All that remained of the dam on 13 March 1928.

CASE STUDY 9.4

The Nile River—Three Cubits between Security and Disaster

Flooding is a natural part of the hydrologic cycle, and as farmers, hydrologists, and geologists know, it can be beneficial. The annual flood on the Nile River, the longest river in the world (depending upon how one measures the Amazon), was a source of life for at least 5,000 years. Eighty percent of the water for these floods came from the faraway Ethiopian Highlands and it replenished the soil moisture and deposited a layer of nutrient-rich silt on the floodplain. The Greek historian and naturalist Herodotus, in 457 B.C., recognized the importance of annual flooding when he spoke of Egypt as "the gift of the Nile." The priests of ancient Egypt used principles of astronomy to calculate the time of the river's annual rise and fall. Records of the height of the flood crests go back to 1750 B.C. (some say to 3600 B.C.), as measured with primitive hydrographs known as Nilometers (⊙ Figure 1). Water height was measured in cubits, one cubit being the distance from one's elbow to the tip of the middle finger, approximately 0.5 meter. In Shakespeare's *Antony and Cleopatra,* Mark Antony notes that

> They take the flow o' th' Nile . . . .
> By certain scales . . . they know
> By th' height, the lowness, or the mean,
>   if dearth
> Or foison follow. The higher Nilus swells,
> The more it promises. As it ebbs,
>   the seedsman
> Upon the slime and ooze scatters his grain,
> And shortly comes to harvest.

And, of course, John Lennon of the Beatles supposedly once pointed out that "denial is not a river in Egypt."

The annual flooding cycle has changed since completion of

⊙ **FIGURE 1** The Nilometers of ancient Egypt were primitive hydrographs used to measure the Nile's flood heights in cubits, a cubit being the distance between one's elbow and the tip of the middle finger (usually about 18 in). The environmental interpretations shown here were used during the time of Pliny the Elder (A.D. 23–79) of Mount Vesuvius fame.

the Aswan High Dam on the Upper Nile in southern Egypt in 1970. The dam's purposes were to increase the extent of arable land, generate electricity, improve navigation, and control the flooding on the Lower Nile, and it is accomplishing all of those things. Now that the flooding is controlled, however, crop maintenance depends upon irrigation. Severe environmental consequences have resulted, including depletion of the soil nutrients that are beneficial to farming (necessitating the use of artificial fertilizers) and the proliferation of parasite-bearing snails in irrigation ditches and canals. *Schistosomiasis* is a human disease found in communities where people bathe or work in irrigation canals containing snails serving as temporary hosts to parasitic flatworms (schistosomes). These revolting creatures, commonly called blood flukes, invade human's skin when they step on the snails while wading in the canals. The flukes live in their human hosts' blood, where the females release large numbers of eggs daily, which cause extensive cell damage. Next to malaria, *schistosomiasis* is the most serious parasitic infection in humans. Its outbreak in Egypt is a result of controls on the river that prevent the generous floods that formerly flushed the snails from canals and ditches each year. Another negative consequence is the loss of a huge shrimp fishery off the Nile delta; the dam traps the nutrients that encouraged the growth of plankton, upon which the shrimp larvae fed. Every dam has a finite useful lifetime that is determined by how fast the lake it impounds fills with sediment. Lake Nasser, behind the Aswan Dam, is rapidly filling with the (precious) sediment that formerly enriched the floodplain of the lower Nile.

period ($M = 1$) would have a recurrence interval of about 100 years on average. This is the so-called 100-year flood.

The *probability* (chance) of a flood of a given rank occurring in any one year is 1/T; this yields 1 percent for the 100-year flood, 2 percent for the 50-year flood, and so

forth. It should be noted, however, that there is no meteorological or statistical "prohibition" against more than one 100-year flood occurring in 100 years. Because of the statistical laws involved, there is (only) a 63 percent chance that the 100-year flood will occur in 100 years and a 50 per-

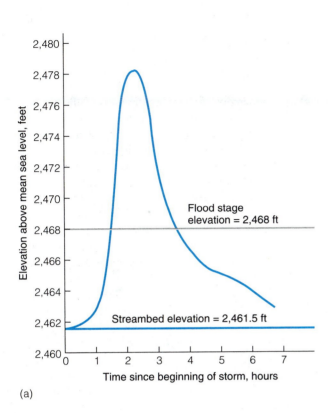

(a)

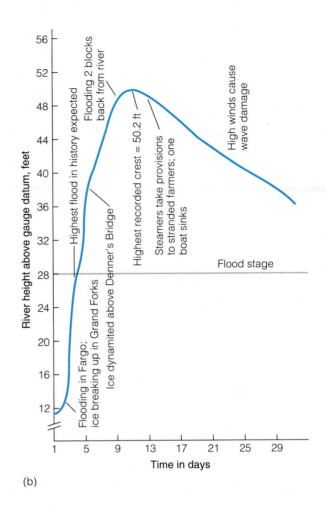

(b)

● FIGURE 9.24 (a) Stage hydrograph for a 5-hour storm in 1975 on Tanque Verde Creek in Tucson, Arizona. Note that the water rose 10 feet above flood-stage elevation and returned to below flood stage in about 2 hours. All the damage was done in a very short time. (b) Stage hydrograph for the great Red River flood of 1897. The waters took about a week to peak at 50 feet above the datum plane and then required almost a month to return to the channel below flood level.

cent chance that it will occur within 69 years (● Figure 9.25). Recurrence intervals are not exact timetables; they merely indicate the statistical probability of a given-size flood. However, knowing the flood discharge that can be expected in a river system every 25, 50, or 100 years enables us to better plan against the flooding that is certain to occur eventually.

Once we have determined recurrence intervals from the historical record, it is possible to construct flood-frequency graphs like the ones in ● Figure 9.26. Such graphs allow us to estimate at a glance how often a flood-level or discharge of a particular magnitude should occur. For example, if we wish to build a warehouse with a useful life of 50 years on the floodplain of the Red River at Grand Forks, we see in Figure 9.26, part a, that the river reaches a reference elevation of about 49 feet every 50 years on average. We would probably want to build the warehouse at or above that ele-

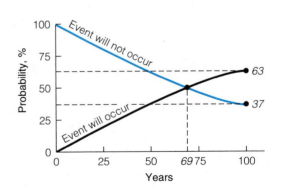

● FIGURE 9.25 Graph of the probability of the 100-year flood. There is a 63 percent chance that it will occur within 100 years and a 50 percent chance of its occurring within 69 years.

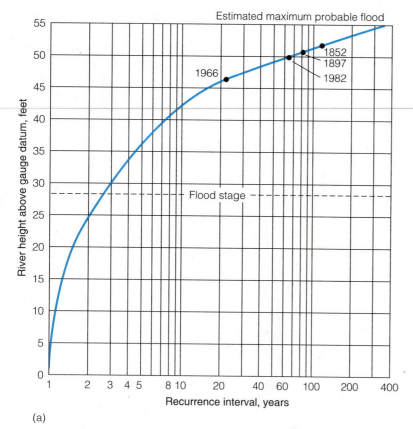

(a)

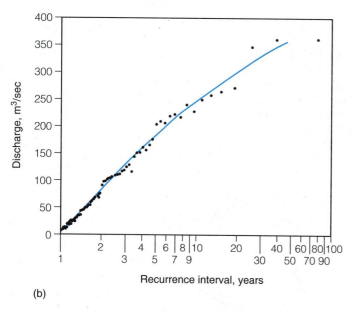

(b)

● **FIGURE 9.26 Flood frequency curves for two rivers, (a) the Red River at Grand Forks, North Dakota, using water elevation as the standard, and (b) the Rio Grande River near Lobatos, Colorado, using water discharge as the measure.**

vation. However, urbanization, artificial levees, and other human-generated changes can alter the amount and timing of runoff and cause higher-discharge flood events to become more common. Consequently, the 100-year flood may become the 70- or 80-year flood. These concerns are discussed later in the chapter.

The Flood of 1993

Forty percent of all the precipitation falling on the contiguous 48 states is collected in the Mississippi River drainage basin. Thus it is not surprising that one of the largest riverine floods attributable to rainfall was the great Mississippi Valley flood of 1993. The flood was the result of a wet spring followed by record-breaking rains in June and July. Above-normal precipitation in the six months preceding the flooding filled reservoirs and produced stream flows that were well above normal. In June and July the cool, dry air of the jet stream dipped abnormally south and collided with moist air from the Gulf of Mexico. This caused heavy rainfall in parts of nine Midwestern states. Large streamflows, exceeding those expected only once every 100 years, were recorded at 45 gauging stations in these states (● Figure 9.27). Flooding along the Mississippi and its tributaries inundated 4 million hectares (10 million acres) in the nine states. At least 50 lives were lost, and damage estimates ranged from 10 to 12 billion dollars.

Most of the rain fell in the upper Mississippi drainage basin, and the flood crests migrated progressively downstream from St. Paul, Minnesota (June 26), to Dubuque, Iowa (July 6), to Quincy, Illinois (July 13), and finally to St. Louis, Missouri (July 19–20; ● Figure 9.28, page 260). Levees were overtopped, and the adjacent floodplains inundated in spite of volunteers' heroic efforts to raise them by sandbagging. The July 19 crest at St. Louis was 46 feet above the datum plane, 3 feet higher than ever before recorded. The next day the river crested at 47.1 feet, setting another record. On August 1, the river surged again to a record 49.4 feet. The Mississippi River is joined by the Missouri and Ohio Rivers at St. Louis and Cairo, Illinois, respectively. Because the Mississippi is deeper and wider below Cairo, it was able to accommodate the flood discharge there.

At the peak of flooding the Mississippi River valley was referred to as the sixth Great Lake. Flooding was most severe where artificial levees failed and along stretches of the river where natural levees had not been raised artificially. Rivers flood naturally, and the only way to reduce the losses they cause is to limit floodplain development. The head of Wisconsin's floodplain program has said, "If we don't [limit development], sooner or later the river will take it back." Many hydrologists, geologists, and planners share this view.

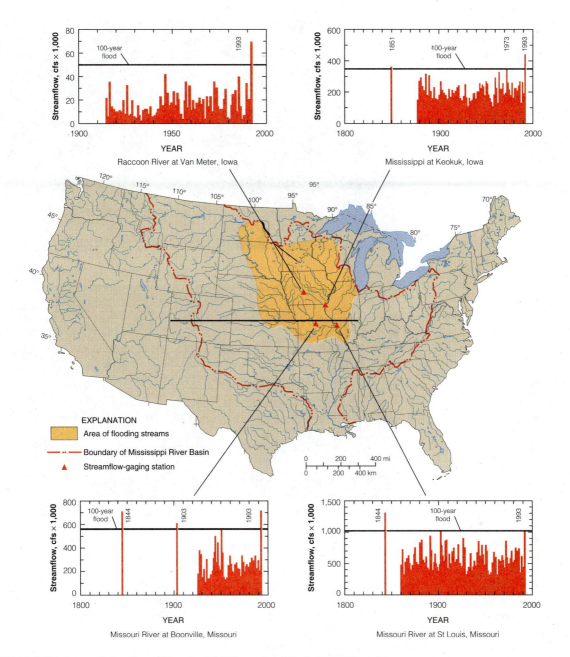

● **FIGURE 9.27 Area of the Great Flood of 1993 on the Mississippi River. The 100-year flood levels were met or exceeded at the four gauging stations shown. Note that the estimated flood of 1844 exceeded the 1993 event at St. Louis.**

MITIGATION

The construction of dams, retaining basins, floodwalls, and artificial levees and other means of altering river channels are time-honored methods of flood control. All of these are attempts to keep floodwaters within the river channel or to store water for slow release at a later time. Unfortunately, these engineering works tend to become obsolete, because population and property values adjacent to and below flood–control structures increase faster than the land can be protected. That is, floodplain development occurs in response to speculation that flood control has been achieved or is not far off. For example, many artificial levees were built along the Mississippi River to protect farmland. The floodplain was later urbanized, and the existing levees, though adequate to protect farmland, are not adequate for protecting homesites (● Figure 9.29, page 261).

An artificial levee is built upward upon a natural levee or riverbank so as to increase the amount of water the channel can accommodate and thus prevent the natural periodic

● FIGURE 9.28 Satellite photos of the confluence of the Mississippi, Illinois, and Missouri Rivers during *(a)* the drought of 1988 and *(b)* the flood of 1993. The top of the photos is north. The Illinois River flows south to the Mississippi, joining it at Grafton, Illinois. The Missouri River flows eastward across Missouri and empties into the Mississippi just north of St. Louis.

(a)

(b)

flooding on adjacent land. When streamflow is high, the area's would-be floodwater flows downstream to areas that have lower natural levees or no levees, which then experience greater flooding than they would if the artificial levee had not been built upstream. Some of these downstream communities will build their own artificial levees to prevent this in the future. Then areas without artificial levees downstream from those levees are subject to flooding, so more levees are built there, and so it goes. Artificial levees are thus self-perpetuating, and the Mississippi River today is largely

contained within them. Even at nonflood river levels, the water surface is above the adjacent floodplains, so the danger of flooding is ever present.

Riverside land adjacent to the old French Quarter of New Orleans, known locally as Jackson Square and the Place du Armes, is extremely valuable tourist property. The natural levees on the Mississippi River here were about 2.1 meters (7 ft) above sea level *(a.s.l.)* during settlement times in the 1700s, and natural flood stages probably never exceeded 3.0 meters a.s.l. However, confinement of the river upstream by artificial levees raised the flood level to 6.5 meters (21 ft) in modern times, and artificial levees were constructed. Due to the riverside real estate's historic and economic value, it was felt that additional protection should be provided as a safety factor—such as that provided by extending the levees to an elevation of 7.4 meters a.s.l. (24 ft). Because levees constructed of fill require a lot of land area (to provide a slope of 20 meters horizontal to 1 meter vertical), it was decided to construct a floodwall just landward of the existing levee's crest instead. The floodwall has gates that close during flooding events (● Figure 9.30). The 13-kilometer-long (8-mi) wall looks innocent enough, but

(a)

(b)

(c)

● FIGURE 9.29 (left) Artificial levees protect land from the river and the sea and sometimes need protection themselves. *(a)* A levee with a road on the top protects an industrial area on the floodplain northeast of New Orleans. *(b)* A levee protects homes from the waters of Lake Pontchartrain. *(c)* A damaged levee on the Mississippi River was protected by snowfencing and sandbags.

it is built of 2-foot-thick reinforced concrete resting on a continuous wall of wooden piles driven 7.6 meters (25 ft) into the ground. The "sheet" piles in turn rest upon concrete piles that extend down another 15 meters (almost 50 ft). The sheet piles prevent underseepage of water during floods. New Orleans's solution to the flooding problem has served the city well for some time, but eventually the river will win the battle and reclaim its cypress swamps and floodplains, or it will shift its course to another distributary.

For a long time dams and flood-control basins were considered the ultimate solution to the flood problem. The U.S. Army Corps of Engineers, at the request of Congress, has supervised construction of 500 large dams within the past century, one of the newest being the Lower Granite Dam on the Snake River near Pullman, Washington. More than 10,000 dams now control U.S. rivers, half of them more than 15 meters (50 ft) high. Although some projects are regarded as successful, most of them have served to increase the potential risk to human life, mainly because development of the floodplains below the dams has followed. Under certain statistically predictable meteorological conditions, flooding will occur, dam or no dam, because water must be released from a full reservoir. More than $35 billion in federal funds was spent on dams between 1960 and 1987, most of them water-storage or flood-control dams. Even those dams built to provide hydroelectric power provide some flood protection.

It appears that constructing dams on rivers today is unacceptable. Their costs far exceed the benefits derived. They

● FIGURE 9.30 Floodwalls are built to protect highly valued land. The floodwall here protects New Orleans' historic French Quarter.

● **FIGURE 9.31** Fluvial flood-hazard areas as defined by the U.S. government. No building is allowed in the regulatory floodway, and only protected or raised structures are allowed to be built in the fringe area.

Regulatory floodway—Kept open to carry flood water: no building or fill permitted.

Regulatory floodway fringe—Use permitted if protected by fill, flood-proofed, or otherwise protected.

– – – Regulatory flood limit—based on technical study; outer limit of the floodway fringe.

– . – Standard project flood (SPF) limit—Area subject to possible flooding by large floods.

←Regulatory floodplain→
←Floodplain (SPF)→

devour huge amounts of otherwise usable land for reservoirs, which displaces families and in some cases entire cultures. Their lifetime is limited because of sedimentation in their reservoirs, and the threat of catastrophic failure is always present, as exemplified by St. Francis Dam (Case Study 9.3), the Vaiont Dam (Chapter 7), and others.

Other methods of reducing the impact of flooding on humans include flood insurance, flood-proofing, and flood-plain management. The National Flood Insurance Act of 1968 provides federally subsidized insurance protection to property owners in flood-prone areas. The Federal Insurance Administration chose the 100-year-flood level as the **Regulatory Floodplain** on which to establish the limits of government management and set insurance rates. Maps have been developed that delineate Regulatory Floodplains in literally thousands of communities in the United States. The subsidized flood insurance is not available on new homes within the regulatory floodway (●Figure 9.31). Pre-1968 houses in the floodway were "grandfathered" (allowed to stay and to be insured under the government program), but they cannot be rebuilt at that site with flood-insurance money if they are more than 50 percent damaged by flooding.

Unfortunately, there are serious problems with the 1968 act. Because the insurance is government-subsidized, owners need not assume the full insurance risk for homes in flood-prone areas. This subsidy

serves to encourage *more* building in flood-prone areas, because the land is cheap and the risk is subsidized. Also, as with much well-meaning federal legislation, enforcement is at the local-government level and thus varies considerably from place to place.

Three methods of flood-proofing are (1) raising structures above the 100-year-flood level by artificial filling with soil, (2) building walls and levees to resist floodwaters, and (3) using water-resistant building materials. Good flood-plain management limits the uses of floodplains in chronically flood-prone areas. As noted earlier, alluvial fans in Southwestern deserts are subject to flash floods and debris flows from intense storms. Nonetheless, they are extensively developed in resort areas of the Southwest such as Palm Springs, California, although they are subject to strict regulation. Fans present a special problem, because the chance of flooding near the fan's apex does not determine the probability or depth of flooding below that point. Flood intensities decrease downfan, from life-threatening channelized debris flows near the apex, to a zone where streams rapidly change channels in a "braided" drainage pattern, to the area of little hazard at the toe of the fan (●Figure 9.32).

The building pads of homes built on alluvial fans near Palm Springs are required to be raised artificially several feet above the natural ground so as to locate the structure above possible floodwater lev-

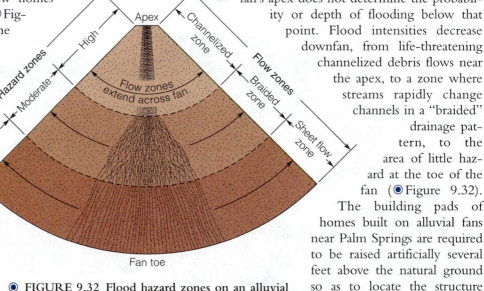

Apex

Channelized zone

Flow zones

Braided zone

Sheet flow zone

Hazard zones

High
Moderate
Low

Flow zones extend across fan.

Fan toe

● **FIGURE 9.32** Flood hazard zones on an alluvial fan. Hazards exist across the full width of the fan.

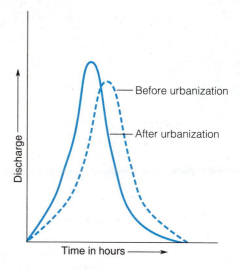

● **FIGURE 9.33 Hypothetical hydrographs of flooding patterns before and after urbanization. After urbanization the maximum flood stage occurs sooner and is higher.**

els. When built in this manner, they qualify for federally subsidized flood insurance.

Urban development presents special problems, because it alters the hydrology of a watershed. A conceptual hydrograph can be used to illustrate the fact that because of urbanization, floods peak sooner and their total runoff and peak flow are greater (● Figure 9.33). To geologists and hydrologists, urbanization represents the replacement of permeable soil with impermeable pavement, which produces the observed greater flooding and damage. It has been found that when 50 percent of a drainage basin is urbanized, the frequency of overbank flows (floods) is four times what it was before urbanization.

Development has literally poured onto floodplains in the United States, and annual flood losses have increased accordingly. Annual flood damage (in constant dollars) almost tripled between 1950 and the mid 1990s to an average of $3 billion per year. Floods are natural and inevitable, as demonstrated in this chapter. A congressional report perhaps summed it up best in noting that "Floods are an act of God. Flood damages result from acts of men."

Flood Planning

The following flood and flood-control information is useful in the planning process:

- Floods are the most consistently destructive natural hazard in the United States.
- Flooding cannot be controlled completely, but its damaging effects can be mitigated by restricting floodplain development or by government acquisition of floodplain lands.
- Flood damage costs have been increasing at a rate of 5 percent annually over the past several decades in real dollars.
- Urbanization increases flood peaks on small streams by two to six times.
- Use of the 100-year-flood standard for regulation is effective and should be continued, and the 500-year flood

should prevail for such critical facilities as hospitals, water-supply facilities, and power plants.

Two towns on the Illinois River, both heavily damaged during the 1993 floods, have totally different flood-planning options. Liverpool, Illinois, a town of 200 residents, has opted to build a levee to protect it against future flood events. Construction of a 1.4-kilometer-long (0.87-mi) levee at a cost of $2.1 million should solve Liverpool's flood problem. At Kampsville to the south, a levee is not feasible for a variety of reasons, and since 1986 the federal government has been buying homes and businesses on the floodplain there. Between 1986 and 1993 the government spent $1.1 million buying 60 homes, but as many remained and they were hit hard in 1993. One of the problems, residents say, is that the government does not offer them enough money for their property and its flood insurance allows them to rebuild after each flood. Thus, the government pays whether or not it buys floodplain property. However, as noted earlier, if a home is 50 percent damaged, an owner is not allowed to use flood-insurance money to repair or rebuild it. This has probably forced the remaining Kampsville residents to sell out. This example illustrates the problem with federally subsidized floodplain insurance.

Don Barnett, the mayor of Rapid City, South Dakota, at the time of its disastrous 1972 flood, said, "Elected public officials must give the same attention and priority to their drainage problems as they give to their police and fire problems. In the history of Rapid City, perhaps 35 people have died in fires and another 35 have been killed during the commission of crimes, but in just two hours, 238 died in a flood."

GALLERY

Gorgeous Gorges and Natural Bridges

As we have seen, river erosion forms valleys. When a river has more energy than is necessary to transport its sediment load, it carves its valleys deeper and deeper. In areas of uplift and steep gradients, streams will cut steep-sided canyons, such as the Grand Canyon of the Colorado River in Arizona (⊙ Figure 9.34), or almost vertical-sided gorges such as the Royal Gorge of the Arkansas River in Colorado (⊙ Figure 9.35).

The first sighting of the Grand Canyon by Europeans is credited to the Coronado expedition of 1540, and subsequent Spanish explorers believed "the awesome abyss," as it is sometimes called, was formed by a great earthquake. Established as a national park in 1919, Grand Canyon is visited by more than two million people a year. The canyon is the product of a dry climate, which preserves the beautiful stairstep topography, and a river with a steep gradient and great erosive power. The precanyon river flowed westerly, and it has remained essentially in place, forming the canyon by incessant downward erosion as the plateau has slowly been uplifted and tilted southward. This has taken place during the last 5 million years, mostly during late Miocene and Pliocene time, and it explains the west-flowing river on the south-sloping plateau.

Rainbow Bridge, just north of Utah's southern border, is the world's largest natural bridge (⊙ Figure 9.36). The colorful, rainbow-shaped bridge rises 88 meters (290 ft) above the creek below and is 85 meters (278 ft) long. The

Navajos regard it as sacred because they consider atmospheric rainbows to be guardians of the universe. Before construction of the Glen Canyon Dam on the Colorado River, which formed Lake Powell and provided boat access to the site, this was one of the most inaccessible places in the 48 states. Its existence was not even known by non-Native Americans until 1909. The bridge is composed of red sandstone, mostly of wind-blown origin, with dark vertical streaks of "desert varnish," a film of iron and manganese oxide. Bridge Creek flowing off Navajo Mountain 8 kilo-

⊙ FIGURE 9.34 The Grand Canyon downstream from Marble Canyon. Flat-lying Paleozoic sedimentary rocks form near-vertical cliffs in this part of the canyon.

meters (5 mi) to the southeast eroded deep meanders into the red sandstone as it flowed across the uplifted Colorado Plateau (⊙ Figure 9.37, part a). The creek cut downward until it encountered the hard Kayenta Formation at the base of the bridge. Unable to erode downward easily, the stream cut laterally. It undercut the canyon walls and then cut through the thin wall ("neck") of the meander loop, forming the bridge (Figure 9.37, part b).

◉ FIGURE 9.36 Rainbow Bridge National Monument, Glen Canyon, Utah.

◉ FIGURE 9.35 (above) Royal Gorge of the Arkansas River near Canon City, Colorado. The suspension bridge that spans it is 1,053 feet above the river. Zebulon Pike, for whom Pike's Peak was named, is believed to have been among the first Europeans to see the gorge.

◉ FIGURE 9.37 Geologic origin of natural bridges. *(a)* A meandering stream cuts downward into rock, and the meanders become very sinuous. *(b)* The stream cuts through a meander's thin wall, or neck, leaving an arch.

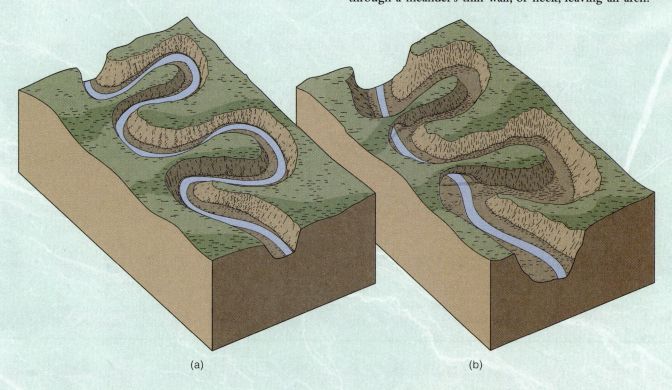

(a)

(b)

SUMMARY

Water Resources

HYDROLOGIC CYCLE The continuous flow of water in various states through the environment.

WATER BUDGET Varies with geologic location, but it can be generalized for the 48 contiguous states.

1. *Precipitation* supplies about 75 cm/y.
2. *Evapotranspiration* returns 55 cm/y back to the atmosphere by evaporation and transpiration.
3. *Runoff into oceans and infiltration into the ground* account for the remaining 20 cm/y.

USES

1. Consumptive uses rule out immediate reuse. Agricultural irrigation is consumptive use.
2. Nonconsumptive uses return water to the surface flow via storm drains, sewage treatment plants, and rivers. Household use is nonconsumptive.

SOURCES Rivers and lakes provide 80 percent of U.S. offstream water use. In arid regions where consumption is greater than the local supply, water must be imported or underground water must be "mined" (defined as extraction > recharge).

River Systems

DRAINAGE BASIN The fundamental geographic unit or tract of land that contributes water to a stream or stream system. Drainage basins are separated by divides.

DISCHARGE The amount of water flowing in a stream channel depends upon the amount of runoff from the land, which in turn depends upon rainfall or snowmelt, the degree of urbanization, and vegetation.

EROSION The erosive power of a stream is a function of its velocity; the greater the velocity, the greater the erosion. Stream velocity is determined by discharge, channel shape, and gradient.

BASE LEVEL The lowest level to which a stream or stream system can erode. This is usually sea level, but there are also temporary base levels such as lakes, dams, and waterfalls.

GRADED STREAM A stream that has reached a balance of erosion, transporting capacity, and the amount of material supplied to it. This condition is represented by a concave-upward profile of equilibrium.

Stream Features

ALLUVIUM Sediment deposited by a stream, either in or outside the channel.

ALLUVIAL FAN A buildup of alluvial sediment at the foot of a mountain stream in an arid or semiarid region.

DELTA Forms where a sediment-laden stream flows into standing water. An example is the delta of the Nile River.

FLOODPLAIN A low area adjacent to a stream that is subject to periodic flooding and sedimentation; the area covered by water during flood stage.

MEANDERS, OXBOW LAKES, AND CUTOFFS Flowing water will assume a series of S-shaped curves known as meanders. The river may cut through the neck of a tight meander loop and form an oxbow lake.

Flooding and Flood Frequency

FLOODS Upland floods come on suddenly and move with great energy through narrow valleys. Upland floods are often "flash" floods; the water rises and falls in a matter of a few hours. Riverine floods, on the other hand, inundate broad adjacent floodplains and may take many days, weeks, or even months to complete the flood cycle.

HYDROGRAPH A graph that plots measured water level (stage) or discharge over a period of time.

RECURRENCE INTERVAL The length of time (T) between flood events of a given magnitude. Mathematically, $T = (N + 1) \div M$, where N is the number of years of record and M is the rank of the flood magnitude in comparison with other floods in the record. This enables engineers and planners to anticipate how often floods may occur and to what elevation the water may rise.

FLOOD PROBABILITY The chance that a flood of a particular magnitude will occur within a given year based on historical flood data for the particular location. Using the calculation for T established for the recurrence interval, probability is calculated as $1/T$. Thus the 100-year flood has a 1 percent chance of occurring in any one year.

Mitigation

STRUCTURES Dams, retaining basins, artificial levees, and raising structures on artificial fill are common means of flood protection.

FLOOD INSURANCE The National Flood Insurance Act of 1968 provides insurance in flood-prone areas, provided certain building regulations and restrictions are met.

Urban Development

HYDROLOGY Urbanization causes floods to peak sooner during a storm, results in greater peak runoff and total runoff, and increases the probability of flooding. Floods, including coastal flooding during hurricanes, are the greatest natural hazards facing humankind.

KEY TERMS

alluvial fan	drainage basin
alluvium	drainage divide
artificial levee	evapotranspiration
base level (stream)	flood frequency
channel shape	floodplain
Continental Divide	floodwall
cutoff	geologic agents
delta	graded stream
discharge	gradient

hydrograph
hydrologic cycle
hydrology
meander
natural levee

oxbow lake
point bar
recurrence interval
Regulatory Floodplain
wetted perimeter

STUDY QUESTIONS

1. Distinguish between weathering and erosion.
2. How can we prevent floods or at least minimize the property damage caused by flooding?
3. How is the velocity of a stream related to its erosive power, discharge, and cross-sectional area?
4. Sketch the hydrologic cycle.
5. What has been the impact of urbanization on flood frequency, peak discharge, and sediment production?
6. According to the local newspaper, your town on the Ohio River has just experienced a 50-year flood. What is the probability of a repeat of this event the following year? What is the weakness of such statistical forecasts?
7. How does the hydrograph record of a flash flood differ from that of a lowland riverine flood? Where do we find flash flooding, and what precautions should be observed in such areas?
8. What is the source of most of the fresh water used in the United States? Explain the difference between consumptive and nonconsumptive use.
9. What has the federal government done to alleviate people's suffering due to flooding? Do you believe this is a good thing? Does it promote more construction in flood-prone areas? If so, does this construction itself have possible negative impacts? Support your answers with examples and rationales.

10. Consider a stream whose headwaters are 1,500 meters (4,900 ft) above sea level and that flows 300 kilometers (180 mi) to the ocean. What is the average gradient of the stream in m/km and in ft/mi?

FURTHER INFORMATION

BOOKS AND PERIODICALS

Carrier, Jim. 1991. The Colorado: A river drained dry. *National geographic,* June.

Devine, Robert S. 1995. The trouble with dams. *Atlantic monthly,* August: 64–74.

Hays, W. W., ed. 1981. *Facing geologic and hydrologic hazards: Earth science considerations.* U.S. Geological Survey professional paper 1240-B. Washington, D.C.

Leopold, Luna B. 1968. *Hydrology for urban land planning: A guidebook on the hydrologic effects of urban land use.* U.S. Geological Survey circular 554.

Leopold, L.; M. G. Wolman; and John P. Miller. 1964. *Fluvial processes in geomorphology.* San Francisco: W. H. Freeman and Co.

McDowell, Jean, and Richard Woodbury. 1991. The Colorado: A fight over liquid gold. *Time,* July 22.

McPhee, John. 1989. Atchafalaya, chapter 1. *The control of nature.* New York: Farrar, Straus & Giroux.

National Geographic. 1993. Water: the power, promise and turmoil of North America's fresh water. Special edition, October.

Saarinen, T. F.; V. R. Baker; R. Durrenberger; and T. Maddock, Jr. 1984. *The Tucson, Arizona, flood of October 1983.* Washington D.C.: National Research Council, National Academy Press.

U.S. Geological Survey. 1993. Mississippi River flood of 1993. *U.S. Geological Survey yearbook:* 37–40.

Waananen, A. O., and others. 1977. *Flood-prone areas and land-use planning: Selected examples from the San Francisco Bay region, California.* U.S. Geological Survey professional paper 942.

WATER UNDER THE GROUND

*Streams will burst forth in the desert, And rivers in the steppe.
The burning sands will become a pool, And the thirsty ground,
springs of water.*

ISAIAH 41:18

S ome of the water in the hydrologic cycle infiltrates underground and becomes **ground water,** one of our most valuable natural resources (see Figure 9.1). In fact, most of the earth's liquid fresh water exists beneath the land surface (see Table 9.1), and its geologic occurrence is the subject of many misconceptions. For example, it is commonly believed that ground water occurs in large lakes or pools beneath the land. The truth is that almost all ground water is in pore spaces and fractures in rocks. Another misconception is that deposits of ground water can be found by people with special skills or powers, people referred to as *water witches* or *dowsers.* They locate underground water using two wires

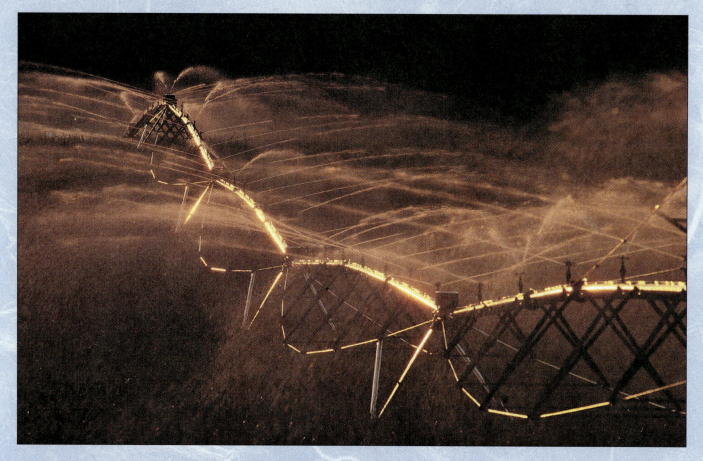

OPENING PHOTO
For decades, ground water from the Ogallala Aquifer has irrigated Nebraska cornfields. Seen here is a center-pivot irrigation system that is connected to a well tapping the aquifer that can yield as much as 4,000 liters per minute (about 1,000 gallons per minute), 24 hours a day.

bent into L shapes, forked branches, or other forked devices called *divining rods*. A dowser locates an underground "tube" or lake of water by walking over the ground, holding the divining rod in a prescribed manner until it jerks downward, pointing to the location of ground water, a mineral deposit, or a lost set of keys. Some dowsers claim they can locate good places to drill water wells using only a map of the land. The divining rod is passed over the map until it jerks downward where presumably a well should be located (X marks the spot). Water dowsers are found throughout the world, particularly in rural areas, and many of them will guarantee finding water. This is one difference between dowsers and ground-water geologists, or **hydrogeologists.** Hydrogeologists will not make this guarantee.

In this chapter we will investigate an important source of public and agricultural water supplies—underground water. The geology of underground water, how it is located and extracted, and its management will be discussed. Finally, we will study ground-water quality and ways that this important resource becomes polluted. The principles and facts conveyed in this chapter should enable you to make an intelligent evaluation of the claims of both water dowsers and hydrogeologists.

WATER SUPPLY

Ground water provides 40 percent of public water supplies and 30 to 40 percent of all water used exclusive of power generation in the United States. It is by far the cheapest and most efficient source of municipal water, because obtaining

it does not require the construction of expensive aqueducts and reservoirs. Thirty-four of the largest 100 cities in the United States depend entirely upon local ground-water supplies; Miami Beach, San Antonio, Memphis, Honolulu, and Tucson are just a few. Ground water provides 80 percent of the water for rural domestic and livestock use and it is the only source of water in many agricultural areas. Consumption by state is shown in ⦿ Figure 10.1. It is not surprising that some of the largest agricultural states—such as Texas, Nebraska, Idaho, and California—are the largest consumers, accounting for almost half of all the ground water produced. Ground water is important even to communities that import water from distant areas, because in almost every case, local ground water provides a significant low-cost percentage of their water supply.

Although it is not generally known, ground water sustains and maintains streamflows during periods when there is no rain. In arid climates where there are permeable water-saturated rocks beneath the surface, ground water seeping into stream channels may be the only source of water for rivers. The fabled oases of the Sahara and Arabian deserts occur where underground water is close to the surface or where it intersects the ground surface, forming a spring or watering hole.

LOCATION AND DISTRIBUTION

Between the land surface and the depth at which we find ground water is the *zone of aeration,* a zone where voids in soil and rock contain only air and water films. The contact

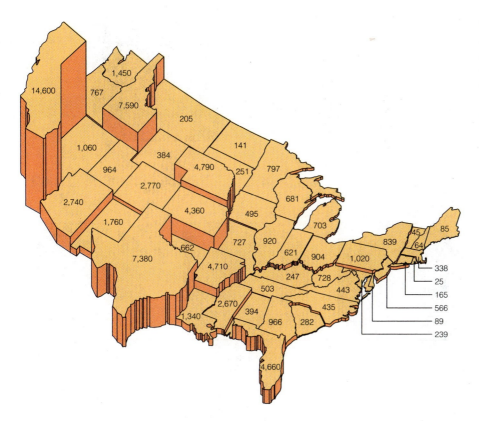

⦿ FIGURE 10.1 Production of ground water in millions of gallons per day per state in 1990. The greatest demand is in states that have extensive irrigated agriculture; California, Idaho, Texas, and Nebraska led the nation in 1990. See Figure 9.5 for surface-water use.

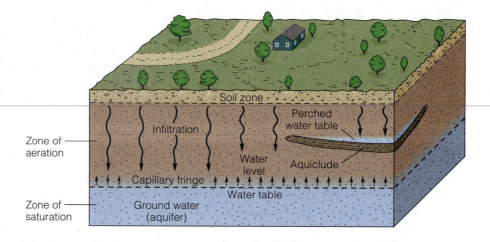

◉ FIGURE 10.2 Schematic cross section of ground-water zones showing a static water table with a capillary fringe. A perched water table exists where a body of clay or shale acts as a barrier to further infiltration.

Soil zone

Zone of aeration

Infiltration

Perched water table

Water level

Aquiclude

Capillary fringe

Water table

Zone of saturation

Ground water (aquifer)

between this zone and the *zone of saturation* below it is the **water table** (◉ Figure 10.2). Above the water table is a narrow moist zone, the *capillary fringe*. In the saturated zone, free (unattached) water fills the openings between grains of clastic sedimentary rock, the fractures or cracks in hard rocks, and the solution channels in limestones and dolomites (◉ Figure 10.3).

Water-saturated geologic formations whose porosity and permeability are sufficient to yield significant quantities of water to springs and wells are known as **aquifers**. A rock's **porosity** (root word *pore*) is a function of the volume of openings, or void spaces, in it and is expressed as a percentage of the total volume of the rock being considered. With-

out such voids, a rock cannot contain water or any other fluid. **Permeability** is the ease with which fluids can flow through an aquifer; it is thus a measure of the connections between pore spaces. **Hydraulic conductivity** is the term hydrologists prefer for the ease with which fluids flow through porous strata, but *permeability* will suffice for our purposes, as it has a long history of use and is easy to remember. **Aquicludes** are rocks or sediments such as shale and clay that lack permeability and hence will not transmit water. In some places aquicludes occur as *lenses* that trap water and prevent it from percolating down to the water table. Such a condition creates a **perched water table** (see Figure 10.2).

◉ FIGURE 10.3 Relative porosity of selected rocks and unconsolidated sediments. The uniform grain size of well-sorted sand creates a packing arrangement with relatively high porosity. The grain sizes in poorly sorted sand range from fine to coarse. This allows small grains to occupy pore spaces among the larger grains, thus reducing relative porosity.

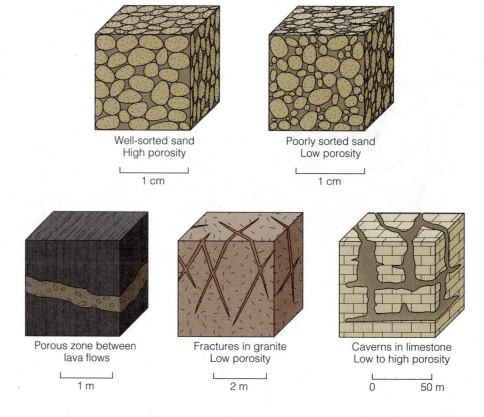

Well-sorted sand
High porosity

1 cm

Poorly sorted sand
Low porosity

1 cm

Porous zone between lava flows

1 m

Fractures in granite
Low porosity

2 m

Caverns in limestone
Low to high porosity

0   50 m

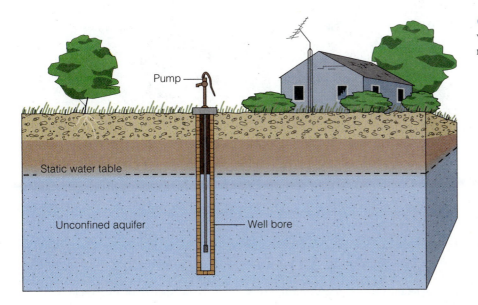

Aquifers are important because they transmit water from *recharge areas*—that is, areas where water is added to the zone of saturation—to springs or wells. In addition, they store vast quantities of drinkable water. In terms of the hydrologic cycle, we can view aquifers both as transmission pipes and as storage tanks for ground water. The Nubian Sandstone in the Sahara Desert, for instance, is estimated to contain 600,000 cubic kilometers (144,000 mi^3) of water, mostly untapped. It is the world's largest known aquifer and is a resource of great potential for North Africa.

Static Water Table

Unconfined aquifers are formations that are exposed to atmospheric pressure changes and that can provide water to wells by draining adjacent saturated rock or soil (● Figure 10.4). As the water is pumped, a **cone of depression** forms

around the well, creating a gradient that causes water to flow toward the well (● Figure 10.5). A low-permeability aquifer will produce a steep cone of depression and substantial lowering of the water table in the well. The opposite is true for an aquifer in highly permeable rock or soil. This lowering, or the difference between the water-table level and the water level in the pumping well, is known as **drawdown.** Stated differently, low permeability produces a large drawdown for a given yield and increases the possibility of a well "running dry." A good domestic (single-family) well should yield at least 11 liters (2.6 gal) per minute, which is equivalent to about 16,000 liters (3,700 gal) per day, although some families use as little as 4 liters per minute (1,350 gal/day). As pointed out, a pumping well in permeable materials creates a small drawdown and a gentle gradient in the cone of depression. Some closely spaced wells, however, have overlapping "cones," and excessive pumping

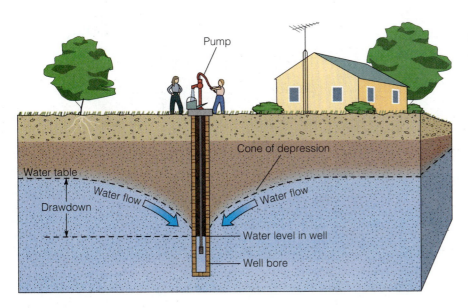

● FIGURE 10.5 When pumping begins, the water table is drawn down an amount determined by the pumping rate and the aquifer's permeability. The development of a cone of depression creates a hydraulic gradient, which causes water to flow toward the well and enables continuous water production. The amount of water the well yields per unit of time pumped depends upon the aquifer's hydrologic properties.

● **FIGURE 10.6** Cones of depression overlap in an area of closely spaced wells. Note the contributions to ground water from septic tanks.

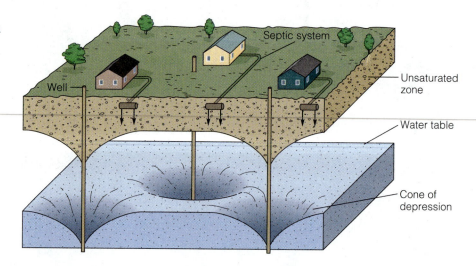

Septic system

Well

Unsaturated zone

Water table

Cone of depression

in one well lowers the water level in an adjacent one (● Figure 10.6). Many lawsuits have arisen for just this reason. Water law is discussed briefly in Case Study 10.1.

The water table rises and falls with consumption and the season of the year. Generally, the water table is lower in late summer and higher during the winter or the wet season. Thus water wells must be drilled deep enough to accommodate seasonal and longer-term, climatic fluctuations of the water table (● Figure 10.7). If a well "goes dry" during the annual dry season, it definitely needs to be deepened.

Streams that are located above the local water table are called **influent,** or "losing," streams, as they contribute to the underground water supply (● Figure 10.8, part a). Beneath influent streams there may be a mound of water above the local ground-water table known as a *recharge mound.* A stream that intersects the water table and which is fed by both surface water and ground water is known as an **effluent,** or "gaining," stream (Figure 10.8, part b). A stream may be both a gaining stream and a losing stream, depend-

ing upon the time of year and variations in the elevation of the water table.

Where water tables are high, they may intersect the ground surface and produce a spring. Springs also are subject to water-table fluctuations. They may "turn on and

CONSIDER THIS . . .

Some hot springs form geysers that are celebrated for their regularity. Old Faithful in Yellowstone National Park is an example. Cold springs, on the other hand, do not generate attention-grabbing geysers, and their flow is often far from regular. Indeed, many springs operate on a sort of stop-and go-basis. How can we explain this intermittent flow?

● **FIGURE 10.7** The effect of seasonal water-table fluctuations on producing wells. A shallow well may go dry in the summertime.

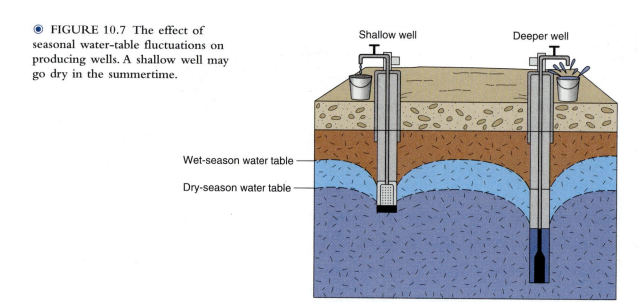

Shallow well

Deeper well

Wet-season water table

Dry-season water table

CASE STUDY 10.1

Ground-Water Law

Ownership of ground water is a property right and has traditionally been regulated as such by individual states. Each state applies one or more of four legal doctrines to ground-water rights, none of which considers the geological relationship between surface-water infiltration and ground-water flow. The doctrines are

- *Riparian doctrine* (Latin *ripa*, "river or bank") holds that landowners overlying well-defined underground "streams" have absolute right to that water. They may exercise that right whenever they wish without limitation.

- *Reasonable-use doctrine* restricts a landowner's right to ground water to "reasonable use" on the land above that does not deprive neighboring landowners of their rights to "reasonable use." This doctrine results in a more equal distribution of water when there is a shortage.

- *Prior-appropriation doctrine* holds that the earliest water users have the firmest rights; in other words, "first-come, first-served."

- *Correlative-rights doctrine* holds that landowners own shares to the water beneath their collective property that are proportional to their shares of the overlying land.

In the San Joaquin Valley of California, for example, surface-water rights are governed by the doctrine of prior appropriation (first-come, first-served), whereas ground-water rights are governed by the doctrine of correlative rights (proportional shares). In the late nineteenth century the Sierra Nevada to the east of the valley and the lower ranges to the west maintained high water tables, and they both contributed water to the flow of the San Joaquin River (◉ Figure 1). By 1966, however, heavy ground-water withdrawals on the west side of the San Joaquin Valley had reversed the regional ground-water flow, so that the river was losing water underground, rather than gaining. The San Joaquin had become an influent stream. Imagine a courtroom scenario in which hundreds of people with first-come, first-served rights to *surface* water sue thousands of people holding proportional shares rights to *underground* water for causing their surface water to go underground.

As explained throughout this chapter, it is impossible to treat ground water and surface water as separate entities. Surface water contributes to underground water supplies, and ground water contributes to streamflow and springs.

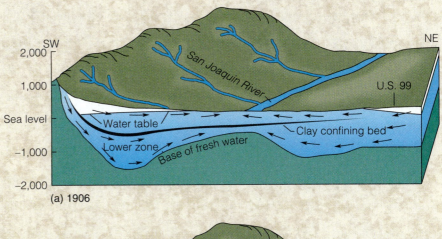

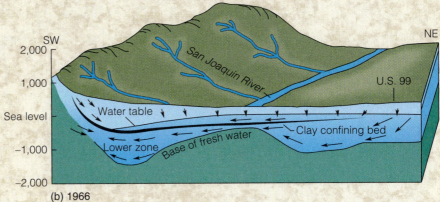

◉ **FIGURE 1** Ground-water flow in the central San Joaquin Valley in *(a)* 1906 and *(b)* 1966. Note that the San Joaquin River was an effluent (gaining) stream in 1906 and an influent (losing) stream is 1966. The change reflects intensive development and extraction of underground water.

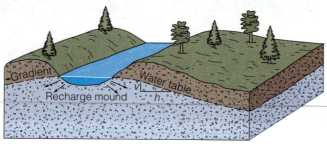

Gradient
Recharge mound
Water table
v
h

(a) Influent (losing) stream

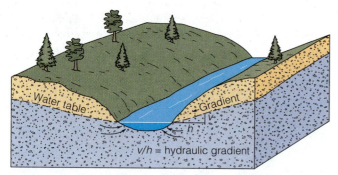

Water table
Gradient
v
h
v/h = hydraulic gradient

(b) Effluent (gaining) stream

⊙ **FIGURE 10.8 Hydraulic gradient** (v/h) **in** *(a)* **an influent (losing) stream that contributes water to the water table, and** *(b)* **an effluent stream that gains water from the water table.**

off" depending upon rainfall and the season of the year. ⊙ Figure 10.9 illustrates the hydrology of three kinds of springs.

Pressurized Underground Water

Pressurized ground-water systems cause water to rise above aquifer levels and sometimes even to flow to the ground surface. Pressurized systems occur where water-saturated permeable layers are enclosed between aquicludes, and for this reason, they are called **confined aquifers** (⊙ Figure 10.10). A confined aquifer acts as a water conduit, very similar to a garden hose that is filled with water and situated such that one end is lower than the other. In both cases a water-pressure difference known as a *hydraulic head* is created. Water gushes

CONSIDER THIS...

As explained in Case Study 10.1, the legal issues surrounding the withdrawal and use of ground water are complex, and several legal doctrines apply to them. From a philosophical viewpoint, which of the four doctrines seems fairest for multiple users of an aquifer? Please explain your choice.

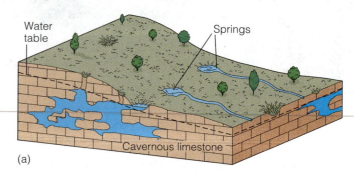

Water table
Springs
Cavernous limestone

(a)

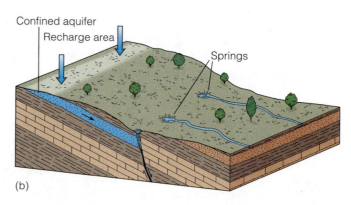

Confined aquifer
Recharge area
Springs

(b)

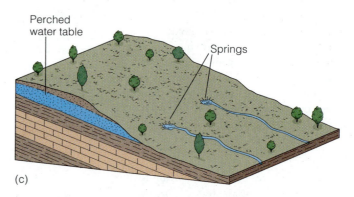

Perched water table
Springs

(c)

⊙ **FIGURE 10.9 Geology and hydrology of three kinds of springs.** *(a)* **Springs form where a high water table in cavernous limestone intersects the ground surface.** *(b)* **Artesian springs. A fault barrier can cause pressurized water in a confined aquifer to rise as springs along irregularities in the fault.** *(c)* **Springs form where a perched water table intersects the ground surface. Such springs are intermittent, because the small volume of perched water is quickly depleted.**

from the low end until there is no longer a pressure difference. At that point the water level stabilizes, becomes *static*.

Wells that penetrate confined aquifers are called *artesian wells*, after the province of Artois, France, where they were first described. Artesian systems are distinguished by a water-pressure gradient called a **potentiometric surface** (see Figure 10.10, part b). This "surface" is the level to which water will rise in a well at a given point along the aquifer. It is important to note that the slope of the water table of an unconfined aquifer also is a potentiometric surface. Underground water moves from areas of high hydraulic head to

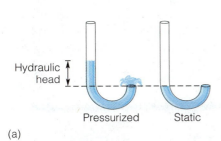

Hydraulic head

Pressurized   Static

(a)

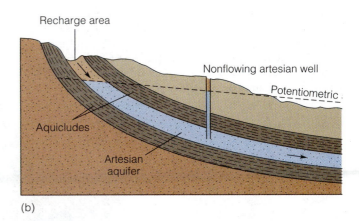

Recharge area

Nonflowing artesian well

Potentiometric

Aquicludes

Artesian
aquifer

(b)

(c)

● FIGURE 10.10 Water under pressure. *(a)* The hydraulic head in a hose lowers as water flows out the low end. Eventually a static condition prevails. *(b)* A confined aquifer, commonly called an *artesian* aquifer. The potentiometric surface is the level to which water will rise in wells at specific points above the aquifer. *(c)* A flowing artesian well; South Fork of the Madison River, Gallatin County, Montana.

areas of low hydraulic head (pressure). Contour maps of the potentiometric surface can be very useful, because they indicate the direction of ground-water flow and the slope of the hydraulic gradient (● Figure 10.11). If an aquifer's hydraulic conductivity can be estimated or measured, the rate and volume of ground-water flow can be determined using Darcy's Law (see Appendix 4).

Artesian wells differ from water-table wells in that the latter depend upon pumping or topographic differences to create the hydraulic gradient that causes the water to flow. Because the energy required to raise water to the surface is the greatest expense in tapping underground water, artesian systems are much desired and exploited. Not all artesian wells flow to the ground surface, because the potentiometric surface lowers as water is extracted; water may rise naturally only part way up the well bore. Nevertheless, a natural rise of water any distance reduces the cost of extracting ground water.

Finally, many bottled water and beverage producers attempt to lead us to believe that deep-well or artesian water is purer than other kinds of water. *Artesian* refers

CONSIDER THIS . . .

Ground-water pollution by surface-water seepage through waste-disposal sites into the water table is a major problem. Where would you locate a facility for solid-waste disposal in the setting of Figure 10.11 to minimize this hazard? Designate the site you choose by relating it to the letters at the corners of the diagram. Also, which three wells would you expect to provide the best-quality water for domestic use in the city? Briefly explain your answers.

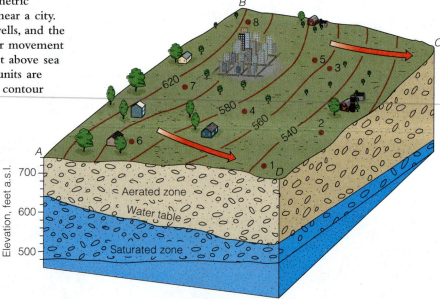

● FIGURE 10.11 Hypothetical potentiometric contour map based upon water-well data near a city. The dots indicate the locations of water wells, and the arrows show the direction of ground-water movement in the saturated zone. Elevations are in feet above sea level, rather than meters, because English units are used almost exclusively on potentiometric contour maps in the United States.

only to an aquifer's hydrogeology and has no water-quality connotation. Do not be misled by advertisements attributing greater purity to artesian or deep-well water. Such waters may or may not have low levels of dissolved contaminants.

 ## GROUND-WATER STORAGE AND MANAGEMENT

In order to manage underground water supplies in a ground-water basin intelligently, the hydrologist or planner must have two measurements:

- the quantity of water stored in the basin, and
- the **sustained yield,** the amount of water the aquifers in the basin can yield on a day-to-day basis over a long period of time.

A ground-water basin consists of an aquifer or a number of aquifers that have well-defined geological boundaries and *recharge areas*—places where water seeps into the aquifer (see Figure 10.10, part b). To determine the amount of usable water in the basin, we rely on the concept of specific yield of water-bearing materials. **Specific yield** is the ratio of the volume of water an aquifer will give up by gravity flow to the total volume of material, expressed as a percentage. For example, a saturated clay with a porosity of 25 percent may have a specific yield of 3 percent; that is, only 3 percent of the saturated porosity will drain into a well. The remaining water in the pore spaces (22 percent) is held by the clay-mineral surfaces and between clay layers against the pull of gravity. This "held" water is its *specific retention*. Similarly, sand with a porosity of 35 percent might have a specific yield of 23 percent and a specific retention of 12 percent. Note that the sum of specific yield and specific retention equals the porosity of the aquifer material.

Thus the total water-storage capacity of an aquifer may be calculated by multiplying the specific yield by the volume of the aquifer determined from water-well data. The process is a bit more complicated for artesian aquifers, but it follows the same procedure. The amount of water available to us in a ground-water basin can be determined fairly accurately if the basic data are available from numerous wells.

As noted, sustained yield is the amount of water that can be withdrawn on a long-term basis without depleting the resource. Sustained yield is more difficult to assess than aquifer storage capacity, as it is affected by precipitation, runoff, and recharge. Nevertheless, it must be at least estimated for intelligent management of the resource.

Ground-Water Mining

When the amount of water withdrawn from an aquifer exceeds the aquifer's sustained yield, an overdraft condition called *ground-water mining* exists (● Figure 10.12). The Ogallala Sandstone makes up about 80 percent of what is collectively known as the High Plains aquifer. These aquifers contain as much water as Lake Huron and underlie 480,000 square kilometers (174,000 mi^2) of Kansas, Colorado, Nebraska, Oklahoma, and Texas. The Ogallala's average thickness is 60 meters (about 200 ft), but thicknesses as great as 425 meters (1,400 ft) have been measured. In general, water levels in this important aquifer are dropping. In Texas, overdrafts have exceeded 65 percent, and overdrafts of 95 percent are known in a few places. In the latter case, 20 times more water was extracted from the aquifer than was recharged. This magnitude of overdraft is a serious problem because much of the water in the aquifer was acquired during wetter glacial climates thousands of years ago.

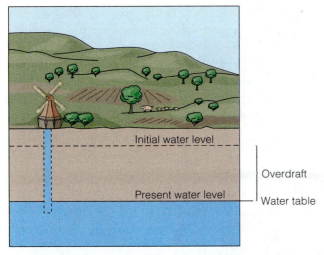

● FIGURE 10.12 The effect of ground-water mining on the water table. The drop from the initial level to the present level is the *overdraft,* the amount of water that has been mined.

The High Plains aquifer was not massively tapped until the 1950s, and by 1980 the water-saturated zone had decreased an average of 3 meters, and as much as 30 meters under some parts of Texas and Nebraska (● Figure 10.13). Center-pivot sprinklers, which irrigate large circular areas, contribute to the ground-water deficits, because many of them are supplied by wells that produce 4,000 liters per minute (about 1,000 gals/ min) 24 hours a day. In addition, government subsidies have encouraged growing water-gulping corn, rather than less-water-intensive wheat, sorghum, and cotton. In addition, a "use it or lose it" mentality has arisen, because one's water rights can be lost if a certain minimum volume of water is not pumped every year from beneath one's property.

In the early 1980s water decline rates decreased because of heavy rain and snow, better water management, and new technologies. New *LEPA* (low-energy, precision application) irrigation nozzles use less water and less energy and decrease evaporation by as much as 98 percent from the amounts lost with spray-type irrigation. In parts of Kansas and Nebraska, meters are now mandatory on irrigation wells to ensure that farmers do not exceed their water allotments. Furthermore, treated wastewater is now being recycled on fields, and rainfall has been induced by cloud seeding in attempts to recharge the aquifer in Kansas.

A 1982 Department of Commerce study found that 6 million hectares (15 million acres) were being irrigated with water drawn from 150,000 wells in the Ogallala and other High Plains aquifers. By the year 2020, according to the study report, about a fourth of the aquifer's water will have been mined at the present rate of withdrawal. A 1995 estimate of the total overdraft of the Ogallala is that it equals one year's flow of the Colorado River. The Ogallala aquifer is certainly the life-blood of High Plains agriculture, and most users of its water are aware that "mining" the resource must be discontinued. Strict management that includes conservative usage, monitoring, and utilization of new technologies is required in attempting to balance its recharge and withdrawals. As a postscript to the Ogallala story, the aquifer's name is derived from that of the Oglala Tribe of the Sioux Nation. Led by Chiefs Red Cloud and Crazy Horse, the tribe fought to retain tribal lands in the late 1800s.

Ground-water mining causes shallow wells to go dry, increases the cost of lifting water to the surface as water levels drop, and may eventually cause ground subsidence (see Chapter 8). Currently ground-water pumping is twice the volume of natural recharge in Arizona's Tucson ground-water basin, and it has been necessary to import water from the Central Arizona Project to end the overdraft. One product of nuclear testing in the 1950s and early 1960s was radioactive tritium. This isotope of hydrogen has been used as a tracer of ground water recharged into the Tucson basin on alluvial fans of the adjacent Catalina and Rincon Mountains. The study indicates that a significant amount of the basin's ground water is very young

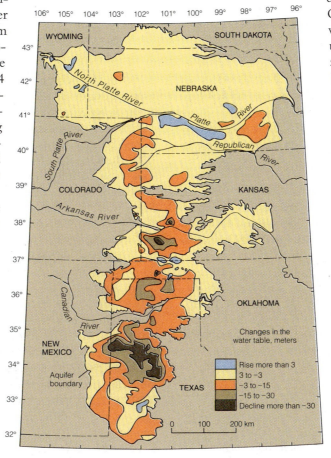

● FIGURE 10.13 Areal extent of the High Plains aquifer and changes in the water table from predevelopment times to 1980.

● FIGURE 10.14 Water-spreading grounds for artificial recharge; Santa Ana River, Southern California. The dam is inflatable and may be raised or lowered to impound or release water as needed.

and has infiltrated since the atom-bomb tests. This does not bode well for bringing the water table back to predevelopment levels in the basin.

The Santa Clara Valley of central California is a classic case of regional overdraft. Water levels there declined 46 meters (150 ft) between 1912 and 1959, and downtown San Jose dropped 2.76 meters (9 ft). As a final example, overpumping in Alabama dropped the potentiometric surface in one aquifer 62 m (200 ft); it is now below sea level more than 80 miles inland from the sea. The solution to such a problem is *artificial recharge,* or "water spreading," which is accomplished by importing or diverting surface water and ponding it where it can percolate into the aquifer (● Figure 10.14). Such a program was initiated in the Santa Clara Valley in 1959, which, in combination with decreased pumping, has restored the water table to 1912 levels. Raising the water table does not bring the land back to its original level, however, because pore spaces are lost as subsidence occurs (see Chapter 8).

Saltwater Encroachment

Aquifers in coastal areas may discharge fresh water into the ocean, creating bodies of diluted seawater. If the water table is lowered by pumping to or near sea-level elevation, however, saltwater invades the freshwater body, causing the water to become saline *(brackish)* and thus undrinkable. The direction of flow between the two water masses is determined by their density differences. Seawater is 2.5 percent (1/40th) heavier than fresh water. This means that a 102.5-foot-high column of fresh water will exactly balance a 100-foot-high column of seawater (● Figure 10.15, part a). Fresh water floats on seawater in the Hawaiian Islands, the Outer Banks of North Carolina, Long Island, and many other coastal areas (Case Study 10.2).

This relationship is known as the *Ghyben–Herzberg lens*— named for the two scientists who independently discovered it and for the lenslike shape of the freshwater body (Figures 10.15, part b and ● 10.16). In theory, the mass of fresh water extends to a depth below sea level that is 40 times the water-table elevation above sea level. If the water table is 2.5 feet above sea level, the freshwater lens would extend to a depth of 100 feet below sea level. Any reduction of water-table elevation would cause the saltwater to migrate upward into the freshwater lens until a new balance is established. If the water table drops below the level necessary for maintaining a balance with denser seawater, wells will begin to produce brackish water and eventually saltwater. Thus it is

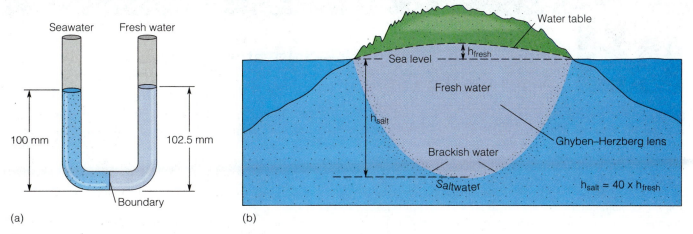

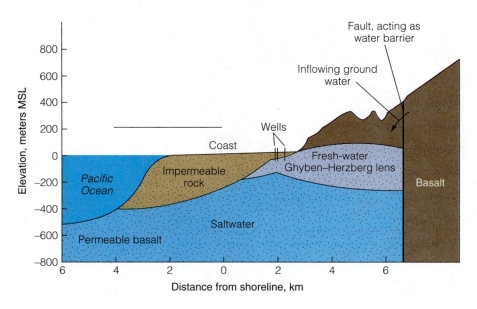

(a)   (b)

⦿ FIGURE 10.15 *(a)* Illustration of the density balance between fresh water and seawater. Because seawater is 2.5% heavier than fresh water, 102.5 millimeters of fresh water is required to balance 100 millimeters of salt water. *(b)* Hypothetical island and Ghyben–Herzberg lens of fresh water floating on seawater. The freshwater lens extends to a depth 40 times the height of the water table above sea level. Not to scale.

important to maintain high water-table levels in coastal zones so that the resource is not damaged by *saltwater encroachment*. This is accomplished by good management combined, when necessary, with artificial recharge using local or imported water (water spreading), or with injection of imported water into the aquifer through existing wells.

 WATER QUALITY

Dissolved Substances

Dissolved in natural waters are varying concentrations of all the stable elements known on earth. Whereas some of these elements and compounds, such as arsenic, are poisonous, others are required for sustaining life and health; common

table salt is one and it occurs in almost all natural waters. Years ago the U.S. Public Health Service established **potable** (drinkable) **water** standards for public water supplies based on purity, which is defined as allowable concentrations of particular dissolved substances in natural waters. These concentrations are expressed in weight per volume, which is milligrams per liter (mg/L) in the metric system. The standards can also be expressed in a weight-per-weight system. For example, in the English system, one pound of dissolved solid in 999,999 pounds of water would be one part per million (ppm). The Public Health Service standards allow only minuscule amounts of pollutants in public water supplies. For example, 1 ppm or 1 mg/L is equal to about 1 ounce of a dissolved solid in 7,500 gallons of water, or 0.0001 percent. The standard for "sweet" water—the

⦿ FIGURE 10.16 Hydrologic cross section through Honolulu, Hawaii, showing saltwater–freshwater relationships. Note that where fresh water is pumped, saltwater comes up below the wells and the water table is lowered. Not to scale.

Long Island, New York—Saltwater or Fresh Water for the Future?

Long Island is an informative case study of the cause and remedies of saltwater encroachment. By far, Long Island's most important source of fresh water is its aquifers, which are estimated to hold 10–20 trillion gallons of recoverable water (Figure 1). This ground water infiltrated underground over centuries before development, and the excess was discharged naturally to the sea. Ground water was exploited as the island was urbanized, causing the water table in Kings County, the county closest to the mainland, to drop below sea level. By 1936 saltwater had invaded the freshwater aquifer (Figure 2, part a). Pumping wells were then abandoned or were converted to recharge wells using imported water, and by 1965 the water tables had recovered to acceptable elevations above sea level (Figure 2, part b).

Meanwhile, in adjoining Queens County, pumping had been increasing and recharge was diminishing due to the construction of sewers and paving of streets. Treated wastewater was thus being carried to the sea, rather than infiltrating underground as it would with cesspools and septic tanks. With lowered water tables, saltwater invaded the aquifers below Queens County (Figure 2), and the State of New York in 1970 asked the U.S. Geological Survey to conduct ground-water studies.

In order to preserve the fresh water below Long Island, it was necessary to determine the *safe yield,* the amount of water that could be withdrawn from Long Island aquifers without causing undesirable results. It was also necessary to achieve a balance between total freshwater outflow and recharge. Several methods of managing the water resource and preventing saltwater encroachment were developed, and they are offered here as examples.

The basic challenge was to offset ground-water withdrawals of 1.7 million cubic meters per day with equal amounts of surface infiltration or injection. A line of injection wells was proposed for replacing the decreased natural recharge. These wells would inject treated sewage effluent or imported water and build a freshwater barrier against saltwater intrusion (Figure 3, part a). Spreading highly treated sewage effluent and natural runoff into recharge basins could be carried out along with injection into the deeper aquifer. In both the injection-well and the spreading-basin alternatives, the treated sewage is purified to drinking-water standards so that it does not affect local shallow wells. Recharge, like injection, builds a freshwater ridge that keeps saltwater at bay. An unusual but viable method is to allow controlled saltwater intrusion so that a true Ghyben–Herzberg lens of fresh water floating on saltwater develops (Figure 3, part b). Treated sewage

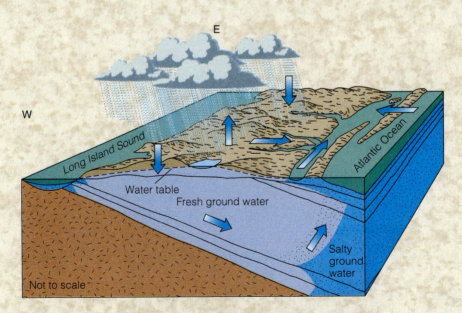

 FIGURE 1 Hydrologic relationship between fresh and salty underground water; Long Island, New York. Arrows indicate direction of water movement in the hydrologic cycle.

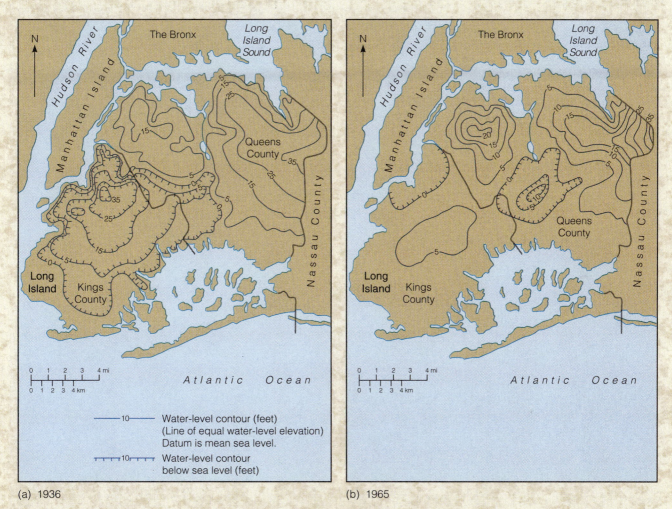

(a) 1936

(b) 1965

⦿ **FIGURE 2** Water-table levels in *(a)* 1936 and *(b)* 1965 in Kings and Queens Counties, Long Island, New York.

effluent would be discharged into the ocean or injected underground as the need arose. This method has the distinct advantage of salvaging much of the fresh water that otherwise would flow through aquifers into the ocean. In addition, this alternative would increase the yield of the aquifer by several hundred million cubic meters per day, although it would decrease the total volume of fresh water in the reservoir.

Each method of balancing outflow and inflow—injection wells, spreading basins, and controlled saltwater intrusion—has advantages and disadvantages. As is the case with many geologi-

cal problems, applying a combination of methods yields the most fruitful and cost-effective solution. Kings and Queens Counties now rely entirely upon importation for their freshwater needs. Public water in Nassau and Suffolk Counties is entirely underground water, and only a very small amount of seawater intrusion occurs in the southwest corner of Nassau County. Injection of tertiary treated wastewater (see Chapter 15) was tested in the 1980s. Unfortunately, the chemical differences of the injected water, natural fresh water, and saltwater caused the injection wells

continued on next page

● FIGURE 3 Methods of balancing freshwater outflow and seawater inflow; Long Island, New York. *(a)* The aquifer is recharged with highly treated wastewater *(T)* using injection wells. This method reverses saltwater intrusion and also improves the outlook for long-term water yield. *(b)* Saltwater is allowed to move inland, and a lens of fresh water forms, floating on the saltwater. This serves to establish a new equilibrium between fresh water and seawater.

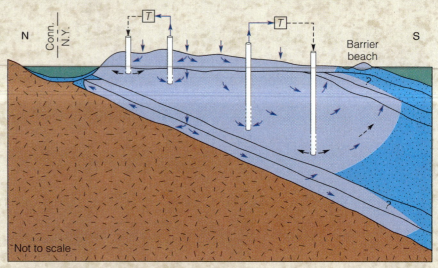

(a) Inject treated water as a barrier.

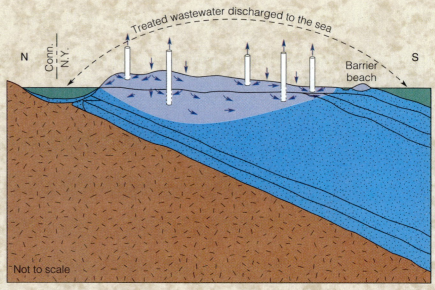

(b) Permit saltwater intrusion.

to clog and require excessive maintenance. Wastewater injection is feasible but at present is too expensive. Currently, there is a network of recharge basins that collect surface-water runoff and allow it to percolate underground into the shallow aquifer (● Figure 4). The counties require developers to dedicate land for recharge basins within their housing or commercial projects. This method of recharge has proven effective in maintaining the balance between seawater and fresh water.

● FIGURE 4 (right) A recharge basin for capturing surface-water runoff; Nassau County, New York.

DISSOLVED SUBSTANCE	STANDARD, (mg/L)	EFFECT
Calcium and magnesium	—	principal cause of water hardness; boiler scale (Ca and Mg salts)
Chloride	250	unpleasant taste; corrodes pipes
Copper	1.0	metallic taste; stains clothes and containers
Fluoride	2.0	may stain tooth enamel
Sulfate	250	bitter taste; laxative effect
Iron	0.3	unpleasant taste
Zinc	5.0	unpleasant taste
Manganese	0.05	unpleasant taste
Total dissolved solids (tds)	500	possible relationship to cardiovascular disease; corrosive to pipes

*Secondary standards are nonenforceable and apply to substances that occur in concentrations below U.S. Public Health Service standards and affect such water qualities as taste and odor. For example, although it is desirable that drinking water contain no iron, sulfate, or zinc, concentrations above these standards are permissible.

preferable quality classification for domestic use—is less than 500 ppm of *total dissolved solids (tds),* and the standard for "fresh" water is less than 1,000 ppm of total dissolved solids. ■Table 10.1 lists the Public Health Service secondary standards, and ■Table 10.2 lists some common water problems and suggestions for correcting them.

The quality of ground water can be impaired either by a high total amount of various dissolved salts or by a small amount of a specific toxic element. High salt content may be due to solution of minerals in local geological formations, seawater intrusion into coastal aquifers, or the introduction of industrial wastes into ground water. (Organic contaminants, mainly animal wastes, solvents, and pesticides, also degrade water. They are discussed in the next subsection.) ■Table 10.3 lists Public Health Service limits for some of the inorganic substances that occur in ground water and impair its safety.

Water with high salt content, particularly bicarbonate and sulfate contents, can have a laxative effect on humans. Dissolved iron or manganese can give coffee or tea an odd color, impart a metallic taste, and stain laundry. Concentrations greater than 3 ppm of copper can give the skin a green tinge (it's not easy being green, as Kermit says; the condition is reversible), and lead is both poisonous and persistent.

■ TABLE 10.2 Common Water Quality Problems

PROBLEM	PROBABLE CAUSE	CORRECTION
Mineral scale buildup	water hardness	water softener or hardness inhibitor (such as Calgon)
Rusty or black stains	iron and/or manganese	aeration/filtration; chlorination; remove source of Fe/Mn
Red-brown slime	iron bacteria	chlorination filtration unit; remove source of slime
Rotten-egg smell or taste	hydrogen sulfide and/or sulfate-reducing bacteria; low oxygen content	aeration; chlorination filtration unit; greensand filters
Salty taste	chloride—seawater or other saline water	reduce pumping to raise water table
Gastrointestinal diseases, typhoid fever, dysentery, and diarrhea	coliform bacteria and other pathogens from septic tanks or livestock yards	remove source of pollution; disinfect well; boil water; chlorinate; abandon well and relocate new well away from pollution source
Petroleum smell or film	fuel oil, diesel fuel, or lubricating oil	remove source of pollution

SOURCE: D. Daly, "Groundwater Quality and Pollution" (Geological Survey of Ireland Circular 85–1, 1985), in J. E. Moore and others, *Ground Water: A Primer,* American Geological Institute, 1994.

TABLE 10.3 Health Effects of Significant Concentrations of Common Inorganic Environmental Pollutants

INORGANIC POLLUTANT	STANDARD, mg/L	HEALTH EFFECTS
Arsenic	0.05	dermal and nervous-system toxicity effects, paralysis
Barium	1.0	gastrointestinal effects, laxative
Boron	1.0	no effect on humans; damaging to plants and trees, particularly citrus trees
Cadmium	0.01	kidney effects
Copper	1.0	gastrointestinal irritant, liver damage; toxic to many aquatic organisms
Chromium	0.05	liver/kidney effects
Fluoride	4.0	mottled tooth enamel
Lead	0.05	attacks nervous system and kidneys; highly toxic to infants and pregnant women
Mercury	0.002	central-nervous-system disorders, kidney disfunction
Nitrates	10.0	methemoglobinemia ("blue baby syndrome")
Selenium	0.01	gastrointestinal effects
Silver	0.05	skin discoloration (argyria)
Sodium	20–170	hypertension and cardiac difficulties
	70.0	renders irrigation water unusable

TABLE 10.4 Water Hardness Scale*

CONCENTRATION OF CALCIUM^{2+}, mg/L	CLASSIFICATION
0–60	soft
61–120	moderately hard
121–180	hard
>180	very hard

*Hardness is expressed here as milligrams of dissolved Ca^{2+} per liter. Dissolved calcium and magnesium in water combine with soap to form an insoluble precipitate that hampers water's cleansing action.

A *persistent pollutant* is one that builds up in the human system because the body does not metabolize it. (The Romans used lead pipes for plumbing, and skeletal materials exhumed from old Roman cemeteries retain high lead concentrations.) Lead concentrations in the water supply greater than about 0.1 ppm can build up over a period of years to debilitating concentrations in the human body. Arsenic is also persistent, and the standards for arsenic and lead are the same—less than 0.05 ppm, preferably absent. Nitrates in amounts greater than 10 ppm are known to inhibit the blood's ability to carry oxygen and to cause "blue baby syndrome" *(infantile methemoglobinemia)* during pregnancy. The element with the greatest impact upon plants is boron, which is virtually fatal to citrus and other plants in concentrations as low as 1.0 ppm.

Dissolved calcium (Ca^{2+}) and magnesium (Mg^{2+}) ions cause water to be "hard." In hard water soap does not lather easily, and dirt and soap combine to form scum. Hard water is most prevalent in areas underlain by limestone ($CaCO_3$) and dolomite ($CaMgCO_3$). A water softener simply exchanges sodium (Na^+) for calcium and magnesium ions in the water using a zeolite (a silicate ion exchanger) or mineral sieve. Water with more than 120 ppm of dissolved calcium and magnesium is considered hard (Table 10.4). Because sodium is known to be bad for persons with heart problems (hypertension) and water softeners increase the sodium content of water, those people on low-sodium diets should not drink softened water.

The good news regarding dissolved ions is that fluoride ion (F^-) dissolved in water reduces tooth cavities when it is present in amounts of 1.0–1.5 ppm. The explanation for this is that tooth enamel and bone are composed of the mineral apatite, $Ca_5(PO_4)_3(F, Cl, OH)$, whose structure allows the free substitution of fluoride F^-, chloride Cl^-, or hydroxyl $(OH)^-$ ions in the mineral structure. In the presence of fluoride, the crystals of apatite in tooth enamel are larger and more perfect, which makes them resist decay. Too much fluoride in drinking water, however—concentrations greater than 4 ppm—can cause children's teeth to become mottled with dark spots. This cosmetic blemish, among children in particular, led to the discovery of the beneficial effect of fluorine and to the establishment of concentration standards for it. Fluoride in natural waters is rare, but Colorado Springs, Colorado, is a well-known area with very few children's dentists. Dissolved fluoride is also believed to slow osteoporosis, a bone-degeneration process that accompanies aging.

Ground-Water Pollutants

The term *water pollution* refers to the introduction of chemical, physical, or biological materials into a body of natural

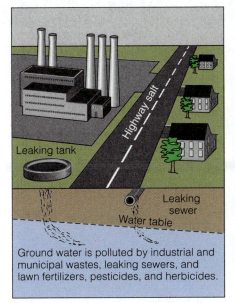
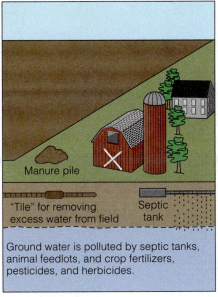
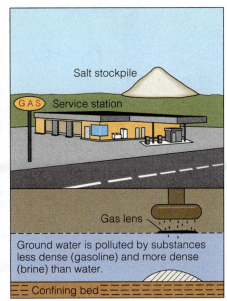

(a) Urban areas

(b) Rural areas

(c) Density effects

⦿ FIGURE 10.17 (a) Urban-area and (b) rural-area pollution sources. (c) Some pollutants float and others sink due to differences in density between ground water and the pollutants.

water that affect its future use. It is more widespread than formerly thought, as revealed by studies by state and federal agencies and by complaints from users. It is a serious problem because thousands of years may be required to flush contaminated water from an aquifer and replace it with clean water. River water, in contrast, is exchanged in a matter of hours or days. **Residence time,** the length of time a substance remains (*resides*) in a system, is an important concept in water pollution studies. We may view it as the time that passes before the water within an aquifer, lake, river, glacier, or ocean is totally replaced and continues on its way through the hydrologic cycle. The average residence time of water in rivers has been found to be a few days; of water in large lakes, several decades; in shallow gravel aquifers, a few days; and in deeper, low-permeability aquifers, perhaps thousands or even hundreds of thousands of years. This knowledge allows us to estimate the approximate time a particular system will take to clean itself naturally and indicates those cases in which a pollutant must be removed from a water body artificially by human means because of a long residence time.

Although most pollution is due to careless disposal of waste at the land surface by humans, some is due to the leaching of toxic materials in shallow excavations or mines. Ground-water pollution occurs in urban environments due to improper disposal of industrial waste and due to leakage from sewer systems, old fuel-storage tanks, wastewater settling ponds, and chemical-waste dumps (⦿ Figure 10.17, part a). In suburban and agricultural areas, septic-tank leakage, fecal matter in runoff and seepage from animal feedlots,

and the use of inorganic fertilizers, such as nitrates and phosphates, and of weed and pest-control chemicals all can serve to degrade both surface and subsurface water (Figure 10.17, part b).

Leaking gasoline-storage tanks are a more recent cause of soil and ground-water pollution. By federal law, all underground tanks now must be inspected for leakage, and if gasoline is found in the adjacent soil, the tanks must be removed and the soil cleaned. (Although the problem is not amusing, the technology created by the need for hydrocarbon cleanup has come to be lightly referred to as *yank-a-tank*.) If gasoline is found floating on ground water, it must be removed (Case Study 10.3). Both the gasoline and the decontaminated ground water are then usable (Figure 10.17, part c). Soils can be cleansed of hydrocarbons biologically with bacteria, a process known as **bioremediation,** or by aeration and oxidation of the hydrocarbon films. (See Case Study 6.4.)

▪ Table 10.5 lists some of the detrimental effects of selected organic pollutants. Water standards for common organic pollutants of water bodies are given in ▪ Table 10.6. The very low limits allowed for the pesticides lindane and endrin indicate their toxicities. Some of these toxic chemicals are also carcinogenic (cancer-causing), which makes them doubly dangerous.

Density differences between ground water and particular pollutants often govern the remedial measures that can be used; for example, gasoline and oil float on ground water, whereas salt brines from industry or agriculture sink to lower levels. In some cases it is possible to skim a layer of

DNAPL and LNAPL

A couple of the newest acronyms in the environmental cleanup business are *DNAPL* (for *dense, nonaqueous-phase liquid* hydrocarbons) and *LNAPL* (for *light, nonaqueous-phase liquid* hydrocarbons). With the discovery of crude oil and the explosion (no pun intended) of its use in the reciprocating engine, numerous refineries and above-ground petroleum-storage-tank farms were developed. All the states face the problem of these tanks leaking and polluting the soils beneath them with refined hydrocarbons, including lubricating oils, diesel fuel, jet fuel, and gasoline—all of them LNAPLs (● Figure 1).

No place in the United States is the problem more obvious than the Los Angeles coastal plain, where there are 16 major refineries and 33 above-ground petroleum-storage-tank farms. Underlying these refineries are several important aquifers, one of them less than 60 meters (200 ft) below the ground surface. Some of the refineries are on the Environmental Protection Agency (EPA) National Priorities List (Superfund sites), and others are regulated by the Resource Conservation and Recovery Act (RCRA) as hazardous waste sites (see Chapters 1, 11, 14, and 15). The environmental concerns at these sites include:

- Hydrocarbons dissolved in the ground water will limit the water's beneficial use.
- LNAPL hydrocarbon pools occurring as "perched" zones serve as continuous sources of contamination for both soil and ground water.
- Soil contamination might convert the soil to a hazardous material.
- The accumulation of hydrocarbon vapors poses potential fire or explosion hazards.

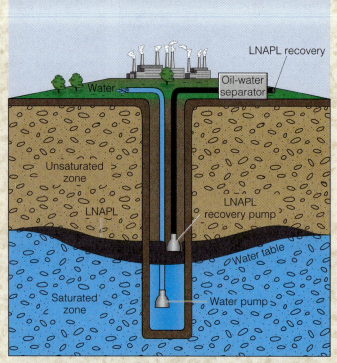

● **FIGURE 2 A two-pump LNAPL-hydrocarbon recovery well. Sensors activate or deactivate pumps so that the upper pump draws only LNAPL hydrocarbons and the lower one draws only water. Recovered oil is reused.**

The combined areal extent of LNAPL pools perched, or floating, on ground water beneath the refineries is estimated as being on the order of 600 hectares (1,500 acres or 2.3 mi²), and the volume is estimated to be between 1.5 and 7.5 million barrels—the size of a small oil field!

The most sophisticated LNAPL-hydrocarbon recovery is accomplished by a recovery well with a two-pump system (● Figure 2). The two pumps are controlled by a series of electrodes set at predetermined depths. The lower, water–recovery pump runs until the drawdown brings the LNAPL–water interface into contact with the electrode just above the lower pump's intake. When the electrode senses the presence of LNAPL hydrocarbons, it shuts down the water pump so that LNAPLs are not ingested and mixed with fresh water. The upper, LNAPL-recovery pump is switched on and off by a probe that measures the electrical resistivity of fluids adjacent to the intake. LNAPLs have high electrical resistivity, and water has a low resistivity. The probe automatically shuts down when it senses fresh water. In this manner, water and hydrocarbons are recovered separately, treated, and then recycled to the benefit of all concerned.

● **FIGURE 1 Leaking oil-storage tanks, West Texas. Such leaks lead to soil and ground-water contamination.**

ORGANIC SUBSTANCE	HUMAN HEALTH EFFECTS	ENVIRONMENTAL EFFECTS
Dieldrin	convulsions, kidney damage	toxic to aquatic organisms
DDT	convulsions, kidney damage	reproductive failure in animals, eggshell thinning
PCBs	vomiting, abdominal pain, liver damage	eggshell thinning in birds, liver damage in mammals
Benzene	anemia, bone-marrow damage	toxic to some fish and aquatic invertebrates
Phenols	death at high doses	decreased phytoplankton productivity
Dioxin	acute skin rashes, systemic damage, mortality	lethal to aquatic birds and mammals

★Insecticides or compounds used in their manufacture.
SOURCE: U.S. Public Health Service; Environmental Protection Agency.

gasoline from the top of the water table. Gasoline or fuel oil that is floating on the capillary fringe may yield vapors that can rise and be trapped beneath or in the walls of buildings. If these fumes contain BTX (*b*enzene, *t*oluene, and *x*ylene), they are carcinogenic. "Air-stripping" is currently the preferred method of removing volatile organic pollutants such as BTX compounds and trichloroethylene (TCE) and tetrachloroethylene (PCE) cleaning solvents from ground water. (The origin of these substances is usually connected to refinery, industrial, or military activities.) The polluted water is brought to the earth surface, where the contained volatile pollutants (TCE or BTX) are gasified (air-stripped) and wasted to the atmosphere with no adverse effects.

In Canada 10,000 metric tons of PCE are used per year, primarily in the dry-cleaning and metal-cleaning industries. An exhaustive study mandated and sponsored by the Canadian Environmental Protection Act (CEPA) indicates that PCE is entering the environment in significant quantities. Due to its volatility (it evaporates readily), it is found mostly in the atmosphere, where it has a relatively short residence time and does not contribute to global warming. The study concludes that the present atmospheric concentration of PCE is not expected to have adverse effects on Canada's aquatic and terrestrial animals, including humans, but it has the potential of harming trees. Local spills, of course, require cleanup and remediation.

 ## CONSERVATION AND ALTERNATIVE SOURCES

Drought in the southwestern United States is a persistent and nagging problem. In 1990–91, for example, following several years of below-average rainfall in California, surface-water reservoirs became depleted. This led to excessive ground-water withdrawals and eventual mining. Remedies that have been proposed include desalinizing seawater, recycling wastewater, and even bringing huge Antarctic icebergs to dry coastal areas to melt and provide fresh water. Voluntary water conservation has been quite successful,

■ TABLE 10.6 Water Standards for Common Organic Pollutants★

SUBSTANCE	MAXIMUM SAFE CONCENTRATION, mg/L
Cyanide	0.05
Lindane	0.004
Endrin	0.0002
2,4-D	0.10
Phenols	0.001

★Insecticides and animal-control compounds.
SOURCE: U.S. Public Health Service; Environmental Protection Agency.

reducing consumption by as much as 15 percent, and even greater reductions have been achieved where rationing is mandatory.

Conservation is every citizen's obligation where a resource is in short supply. ■ Table 10.7 provides some examples of how water can be conserved in the home. Using recycled wastewater, so-called gray water, is the simplest conservation measure after rationing. Recycled shower, bath, and laundry water can be used for irrigating lawns and certain plants, especially when biodegradable soaps and detergents are used. Gray water contaminated with chlorine should not be used, however.

Desalinization of seawater is a viable alternative where no other water source is available. The process is expensive, and the water must be pumped uphill to users, which further increases its cost. In most cases, more energy is required to pump the water than to desalinize it. For this reason, desalinization factories are found only on desert coasts (as in some Middle Eastern nations), on small islands that do not receive much precipitation, and on military installations.

Artificial recharge (water-spreading; see Figure 10.14) of aquifers during heavy-rain years—"banking" water, so to speak—may provide longer-term benefits. This is done by diverting water that ordinarily would flow into rivers into

Hot Springs, Geysers and Other Marvels of Underground Water

In areas of abundant underground water and cooling igneous rocks at shallow depths we may find hot springs (by definition, >37° C). Some of these are mere trickles of warm water, and others are scalding and widespread, as at Yellowstone National Park, in Iceland, and at Rotorua in New Zealand (see Chapters 5 and 13). Where a system of intricate and interconnected underground openings occur, hot springs may periodically erupt with tremendous force as geysers (Icelandic *geysir,* "gush forth"). Perhaps the most famous geyser area in the world is in Wyoming's Yellowstone Park, where Old Faithful erupts every 30 to 90 minutes (◉ Figure 10.18).

A special underground plumbing is required for deep water near its boiling temperature to flash into steam, and not simply to convect upward as it expands (◉ Figure 10.19). The deep water of a geyser system has a higher boiling temperature than the near-surface water, because it is under pressure. Any sudden reduction in that pressure causes the high-temperature, deep water to flash into steam. Such a pressure reduction can result when rising steam bubbles carry surface water out of the system—causing sort of a pre-eruption "burp"—after which the main eruption occurs. After the eruption, the cooled waters seep back underground where they are reheated, leading to a repetition of the eruption cycle.

Most hot-spring waters are highly mineralized and thus they deposit mineral matter as they cool or evaporate. Waters rich in calcium carbonate ($CaCO_3$) deposit *traver-*

◉ **FIGURE 10.18** Old Faithful geyser in Yellowstone National Park, northwest Wyoming, erupts every 30 to 90 minutes.

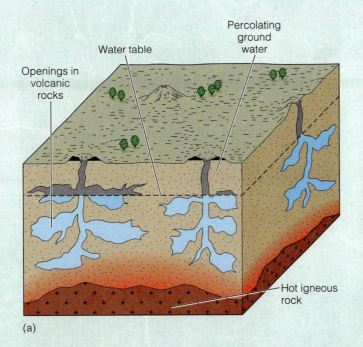

Openings in volcanic rocks

Water table

Percolating ground water

Hot igneous rock

(a)

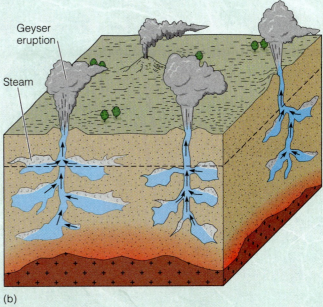

Geyser eruption

Steam

(b)

● FIGURE 10.19 *(a)* Geologic conditions of an active geyser basin. *(b)* Deeper water is trapped in irregular openings and is under pressure. When at its boiling point, a sudden pressure reduction in an individual cavity causes the water to flash into steam and rise to the surface as a fountain of steam and hot water.

tine, the hard, dense variety of calcareous tufa (◉ Figure 10.20), whereas waters rich in silica (SiO_2) deposit a spongy hydrated silica known as *geyserite* (◉ Figure 10.21).

Mineralization by underground water is also responsible for transforming organic material into hard mineral matter —a process known as *fossilization.* The replacement mineral

continued on next page

● FIGURE 10.20 (left) Minerva Terrace at Mammoth Hot Springs in Yellowstone National Park is composed of calcium carbonate ($CaCO_3$), which precipitates as the waters spill out of the spring and cool. Such fresh-water deposits are known as travertine (Italian *tivertino,* "stone of *Tivoli,*" the old Roman name of a town near Rome where extensive deposits are found). It polishes beautifully and is used as exterior and interior facing stone for buildings.

FIGURE 10.22 Petrified wood from Arizona. Woody tissue is replaced cell by cell with mineral matter from ground water.

FIGURE 10.21 (right) Liberty Cap, a large geyserite mound, formed by repeated eruptions and precipitation of hydrated silica ($SiO_2 \cdot nH_2O$) around a geyser vent; Yellowstone National Park, Wyoming.

FIGURE 10.23 Amethyst geode from Brazil. Silica dissolved in underground water percolates into an opening in the host rock and is deposited, forming crystals. Most commonly, geode crystals are varieties of quartz, but other minerals dissolved in ground water may also crystallize.

is typically silica or calcium carbonate, and replacement may take place cell by cell as in the case of petrified wood (Figure 10.22).

Geodes (pronounced "gee-odes") are rounded cavities in rocks that are lined with quartz crystals formed from percolating ground water. They are one of nature's most beautiful geological creations. Geodes from Mexico and Brazil are lined with light-colored agate from which "grow" beautiful purple amethyst or smoky quartz crystals (Figure 10.23). They are produced when silica in aqueous solution is deposited directly into cavities in limestone or volcanic rocks.

ACTIVITY	NORMAL USE	CONSERVATIVE USE
Shower	5 minutes, 30–40 gallons	Reduce shower length; use water-conserving shower head.
Tub bath	full tub, 36 gallons	Use minimal water level (10–12 gallons).
Teeth brushing	tap running, 2 gallons	Wet brush, rinse briefly.
Shaving	tap running, 5 gallons	Fill sink or basin (1 gallon).
Dishwashing	tap running, 30 gallons	Wash and rinse in sink (5 gallons).
Automatic dishwashing	full cycle, 25 gallons	Wash full loads on short cycle (12 gallons).
Hand washing	tap running, 2 gallons	Fill basin (1 gallon).
Toilet flushing	5–7 gallons per flush	Place bottles or bar dam in tank (4–6 gallons).
Washing machine	full cycle with top water level, 30–35 gallons	Wash full loads on short cycle with minimum water level (15–17 gallons).
Outdoor watering:		
hose	5 minutes, 40–50 gallons	Water in the morning (20 gallons).
lawn sprinklers	20 minutes daily, 1,200 gallons per week	Water as needed.

CONSIDER THIS . . .

After the 1959 major earthquake near Hegben Lake, Montana, some 20–50 miles west of Yellowstone National Park, the periodicity of some of the park's geysers changed. Some other geysers stopped erupting, and a few hot springs became geysers. How can this be explained?

manufactured basins surrounded by dikes. This temporarily ponds the running water, giving it time to infiltrate through the streambed and into the aquifer. By this restocking of aquifers when surplus surface water or abundant cheap imported water is available, the people of some drought-prone areas are able to prepare for those episodic droughts that seem to strike all mid-latitude areas sooner or later.

 ## SUMMARY

Ground Water

DEFINED Loosely defined as all water underground, as distinct from surface water; the water in the hydrologic cycle that infiltrates underground.

ZONES
1. The *zone of aeration,* where the voids in rock or sediment are filled with air and water films.
2. The lower *zone of saturation,* where voids are filled with water.

WATER TABLE The contact between the zone of aeration and the zone of saturation.

Aquifers

DEFINED Bodies of porous and permeable rock or sediment that are saturated with water.
1. *Porosity*—The volume of pore spaces in a geologic material, expressed as a percentage of the total volume.
2. *Permeability*—Ease with which fluids flow through a geologic material.
3. *Aquiclude*—A geologic material that lacks permeability and inhibits water flow.

TYPES
1. Unconfined—An aquifer in which the water table is static and that is exposed to atmospheric pressure changes. A slope, or gradient, is required to induce subsurface flowage in a given direction. A well that taps the static water table creates a cone of depression, which provides the necessary gradient for water to flow to the well. Drawdown is the amount of lowering of the water table in a pumping well.
2. Artesian—An aquifer that is confined between aquicludes. The hydraulic head causes water to rise in a well when the aquifer is penetrated. The height to which water will rise at any point along the aquifer is known as the *potentiometric surface.*

Ground-Water Management

DEFINED Managing the amount of water contained in a ground-water basin so that it will produce water in the future.
1. *Sustained yield*—The amount of water an aquifer will yield on a day-to-day basis.
2. *Specific yield*—The amount of water a body of rock or sediment will yield by gravity alone.
3. *Specific retention*—The amount of water held by a body of rock or sediment against the pull of gravity.

Specific yield *plus* specific retention *equals* porosity.

SEAWATER ENCROACHMENT   The invasion of coastal aquifers by saltwater. Management requires maintaining high freshwater tables to balance the denser saltwater.

Water Quality

DEFINED   Purity or drinkability (potability) of water as established by the U.S. Public Health Service. Dissolved constituents are measured in milligrams per liter (volume) or parts per million (weight).

STANDARDS   Less than 500 ppm total dissolved solids for "sweet" water, and limits on ions and molecules that impair the quality of water.

Pollution

DEFINED   Chemical, physical, or biological materials that impair the future use of water and that may be a health hazard.

RESIDENCE TIME   The average length of time a given substance will stay in a system, such as water in a lake or aquifer. Whereas the residence time of human-generated smog in the atmosphere of large cities is on the order of ten days, pollutants may reside in some aquifers for thousands of years.

COMMON POLLUTANTS   Hydrocarbons, carcinogenic BTX solvents (benzene, toluene, and xylene), fertilizers, pesticides, livestock fecal matter, and gasoline.

Conservation and Alternative Sources

Conservation means include the use of treated wastewater for irrigation and recharging aquifers, water-saving irrigation devices, water-spreading during wet years, and voluntary water conservation. Desalinization and importation are the only viable alternative sources at present.

KEY TERMS

aquiclude
aquifer
bioremediation
cone of depression
confined aquifer
drawdown
effluent (stream)
ground water
hydraulic conductivity
hydrogeologist
influent (stream)

perched water table
permeability
porosity
potable water
potentiometric surface
residence time
specific yield
sustained yield
unconfined aquifer
water table

STUDY QUESTIONS

1. Sketch a geologic cross section that shows the ground-water zones and the static water table.

2. Sketch or define a confined aquifer. Explain the potentiometric surface and its importance in production of water from an artesian well.

3. How does topography affect the shape of the static water table?

4. How does one determine the amount of water a well that taps the static water table might produce on a long-term basis?

5. Explain the distinction between effluent and influent streams.

6. Explain the relationship between the amount of drawdown in a water well and the aquifer's permeability.

7. What states are the largest consumers of ground water, and why is their usage so heavy?

8. What is meant by ground-water *mining?* Cite the case history of an aquifer where this has happened.

9. Many beverage manufacturers claim the water in their drinks is purer because it comes from artesian wells or springs. Comment on these claims.

10. What are the Public Health Service standards for total dissolved solids in drinking water? What is the maximum amount of dissolved salts for safe drinking water?

11. Name some organic compounds sometimes found dissolved in ground water that are hazardous to human health.

12. Considering that agriculture is the largest consumer of ground water and that aquifers in many agricultural areas are being mined, how might farmers conserve ground water?

FURTHER INFORMATION

BOOKS AND PERIODICALS

American Institute of Professional Geologists. 1985. *Ground water: Issues and answers.* Arvada, Colo.: A.I.P.G.

Baldwin, H. L., and C. L. McGuinness. 1963. *A primer on ground water.* Washington: U.S. Geological Survey.

Bull, W. B., and R. E. Miller. 1975. *Land subsidence due to ground-water withdrawal in the Los Banos–Kettleman City area, California.* U.S. Geological Survey professional paper 437-E.

Cohen, P.; O. L. Franke; and B. L. Foxworthy. 1970. *Water for the future of Long Island, New York.* New York Division of Water Resources, Department of Environmental Conservation, in cooperation with the U.S. Geological Survey.

Heath, Ralph C. 1983. *Basic ground-water hydrology.* U.S. Geological Survey water-supply professional paper 2220.

Loveland, D. G., and Beth Reicheld. 1987. *Safety on tap: A citizen's drinking water handbook.* Washington: League of Women Voters Education Fund.

Moore, John E.; A. Zaporozed; and James W. Mercer. 1994. *Ground water: A primer.* Environmental awareness series. Alexandria, Va.: American Geological Institute.

Solley, W. B.; R. R. Pierce; and Howard A. Perlman. 1993. *Estimated use of water in the United States in 1990.* U.S. Geological Survey circular 1081.

OCEANS AND COASTS

Man marks the earth with ruin; his control stops with the shore.

LORD BYRON, *CHILDE HAROLD'S PILGRIMAGE* (1812)

Although Lord Byron's insight into human degradation of the planet was prophetic, environmental pollution no longer stops at the shoreline, and no relationship is more important to humans than that between the two most common substances on earth, water and air. The transfer of heat from the ocean to the atmosphere feeds energy-hungry hurricanes. The reverse transfer causes evaporation, which puts moisture into the atmosphere and begins the hydrologic cycle. Heat exchanges between the ocean and the atmosphere create maritime climates and modify day-to-day weather along coasts and adjacent inland areas. Much of the severe weather along the U.S. Gulf Coast is the result of moisture-laden air flowing northward from the Gulf of Mexico mixing with cold air flowing southward from the northern states and Canada.

OPENING PHOTO

"Watch that first step!" would be a good warning to anyone exiting these stilt houses that once rested on the beach. Strong waves reduced the beach level almost four meters during one storm. Westhampton, New York.

Coastal oceans are important because they are the location of beaches and estuaries. Beaches are geologically important because they are the last rampart protecting the land from the sea. In addition, they are unexcelled as areas of recreation and relaxation. **Estuaries** are semienclosed bodies of water where fresh water and saltwater mix, forming brackish waters that are attractive to a host of marine and terrestrial life. A significant estuary is Chesapeake Bay, the site of many important seaports and centers of commercial fishing activities.

According to the 1990 U.S. census, coastal states have been experiencing the largest population growth; 50 percent of the U.S. population can now drive to a coast in less than an hour. Coastal environments are constantly being modified by natural and human intervention, usually with negative impacts on their inhabitants. For example, Hurricane Andrew left thousands of people homeless in 1992. Wetlands are being filled or made into marinas; sewage and solid waste have been dumped into estuaries and coastal waters; and shoreline modifications have eroded beaches and silted in harbors.

Wind-driven waves that arise in the open ocean eventually strike shorelines, generating dynamic processes that constantly alter beaches. Beach erosion is a serious, costly problem in most coastal areas. Armed with knowledge of how waves erode beaches and of the fate of eroded materials, we may be able to mitigate many of the human-induced coastal problems.

 WIND WAVES

Waves in Deep Water and at the Shore

When wind blows over smooth water, surface ripples are created that enlarge with time to form local "chop" (short, steep waves) and eventually wind waves. The factors that determine the size of waves at sea and ultimately the size of the surf that strikes the shoreline are

- the wind velocity,
- the length of time the wind blows over the water, and
- the **fetch,** the distance along open water over which the wind blows (● Figure 11.1).

For a given fetch, wind velocity, and wind duration, wave dimensions will grow until they reach the maximum size for the existing storm conditions. This condition, known as a *fully developed sea*, represents the maximum energy that the waves can absorb for wind of a given velocity. For example theoretical wave heights of 32 meters (100 ft) are possible for a 50-knot (93 km/h, 58 mph) wind blowing for 3 days over a fetch of 2,700 kilometers (1,600 mi). Wind waves will move out of the area of strong atmospheric disturbance as regularly spaced, long waves known as *swell* and eventually become huge breakers on a distant shoreline.

Important dimensions of wind waves are their length, height, and *period*—the length of time between the passage

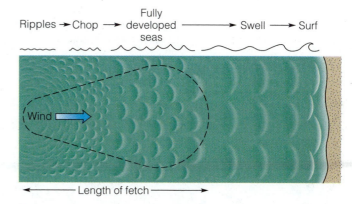

● **FIGURE 11.1 The evolution of wind-generated sea waves from a rippled water surface to fully developed seas (storm waves). Long waves with very regular spacing and height that outrun the storm area are called *swell*.**

of equivalent points on the wave. These dimensions are measures of the amount of *energy* the waves contain; that is, their ability to do work (● Figure 11.2). Although waves with long wavelengths and long periods are the most powerful, both long and short waves are capable of eroding beaches.

The actual water motion in a wind-generated wave describes a circular path whose diameter at the sea surface is equal to the wave height. This circular motion diminishes with depth and becomes essentially zero at a depth equal to half the wavelength ($L/2$), the depth referred to as **wave base.** Wave base is the maximum depth to which waves can disturb or erode the sea floor. (This explains why submersibles at depth are not tossed about during storms as surface ships are.) Although the waveform moves across the water surface with considerable velocity, there is no mass forward movement of water. Such waves are called **waves of oscillation.** Floating objects simply bob up and down with a slight back-and-forth motion as wave crests and troughs pass.

As waves approach a shoreline and move into shallow depths, the sea floor interferes with the oscillating water particles. This friction causes wave velocity to decrease. This in turn causes the wave to steepen; that is, the wavelength shortens and the wave height increases, a phenomenon easily seen by surf watchers. Eventually the wave becomes so steep that the crest outruns the base of the wave, and the crest jets forward as a "breaker." (Figure 11.2). Breakers are **waves of translation,** because the water in them physically moves landward in the surf zone. Where the immediate offshore area is steep, we find the classic tube-shaped *plunging breakers* typical of Sunset Beach and Waimea Bay on the north shore of Oahu (● Figure 11.3). This is because the wave peaks rapidly on a steep sea floor. On a flat nearshore slope, on the other hand, waves lose energy by friction, so they build up slowly, and the crest simply spills down the face of the wave. This latter type are called *spilling breakers*. They are typical of wide, flat

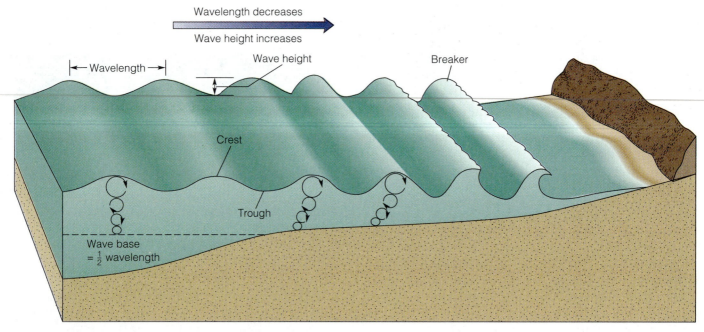

Wavelength decreases

Wave height increases

Wavelength

Wave height

Crest

Breaker

Trough

Wave base = $\frac{1}{2}$ wavelength

● FIGURE 11.2 Motion of water in a wave and the terminology of wave geometry. Note that the orbital water motion decreases with depth until it is essentially nil at a depth of one-half the wave length (*L/2*), a depth called *wave base*. The height on land reached by the broken wave is called its *runup*.

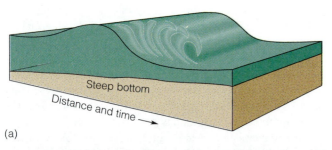

Steep bottom

Distance and time

(a)

(b)

● FIGURE 11.3 *(a)* Plunging (tubular) breakers form on steep offshore slopes. The front face of the wave steepens until the top of the wave "jets" forward and down onto the wave's front slope. *(b)* A tubular plunging wave; Waimea Bay, Oahu, Hawaii.

nearshore bottoms such as that at Waikiki Beach on Oahu (● Figure 11.4). Regardless of the type of wave, *swash,* the water that rushes forward onto the foreshore slope of the beach, returns seaward as *backwash* to become part of the next breaker. A common misconception is the existence of a bottom current called *undertow* that pulls swimmers out to sea. The strong pull seaward one feels in the surf zone is simply backwash water returning to be recycled in the next breaker.

Wave Refraction

Where waves approach a shoreline at an angle, they are subject to **wave refraction,** or bending. Waves may be seen to slow first at their shallow-water ends while their deeper-water ends move shoreward at a higher velocity. Thus a fixed point on a wave crest follows a curved path to the shoreline (● Figure 11.5).

Because waves seldom approach parallel to the shore, some water is discharged parallel to the beach as a weak current within the surf zone called a **longshore current.** The greater the angle at which the wave approaches the shoreline, the greater the volume of water that is discharged into the longshore-current stream. This current is capable of moving swimmers as well as sand grains down the beach. Sand becomes suspended in the turbulent surf and swash zone and moves in a zigzag pattern down the beach's foreshore slope (see Figure 11.5). Beaches are dynamic; the sand grains present on the foreshore today will not be the same ones we see there tomorrow. Beach-sand transport parallel to the shore in the surf zone is known as **littoral drift,** and

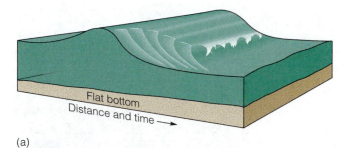

(a)

(b)

● FIGURE 11.4 *(a)* Spilling breakers form on gently sloping offshore areas. The forward face of a spilling wave steepens, and the crest spills down the front of the wave. *(b)* A spilling-wave front forms a "wall" across which the surfer glides, much as a skier moves across a slope, rather than straight downhill.

its rate is usually expressed in thousands of cubic meters or yards of sand per year.

Another consequence of wave refraction is the straightening of irregular shorelines. This occurs because the concentration of wave energy at points causes shoreline erosion, and the dissipation of wave energy in embayments allows sand deposition (● Figure 11.6).

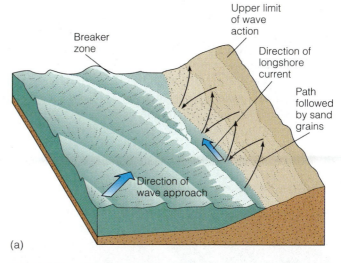

(a)

(b)

● FIGURE 11.5 *(a)* Refraction, or bending, of wind waves and the generation of a longshore current parallel to the beach. The longshore current transports beach sand, resulting in a dynamic process called *beach drifting*. *(b)* Wave refraction along a relatively straight shoreline near Santa Barbara, California.

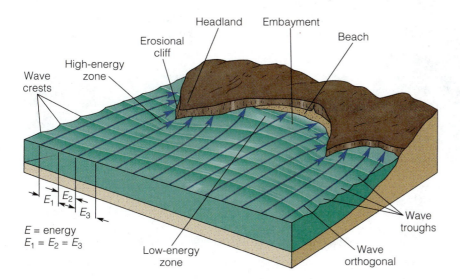

● FIGURE 11.6 Wave refraction leads to concentration of energy on points and dissipation in embayments. The arrowed lines, called *orthogonals,* are constructed perpendicular to the wave crest and show the wave path. E_1, E_2, and so on, represent equal parcels of wave energy and, therefore, equal ability to erode. Note that energy is focused toward the points and spread out in the bays.

Longshore currents are likened to rivers on land, with the beach and the surf zone constituting the two banks. Just as damming a river causes sedimentation to occur upstream and erosion to occur downstream, so does building structures perpendicular to the beach. Such a structure slows or deflects longshore currents, causing sand accretion upstructure and erosion below it (◉ Figure 11.7). **Groins** (French *groyne,* "snout") are constructed of sandbags, wood, stone, or concrete, ostensibly to trap sand and create a wider beach. For groins of equal length and design, the severity of the erosion–deposition pattern is a function of the strength

◉ **FIGURE 11.8** Groin field at Cape May, New Jersey. Note the bread-knife, or sawtooth, appearance that has resulted from sand accretion and erosion above and below the groins.

of the waves and the angle at which they strike the shoreline. Most groins are short and are intended to preserve or widen a beach in front of a private home or resort. Unfortunately, because this causes neighbors' beaches below the groin to narrow, many of those neighbors build groins to preserve and widen *their* beaches, and so on. Like artificial levees on a river, groins tend to generate more groins, and we generally see fields of them along a given stretch of beach instead of just one (◉ Figure 11.8).

Just as rivers overflow their banks, longshore currents may overflow through the surf zone, particularly during periods of big surf. Overflow (really *outflow* to the open ocean) occurs because the water level within the surf zone is higher than the water level outside the breakers, making for a hydraulic imbalance. The higher water inshore flows seaward to lower water offshore as a **rip current** through a narrow gap, or *neck,* in the breakers (◉ Figure 11.9).

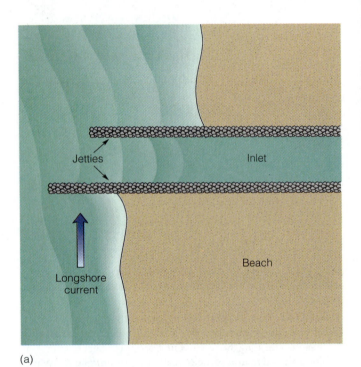

(a)

(b)

◉ **FIGURE 11.7** *(a)* Accretion and erosion of beach sand at jetties constructed to improve an inlet to a navigable lagoon. *(b)* This exact condition is seen at Boca Raton, Florida.

CONSIDER THIS . . .

After a few hours of frolicking in the surf, you find you have drifted along the shore to a spot where the waves aren't breaking. Suddenly, you find yourself being drawn seaward by an invisible hand, no matter how vigorously you swim toward the beach. What is happening? What should you do, and why?

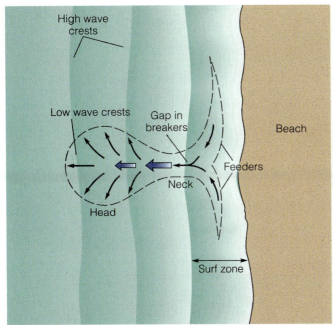

 FIGURE 11.9 A rip current is characterized by relatively weak *feeder currents,* a narrow *neck* of strong currents, and a *head* outside the surf zone, where the rip dissipates.

Because the location of rip currents, erroneously called rip *tides,* is controlled by bottom topography, they usually occur at the same places along a beach. Rip currents can become a problem to swimmers under big-surf conditions, when velocities in the neck may reach 4–5 knots. Awareness can save your life, so be alert to a gap in the breakers, white water beyond the surf zone, and objects floating seaward—all of which are evidence of a strong rip current

flowing out to sea (Figure 11.10). Should you find yourself in a rip current, swim parallel to the beach until you are out of the narrow neck region of high velocity.

SHORELINES

Where the sea and land meet we find *shorelines,* some of which are shaped primarily by erosional geologic processes and others by depositional geological processes. Shorelines that are exposed to the open ocean and high-energy wave action are likely to have *erosional* features such as sea cliffs or broad, wave-cut platforms. Shorelines that are protected from strong wave action, by offshore islands or by the way they trend relative to the direction of dominant wave attack, are characterized by *depositional* features such as sand beaches.

Technically, the shoreline is the distinct boundary between land and sea that changes with the tides, whereas the *coast* is the area that extends from the shoreline to the landward limit of features related to marine processes. Thus, a coast may extend inland some distance, encompassing estuaries and bays. Human use of the shore is most intensive where there are broad sand beaches. Poorly designed shoreline engineering works have caused intensive beach erosion and even total loss of some beaches. Thus, human-induced beach erosion can have considerable impact on the shore environment and on populations that use beaches.

Beaches

In addition to their usefulness as recreational areas, beaches also serve to protect the land from the erosive power of the sea. They are composed of whatever sediment rivers and waves deliver to them (Figure 11.11). For example, in some places in Hawaii, black sand forms from weathered

(a)

(b)

 FIGURE 11.10 *(a)* Small rip currents. Note the gap in the breakers and the white water seaward of the surf zone. *(b)* Signs warn swimmers about dangerous currents in the surf at some beaches.

● FIGURE 11.11 Beach materials. *(a)* Black sand beach; American Samoa. *(b)* Coral sand; Caribbean. *(c)* Mineral-grain beach sand; Santa Barbara, California. *(d)* Cobble, or shingle, beach; Oceanside, California.

basalt and also from hot lava that disintegrates when it flows into the sea. In contrast, reef-fringed oceanic islands are characterized by beautiful white coral sand and shell beaches. Mainland beach materials are largely quartz and feldspar grains derived from the breakdown of granitic rocks. *Shingle* beaches are composed of gravels; they typically occur where the beach is exposed to high-energy surf. Amusing or not, tin cans from a local dump once formed a beach at Fort Bragg, California.

TYPES OF BEACHES. Four types of beaches can be recognized, according to the geologic setting where they are found:

- coastal-plain, or mainland, beaches,
- pocket beaches,
- barrier islands, or barrier beaches, and
- sand spit beaches (● Figure 11.12).

Each of these has its own peculiar stability problems, but all are influenced by seasonal differences in wave action. As a rule, winter beaches are narrow or nonexistent, because high winter-storm waves erode sand from the beach berm and deposit it in offshore sandbars parallel to the beach (● Figure 11.13). You may have experienced this while wading in waist-deep water that abruptly shallowed as you walked onto

CONSIDER THIS . . .

Your favorite beach disappears during a large winter storm. Will you have to find a new beach for next summer? If not, will the summer beach look any different and, if so, how will it be different? Explain your answer.

(a)

(b)

(c)

(d)

● FIGURE 11.12 Examples of the four types of beaches. *(a)* Coastal plain beach; Huntington Beach, California. Note that the high angle of wave approach produces a strong longshore current along the beach from top to bottom in the photo. *(b)* Pocket beaches on a cliffed shoreline; Laguna Beach, California. *(c)* Barrier beach; New Jersey. Note the groin field and accretion of sand on the upcurrent side of the structures. *(d)* The sand spit that encloses San Diego Bay is the result of a huge volume of sand transported northward from the Tijuana River in Mexico. Naval Air Station Coronado is at the south entrance to the bay, and rock-supported Point Loma forms the north entrance.

a sandbar. With the onset of long, low summer waves, the sand in the offshore bar gradually moves back onto the beach, and the beach berm widens. Thus, the natural annual beach cycle is from narrower and gravelly in the winter, to wider and sandy in the summer.

Barrier islands form the most extensive and tenuous beach system in the United States (● Figure 11.14). They extend southward from Long Island and Coney Island, New York, to Atlantic City, continuing as the Outer Banks of North Carolina (the location of Kitty Hawk) and then

● FIGURE 11.13 (right) *(a)* Winter and summer beach profiles and *(b)* some beach terminology.

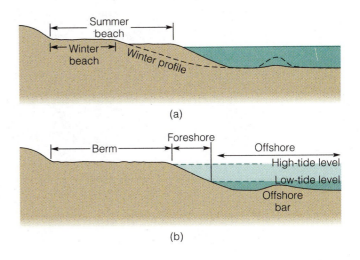

(a)

(b)

● FIGURE 11.14 (left) Satellite photo of the barrier island system off Long Island and New Jersey.

down the coast of Florida to Daytona Beach, Cape Canaveral, Palm Beach, and Miami Beach. The barrier system begins again on the west coast of Florida and continues to the Padre Islands in Texas, interrupted only by the Mississippi River delta. These barrier beaches are transient features; formed by wave action, they are reoriented by wave action and frequently overtopped by storm surges. They are separated from the mainland by shallow lagoons and marshland over which they migrate as large storms overwash the islands, transporting sand from their seaward sides to their landward sides.

Hog Island, Virginia, is a barrier island that was a popular hunting and fishing spot of the rich and famous in the late 1800s (● Figure 11.15). Its town of Broadwater had 50 houses, a school, a cemetery, and a lighthouse. After a 1933 hurricane inundated the island, killing its pine forest, Broadwater was abandoned. Today Broadwater is under water a half-kilometer offshore. Typical shoreward migration rates for an Atlantic-coast barrier island are 2 meters per year, but extraordinary displacements have been reported. Hatteras Island off mainland North Carolina migrated 500 meters (1,550 ft) landward between 1870 and 1990 (4.2 m/y; more than 14 ft/y). The lighthouse at Cape Hatteras, a notorious graveyard for ships, was built almost a half-kilometer landward from the shoreline in 1870. It now stands on a point in the water and is protected by sandbags.

Spits, such as Cape Cod, Massachusetts, are elongated masses of sand attached at one end to the mainland. Spits

● FIGURE 11.15 The south end of Hog Island, a Virginia barrier island, migrated almost 3 kilometers (1.9 mi) landward between 1852 and 1963.

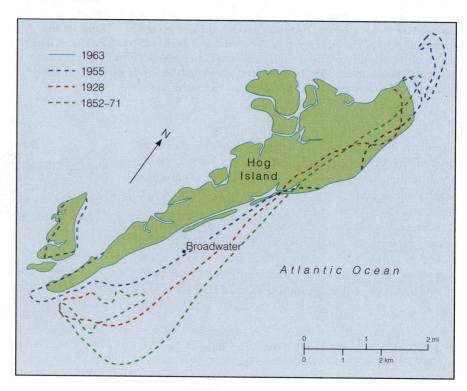

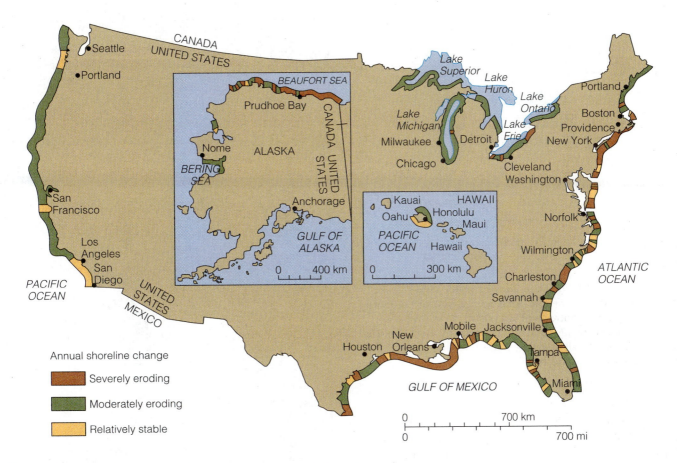

Annual shoreline change

■ Severely eroding

■ Moderately eroding

■ Relatively stable

◉ **FIGURE 11.16** Present coastal erosion in the United States. Note that most of the shorelines are eroding. Uncolored coastal areas on the map represent shorelines for which data are not available.

may be formed by longshore currents when they transport and deposit sand across the mouths of bays or estuaries. Newport Beach, California is a spit deposited across Balboa Bay, an estuary of an abandoned Pleistocene river channel. Coronado Island, a sand spit that protects San Diego Bay, formed from huge volumes of sand delivered from the south by Mexico's Tijuana River (Figure 11.12, part d).

BEACH EROSION—THE RIVER OF SAND GONE DRY. A beach is stable when the sand supplied to it by longshore currents replaces the amount of sand removed by waves. Where the amount of the supplied sand is greater than the available wave energy can remove—typically near the mouth of a river—the beach widens until wind action forms sand dunes. Significant areas of coastal dunes occur in Oregon, on the south shore of Lake Michigan in Indiana, at El Segundo in California, near the Nile River in Egypt, and on the Bay of Biscay on France's west coast, to name only a few. The more usual case is for wave action and longshore currents to move more sand than is supplied, and here we find eroding beaches or no beaches at all. All 30 of the U.S. coastal states have beach erosion problems of varying magnitudes resulting from both human activities and natural causes (◉ Figure 11.16).

Inasmuch as sand comes to beaches by way of rivers, we should look to the land for causes of long-term beach erosion. Water-storage and flood-control dams trap sand that would normally be deposited on beaches. The need for upland flood protection is clear, but it can conflict with other needs, including the need for well-nourished beaches. For instance, 80 percent of Southern California's largest watershed, that of the Santa Ana River, is behind a dam (◉ Figure 11.17). The dam was built to control floodwaters and to provide slow release for water spreading downstream. As a result, the normal sand supply to the affected beaches is diminished, and they are eroding. Urban growth also increases the amount of paving, which seals sediment that might otherwise be eroded into local rivers and ultimately to the ocean.

Significant natural causes of beach erosion are protracted drought and rising sea level. Little runoff occurs during drought conditions, and therefore, little sand is delivered to beaches. This is true even for unmodified watersheds without dams. The impact of drought on beach nourishment is less where the watershed is behind a dam. Dams trap sediment so that even with normal rainfall sand is prevented from reaching the beach.

A rise in sea level could have catastrophic results, as indicated by the map of coastline changes in the eastern United

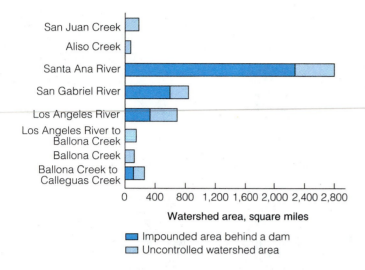

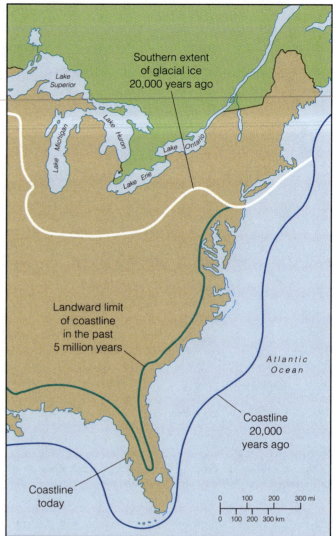

● FIGURE 11.17 Drainage areas of major sand-contributing streams behind dams in a region of Southern California. Flood-control and water-storage dams have greatly reduced the amount of sand that streams contribute to the beaches. Small watersheds are uncontrolled and also contribute sand to local beaches.

● FIGURE 11.18 Position of the coastline in the eastern United States in late Pleistocene time (20,000 years ago), Miocene time (5 million years ago), and today.

States over the past 20,000 years (● Figure 11.18). On a shorter time scale, it has been estimated that as the earth warms over the next century, sea level will rise 1 to 3 meters (up to 10 ft) due to thermal expansion and melting of glacial ice. On flat coastal plains where slopes are as gradual as 0.2–0.4 meters per kilometer, a sea-level rise of 2.54 centimeters (1 in) would cause the shoreline to retreat between 62 and 125 meters (200–400 ft). If the coast is subsiding, this shoreline migration landward would be magnified. On the other hand, tectonic uplift along parts of the Alaska and California coasts is faster than sea-level rise, which makes for a net shoreline retreat seaward. Careful tidal gauge measurements at New York City, where the shoreline is tectonically stable, indicated a sea-level rise of 25 centimeters (almost 10 in) between 1900 and 1970. If that trend continues, plan to take a boat to the Statue of Liberty's big toe and to say goodbye to Long Island, Miami Beach, and much of Texas's Galveston County by the middle of the next century.

LOCAL EROSION. Long-term beach erosion is aggravated locally by poorly conceived engineering works, such as *breakwaters* built to create quiet water for safe yacht anchorage, and long **jetties** built to prevent sedimentation in harbor or river inlets. Such structures cut off the littoral flow, causing accretion of sand in their immediate vicinity and erosion on their downcurrent side. Rates of erosion and accretion are calculated in cubic yards of sand per year, and it is a rule of thumb that a cubic yard of sand equals a square foot of beach. Thus, the littoral drift rates given in

■ Table 11.1 can be roughly translated into gains and losses of beach area. However, you can bet that where beach sand has accreted and a beach has widened, it has occurred at the expense of some beach downcurrent.

Ocean City, Maryland, is at the south end of Fenwick Island, one of the chain of barrier islands extending down the U.S. east coast from New York to Florida. Long jetties were constructed there in the 1930s to maintain the inlet between Fenwick and Assateague Island to the south. As a result, the beach has widened at Ocean City—because the north jetty blocked the southward longshore transport of sand—and the Assateague Island beach has been displaced 500 meters (1,650 ft) shoreward due to sand starvation (● Figure 11.19). Even though the Ocean City beach had widened, a strong storm in March 1962 inundated all but the highest dune areas, causing $7.5 million worth of damage.

TABLE 11.1 Rates of Littoral Drift at Selected U.S. Locations

LOCATION	DRIFT RATE, CUBIC YDS/Y	DIRECTION
West Coast (all California)		
Newport Beach	300,000	south
Port Hueneme	290,000	south
Santa Barbara	250,000	south
Santa Monica	160,000	south
Anaheim Bay	90,000	south
East Coast		
Fire Island Inlet, N.Y.	350,000	west
Rockaway Beach, N.Y.	260,000	west
Sandy Hook, N.J.	250,000	north
Ocean City, N.J.	230,000	south
Palm Beach, Fla.	130,000	south
Great Lakes		
Waukegan, Ill.	43,000	south
Racine Co., Wis.	23,000	south

SOURCES: U.S. Army Corps of Engineers and other sources.

(a)

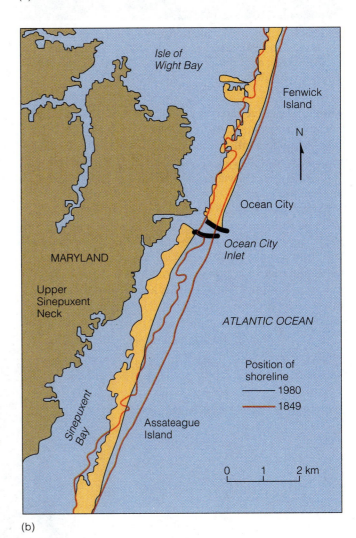
(b)

In the 1920s Santa Barbara's civic leaders decided to mold their community into the "French Riviera" of California. In spite of unfavorable reports by the Corps of Engineers, city officials authorized construction of a long, L-shaped breakwater for a yacht anchorage in 1929 (● Figure 11.20). The predominant drift direction here is southeasterly, and as the beach west of the breakwater grew wider, near-record erosion occurred on the beaches east of the structure. Eventually, the southeast-moving littoral drift traveled around the breakwater, forming a sandbar in the harbor entrance. The bar built up, diminishing the size of the harbor and creating a navigational hazard that remains to this day. When the Santa Barbara city fathers asked Will Rogers, the noted American humorist, what he thought of the harbor in the 1930s, he is alleged to have replied, "It would grow great corn if you could irrigate it." The drift rate here is about 250,000 cubic yards per year, which necessitates dredging sand from the harbor and pumping it to the beach about a mile downdrift. Summerland Beach, about 16 kilometers (10 mi) downdrift, retreated 75 meters (250 ft) in the decade following construction of the breakwater.

As a postscript, the City of Santa Monica to the south built a breakwater parallel to the shoreline in hopes of avoiding the kinds of problems experienced in Santa Barbara. The parallel breakwater created a *wave-energy shadow* on its land-

● **FIGURE 11.19** Jetties built at Ocean City inlet in the 1930s cut off the supply of sand from the north. The north end of Assateague Island is now 500 meters shoreward of the south end of Fenwick Island. *(a)* Recent aerial photo of the inlet, looking north. *(b)* Map of the islands, 1849 and 1980.

● FIGURE 11.20 The Santa Barbara, California, breakwater in 1988. Sand accumulation in the harbor is clearly visible, as is the dredge used to pump sand to the beaches downdrift.

● FIGURE 11.21 Beach at Santa Monica, California, about 1988. Note the sand accumulation in the "wave shadow" of the breakwater. The breakwater is the line of whitewater parallel to the shoreline.

ward side, causing sand to accumulate behind the structure but no serious downdrift erosion (● Figure 11.21). However, dredging is still required to keep the sand from building out to the breakwater and filling in the harbor.

Communities along the shores of the Great Lakes have erosion problems, also. Early in this century, protective structures were built on the Lake Michigan shoreline. Many of these have now deteriorated, particularly during the record high lake-level of 1984 (582.9 feet above mean sea level) (● Figure 11.22). Lake levels began dropping around 1960, and there was little concern at the time about shoreline erosion. Since 1964 (record low of 576.6 feet above mean sea level), levels have been rising, and there is increasing public pressure to put more protective structures on the

shoreline (see Mitigation of Beach Erosion later in this section). A large volume of federal shoreline legislation affecting all coastal states is sponsored by U.S. congresspersons and senators who represent the Great Lakes area.

Sea Cliffs

Most people enjoy ocean views, particularly from atop a 100-foot-high sea cliff. Unfortunately, such cliffs are eroded both by wave action at the toe and by mass-wasting processes on the face and the top. An **active sea cliff** is one whose erosion is dominated by wave action. This results in a steep cliff face and little rock debris at the base (● Figure 11.23, part a). The erosion of an **inactive sea cliff** is dom-

● FIGURE 11.22 Heavily engineered shoreline of Lake Michigan at Chicago. Note that the groins have trapped sand, causing the beach to widen.

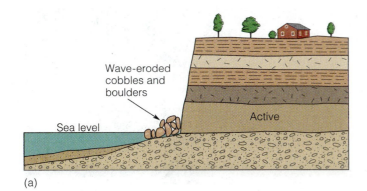

(a)

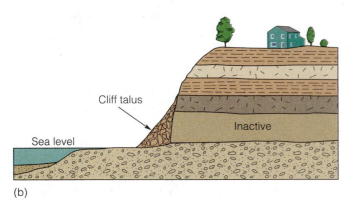

(b)

◉ **FIGURE 11.23** *(a)* Active and *(b)* inactive sea cliffs are distinguished by their dominant eroding agent and shape.

inated by running water and mass-wasting processes. Its slopes are much gentler, and angular debris accumulates at the toe (Figure 11.23, part b). Note that the nature of the debris at the foot of a sea cliff can aid in determining if it is active or inactive. Wave erosion produces rounded cobbles, whereas mass-wasting debris (talus) is angular. Rates of cliff erosion vary from a few centimeters to tens of centimeters per year, depending on the geology, climate, tectonic setting, and relative sea-level changes.

It is standard practice to consider 50 years as the lifetime of a house or other structure. Therefore, a structure should be set back from the top of a cliff at least 50 times the annual rate of cliff retreat. In all too many cases, the rates are not known or people choose to ignore them, with inevitable results (◉ Figure 11.24). Orrin Pilkey of Duke University (see Further Information) has developed a checklist for people to use when they are considering buying property on cliffs and bluffs. Pilkey's list includes such items as looking for evidence of failure at the toe of the cliff, determining rock hardness, and learning the history of cliff recession in the area from local land records and photographs.

Mitigation of Beach Erosion

Most coastal engineering works are built to protect the land from the sea. This battle against the sea is older than the tale of the Danish king Canute who commanded his men to

(a)

(b)

◉ **FIGURE 11.24** *(a)* The white cliffs of Dover at Brighton, England. These easily eroded chalk cliffs have required construction of a massive seawall–walkway at the toe of the cliff and groins to preserve what little sand occurs there. The setback of the houses from the edge of the cliff indicates building officials' recognition that the cliffs are susceptible to wave erosion. *(b)* Drastic undercutting by wave action has left this house in San Diego County, California, in a precarious position. The near-vertical cliff and the rounded cobbles at the base indicate that this is an active sea cliff (see Figure 11.23, part a).

whip the ocean in order to keep back the rising tide. It is perhaps presumptuous to think we can build works that will stop the rise of sea level or withstand enormous storm surges, but that does not keep us from trying. Although the sea will ultimately win the battle, we should attempt to build protective works that conform to nature as much as possible and that may outwit her for awhile.

Rock revetments and concrete seawalls are built landward of the shoreline and parallel to the beach not to protect the beach, but rather, to protect the land behind them (◉ Figure 11.25). Commonly they reflect wave energy downward, where it removes sand and undermines the

● FIGURE 11.25 Seawall at Palm Beach, Florida. Large waves reflecting from the wall remove beach sand from in front of the wall.

● FIGURE 11.26 This breakwater at Venice, California, was built so close to the beach that sand deposited behind it has connected the breakwater to the beach, forming a tombolo.

structure. There is a saying among coastal engineers and geologists that goes something like "Do you want a revetment, or do you want a beach?" Because waves reflecting off hard shoreline structures remove the beach material in front of them, the best-designed rock revetments have sufficient permeability to absorb some of the wave energy. Another problem is that wave action erodes around the ends of revetments. When that happens, they must be extended to prevent damage to structures located there.

Groins are constructed to trap sand for the purpose of widening beaches. They interrupt sand flow on their upcurrent sides and deprive downcurrent beaches of sand. Once sand has built out to the end of a groin, it flows around the tip of it and begins filling in the eroded area downcurrent. Frequently another groin is built to widen *that* beach, and the cycle is repeated. The result is a beach with groins every few hundred feet and a shoreline with a scalloped or bread-knife appearance. The northern New Jersey shore and Miami Beach are classic textbook examples of multiple-groin shorelines. Permeable groins are often constructed today. They allow some sand to filter through them for nourishing downcurrent beaches.

Jetties are the real culprits of beach erosion, because they are very long and they usually occur on both sides of a harbor entrance or where a river flows into the sea. They have caused entire beaches and whole communities to disappear. The jetties at Ocean City, Maryland, have already been cited. Jetties at the U.S. Navy's Port Hueneme on the Cal-

ifornia coast completely blocked the longshore current, leading to the disappearance of 225 meters (over 700 ft) of beach width to the south.

Breakwaters intended to create quiet water for yacht sanctuary can create unintended environmental problems. We have already seen the results of breakwater construction at Santa Barbara and Santa Monica in California. (Figures 11.20 and 11.21). Venice Beach, California, a community with a reputation for the unusual, built a parallel breakwater. It was constructed so close to shore that the sand deposited in the wave "shadow" behind the breakwater has connected the breakwater to the land. This sand feature is called a **tombolo** (● Figure 11.26).

At present, the designs of most long structures to be built perpendicular to the shoreline incorporate permanent *sand-bypassing works*. These features allow for the pumping of sand from the upcurrent side of the structure to nourish the beaches downcurrent. In addition, downcurrent beaches can be artificially nourished by pumping in sand from off-

CONSIDER THIS . . .

Several states have outlawed hard protective structures between beaches and the land. Why did they do this? Should other states, including those on the Great Lakes, enact similar legislation?

● FIGURE 11.27 Artificial beach nourishment by dredging from an offshore source; Redondo Beach, California. Typically, about 15 percent of dredged artificial beach fill is too fine-grained to be retained and is lost back to the sea.

(a)

(b)

◉ FIGURE 11.28 *(a)* King Harbor Marina, Redondo Beach, California, during a damaging storm in January 1988. *(b)* Model of King Harbor built after the storm to aid in assessing potential damage from waves of varying height and source direction in order to make design improvements.

shore areas or by transporting it to them from land sources (◉ Figure 11.27). Unfortunately, artificially nourished beaches are generally short-lived (see Case Study 11.1). One reason is that the imported sand is often finer than the native beach sand and is thus easily removed by the prevailing waves. The other reason is that artificial beaches usually have a steeper shoreface and are thus more subject to direct wave attack than is a gently sloping, natural shore, which dissipates wave energy. Artificial beach nourishment is expensive and it is most commonly done where a government project caused the erosion problem.

Because theory is lacking, thorough shoreline-impact studies using models should be performed for all large projects. Such a model is an exact replica of the project at a scale of about 1:100 (◉ Figure 11.28). Unfortunately, the models are usually built and studied after structures have created problems, rather than before.

Shoreline Construction

Buildings that may be periodically subject to wave attack need special foundation considerations. The usual technique is to build the structures on piles that extend through the beach sand to firm bedrock or other hard geologic layer (◉ Figure 11.29). Thus if sand in front and under the structure is removed, the building will remain standing (◉ Figure 11.30, page 312). Pile foundations are often used in conjunction with solid wooden bulkheads on the beach side of the structure that reflect wave energy. These must be designed to withstand anticipated wave

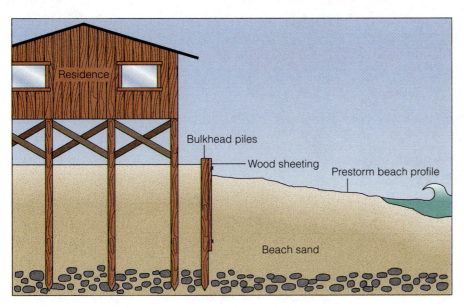

◉ FIGURE 11.29 Recommended construction for beach homes. The foundation rests on piles driven through the sand to firm bedrock or a hard layer. A wooden bulkhead prevents or minimizes the erosive effect of waves on the pile foundation and sand beneath the house.

The Care and Feeding of Undernourished Beaches

Most shorelines in the world's developed nations are eroding. In the United States erosion control is managed by the U.S. Army Corps of Engineers as part of their mission to maintain navigable waters. They usually address the problem by installing "hard" shoreline-stabilization structures such as seawalls or by using the "soft" solution of artificial beach nourishment. A third solution, not advocated by the corps for a variety of engineering and political reasons, is to relocate any structures that are present on the beach and allow the shoreline to establish its own equilibrium. Because of the need to preserve recreational beaches, the use of hard shoreline protection is now being restricted, and New Jersey, North Carolina, South Carolina, and Maine have made hard stabilization illegal. This leaves artificial beach replenishment, which does not have a good longevity record, as the option of choice in many places. And this is the focus of controversy between the engineers who advocate it and the many geologists who question its effectiveness.

More than 90 beaches were replenished by 270 separate pumping contracts along East Coast barrier-island shores between the early 1960s and the early 1990s. Studies by Orrin Pilkey at Duke University from 1988 to 1992 show that the life spans of these replenished beaches are highly variable, but that most of them are very short lived. At Miami Beach, a 16-kilometer (10-mi) $60 million artificial beach was still largely in place after 15 years (⦿ Figure 1). This is in sharp contrast to the experience at Ocean City, Maryland. In January 1989, Ocean City's mayor extolled the virtues of the community's new $12

(a)

(b)

⦿ FIGURE 1 *(a)* Miami Beach had little beach width in the early 1970s. *(b)* Miami Beach in 1981, after what has proven to be a successful artificial sand-replenishment project.

● FIGURE 2 Replenishment history of Wrightsville Beach, North Carolina, showing the cumulative amount of sand placed on the beach from 1965 to 1991. The steady rise of the curve indicates that regular additions of sand were required to maintain the beach and that no steady-state, or equilibrium, condition had been reached by 1991.

(a)

(b)

● FIGURE 3 Wrightsville Beach *(a)* after one of its many replenishment projects (note width of beach berm) and *(b)* three years later. The beach was replenished and destroyed again in 1996 by Hurricane Fran.

million replenished beach. Two months later, the winter's first big storm had removed much of the new beach, some of it clear back to its prereplenishment location (see Figure 11.19). Essentially no replenished beaches north of Florida have lasted more than five years, because the newly placed beach fill is quickly removed by storm waves. North Carolina's Wrightsville Beach, one of the most extensively replenished beaches in the United States, has been replaced ten times (● Figure 2). ● Figure 3 shows Wrightsville Beach just after one of these replenishments and then three years later. According to Pilkey, the main problem with artificial replenishment is many coastal engineers' strong belief in the theoretical "textbook" explanation of sand transport on a shoreline. That explanation is based on many questionable assumptions about the strength of future storms, the rate of sea-level rise, and coastal-zone development.

Artificial beach replenishment is expensive and temporary; it may be that no nation can afford it for long. A growing number of geologists believe the only feasible long-term solution is relocation, moving out whatever buildings are there—or simply letting them fall into the sea. This would preserve the beach, save shoreline stabilization costs, and preserve the buildings if moved. By now it is clear that the shoreline-erosion crisis is human-made; if no one lived next to the shore there would be no problem. This alternative has merit and many strong advocates. It is politically unpopular, however, as it could result in land loss, and it could also be very costly.

FIGURE 11.30 A house constructed as in Figure 11.29 but without a bulkhead; Oceanside, California. Wave action has removed 3–4 meters of sand from beneath the deck, making it a stilt structure.

FIGURE 11.31 A massive bulkhead protects a hotel in Ft. Lauderdale, Florida. Note that erosion has bypassed the end of the bulkhead and eroded the adjacent beach. Sandbags are required to prevent waves from undermining the side of the hotel.

forces, however; if they should break up, their large timbers could become projectiles and battering rams that would do further damage (Figure 11.31).

Estuaries

Estuaries (Latin *aestuarium,* "tidal") are semienclosed basins in which river water (fresh water) mixes with tidally influenced seawater. The river water flowing into the estuary "floats" on the denser seawater, which is constantly moving landward and seaward with the tide. The mixing of fresh water and saltwater generally occurs below the water surface, and the resulting mixture is dilute, or brackish, water. There are almost 900 estuaries along U.S. coasts, covering 68,000 square kilometers (27,000 mi²). Based upon their geology, four major types of estuaries are recognized:

- *Coastal-plain estuaries* form where the land subsides or sea level rises and submerges a river valley. Chesapeake Bay (Figure 11.32) is an example of this type, as are Delaware Bay, the Hudson River, the Mississippi River, the Thames River in England, and the Seine River in France.
- *Fiords* are drowned U-shaped valleys formed by glaciers (see Chapter 12). They are common in Scandinavia, Greenland, Alaska, British Columbia, and New Zealand.
- *Fault-block estuaries* (grabens; see Chapter 4) form on tec-

tonically active coasts such as the southern part of San Francisco Bay.
- *Bar-formed estuaries* occur as lagoons between barrier islands and the mainland (Figure 11.33). Examples are Pamlico Sound within the Outer Banks of North Carolina and Laguna Madre in Texas.

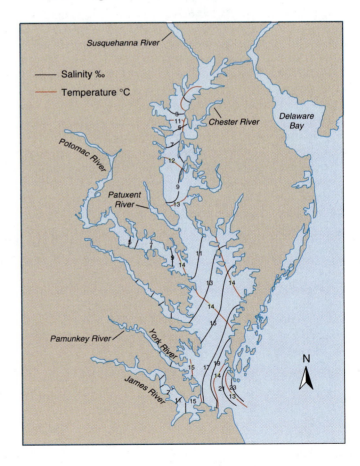

FIGURE 11.32 (right) Chesapeake Bay is the drowned valleys of the Susquehanna, Potomac, and James Rivers and many smaller watersheds. It is 300 kilometers long and 65 kilometers wide (180 mi × 40 mi). Spring salinities in the estuary are shown in parts salt per thousand parts seawater. Normal seawater has a salinity of 34‰.

⦿ FIGURE 11.33 A barrier island with an estuary and marshlands on its west *(left)* side and a smooth seaward profile.

Estuaries and the marsh wetlands on their shores are habitats for flourishing biological communities. The waters of estuaries have high nutrient contents that stimulate phytoplankton (plant) growth, which in turn supports a rich and diverse fauna. Estuaries are the habitat of a host of shellfish—including blue crab and common shrimp—and the breeding place of many fish species. The tidal marshes on their shores support a variety of waterfowl. It is estimated that 95 to 98 percent of the important commercial fish species are spawned in estuaries.

Urbanization around estuaries has placed conflicting demands on their waters. For example, the fishing and shellfish industry that the Hudson River estuary (⦿ Figure 11.34) once supported has drastically declined since 1960. The Hudson is now the emergency water supply for New York City and a major transportation and recreational boating waterway. It is also the receptacle for the sewage of 16 million people—about a third discharged from Manhattan Island, another third contributed from New Jersey, and the remainder coming from Brooklyn, Queens, and the Bronx. Because of the Clean Water and Ocean Dumping Acts of 1972, no raw sewage or sludge (solids) may be discharged into the river, but discharges of treated sewage have resulted in elevated bacteria levels, the presence of heavy metals and organic compounds in bottom sediments, and a reduced dissolved-oxygen content in the water column. Sewage sludge from the surrounding area is disposed of at sea (see Chapter 15). The aesthetic amenities and resources of this estuary have been severely degraded by human activities.

Further estuary and wetland degradation occurs when channels are dredged in them to accommodate large ships and when they are filled in to build highways and other structures. Such multiple use has occurred at San Francisco Bay, where only 20 percent of the original 90,000 hectares (200,000 acres) of marshland remains today. It has been estimated that whereas marshlands return $12,000 per hectare

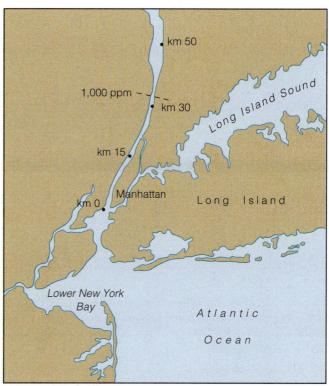

⦿ FIGURE 11.34 Geography of the Hudson River. Brackish water is found as far north as kilometer 30.

per year in their role of supporting fisheries, one hectare of marshland may put a million dollars into a developer's pocket.

 HURRICANES AND COASTAL FLOODING

Andrew, Beulah, Camille, Opal, Gilbert, and *Hugo* are innocent-sounding names, but onshore surges of water produced by hurricanes with such names are our greatest natural hazard. Hurricanes (called *typhoons* in the North Pacific Ocean and *cyclones* in the South Pacific and Indian Oceans) get their energy from the equatorial oceans in the summer and early fall months when seawater is warmest, usually above 25° C (about 80° F). Hurricanes begin as tropical depressions when air that has been heated by the sun and oceans rises, creating reduced atmospheric pressure, clouds, and rain. As atmospheric pressure drops, the warm air moves toward the center of low pressure, the eye of the impending hurricane. When the rising air reaches higher elevation it releases heat, causing it to rise even faster and creating even lower atmospheric pressure and greater wind velocities. Wind blowing toward the center of low pressure is given a counterclockwise rotation by the Coriolis force in the Northern Hemisphere (⦿ Figure 11.35; see also Chapter 12).

Thus, what begins as a localized tropical disturbance can grow to be 800 kilometers (500 mi) across, with wind

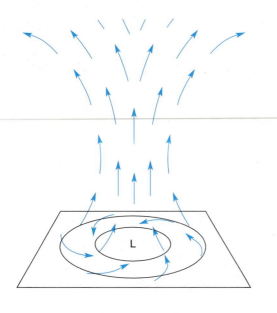

(a)   (b)

◉ FIGURE 11.35 *(a)* Wind-flow pattern associated with low-pressure centers in the Northern Hemisphere. *(b)* Satellite photograph of Hurricane Gilbert practically filling the Gulf of Mexico, 15 September 1988. Gilbert's rotation and central eye are clearly visible.

velocities far in excess of 120 kilometers per hour (74 mph), the lower velocity limit for hurricane designation. Hurricanes travel as coherent storms with velocities of 10 to 35 knots (12–40 mph), but their exact course and where they will strike land is not completely predictable (◉ Figure 11.36). If we divide a storm system in the North Atlantic into four equal quadrants, we find the strongest winds in the northeast quadrant (designated as quadrant I), because the cyclonic winds and the storm system are moving in the

same direction there (◉ Figure 11.37). When a storm makes a landfall, the greatest wind damage usually occurs in this sector.

A "hill" of water called a **storm surge** is pushed ahead of the hurricane that may cause sea level to rise many meters above the normal high-tide level. The reduction in atmospheric pressure associated with a hurricane also causes sea level to rise—like fluid rising by suction in a soda straw. Average sea-level atmospheric pressure is 1,013 millibars (or

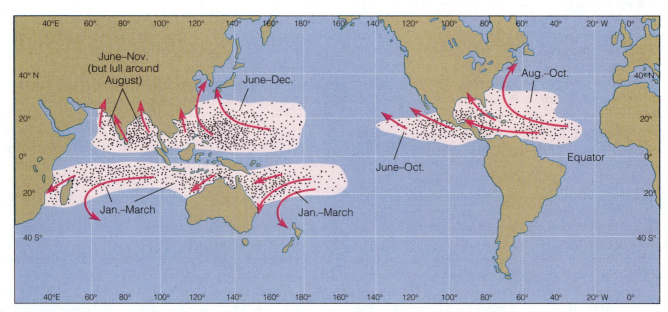

◉ FIGURE 11.36 Where hurricanes, typhoons, and cyclones originate and their typical paths. Each dot represents a storm that ultimately reached hurricane intensity.

CONSIDER THIS...

"Why sweat a swamp?" is what some developers say about the fuss over preserving estuaries and marshlands. Is this attitude justified, or are these features valuable to the environment?

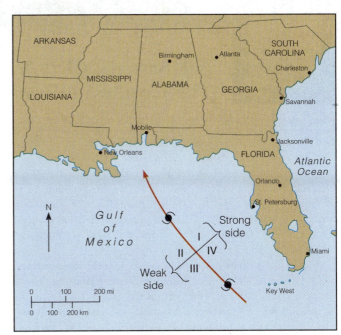

● **FIGURE 11.37** Northern Hemisphere hurricane quadrants. The highest winds occur on "the strong side."

29.92 inches of mercury), and the rise in sea level is about 1 centimeter for every millibar of pressure drop. Hurricane Camille in 1969, the strongest storm to hit the U.S. mainland, had a near-record low barometric pressure of 905 millibars (26.61 inches of mercury). Hurricane surge and winds devastated the entire Gulf Coast from Florida to Texas.

Twenty-five residents of the Richelieu Apartments in Pass Christian, Mississippi, on the state's south coast were enjoying a hurricane party when the police chief came to alert them of the danger. They laughed when he requested the names of their next of kin because they did not appear to be willing to take his advice and evacuate. The storm surge at Pass Christian reached 7 meters (23 ft) above mean sea level, and only two people survived the party. One of them was rescued from a floating sofa 8 kilometers (5 mi) from the apartment complex.

Camille killed 130 people and flooded 40,000 homes. Three thousand homes simply disappeared, and property damage exceeded $1.3 billion. Yet real estate was resold, and new homes were built on the same sites where homes had been removed by the storm. Camille was rated as a category-5 hurricane, the highest category on the Saffir–Simpson Scale

(■ Table 11.2). However, neither Hurricane Camille nor any other natural disaster to date has caused as much dollar-damage as Hurricane Andrew caused in 1992 (● Figure 11.38 and Case Study 11.2).

Some hurricanes veer eastward and follow a track up the U.S. East Coast, causing flooding and beach erosion from the Carolinas to Maine. In 1989 Hurricane Hugo struck the coast just north of Charleston, South Carolina, with sustained winds of more than 215 kilometers per hour and a storm surge of 6 meters (20 ft; ● Figure 11.39). Complete

■ TABLE 11.2 Saffir–Simpson Hurricane Scale

| CATEGORY | WIND VELOCITY | | DAMAGE | EXAMPLES |
	km/h	*mph*		
1	119–153	74–95	minimal	Agnes, 1972 Juan, 1985
2	154–178	96–110	moderate	David, 1979 Bob, 1991
3	179–210	111–130	extensive	Betsy, 1965 Frederic, 1979
4	211–250	131–155	extreme	Hugo, 1989 Andrew, 1992 Opal, 1995
5	>250	>155	catastrophic	Labor Day, 1935 Camille, 1969

SOURCE: Coastal Weather Research Center, University of South Alabama, Mobile.

A Disaster Called Andrew—
The Costliest in History

Hurricane Andrew, in August 1992, killed 25 persons in Florida, Louisiana, and the Bahamas and did an estimated $20 billion worth of damage. Andrew cut a 45-kilometer-wide (28-mi) swath through South Florida that destroyed 63,000 homes in and around Homestead and left a quarter-million people homeless (◉ Figure 1). Although Andrew was classified as a strong category-4, it is probable that gusts exceeded 300 kilometers per hour (186 mph), and the damage did indicate category-5 winds (see Table 11.2).

The storm system developed from a tropical low over the eastern Atlantic Ocean near the Cape Verde Islands, but what was to become Andrew began just south of the Sahara Desert in the second week of August. Once this low-pressure disturbance left Africa and hit the Atlantic Ocean, it acquired heat, moisture, and cyclonic rotation and increased in wind velocity. Moving northwestward, the storm was recognized as a hurricane by August 22. Normally these so-called Cape Verde storms curve northward and pose a threat to the Carolinas and points farther north on the U.S. East Coast. However, a ridge of high pressure from Canada covered the entire coast and acted to prevent Andrew from recurving to the north. So Andrew moved westward at a brisk 25–30 kilometers per hour (16–9 mph) with lit-

tle deviation, always probing for a breach in the high-pressure system in order to turn north. Meanwhile, atmospheric pressure in the eye was dropping, and Andrew's wind velocities were increasing as he headed for South Florida. During the hurricane watchers' last reconnaissance flight before landfall on August 24, the Air Force spokesperson reported a peak wind velocity of 320 kilometers per hour (196 mph).

Once a hurricane is over land, it usually loses its punch due to friction with terrain and loss of heat. Because South Florida is so flat, however, Andrew lost little strength in his devastating journey across the state. After crossing the narrow peninsula, the storm again intensified over water and maintained its strength as it headed for Louisiana. It struck just south of Morgan City the night of August 25, with peak winds of more than 200 (125 mph) kilometers per hour, and dumped torrents of rain on Louisiana, Mississippi, and Alabama for the next 48 hours.

In Hurricane Andrew we see the incredible energy that the atmosphere can acquire from the oceans. Fortunately, not all tropical disturbances grow to hurricane force, and not all hurricanes are as devastating as Andrew, the costliest natural disaster in U.S. history.

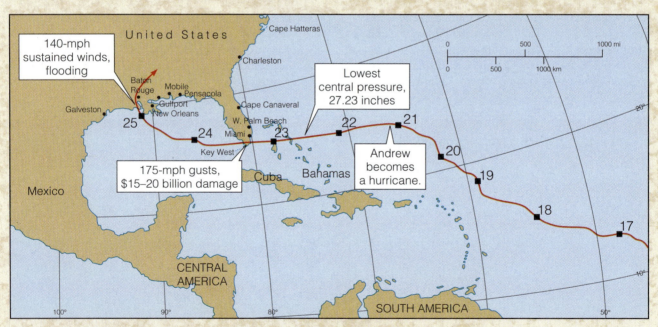

◉ FIGURE 1 The path of Hurricane Andrew during the last 9 days of its life, 17–25 August 1992. All positions are as recorded at 10 P.M. CDT. Although the storm was spawned in North Africa during the second week of August, it did not reach hurricane proportions until it acquired heat and moisture from the Atlantic Ocean.

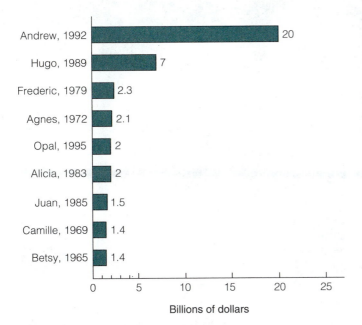

Andrew, 1992 — 20
Hugo, 1989 — 7
Frederic, 1979 — 2.3
Agnes, 1972 — 2.1
Opal, 1995 — 2
Alicia, 1983 — 2
Juan, 1985 — 1.5
Camille, 1969 — 1.4
Betsy, 1965 — 1.4

Billions of dollars

⦿ FIGURE 11.38 Costliest hurricanes in U.S. history. Losses are reported as dollar-values at the time of their occurrence.

television coverage of this storm provided visual evidence that a 3-meter-high barrier island cannot withstand the force of a 6-meter storm surge. A total of 29 people died in South Carolina, and damage totaled $5.9 billion.

The 1995 hurricane season in the North Atlantic Ocean was the worst in more than 60 years. Wet weather in northwest Africa spawned slow-moving disturbances, some of which grew into tropical storms and then became hurricanes over the ocean. Some of these hurricanes struck the U.S. Gulf and East Coasts. Opal was the worst of these and she made a direct hit on coastal Alabama and Florida in October. "She" caused $2 billion in damage and was entered in the record book as one of the costlier storms in U.S. history (Figure 11.38). Many resort hotels and private beach homes were severely damaged or totally destroyed by the 6-meter (20-ft) storm surge and high winds. Afterward one expert commented, "You guys [Alabama and Florida] are in hurricane alley, and the message should be clear. Do not rebuild close to the beach." It is doubtful that the message has been heard, however. Rebuilding began almost immediately on the old foundations.

Coastal flooding induced by hurricanes or typhoons is the greatest natural hazard to humans. Nowhere has this been shown more vividly than in flood-prone Bangladesh, where a third of the country lies less than 20 feet above sea level. In 1970 a half-million people lost their lives there in a massive storm surge on the low-lying delta of the Ganges River. In April and May 1991, the area was struck by another fast-moving wall of water more than 6 meters high. A string of islands along the delta margin were inundated, and a reported 100,000 people died this time (see Figure 9.11). Winds of 250 kilometers per hour (155 mph) were

(a)

(b)

⦿ FIGURE 11.39 (a) Homes at beautiful Folley Beach, North Carolina, before Hurricane Hugo. Note the rock revetment facing the beach to prevent sand erosion and damage to structures. (b) The area had five fewer houses after Hurricane Hugo, but the palm trees and the rock revetment remained. Three houses in the background were damaged by battering-ram action of the demolished first-row structures. This is the hard way to become an owner of beachfront property.

reported on the low-lying islands of sand with names like *Chittagong, Cox's Bazar,* and *Sandwip.* Testimony to the storm's power was the recovery of a man's body 20 kilometers (12 mi) inland from his village. If the storm had hit the U.S. coast, it would have been classified a category-5 event, as Camille of 1969.

Natural disasters such as flooding, earthquakes, and massive debris flows are most devastating to Third World peoples. One explanation is that communication and warning systems are virtually nonexistent in emerging nations. Fortunately, Indian forecasters projected the 1991 storm path far enough in advance to evacuate three million people. In contrast, a weaker cyclone 21 years earlier killed five times as many people. Another explanation for the large death tolls is that overpopulation in many of these areas forces people

Wonders of the Coastal Zone

The interaction of the ocean and the atmosphere at shorelines creates interesting geologic features, some of which become tourist attractions. **Stacks,** outlying remnants of retreating sea cliffs, are picturesque features along all cliffed shorelines. None are more picturesque than the "Twelve Apostles," a must for tourists who visit the south coast of Australia (Figure 11.41) Sometimes a stack or small island close to a shore creates a wave-energy

● **FIGURE 11.40** A tombolo and tied island near Friday Harbor, San Juan Islands, Washington State.

● **FIGURE 11.41** The "Twelve Apostles" sea stacks on the south coast of Australia near Apollo Bay.

to settle on terrain that is subject to life-threatening natural hazards. Bangladesh, occupying a land area the size of Wisconsin, has 115 million people, about half the U.S. population, and most of them are desperately poor. Even in areas where people were warned of the 1991 storm, many people were reluctant to leave their few possessions to looters.

 ## SUMMARY

Coastal Ocean

IMPORTANCE
1. Impact on local climate and weather
2. Estuaries
3. Beaches

Waves

CAUSE Wind blowing over the water surface creates ripples, then stormy irregular seas, and finally regularly spaced swell that outruns the source area.

FACTORS OF SIZE Wave size depends upon wind velocity, the length of time wind blows over the water, and fetch. A fully developed sea has the highest waves attainable for the three variables.

MOTION Water parcels move in circular paths to a depth of half the wavelength, $L/2$, called *wave base*. This causes movement of the waveform known as *waves of oscillation*. Water physically moves shoreward when waves break in the surf zone. These breaking waves are called *waves of translation*.

TYPES OF BREAKERS Steep offshore slopes develop plunging breakers, and gentle slopes produce spilling breakers.

WAVE REFRACTION Waves generally approach the shore at some prevailing angle. This results in longshore currents that move sand along the beach. The transported sand is known as *littoral drift*. Groins act as dams to littoral drift, causing sand to be deposited upcurrent

● FIGURE 11.42 A revetment's half-ton to one-ton stone blocks were lifted and hurled through the walls and windows of a living space they were intended to protect; Hurricane Hugo, 1989.

● FIGURE 11.43 Delta House Movers appeared to have added boat-moving services after Hurricane Camille in 1969; Gulfport, Mississippi.

shadow—as a parallel breakwater does (see Figure 11.26)—and becomes connected to the mainland by a sandbar deposited in its lee. Such a sand bridge is known as a *tombolo,* and the system is called a *tombolo and tied island.* The Rock of Gibraltar on the south coast of Spain is such a system, but literally thousands of less famous and just as pho-togenic examples exist (● Figure 11.40). As sort of a defiant parting shot, Hurricane Hugo left one family a reminder of the weakness of human efforts to hold back the sea in comparison to nature's power to claim the land (● Figure 11.42). Hurricane Camille also contributed to our appreciation of nature's forces (● Figure 11.43).

and eroded downcurrent. Wave refraction concentrates wave energy on points and dissipates it in bays, thus tending to straighten an irregular shoreline.

Beaches

DEFINED Narrow strips of shore that are washed by the waves or tides, usually covered by sand or pebbles.

TYPES
1. Beaches backed by coastal plains
2. Pocket beaches along cliffed shorelines
3. Barrier islands (barrier beaches)
4. Spits

Beach Erosion

DEFINED Less sand is supplied by littoral currents than is removed by wave action.

NATURAL CAUSES
1. Drought
2. Rising sea level

HUMAN CAUSES
1. Dams that impound sediment
2. Groins, jetties, and breakwaters that impound sediment
3. Hard structures such as seawalls and revetments

MITIGATION Design of groins, breakwaters, and jetties can incorporate features that allow sand to pass through or around them. Seawalls, revetments, and artificial nourishment are only temporary solutions. Model studies of existing structures and shorelines help us to better understand the problem and thus to effect remedies.

Sea Cliffs

DEFINED Cliffs formed by wave action.

TYPES Active and inactive.

EROSION Annual rates between a centimeter and tens of centimeters.

MITIGATION Structures' optimal (safest) setback from the edge of a cliff can be determined when the erosion rate is known.

Estuaries

DEFINED Basins open to the sea in which fresh water mixes with seawater.

TYPES
1. Coastal-plain estuaries, drowned river valleys (Chesapeake Bay)
2. Fiords, drowned glacial valleys (Alaska)
3. Fault-block estuaries (south San Francisco Bay)
4. Bar-formed estuaries landward of barrier beaches (Laguna Madre, Texas)

IMPORTANCE Estuaries and their surrounding wetlands are areas of high biological productivity for marine life and waterfowl.

HUMAN IMPACT Used for dumping sewage, commercial fishing, transportation, and recreation. Structures are built on filled marshland.

Coastal Flooding

HURRICANES Storms generated at sea characterized by counterclockwise winds greater than 120 kilometers/hour (74 mph). They travel across the sea at velocities of 10–35 knots.

FLOOD DANGER A storm surge is a "hill" of water due to high winds and low atmospheric pressure. It may be many meters above normal high-tide levels. Storm surges on low-lying deltas such as the Ganges River delta in Bangladesh are the greatest natural hazard to human life.

CASE HISTORIES Hurricanes Camille (1969), Hugo (1988), Andrew (1992), and Opal (1995) are described in this chapter.

◻ KEY TERMS

active sea cliff	rip current
barrier island	spit
estuary	stack
fetch	storm surge
groin	tombolo
inactive sea cliff	wave base
jetty	wave of oscillation
littoral drift	wave of translation
longshore current	wave refraction

◻ STUDY QUESTIONS

1. Why do submarines and submersibles not experience severe storms at sea?
2. What is a "fully developed sea," and what variables determine this condition?
3. What things are endangering the existence of estuaries and marshlands as a natural resource?
4. What shoreline structures impede littoral drift and cause beach erosion? How may these structures' impact be mitigated so as to minimize damage?
5. Where and when do hurricanes originate, and how do they obtain their energy? Distinguish a hurricane from a typhoon and a cyclone.
6. Name two types of breakers and explain what causes one or the other to occur along a given shoreline.
7. How does the width of a beach vary naturally over the year? Where is the sand stored that is removed from the beach in winter and returned to the foreshore in summer?
8. What is a storm (hurricane) surge, and why is it such a danger in low-lying coastal communities?
9. How do longshore currents develop, and how do they impact beaches *and swimmers* in the surf zone?
10. How can hazardous currents in the surf zone be recognized?

◻ FURTHER INFORMATION

BOOKS AND PERIODICALS

Bascom, Willard. 1964. *Waves and beaches.* Garden City, N.Y.: Doubleday.

Denison, Frank E., and Hugh S. Robertson. 1985. Assessment of 1982–83 winter storms damage, Malibu coastline. *California geology,* September.

Griggs, G., and L. Savoy, eds. 1985. *Living with the California coast.* Durham, N.C.: Duke University Press.

Kaufman, W., and Orrin Pilkey. 1979. *The beaches are moving.* Garden City, N.Y.: Anchor Press/Doubleday.

The Open University. 1989. *Waves, tides, and shallow-water processes.* New York: Pergamon Press.

Pilkey, O. H.; W. Neal; and O. H. Pilkey, Sr. 1978. *From Currituck to Calabash.* Research Triangle Park, N.C.: North Carolina Biotechnology Center.

Pilkey, Orrin H. 1991. Coastal erosion. *Episodes* 14, no. 1 (March).

Pilkey, Orrin H., and others. 1993. The concept of shoreface profile of equilibrium: A critical review. *Journal of coastal research* 9, no. 1, pp. 255–278.

Pipkin, B.; H. Robertson; and R. S. Mills. 1992. Coastal erosion in Southern California: An overview. In Pipkin, B. W., and Richard Proctor, eds., *Engineering geology practice in Southern California.* Belmont, Calif.: Star Publishing Co., pp. 461–483.

Shepard, F. P., and H. R. Wanless. 1971. *Our changing coastlines.* New York: McGraw-Hill.

Williams, J. S.; K. Dodd; and Kathleen Gohn. 1990. *Coasts in crisis.* U.S. Geological Survey circular 1075.

Woods Hole Oceanographic Institute. 1981. The coast. *Oceanus* 23, no. 4.

_____. 1993. Coastal science and policy I. *Oceanus* 36, no. 1.

EXTREME CLIMATES AND CLIMATE CHANGE

We see changes in the climate record that are at least heading in the same direction that one would anticipate from a greenhouse world.

THOMAS R. KARL, CLIMATOLOGIST, 1995

The greenhouse effect is the atmospheric process that makes the earth's climate habitable. In this chapter we examine the current human–induced modification to the greenhouse effect and consider the ways in which this modification may be changing the climate. Also, we examine the geologic processes of two climatic extremes: extreme cold and extreme dryness. The natural processes acting in these environmentally sensitive areas are responsible for a variety of desert, coastal, alpine, polar, and subpolar landforms. Together, these regions of extreme climate constitute about 40 percent of the earth's land area. As the global population grows, humans are forced to occupy such less-desirable lands. We will see how the landforms of extreme climates can impact people who occupy those regions and examine

OPENING PHOTO
Icebergs and slushy ice currently mark the breakup of the Columbia Glacier, near Valdez, Alaska. In 1981 the glacier began a state of catastrophic retreat, its first major change in behavior since 1794, when it was first observed and mapped by Captain George Vancouver.

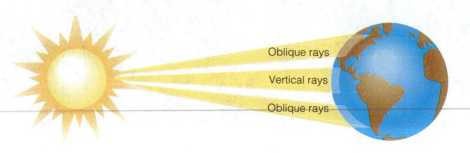

Oblique rays
Vertical rays
Oblique rays

them for clues about climate changes of the past and insights about possible climate changes in the future. Continuing rapid advances in science and technology have reduced climate's influence on human activity, but the continuing importance of climate, and of changing climate, cannot be denied.

THE REASONS FOR CLIMATES

Winds and Atmospheric Circulation

Wind is movement of air from a region of high pressure to a region of low pressure due to unequal heating of the earth's surface. Because the sun's rays strike the earth more directly at the equator than at other places, more radiation (heat) per unit area is received there (● Figure 12.1). Solar heating at the equator warms the air, and it rises, creating a zone of low atmospheric pressure known as the **equato-**

rial low. In polar regions, the sun rays strike the earth at a low angle, resulting in less-intense heating per unit area. There the colder, drier, denser air creates a zone of high pressure known as the **polar high** (● Figure 12.2, part a). The cold, dry, polar air sinks and moves laterally into areas of lower pressure at lower latitudes. Upper-level air flow from the equatorial regions moves toward the poles. When it reaches approximately 30 degrees north and south of the equator, it sinks back to the earth's surface as dense, dry air to form high-pressure belts known as the **subtropical highs,** or the **horse latitudes.** (Figure 12.2, part b). Winds, then, are driven by differences in air pressure caused by unequal heating, but as they move they do not follow straight paths. Instead, they are deflected by the Coriolis effect.

The **Coriolis effect** is caused by the rotation of the earth. A simple analogy of the Coriolis effect is to consider two people playing catch on a merry-go-round (● Figure

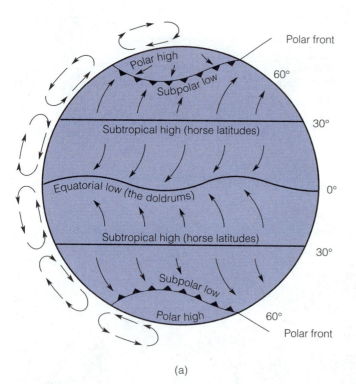

(a)

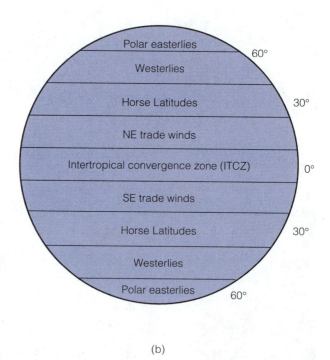

(b)

● **FIGURE 12.2** *(a)* Generalized circulation of the earth's atmosphere and surface air-pressure distribution. *(b)* Atmospheric pressure zones and surface wind belts.

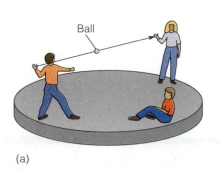

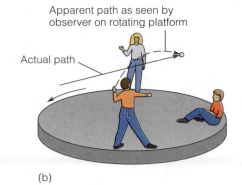

FIGURE 12.3 **The Coriolis effect in a game of catch.** *(a)* The path of the ball is straight when the game is played on a stationary platform, and an observer perceives it as straight. *(b)* On a rotating platform the ball still follows a straight path, but to an observer on the platform, the path appears to be curved. This is the Coriolis effect.

12.3). If the merry-go-round is stationary, the ball will follow a straight path. An apparent paradox occurs when the merry-go-round is turning, however. As seen by stationary observers not on the merry-go-round, the thrown ball still travels in a straight path, but to the people on the turning merry-go-round, the ball appears to follow a curved path. The reason for this paradox is that the merry-go-round and its passengers are turning below the ball as it travels from one person to the other. To the people on the merry-go-round, the path of the ball appears to be deflected as though some force is acting on it. The curve will be to the right if the merry-go-round is turning counterclockwise, the same direction the earth rotates if observed from above the North Pole. The deflection of freely moving objects also occurs on the earth. Rifle bullets, ballistic missiles, ocean currents, and winds that travel long distances all appear to follow curved paths because of the earth's rotation. The path is deflected to the right in the Northern Hemisphere and to the left in the Southern Hemisphere.

The rising warm air in the equatorial belt expands and cools as it rises. Cooling reduces the air's ability to hold moisture, resulting in the formation of clouds and heavy precipitation. For this reason, the equatorial low-pressure region—called the *intertropical convergence zone (ITCZ)*, and sometimes *the doldrums*—is one of the rainiest parts of the world. We generally do not find deserts there. Air also rises at latitudes of approximately 60° north and south, where cold polar surface air and mild-temperature surface air converge but tend not to mix. At this boundary, called the **polar front,** surface air rises, forming a zone of low pressure called the **subpolar low.** It is along the polar front that rising air causes storms to develop. In the subtropics, in contrast, cool, dry air sinks, forming high-pressure regions. As the air sinks, it is compressed, which causes its temperature to increase. This warming enables the air to absorb water and precludes the formation of clouds. This is the reason that the climate belts of subtropical highs, from about 30° to 35° north and south of the equator, the so-called horse latitudes, experience warm, dry winds with generally clear skies and sunny days. It is here that we find most of the world's hot deserts.

CONSIDER THIS . . .

The old expression "caught in the doldrums" is sometimes used to describe a person who lacks motivation and is listless. What is the relationship of this expression to the traditional term for the ITCZ?

Oceanic Circulation

Unequal heating of the earth's surface by the sun moves the atmosphere, and the resulting winds drive the surface currents of the oceans. Each major ocean basin is distinguished by currents that rotate clockwise in the Northern Hemisphere and counterclockwise in the Southern Hemisphere in accordance with the Coriolis effect (Figure 12.4). In both hemispheres, these currents carry heat from the equator toward the poles along the west sides of the basins and, in turn, carry cooled water from polar latitudes back toward equatorial regions along the east sides of the basins. This transfer of heat occurs in all oceans. For example, the California Current cools the west coast of North America, imparting mild summer temperatures to San Diego. In contrast, the warm Gulf Stream on the east coast imparts hot, humid summers to Savannah, Georgia, at the same latitude.

A major oceanic process affecting coasts in the "tradewind" belts north and south of the ITCZ (Figure 12.2, part b) is **upwelling.** Upwelling occurs when winds blowing from the land to the sea *(offshore winds)* sweep surface water away from the shore, creating a condition of hydraulic instability. The displaced nearshore water is replaced by nutrient-rich, cold waters from depths in excess of 200 meters (650 ft; Figure 12.5). During upwelling, coastal waters nearshore are colder than offshore water, which further moderates local weather. Conditions for coastal upwelling are most commonly found along the western sides of continents in the trade-wind belts, as, for example, along the coasts of Peru and California. Upwelling brings

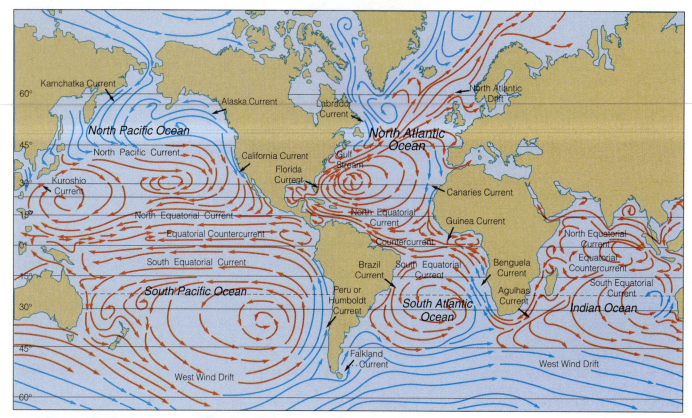

⦿ **FIGURE 12.4 Major wind-driven oceanic surface currents. Warm currents
are shown in red; cool currents in blue. Currents rotate clockwise in the
Northern Hemisphere and counterclockwise in the Southern Hemisphere.**

dissolved phosphates and nitrates to the surface, which
leads to the proliferation of marine plankton, which in
turn promotes large fish populations and gives rise to commercial fisheries. However, if the trade winds break down
or weaken and upwelling ceases, major climatic upsets such
as El Niño may occur, causing floods and severe storms in
some areas (Case Study 12.1). The combined effects of the
general circulation of the atmosphere and ocean currents,
along with other factors, produce deserts.

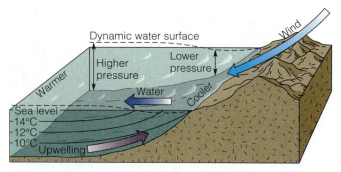

⦿ **FIGURE 12.5 The mechanics of wind-driven upwelling
along a coastline. Upwelled nearshore water is cooler than
offshore surface water, as indicated by the temperature
contours. Upwelling brings nutrients such as phosphates
and nitrates to surface waters, which promotes plant
plankton blooms and more marine life.**

🌀 DESERTS

Kinds of Deserts

A **desert** (Latin *desertus,* "deserted, barren") is defined as a
region with mean annual precipitation of less than 25 centimeters (10 in), with a potential to evaporate more water
than falls as precipitation, and with so little vegetation that
it is incapable of supporting abundant life. Notice that high
temperatures are not necessary for a region to be called a
desert; only low precipitation, a dry climate, and limited
biological productivity. On this basis we can identify four
kinds of desert regions: polar, subtropical, mid-latitude, and
coastal (■ Table 12.1).

- **Polar deserts** are marked by perpetual snow cover, low
precipitation, and intense cold. The most desertlike polar
areas are the ice-free dry valleys of Antarctica. The
Antarctic continent and the interior of Greenland are true
deserts. They are very dry, even though most of the fresh
water on earth is right there in the form of ice.
- **Subtropical deserts** (also called *trade-wind deserts*) are the
earth's largest dry expanses. They lie in the regions of
subsiding high-pressure air masses in the western and central portions of continents. They receive almost no precipitation and have large daily temperature variation. The
Saharan, Arabian, Kalahari, and Australian deserts are of
this type.

- **Mid-latitude deserts,** at higher latitudes than the subtropical deserts, exist primarily because they are positioned deep within the interior of continents, separated from the tempering influence of oceans by distance or a topographic barrier. The Gobi Desert of China, characterized by scant rainfall and high temperatures, is an example. Marginal to many middle-latitude deserts are *semiarid grasslands,* extensive treeless grassland regions. Drier than prairies, they are especially sensitive to human intervention. In North America and Asia the mid-latitude and subtropical arid regions merge to form nearly unbroken regions of moisture deficiency. Also in the midlatitudes are mountain ranges that act as barriers to the passage of moisture-laden winds from the ocean, a situation leading to the formation of **rain-shadow deserts.** As water-laden maritime air is drawn across a continent, it is forced to rise over a mountain barrier, which causes the air to cool and to lose its moisture as rain or snow on the windward side of the mountains. By the time the air descends on the mountains' leeward side, it has lost most of its moisture, and there we find desert conditions (◉ Figure 12.6). The Mojave Desert and the Great Basin, leeward of the high Sierra Nevada, are examples of rain-shadow deserts. Death Valley, the lowest and driest place in North America, is in the rain shadow of the Sierra Nevada. The Sierra Nevada receives heavy precipitation, and Mount Whitney, its highest point and the highest point in the conterminous 48 states, is a scant 100 kilometers (60 mi) from arid Death Valley.
- **Coastal deserts** lie on the coastal side of a large land mass and are tempered by cold, upwelling oceanic currents (Figure 12.5). Their seaward margins are perpetually enshrouded in gray coastal fog, but inland, where the air temperatures are higher, the fog evaporates in the warmer dry air. The Atacama Desert in Chile, which may be the driest area on earth, is a rain-shadow desert as well as a coastal desert. The Andes on the east block the flow of easterly winds, and air from the Pacific Ocean on the west

■ TABLE 12.1 Classification of Deserts		
TYPE	CHARACTERISTIC LOCATION	EXAMPLES★
Polar	region of cold, dry, descending air with little precipitation	ice-free, dry valleys of Antarctica
Subtropical	a belt of dry, descending air at 20°–30° north or south latitude	Sahara, Arabian, Kalahari, Australia's Great Sandy and Simpson
Mid-latitude	deep within continental interiors remote from influence of an ocean	Gobi, Takla Makan, Turkestan
Coastal	coastal area in mid-latitudes where a cold, upwelling oceanic current chills the shore	Atacama, Namib
Rain-shadow	leeward of a mountain barrier that traps moist ocean air (Figure 12.6)	Mojave, Great Basin

★See Figure 12.8

is chilled and stabilized by the Humboldt Current, the ocean's most prominent cold-water current.

Collectively, mid-latitude deserts and semiarid grasslands cover about 30 percent of the earth's land area and constitute the most widespread of the earth's geographic realms.

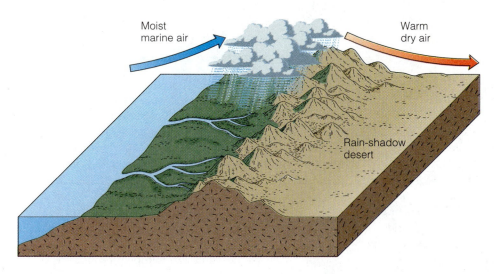

◉ FIGURE 12.6 The origin of rain-shadow deserts in middle- and high-latitude regions. Moist marine air moving landward is forced upward on the windward side of mountains, where cooling causes clouds to form and precipitation to fall. The drier air descending on the leeward side warms, producing cloudless skies and a rain-shadow desert.

Eco-catastrophe—A Day in the Life of El Niño

Certain impressive natural phenomena are the result of direct ocean-atmosphere interaction. Hurricanes are an example, as is El Niño, a flow of warm surface water that has an ecological domino effect with almost worldwide climatic implications. First recognized in coastal regions of Ecuador and Peru, it is called *El Niño* (Spanish, "the boy"), referring to the Christ Child, because it typically occurs just after Christmas.

This part of South America is dominated by coastal upwelling produced by the southeast trade winds blowing surface water offshore. This upwelled, cold, nutrient-rich water, known as the Peru Coastal Current, forms the base for one of the largest marine fisheries in the world, primarily anchoveta and tuna. The rich nutrients, mostly phosphates and nitrates, are converted to energy by microscopic marine phytoplankton (plant floaters). These are eaten by tiny zooplankton (animal floaters), which are eaten by small fish such as anchoveta, which are in turn eaten by large fish such as tuna. Although the food chain is actually more complicated than this, it exists only because of fertile waters that stimulate the phytoplankton growth that is the foundation of the chain. As a bonus, the anchoveta support a dense marine bird population, whose droppings create the phosphate-rich guano deposits that are harvested for fertilizer.

Periodically, every 3 to 7 years on average, the trade winds weaken, upwelling stops, and the warm, dilute Equatorial Countercurrent moves southward along the coast (◉Figure 1). Mass mortality of the anchoveta results, and the normal 10-million-ton catch, 15 percent of the world fish catch, is reduced by half. Most of the anchoveta crop is ground into protein-rich fishmeal for use as cattle and poultry food. When the higher-priced alternative, soybeans, must be used, the prices of chicken and beef rise dramatically. The rotting fish in bays and harbors generate hydrogen sulfide (the source of the rotten-egg odor in the chem lab), which combines with moisture in the atmosphere to form "acid fog." This fog so corrodes and tarnishes bright metal that it is called the *Callao Pintor* (Spanish, "painter"), for the seaport city of Callao, Peru, where it occurs. The seabirds migrate or die, which cuts off the supply of guano and results in low crop yields and local hunger. Because the trade winds normally keep moisture-laden clouds offshore and the

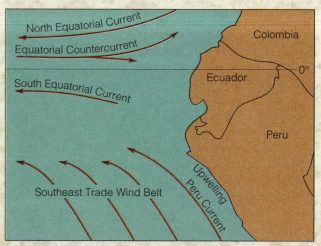

◉ **FIGURE 1** Winds, currents, and pressure centers under normal atmospheric conditions in the east equatorial Pacific Ocean.

Andes block rain clouds from the east, parts of this coast are the driest areas in the world. During El Niño, however, without the trade winds, moisture-laden clouds advance onto the land and rain storms cause flooding.

El Niño events occured in 1972–73 and 1976–77. The "Super El Niño" of 1982–83 made *El Niño* a household word in many parts of the world. Although it had been suspected for many years, the 1982–83 event firmly established El Niño as a worldwide ocean-atmosphere phenomenon. Those years it caused droughts in Africa, Australia, and Indonesia; flooding in Tahiti, California, Ecuador, and Peru; and impressive waves and storms on the California and east coasts of the United States (◉Figure 2). The normal pressure gradient that maintains the southeast trade winds is from high pressure in the eastern Pacific to low pressure in the western Pacific. Periodically this east to west flow breaks down and reverses itself, causing equatorial winds to blow from west to east, pushing a surge of warm water

Desertification

Deserts resulting from human intervention may be called *deserts of infertility.* Such deserts are generally variants of middle-latitude deserts, but they are not controlled by climate as much as by human misuse. An understanding of the causes of this class of desert is important, because much of humankind's future in many parts of the world is dependent on the proper and careful use of arid and semiarid lands.

Human activities, grazing of domestic animals, cutting trees and brush for fuel, certain farming practices, and

other uses beyond the carrying capacity of the land can so deplete the natural vegetation of semiarid grasslands bordering arid lands that true deserts expand into these regions. The expansion of deserts into formerly productive lands because of excessive human use is called **desertification,** and it is occuring at alarming rates in countries that can least afford to lose productive land. Globally, deserts are estimated to be advancing by about 70,000 square kilometers (27,300 mi^2) per year, an area a bit larger than that of the state of West Virginia, and they are expanding over grazing lands, villages, and agriculturally productive lands.

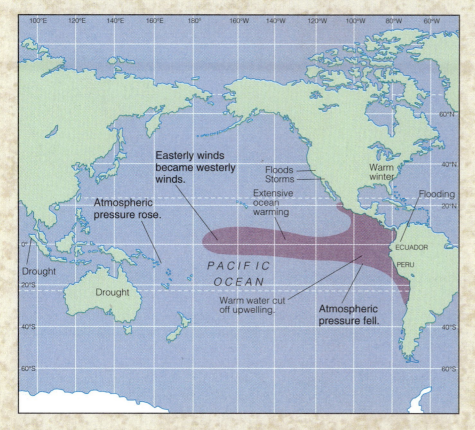

called a *Kelvin wave* toward the Americas. This atmospheric pressure reversal in the southwestern Pacific Ocean, called the *Southern Oscillation,* is the trigger for El Niño. We now call the El Niño-Southern Oscillation relationship the *ENSO event* because of the connection between seemingly unrelated happenings on opposite sides of the Pacific Ocean. The span 1990–1995 is the longest El Niño in 125 years of record, and is cited as evidence of global change.

A broad band of cool water called *La Niña* ("the girl") periodically appears along the equator in the Pacific Ocean. It has been linked to drought conditions in the Northern Hemisphere, such as the one in the United States in 1988 that did more crop damage than the Great Flood of 1993, and it may serve as an indicator of future droughts. It has also been suggested that warm oceanic temperatures cause the jet stream—the high-velocity, upper-level winds that are closely linked to surface weather—to shift northward. This northward shift then allows dry, high-pressure conditions to settle on the Midwest. This model's reliability as a forecasting tool is yet to be tested.

In the past few decades millions of people have died of starvation or have been forced to migrate because of drought in these misused lands where malnutrition is a way of life. All desert regions of the world are expanding (● Figures 12.7 and ● 12.8), but beginning in the 1970s the expansion of the Sahel region of Africa became especially severe.

Although natural long-term climate variations cause gradual expansion and contraction of deserts, recent human overuse of semiarid lands has greatly accelerated the rate of desertification. The clearing of natural vegetation from semiarid lands for cultivating crops and grazing domestic animals seemed necessary for feeding increasing populations in places like the Sahel, the 300- to 1,500-kilometer (180- to 940-mi)-wide belt along the southern edge of the Sahara Desert. Semiarid grasslands on the margins of true deserts are part of a delicately balanced ecosystem that is adapted to natural climate variations. The vegetation indigenous to these marginal lands is tolerant of the occasional droughts that are typical of these climates, the plants being able to survive the rigors of a few dry years. Cultivated crops that replaced the natural vegetation, however, lack drought tol-

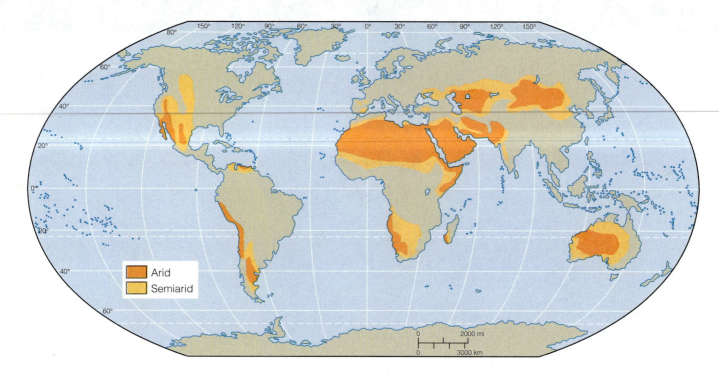

● FIGURE 12.7 Distribution of the earth's arid and semiarid regions.

erance, and for this reason, crop failure is common. The soil is left bare and susceptible to erosion by wind and rain.

Reversing desertification is a desirable goal. The best program for reversing the trend is to halt exploitation immediately and to adopt measures such as planting trees, developing and properly using water resources, and estab-lishing ecological protective screens that will help reestab-lish natural ecosystems. It might take many centuries for misused semiarid regions to recover fully, but some natural recovery occurred very quickly in Kuwait as an unexpected consequence of the 1991 Gulf War. Thousands of unex-ploded land mines and bombs are scattered across the desert

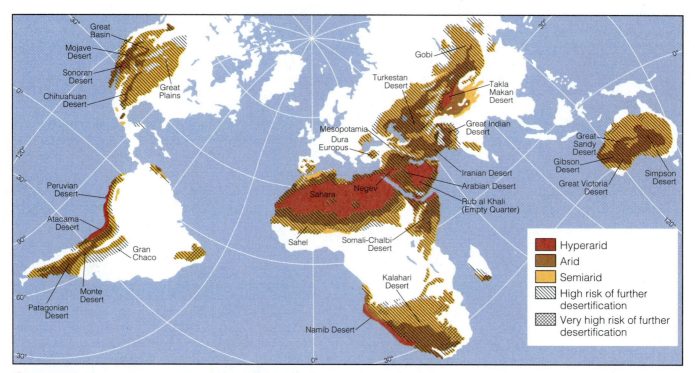

● FIGURE 12.8 Arid and semiarid regions of the world currently experiencing desertification.

● FIGURE 12.9 The Devil's Cornfield, a deflation basin that has been lowered to the capillary fringe at the top of the water table; Death Valley National Park, California. The rugged plants growing here—arrow weed, which resemble shocks of corn—are able to tolerate heat and saline soil, and their roots can withstand the effects of exposure to episodic sandblasting.

● FIGURE 12.10 Desert pavement, an interlocking cobble mosaic that remains after winds have removed the surficial fine-grained sediments. The desert floor is left armored with a tightly interlocked mosaic of gravel. An unsorted mixture of coarse- and fine-grained sediments underlies the armor.

in the western part of the country. These have discouraged off-road driving by hunters, who formerly entertained themselves by shooting the region's desert animals for recreation. The native vegetation is becoming re-established, the desert is now resembling a U.S. prairie, and the number of birds has increased immensely since before the war.

Wind as a Geologic Agent in Deserts

Wind erodes by deflation and by abrasion. **Deflation** occurs when things are blown away. This produces deflation basins where loose, fine-grained materials are removed, sometimes down to the water table (● Figure 12.9 and Case Study 12.2). *Desert pavement,* an interlocking cobble "mosaic," results where winds have removed fine-grained sediments from the surface (● Figure 12.10).

Abrasion occurs when mineral grains are blown against each other and into other objects. Among the wind-abraded geological features are fluted and grooved bedrock outcrops and **ventifacts,** cobbles or boulders bordering a dune field that have been sandblasted and polished by wind action.

Migrating dunes—small hills of migrating, wind-blown sand—and drifting sand, which are common in arid regions, can adversely affect desert highways and communities, but they are not limited to arid regions. Strong sea breezes in humid coastal regions and human activities also can cause dune migration (see Chapter 11 and ● Figure 12.11). The problems of migrating sand have been mitigated by building fences, by planting windbreaks of wind- and drought-tolerant trees such as tamarisk (● Figure 12.12), and by oiling or paving the areas of migrating sand. Dune stabilization in

● FIGURE 12.11 Coastal dune field near Pismo Beach, California. In the 1960s these dunes began migrating leeward onto agricultural lands due to the destruction of stabilizing plant cover by uncontrolled off-road vehicle use in the dune field.

● FIGURE 12.12 Windbreaks of salt-tolerant *tamarisk* (salt cedar) keep drifting sand from burying railroad tracks; Palm Springs area, California.

Deflation Basins, Winds, and the North Africa Campaign of World War II

The Qattara Depression in northwestern Egypt is an immense, hot wasteland of salt marshes and drifting sand that covers an area of about 18,000 square kilometers (7,200 mi²) at an elevation nearly 135 meters (443 ft) below sea level (Figure 1). Its origin is partly due to deflation by wind. Wind scour has lowered the land surface to the capillary fringe at the top of the water table. This has led to evaporation of the saline water and crystallization of the salt as a sparkling white crust on the depression floor.

This impressive and forbidding quagmire played an important role during World War II in a decisive Allied victory in 1942. The impassable trough served to block German Field Marshal Erwin Rommel's Afrika Korps' attempt to turn the flank of the British Eighth Army led by General Bernard Montgomery. The geography forced Rommel to attack the Eighth Army's heavily fortified positions on El Alamein, a 64-kilometer (40-mi)-wide strip of passable land between the Mediterranean on the north and the hills to the south that mark the edge of the

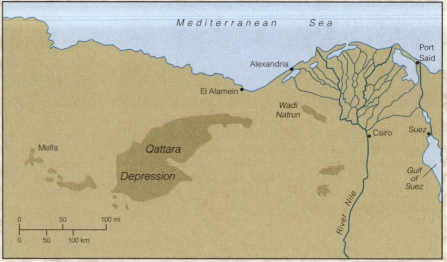

● FIGURE 1 Locations of the Qattara Depression and El Alamein. The decisive battle between the forces of Montgomery and Rommel was fought north of the Qattara Depression.

impassable Qattara (Figure 1). Rommel outran his lines of supply, sustained heavy losses, and was forced into retreat. This pushed him out of Egypt, and eventually westward across Libya.

Throughout the two-year North Africa campaign, the Allied and Axis forces had a common enemy: the *khamsin* (Arabic, "wind of the fifty days"), the searing hot winds that blow out of the Sahara (Figure 2). These winds raised temperatures as much as 19°C (35°F) in a few hours and sent sand grains swirling across hundreds of square kilometers. The onslaught of a khamsin dictated the actions of men and machines, preempting military tactics and strategies. Airplanes had to be grounded, and battlefield action stopped for days, as the sun was blotted out and visibility dropped to zero. Soldiers wore tight-fitting goggles to protect their eyes from the clouds of stinging sand and gasped for air through gas masks or makeshift cloth masks. Winds of hurricane strength that tore up camps and turned over vehicles were common. Swirling clouds of sand generated static electricity that rendered the magnetic compasses needed for navigating across the desert useless. Storm-generated static electricity during one khamsin detonated and destroyed an ammunition dump.

● FIGURE 2 (left) British soldier watching an approaching dust storm generated by a *khamsin* wind near El Alamein during the North Africa campaign, 1942.

humid coastal areas of Europe, the eastern United States, and the California coast has been accomplished by planting such wind-tolerant plants as beach grass *(Elymus)* and ice plant "Hottentot fig" *(Carpobrotus).*

Arid regions of the world have some intriguing and bizarre natural features, as shown in ⦿ Figure 12.13. Among the curiosities produced by wind are teetering rocks that have been nearly severed at the base by wind abrasion and playa rock scrapers, rocks that leave complicated tracks on mud-surfaced playas as they are pushed along by high winds.

GLACIATION

About 97 percent of the earth's water is in the oceans, and three fourths of the remainder (2.25%) is in **glaciers.** A

⦿ FIGURE 12.13 Desert oddities, *(a)* Mushroom rock; Death Valley National Park, California. Rigorous wind erosion in the abrasion zone just above the ground surface can produce unusual landforms such as this. *(b)* Remnants of a 1920s plank road, an early attempt to maintain a road in a region of drifting sand; Algodones dune field between El Centro, California, and Yuma, Arizona. *(c)* Playa lake scraper; Racetrack Playa, Death Valley National Park, California. The boulder's track is preserved on the mud-cracked surface. The boulder, 25 cm (11 in) at its widest, moved or was moved during a winter wet period due to just the right combination of rain-soaked mud and gusty high wind. *(d)* Devil's Golf Course, Death Valley National Park, California, the lowest elevation in the Western Hemisphere. The salt-encrusted surface has formed at the capillary fringe on the top of the underlying water table.

(b)

(c)

(a)

(d)

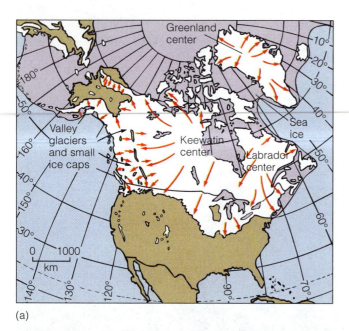

(a)

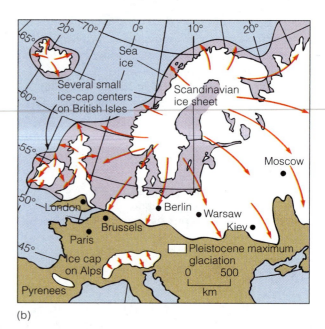

(b)

⊙ FIGURE 12.14 Major centers of ice accumulation and the maximum extent of Pleistocene glaciation in *(a)* North America and *(b)* Europe.

tenth of the earth's land area is covered by glaciers at present, the same amount that is cultivated globally. During the Pleistocene Epoch, which radiometric dating places as beginning 1.6 million years ago and lasting until about 10,000 years ago, as much as thirty percent of the earth was covered by glaciers (⊙ Figure 12.14). Today's glaciers store enormous amounts of fresh water, more than exists in all of the lakes, ponds, reservoirs, rivers, and streams of North America. Farmers in the Midwest grow corn and soybeans in soils of glacial origin, Bunker Hill in Boston is a glacial feature, and the numerous smooth, polished, and grooved bedrock outcroppings in New York City's Central Park were eroded by a glacier that flowed out of Canada (as explained in Case Study 12.4 later in the chapter). Glaciers in Norway, Switzerland, Alaska, Washington state, Alberta, and British Columbia exert a vital effect on regional water supplies during dry seasons. They do this by supplying a natural base flow of meltwater to rivers that helps to balance the annual and seasonal variations in precipitation. Meltwater from glaciers serves as the principal source of domestic water in Switzerland. The Arapahoe Glacier is an important water source for Boulder, Colorado, so much so that Boulder residents take pride in telling visitors that their water comes from a melting glacier. (Actually, the glacier is only one part of a drainage basin of several tributaries that together supply the city's water.)

The steady water supply provided by melting glaciers during long periods of summer drought is widely used as a water resource for generating hydroelectric power in many temperate regions of the world. Utilization of glacial meltwater is highly developed in Norway and Switzerland, for example. Switzerland's hydropower stations, built mainly in the 1950s and 1960s, use water from melting snow and glacial ice. Dams store the meltwater in the summer, when

electrical consumption is low in that country, and in winter, when demand is highest, half of the country's electrical production is generated by water released from the reservoirs.

Origin and Distribution of Glaciers

Glaciers form on land where for a number of years more snow falls in the winter than melts in the summer. Such climatic conditions exist at high latitudes and at high altitudes. The pressure increases that occur as the thickness of the snowpack increases cause the familiar six-sided snowflakes to change into a coarse, granular snow called **firn** ("corn snow" to skiers). Continued accumulation increases the pressure and causes much of the trapped air to be expelled from the firn. Recrystallization into larger crystals occurs, and eventually the dense crystals of "blue" glacial ice form.

Once the ice attains a thickness of about 30 meters (100 ft), it begins to deform as a viscous fluid, flowing due to its own weight. It moves downslope if in mountains, or radially outward if on a relatively flat surface. This kind of behavior of an apparently solid material, described as *plastic flow,* takes place in the lower part of the glacier called the **zone of flowage** (⊙ Figure 12.15). The upper, surficial layer of ice behaves in a more familiar manner as a brittle solid, often breaking into a jagged chaotic surface of *crevasses*—ominous deep cracks—and *seracs*—towering prominences of unstable ice. It is the deformation by plastic flow that distinguishes glaciers from the perennial snow fields that persist at higher elevations in many mountain ranges. Glaciers are more than just masses of frozen water. Although they are largely ice, they also contain large quantities of meltwater and vast amounts of rock debris acquired from the underlying bedrock or mountains where they

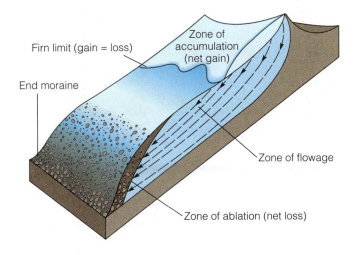

Firn limit (gain = loss)

Zone of accumulation (net gain)

End moraine

Zone of flowage

Zone of ablation (net loss)

● **FIGURE 12.15** (left) Typically, about two thirds of a glacier lies at elevations above the firn limit in the zone of accumulation, where more snow falls in winter than melts in summer. Below the firn limit is the zone of ablation, where glacial ice and the entrained rock debris are exposed by melting in the summer. The ridge of rock debris formed at the edge of the melting ice is an end moraine.

originated and across which they have moved on their journey downslope. There are several types of glaciers, a classification of which appears in ■ Table 12.2.

Glacier Budget

When the rate of glacial advance equals the rate of wasting, the front (terminus) of the glacier remains stationary, and the glacier is said to be in equilibrium. Glaciers **ablate,** or waste away, by melting or, if they terminate in the ocean,

by calving. **Calving** is the breaking off of a block of ice from the front of a glacier that produces an iceberg. If the rate of glacier advance exceeds the rate of melting, the terminus advances. Conversely, if the rate of advance is less than the rate of wasting, the terminus retreats. It is important to realize that glacier ice always flows toward its terminal or lateral margins, irrespective of whether the glacier is advancing, is in equilibrium, or is retreating. Glaciers move at varying speeds. Where slopes are gentle, they may creep along at rates of a few centimeters to a few meters per day; where slopes are steep, they may move eight to ten meters (26–32 ft) a day (● Figure 12.16). Some glaciers, under exceptional conditions, may episodically accelerate with speeds up to 100 meters (328 ft) a day, a condition known as *surging* (see Case Study 12.3).

■ TABLE 12.2 Classification of Glaciers		
TYPE	**CHARACTERISTIC LOCATION**	**EXAMPLE**
Valley (alpine)	mountain valleys	Vaughan Lewis Glacier, Alaska
Ice-field	extensive mountain glaciers that bury many adjoining valleys, leaving only the highest peaks and ridges rising above the ice	Juneau Ice Field, Alaska
Ice-cap	relatively smooth land surfaces in high latitudes or mountain summits where ice flows radially outward	Vatnajökull, Iceland
Piedmont	at the foot of mountain ranges where several valley glaciers emerge, flow laterally, and coalesce	Malaspina Glacier, Alaska
Continental (ice-sheet)	vast land areas	Greenland and Antarctica

● **FIGURE 12.16** Scientists use a jet of hot water to drill a borehole in a glacier. A stake will be placed in the hole that will enable scientists to measure the glacier's movement and summer ablation.

found Icy Strait, at its entrance, choked with ice and Glacier Bay only a slight dent in the ice-cliffed shoreline. By 1879, when John Muir visited the area, the glacier, by then named the "Grand Pacific," had retreated nearly 80 kilometers (50 mi) up the bay. The Grand Pacific Glacier had retreated another 24 kilometers (15 mi) by 1916, and today a 104-kilometer (65-mi)-long fiord occupies the area that just 200 years ago held a valley glacier that was as much as 1,220 meters (4,000 ft) thick (● Figure 12.17).

Glacial Features

Glaciers carry all manner of rock debris that is deposited directly by melting ice as unsorted rock debris, or **glacial**

Probably the best-documented and most dramatic retreat of a glacier is found at Glacier Bay National Park, Alaska. When George Vancouver arrived there in 1794, he

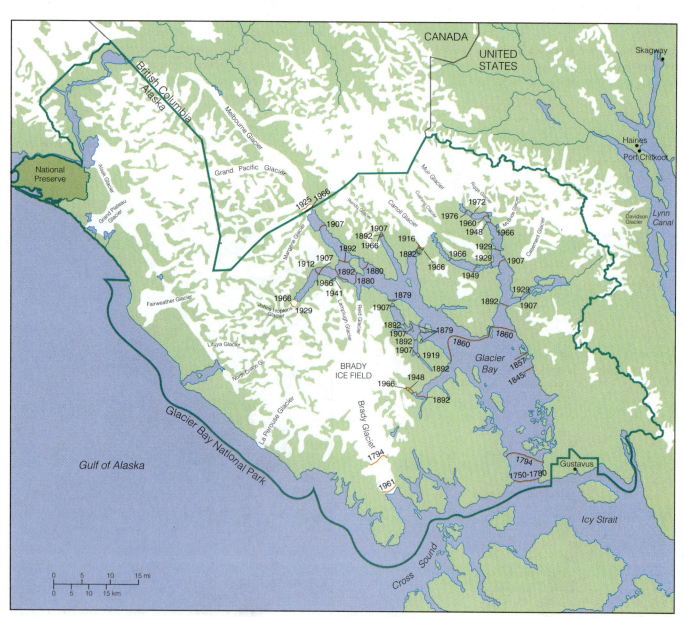

● **FIGURE 12.17** Map of Glacier Bay National Park, Alaska, showing the terminal positions of the retreating Grand Pacific and other glaciers from about 1750 to the present.

When Glaciers Surge It's A Moving Experience

A rare but spectacular type of glacier movement is **surging**. A surge begins when ice in the upper part of the glacier reaches some critical thickness. Subsequently the ice surface breaks up into thousands of crevasses, which rapidly spread downglacier, and then the ice movement accelerates from perhaps only a few centimeters a day to as much as 100 meters (328 ft) a day. If the surge reaches the glacier front, the glacier advances dramatically. Surging may cause a glacier to advance several kilometers within a few weeks and be accompanied by a spectacular release of meltwater that causes great damage downvalley. Alaska's Bering Glacier—the earth's largest surging glacier—began surging in May 1993. After a minor retreat for seven months, it resumed surging in April 1995 and was still surging in March 1996. The surge advanced with maximum rates of 7–10 meters a day during the winter of 1993–94, pushing the glacier terminus forward by about 12 kilometers (7 mi) and producing a large number of icebergs, several of them more than a half-kilometer (1,640 ft) long. The U.S. Geological Survey is monitoring the surge using aerial and field observations, paying special attention to iceberg production. As with the icebergs produced by the Columbia Glacier in the 1980s, any icebergs entering the Pacific Ocean could pose a threat in the oil-tanker lanes leading to and from nearby Prince William Sound.

Surges are generally not considered climatically induced events, and it is important to understand them in order to predict surges that could endanger installations or people. Such is the case of another Alaskan glacier, the Black Rapids Glacier, which is being monitored because a surge by it could threaten a major highway and the Alaska Pipeline. That the concern is justified was demonstrated by its surge in 1936–37. The owners of a lodge on the Richardson Highway correctly perceived that the glacier's 3-kilometer (1.8-mi)-wide front was moving toward them and notified authorities. The glacier was found to be advancing at rates of up to 65 meters (213 ft) a day. Fortunately, the surge halted a short distance from the highway; had it continued, it would have destroyed the lodge, dammed a major river, and cut through the highway.

Surges of the Plomo Glacier in the Argentinean Andes have led to the formation of ice-dammed lakes. When one of these lakes bursts through the ice dam, the resulting flood of water can create destruction downvalley. Several deaths were caused by the Plomo surge of 1934, and 13 kilometers (8 mi) of the Mendoza–Santiago railway and seven bridges were washed out. Similar events in the Karakoram Mountains on the border of Kashmir and China have caused extensive damage and loss of life.

Why glaciers surge is still uncertain, but research on the 1982–83 surge of Alaska's Variegated Glacier has provided new information on surging behavior. Located at the northern end of the Alaskan panhandle, the Variegated Glacier (⊙ Figure 1) flows off a shoulder of the St. Elias Range. It is 20 kilometers (12 mi) long and is normally a well-behaved tongue of ice, flowing at a rate of about 0.1–0.2 meters (0.33–0.67 ft) a day. For some time before the onset of the surge, studies revealed considerable thickening in the upper reaches of the glacier and thinning in the lower part. The consequent steepening of the glacier increased the stress at the base of the ice in the upper area, which gradually closed the subglacial channels that carried away meltwater. Squeezed out of the channels, the water was forced to spread laterally as a layer across the entire glacier bed. This buoyed up the glacier slightly, which reduced the friction and allowed it to slide faster. Rapid movement began in the upper area, and as the faster-moving upper ice pressed on the slower-moving downglacier ice, stresses began transferring to the lower, thinner part of the glacier, and the surge spread downglacier as a wave. The surge's velocity tripled in its upper reaches and doubled in the middle. After the stresses were relieved, the subglacial channels began to reopen, the subglacial reservoir of water emptied in a concluding flood, and the glacier dropped down on its bed and stopped sliding.

SOURCE: Michael Hambrey and Jürg Alean, *Glaciers* (New York: Cambridge University Press, 1992); and Barclay Kamb and others, "Glacier Surge Mechanism," *Science* 227 (1 February 1985), pp. 469–479.

⊙ **FIGURE 1** During the surge of the Variegated Glacier, the glacier's accelerated motion was measured using surveying equipment. The umbrella was used to protect the heat-sensitive electronic equipment from the heat of the sun.

● FIGURE 12.18 Recessional moraines; Coteau des Prairies, near Aberdeen, South Dakota.

till. **Moraines** are landforms composed of till and named for their site of deposition. **End moraines** (also called *terminal moraines*) are deposited at the ends of the melting glaciers. **Recessional moraines** are series of nested end moraines that record the stepwise retreat, or meltback, at the end of an ice age (● Figure 12.18).

After melting, continental and ice-cap glaciers leave a subdued, rounded topography with a definite "grain" that indicates the direction of glacial movement. They may also leave behind numerous **kettle lakes,** water-filled, bowl-shaped depressions without surface drainage (● Figure 12.19). For example, there are about a dozen kettle lakes in northern Indiana's Valparaiso moraine. These lakes are believed to have formed when large blocks of stagnant ice that were wholly or partly buried in the deposits left behind by retreating glaciers melted (● Figure 12.20). End moraines now form the hilly areas extending across the Dakotas, Minnesota, Wisconsin, northern Indiana, Ohio, and all the way to the eastern tip of Long Island, New York

(Case Study 12.4). Some Chicago suburbs have names that reflect their location on high areas of hilly moraines: Park Ridge, Palos Hills, and Arlington Heights, for example. In addition, poor drainage or deranged drainage characterizes many areas that have undergone continental glaciation.

Following a period of glaciation, a **fiord** is formed when the seaward end of a coastal U-shaped valley, eroded by a valley glacier, is drowned by an arm of the sea (● Figure 12.21).

● FIGURE 12.19 Kettle lakes at Kewaskum, Wisconsin.

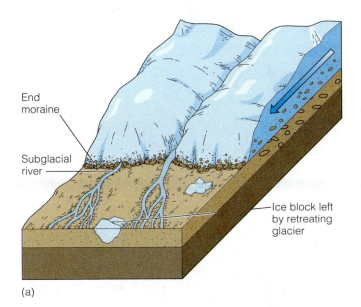

(a)

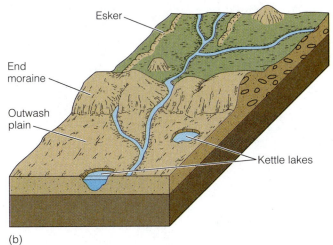

(b)

 FIGURE 12.20 Formation of end moraines, kettle lakes, and eskers *(a)* during glaciation and *(b)* their appearance after glaciation.

THE EFFECTS OF GLACIATION

Although the great ice sheets of the Pleistocene Epoch seem remote in time and modern glaciers seem remote in distance, both have impacts on modern society that are far from trivial.

Distribution of Soils

Farming and forestry generally depend upon the distribution and quality of Pleistocene deposits. Some of the best agricultural soils in the world lie on sediments deposited in former glacial lakes and on glacio-eolian (loess) deposits (see Chapter 6). As Pleistocene glaciers expanded and moved over western Europe and North America, they scraped off soils from the underlying land and redeposited them as moraines of glacial till where the ice melted. Today some

● **FIGURE 12.21** A fiord, Gastineau Channel near Juneau, Alaska.

till-covered areas are excellent forestland and excellent-to-poor farmland (● Figure 12.22, page 340).

Ground-Water Resources

Glacial outwash is sediment that was deposited by meltwater rivers originating at the melting edge of a glacier. The great Pleistocene ice sheet that extended out of Canada into the northern Mississippi River valley released copious amounts of sediment-laden meltwater, resulting in vast sheets of outwash sediments. Some of these glacial-outwash deposits serve as important ground-water aquifers in the end-moraine regions of the Midwest. Moraines associated with the outwash are poor aquifers due to the low porosity and permeability of the clays they contain. Thus geophysical exploration, test-drilling, and careful mapping are necessary in order to delineate the aquifers of porous and permeable sands within the buried stream channels. Glacial outwash can serve as an important source of sand and gravel (● Figure 12.23, page 340).

Sea-Level Changes

The transfer from the oceans to the continents of the millions of cubic miles of water that became the enormous Pleistocene continental ice sheets caused sea level to lower about 100 meters (330 ft). This exposed extensive areas of continental shelves. Consider that a drop of this amount today would increase the area of Florida real estate by more than 30 percent, would leave New Orleans high and dry, and would cause the Panama Canal to drain. Conversely, if all the polar ice were to melt, sea level would rise on the order of 40 meters (130 ft), and London, New York, Tokyo, and Los Angeles would be mostly under water. As sea level dropped during the Ice Ages, the rivers in coastal regions began downcutting across the exposed continental shelves as they adjusted to the new conditions. When sea

Continental Glaciers and the New York Transit Authority

A superb example of a major continental end moraine, the Harbor Hill moraine, extends along Long Island's north shore from Orient Point southwestward through Queens and into Brooklyn, where the rolling ground surface in Prospect Park betrays the glacial origin of the park grounds. The last pulse of the advancing Canadian ice sheet reached this point about 20,000 years ago, leaving the moraine as a kind of glacial footprint. South of the Harbor Hill moraine is the older and less obvious Ronkonkoma moraine. Much of the Harbor Hill moraine on Long Island is inaccessible because it is on private property, but from Port Jefferson to Orient Point, the moraine is easily visible, because its edge is hugged by state highways 25, 25A, and 27 (⦿ Figure 1). Sagamore Hill National Historic Site,

variety of rocks eroded and carried here from distant areas far north of Long Island. At Prospect Park in Brooklyn, which is easily reached by subway, you may stroll along the undulating ground surface of the moraine across terrain described as "knob-and-kettle topography," the *kettles* being depressions that mark former locations of chunks of glacial ice that were buried in the sand and gravel of the moraine and left behind when the ice retreated. Eventually the trapped ice blocks melted, and the ground surface collapsed into the rounded depressions. The moraine continues southwest from Prospect Park to Staten Island, where the moraine gravels form Todt Hill, the highest point in the region at 125 meters (410 ft), and then on into New Jersey and Pennsylvania.

⦿ **FIGURE 1** Locations of the Harbor Hill and Ronkonkoma moraines and other features mentioned in the case study.

at Oyster Bay, Long Island, the family home of Theodore Roosevelt, twenty-sixth President of the United States, sits on a recessional moraine that is younger than the Harbor Hill. The cliffs along the bay at Garvies Point, next to the Nassau County Museum of Natural History at Glen Cove, expose a 1–2 meter (3–7 ft)-cover of glacial till overlying Cretaceous sands and clays. The most striking evidence of Pleistocene glaciation are the **erratic** boulders, large rock fragments on the beach and near the museum, carried here by the glacial ice and deposited with the till (⦿ Figure 2). These erratics consist of granites and gneisses, unlike any bedrock that crops out nearby, which illustrates the

When the Harbor Hill moraine extended across the area that is now the entrance to New York Harbor, it dammed the meltwaters coming down the valley of the modern Hudson River, forming a lake. Eventually the lake waters broke through the moraine dam and eroded the channel to the sea that is now called the Verrazano Narrows. Passing near the Verrazano Narrows Bridge on the New York Transit Authority's Staten Island Ferry or driving across the bridge, one might ponder whether the breakthrough of the moraine was a slow process or a catastrophic one some 20,000 years ago.

In addition to Pleistocene glaciation being responsible for the

⊙ FIGURE 2 Glacial erratic boulders on the beach at Garvies Point, Glen Cove, Long Island, New York.

(a)

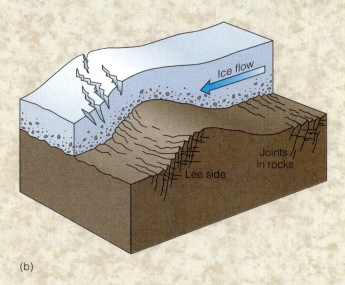

(b)

⊙ FIGURE 3 *(a)* A smoothed roche moutonnée (rock knob) in Central Park, New York City. *(b)* The manner of ice flow and the influence of jointing on the development of a roche moutonnée. The glacier flows upward and over the unjointed part of the rock, smoothing and polishing it to form the gentle side, and then plucks out and removes blocks from the jointed part of the rock, forming the steep side.

development of the Long Island landscape, it also set the stage for the region around Hempstead to become the cradle of American aviation. Before urbanization, the moraines and till of Long Island, like much of New York State, were heavily forested. However, the glacial outwash plain south of the moraines supported a mat of grasses so dense that it excluded trees. It was this treeless plain around Hempstead that Glenn Curtiss and other aviation pioneers recognized as the most suitable site for a landing field for their early flying exploits. Thus, the fledgling aircraft industry got its start here about 1910, and Mitchell and Roosevelt airfields were built soon afterward; it was from Roosevelt Field that Charles Lindbergh began his epic solo transatlantic flight in 1927.

Central Park in Manhattan, accessible by bus or subway, reveals its share of glacial footprints. There are numerous *roche moutonnée* (French, "rock sheep"), rounded knobs of bedrock so sculptured by a large glacier that their long axes are oriented in the direction of ice movement. Some of these show *glacial grooves* that were rasped out by boulders. An especially good set of grooves and glacial polish is found on a large roche moutonnée in the southwestern part of the park immediately south of Heckscher baseball fields numbers 3 and 4 (⊙ Figure 3). The orientation of the grooves shows that the glacier scraped across Manhattan Island from the northwest to the southeast.

● FIGURE 12.22 Rocky soil developed on glacial till; Adirondack Mountains, New York.

● FIGURE 12.23 Part of a vast plain, underlain by Pleistocene glacial outwash near Elgin, Illinois. Outwash sediments, deposited along the margins of the melting continental ice sheets, are spread for thousands of square kilometers across the north-central United States. The ponds in the photograph are the excavations remaining from sand and gravel mining operations.

level rose to its present level, about 5,000 years ago, the mouths of the coastal rivers were drowned, forming what are now harbors and estuaries. This last rise of sea level began about 18,000 years ago, a mere twitch in the scale of geologic time.

Isostatic Rebound

The Ice Age continental glaciers and ice caps, with thicknesses of two miles or more, were so massive that the crust beneath them was depressed by their weight. In the time since the ice retreated at the end of the last ice age, 8,000 to 11,000 years ago, the land regions exposed by the retreating ice sheets in Scandinavia, Canada, and Great Britain have risen due to the geologically rapid removal of the great load of ice. Uplift of the crust due to unweighting is caused by **isostatic rebound,** the slow transfer by flowage of mantle rock to accommodate uplift of the crust in one area that causes subsidence in another area. Evidence of isostatic

rebound in deglaciated areas is seen in the raised beaches around Hudson Bay, the Baltic Sea, and the Great Lakes (●Figure 12.24). Careful surveying in Europe shows that as Scandinavia and Great Britain have been rising, a corresponding subsidence of Western Europe has occurred that is especially pronounced in the Netherlands. The uplift and subsidence in these regions remain active. Above the modern shorelines of the Gulf of Bothnia in Finland and Sweden are raised beaches that show a maximum uplift of 275 meters (900 ft). It is estimated that an additional 213 meters (700 ft) of rise will occur before equilibrium is reached. The rise is so rapid in some places in Scandinavia that docks used by ships are literally rising out of the sea. Uplift in Oslo Fjord, Norway, is about 6 millimeters (¼ in) per year,

● FIGURE 12.24 Isostatically raised beach ridges at Richmond Gulf, Hudson Bay, Quebec, Canada. The highest ridge in the photo lies about 300 meters (984 ft) above sea level and is about 8,000 years old.

which adds another 60 centimeters (2 ft) of elevation per century. Sea-level mooring rings the Vikings used for tying up their dragon boats a thousand years ago are now 6 meters (20 ft) above sea level. Not even the tallest Viking would be able to moor his boat to those rings today. Similar uplifting is recorded along the northern shores of Hudson Bay, where the average uplift is about a meter (3 ft) per century.

Uplifting in the Great Lakes region is slow but continuing and eventually it may cause problems for the residents of Chicago and elsewhere in Illinois. The rebound threatens to cause a drainage change that will force more of Lake Michigan to drain southward into the Mississippi River system (◉ Figure 12.25).

The rapid unloading of the crust at Glacier Bay, Alaska, in the last 200 years has resulted in measurable rebound in the lower parts of the bay. Small islands are slowly emerging, and existing islands are rising at a rate of about four centimeters (1½ in) per year.

Human Transportation Routes

Beginning in 1980, the Columbia Glacier on Prince William Sound, Alaska, began a catastrophic retreat by calving thousands of icebergs each year, some of them as large as houses. The glacier had been in equilibrium at much the same position for at least 200 years—it was first observed in the late eighteenth century—when it began to show evidence of retreat in the late 1970s. Fortunately, the moraine

shoal at its terminus is sufficiently shallow to trap the larger icebergs and prevent them from drifting into Prince William Sound, where they would pose a major threat to the supertankers passing through these waters to and from the oil-loading terminal at Valdez. Because the terminus of the glacier had retreated some 6 kilometers (3.6 mi) or more by 1992, it appears that a new fiord is forming.

Juneau, Alaska, is the only state capital in the United States without highway access, even though it is about 160 miles (by air) west of the nearest highway in Canada. A highway connecting Juneau with Canada would be desirable, but it is not to be, because the advancing Taku and Hole-in-Wall Glaciers threaten to close off the only possible route along the west shoreline of Taku Inlet (◉ Figure 12.26).

Whereas glaciers impede transportation in some areas, in other areas they have served to facilitate it. In wet boggy regions in the early days of New England, the crests of **eskers** provided good travel routes. Eskers are long, winding, steep-sided ridges of stratified sand and gravel deposited by subglacial or englacial streams that flowed in ice tunnels in or beneath a retreating glacier. Hence, especially in Maine, the eskers were called *horsebacks* (Figure 12.20, part b) and ◉ Figure 12.27).

Pleistocene Lakes

The cooler climates of the Pleistocene brought increased precipitation, slower evaporation, and the runoff of glacial

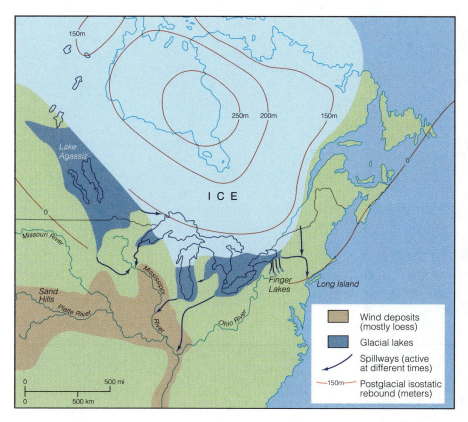

◉ FIGURE 12.25 Locations of maximum postglacial (after glaciation) isostatic uplift in the Hudson Bay region and of land features resulting from the retreat of the late Pleistocene ice sheet about 10,000 years ago. Proglacial lakes (temporary, ice-marginal lakes) overflowed at different times through various spillways. One of the arrows east of the Finger Lakes represents the spillway pictured in Figure 12.30. The spillway draining Lake Agassiz in South Dakota is pictured in Figure 12.31. Windblown loess was deposited adjacent to major glacial meltwater systems. Waters from Lake Michigan drained southwestward into the Mississippi River at the time, which may occur again as isostatic uplift continues.

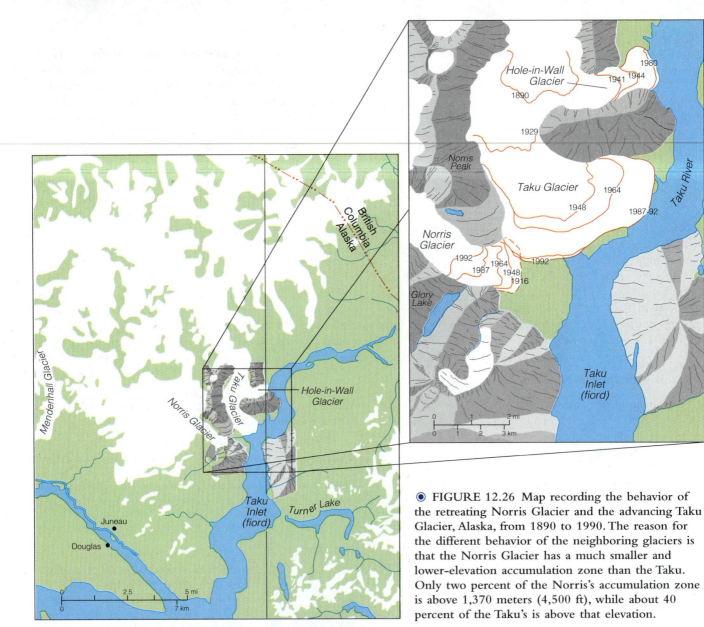

● FIGURE 12.26 Map recording the behavior of the retreating Norris Glacier and the advancing Taku Glacier, Alaska, from 1890 to 1990. The reason for the different behavior of the neighboring glaciers is that the Norris Glacier has a much smaller and lower-elevation accumulation zone than the Taku. Only two percent of the Norris's accumulation zone is above 1,370 meters (4,500 ft), while about 40 percent of the Taku's is above that elevation.

meltwater. Lakes dotted the basins of North America. Accumulations of water, mostly in valleys between the fault-block mountain ranges in the Great Basin, formed hundreds of lakes far from continental glaciers (● Figure 12.28). Today many of these valleys contain aquifers that were probably charged by waters of the Ice Age lakes. California had dozens of lakes, now mostly dried up; among them were Owens Lake, Lake Russell (the ancestor of Mono Lake, ● Figure 12.29), and China Lake. Lake Manley covered the floor of what is now Death Valley National Park, being named after the Manley party that suffered great hardships there in 1849, and Rogers Dry Lake, near Landcaster, California, is the location of Edwards Air Force Base and a landing site for the space shuttle. The Great Salt Lake

● FIGURE 12.27 (left) Esker with a road on its crest; Malingarna area, southern Sweden.

in Utah is all that remains of the once-vast Pleistocene Lake Bonneville. The Bonneville Salt Flats, an area that is ideal for setting automobile speed records, formed by evaporation of the ancient Pleistocene lake. Further east, the Great Lakes and the Finger Lakes of New York State are of glacial origin. In the Pleistocene Epoch, when drainage to the north from the Great Lakes and the Finger Lakes was blocked by ice, these lakes drained through various spillways at different times (Figure 12.25 and ◉ Figure 12.30). Major disruptions in drainage produced large temporary ice-marginal lakes, called **proglacial lakes,** which were dammed partly by moraines and partly by glacial ice.

Ice Age lake sediments are the basis of the rich soils of the northern United States and southern Canada. The rich wheat lands of North Dakota and Manitoba, for example, formed as sediments in Pleistocene Lake Agassiz (Figure

CONSIDER THIS . . .

If there had been no large-scale Pleistocene continental glaciation in North America, what differences might we find in today's Midwestern states? (*Hint:* Carefully examine Figures 12.14, part a and 12.15.)

12.25). Centered in present-day Manitoba, the lake was the largest known North American lake; its area of more than 518,000 square kilometers (200,000 mi²) was more than twice that of the five present-day Great Lakes combined. (It was named after Louis Agassiz, the nineteenth-century sci-

◉ FIGURE 12.28 Pleistocene pluvial lakes in the western United States.

● FIGURE 12.29 Terraces marking the former shorelines of Pleistocene Lake Russell, east-central California. The highest shoreline is approximately 185 meters (600 ft) above the level of the modern Mono Lake.

● FIGURE 12.30 Typical subdued terrain resulting from continental glaciation at Lake George, New York. The lake is dammed by an end moraine. The break in the ridge on the right marks the channel of the ancient river that carried meltwater southward from the Pleistocene glacier. See Figure 12.25.

entist who fostered the theory of the Ice Ages.) The lake formed when the great Canadian ice sheet blocked several rivers flowing northward toward Hudson Bay. For a time, a large spillway channel carried outflow southward from the lake, forming what is now the 200-foot-deep Browns Valley on the border between South Dakota and Minnesota (●Figure 12.31). The many remnants of Lake Agassiz include Lake Winnipeg and Lake of the Woods. Northwest of this area are Great Bear Lake, Great Slave Lake, and Lake Athabaska, descendants of other large ice-marginal lakes in the present-day Northwest Territories, Alberta, and Saskatchewan.

GLOBAL CLIMATE CHANGE

In the past century it has come to be recognized worldwide that Pleistocene—commonly referred to as *the Ice Age*—glacial deposits of different ages exist. The oldest deposits, weathered and gullied from erosion, have been found beneath younger ones, which are fresher-appearing, less weathered, and less eroded. By applying the Laws of Superposition and Cross-cutting Relationships to these observations, it is clear that multiple stages of glaciation occurred during the Pleistocene Epoch. It is now generally accepted that there have been at least four major glacial stages in

● FIGURE 12.31 Big Stone Lake now occupies part of Lake Agassiz's spillway channel. It is in Browns Valley on the boundary between western Minnesota and northeastern South Dakota. See Figure 12.25.

North America and Eurasia. Independently, oceanographers studying fossil shells contained in cores of sea-floor sediments from three oceans have determined that the oceans experienced twenty or more periods of deep cooling with intervening warm periods during the Pleistocene. Because evidence of glaciations on land is easily destroyed by later ice advances, it is assumed that the continental record is probably incomplete.

Global temperatures during Pleistocene Ice Age averaged only about 4° to 10° C (7° to 18° F) cooler than the temperatures at present. The lowering of the average Ice Age temperature was not as significant as the range of seasonal extremes; winters were much colder, but summers were moderate. Consequently, climate zones shifted toward the equator, accompanied by similar shifts of plant and animal communities. The Rocky Mountains, the Sierra Nevada, the Alps, the Caucasus, and the Pyrenees were topped with ice fields similar to the modern ice field in the mountains near Juneau, Alaska. Small mountain glaciers existed in high mountains of the midlatitudes much closer to the equator than any similar modern glaciers. Even the summit of Hawaii's Mauna Loa volcano, only 19° north of the equator, shows evidence of glaciation. It also was cooler south of the equator, but because the landmasses in the Southern Hemisphere's middle latitudes are small or narrow, extensive glaciation was limited to the South Island of New Zealand and the southernmost part of the Andes in South America. Antarctica and Greenland were ice-covered much the same as they are today.

Causes of Ice Ages

Once scholars recognized that cyclic global climate changes were responsible for the multiple stages of glaciation, they began to seek answers to questions about underlying causes. An early answer was provided in the late nineteenth century by Scottish geologist John Croll. Serbian astronomer Milutan Milankovitch expanded upon Croll's work with elegant mathematics in 1941. Both of these scientists recognized that the cyclic pattern of the Pleistocene ice ages could result from periodic minor changes in the earth's rotation and revolution around the sun, with the cold periods of ice advance occurring when three variations in the earth's rotational and orbital elements were synchronous. The three variations are

1. the **obliquity,** or tilt, of the earth's axis;
2. the **eccentricity,** or shape, of the earth's orbit; and
3. the **precession,** or wobble, of the earth's axis (◉ Figure 12.32).

The earth's *obliquity* relative to the sun has been found to vary between 21.2° and 24.5° over a 41,000-year cycle. This slight change causes variations in the temperature range between summer and winter. When the earth is at the minimum tilt, sunlight hits the polar regions at a higher angle, and the global seasonal temperature variation decreases. Every 100,000 years, the orbit of the earth changes from nearly cir-

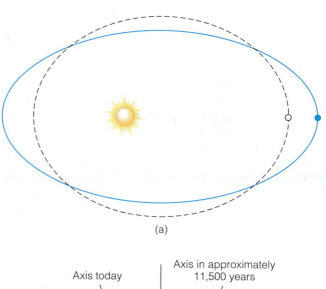

(a)

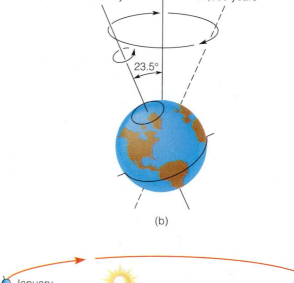

(b)

(c) Conditions now

(d) Conditions in about 11,500 years

◉ **FIGURE 12.32** Geometry of variation in the earth's orbital eccentricity, axial precession, and axial obliquity. *(a)* The orbital eccentricity varies from nearly circular *(dashed line)* to elliptical *(solid line)* to nearly circular and back again over a period of about 100,000 years. *(b)* The earth's axis slowly precesses, like a spinning top tracing out a cone in space, taking about 23,000 years for a complete cycle. *(c)* Conditions today. The earth is closest to the sun in January while the Northern Hemisphere is experiencing winter. *(d)* Conditions in about 11,500 years. The earth will be closest to the sun in July while the Northern Hemisphere is experiencing summer.

cular to more *eccentric*. Seasonal variation in temperature is greater when the orbit is more eccentric. The axis of the earth, which now points to Polaris, the North Star, *precesses* like a spinning top when its axis is disturbed. This precession causes the axis to trace out a cone in space, taking 23,000 years to trace out a complete cone. Consequently, the Northern and Southern Hemispheres "trade" their respective winter and summer seasons every 11,500 years.

Each of these periodic factors alone could cause the limited global cooling that might lead to a "little ice age." More important is the interaction of the three cycles, however. When their cooling effects coincide, there is a sufficient change in the global heat budget to trigger glacial-interglacial episodes, though not enough to explain fully the 4° to 10° C cooling required for the major glacial cycles of the Pleistocene Epoch. In other words, the rotational and orbital factors adequately explain the *timing* of the glacial-interglacial cycles, but not the amount necessary to cause the *changes in global temperature*. There must be other factors.

In order for glaciers to exist, the continents must be in the middle to higher latitudes, and high-standing mountains are "desirable." Plate tectonics can account for both the positioning of continents at "appropriate" latitudes and the uplifting of vast continental areas. Such repositioning of continents and higher elevations of land areas alters the circulation patterns of both the atmosphere and oceans, which leads to long-term climate change. In addition, the higher the continental land masses, the greater is the rock surface area exposed to chemical weathering by atmospheric gases. As the rocks weather, they combine with water and atmospheric carbon dioxide (CO_2), which reduces the amount of the gas in the atmosphere. Further, input of CO_2 from volcanic activity is reduced when, at the end of a tectonic cycle, the volcanic fires at oceanic ridges cool down and the ridges become less active. Because CO_2 is the major greenhouse gas, these reductions promote global cooling. (In the earlier phases of the cycle, the ridges are active, which increases the CO_2 input to the atmosphere, and sea level is high, which reduces weathering. These two factors reinforce each other to increase atmospheric CO_2.) The relationship of CO_2 concentrations in the atmosphere, oceans, and biosphere is enormously complicated, and several feedback mechanisms exist that are not understood (Case Study 12.5). For example, diatoms, the microscopic oceanic plants that can be thought of as the "grass of the sea," may be part of a feedback loop in the biosphere that affects climate. Diatoms may step up productivity when there is excess atmospheric CO_2, thus drawing down the CO_2 content and cooling the climate.

Glacial ice provides evidence of cyclic changes in atmospheric chemistry, temperature, and the amount of atmospheric dust corresponding to cooling and warming periods during the Pleistocene Epoch. As snow accumulates and packs down into ice, it carries information with it that gets locked in the glaciers. The relative abundance of two slightly different oxygen atoms—the oxygen-18 to oxygen-16 isotope ratio, symbolized $^{18}O/^{16}O$—in the ice

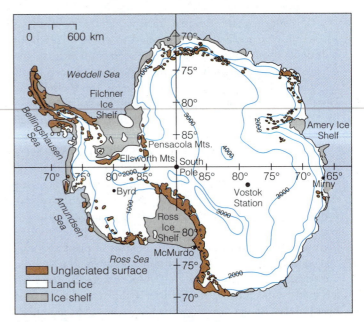

(a)

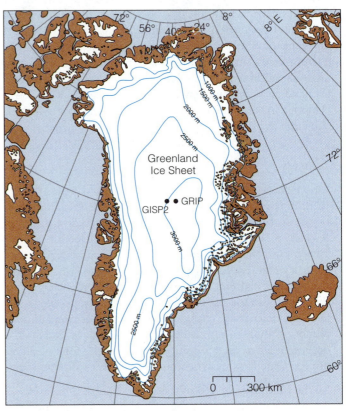

(b)

● FIGURE 12.33 The earth's two continental glaciers. *(a)* The largest glacier complex covers practically all of Antarctica, a continent that is more than half again as large as the 48 states. The ice sheet's thickness averages about 2,160 meters (7,085 ft), with a maximum of 5,000 meters (16,500 ft). *(b)* Greenland's ice sheet has a maximum thickness of 3,350 meters (10,988 ft). Ice thicknesses are contoured in meters. Vostok, GRIP, and GISP2 are the sites of core drilling projects discussed in the text.

molecules indicate what the temperature was when the snow fell. This is because when sea water evaporates from a cold ocean and precipitates on land to form glaciers, water containing the lighter isotope, ^{16}O, is more readily evaporated than is water containing the heavier ^{18}O. Thus, the oceans come to contain more of the heavy isotope, and glaciers to contain more of the light isotope. Furthermore, each year's snowfall forms an easily recognized line in glacial ice. When core samples of glaciers are examined, these lines act like tree rings, allowing scientists to determine the years, and the $^{18}O/^{16}O$ ratios allow each year's temperature to be estimated. Together, these data allow a generalized overview and time scale of global climate change.

The bubbles of ancient atmosphere trapped in glacial ice obtained from 3,348-meter (10,981-ft)-deep cores drilled at Vostok Station in Antarctica, and the Greenland Ice Sheet Project 2 (GISP2) (◉ Figure 12.33, at left) provide a 400,000-year record of unique information about variations in the atmosphere's temperature and composition. The ice-core record and correlative deep-sea-sediment cores show that ice ages are marked by climatic jiggles—intricate cycling within the larger cycles of glacial advance and retreat due to orbital forcing by the Croll–Milankovitch effect. Peculiarities in the lower part of the GISP2 ice core,

presumably representing atmospheric and climatic conditions existing about 120,000 years ago, imply abrupt reversals every 2,000 or so years between glacial climate and warm interglacial climate, with the change occurring in as little as three years. Another study, the Greenland Ice-Core Project (GRIP), drilled about 11 kilometers (7 mi) away from GISP2, provided similar information. The climate interpretations of the upper part of the Vostok core and the two Greenland cores agree, but the wild changes portrayed by the Greenland cores are not seen in the Vostok core. The analysis is far from complete, and there is some evidence that the GISP2 and GRIP cores have been disturbed by intermittent melting and by layering distortions from nearby bedrock. Thus, it is believed that the Vostok record is the only undisturbed ice core that records a full glacial-interglacial cycle—that is, the conditions during the last interglacial period and the glacial period that preceded it (◉ Figure 12.34). The amounts of methane (CH_4) and CO_2, the principal greenhouse gases, in all three cores are lower in the layers accumulated during glacial stages and higher in the layers representing warm interglacial stages. The synchroneity of increasing and decreasing amounts of greenhouse gases with increasing and decreasing average temperature suggests that these changes are related. This

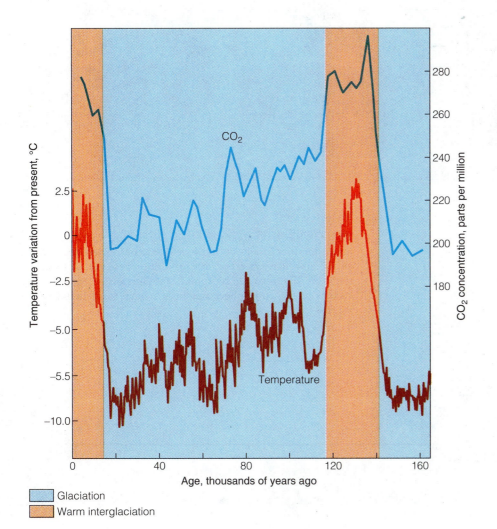

Glaciation
Warm interglaciation

◉ **FIGURE 12.34 Variation in atmospheric CO_2 and air temperature during the last 160,000 years as determined from air bubbles trapped in the Vostok, Antarctica, ice core. The Holocene, the warm interglacial period of the past 9,000 to 10,000 years, was preceeded by more than 150,000 years of generally colder temperatures. Compare this record with the one in Figure 1 of Case Study 12.7 later in the chapter.**

CASE STUDY 12.5

The Carbon Cycle

Complex interactions of the earth's geospheres—the pedosphere, the atmosphere, the hydrosphere, the biosphere, and the lithosphere—produce climate. The interactions are difficult to analyze and they are not well understood. By extending understandings about today's climate into the remote geological past, however, geologists have been able to shed some light on the workings of the climate system. They have reconstructed ancient climates by studying sedimentary layers and deducing their depositional environments, by studying fossil animals and plants, and by carrying out isotope studies that yield information on past temperatures of the earth's oceans and surface. Such *paleostudies* have laid important groundwork for understanding the earth's carbon cycle.

Presently, the earth's reserves of fossil fuels are being burned at a prodigious rate, and the large amount of carbon dioxide (CO_2) being released to the atmosphere by the burning appears to be impacting our planet and its inhabitants. Carbon, the essential element that distinguishes living and once-living matter from other matter, is involved in the changing atmospheric composition that may be causing long-term climate change. Carbon exists in various forms in each of the geospheres:

- The biosphere is the system of all living and some decaying dead organisms, each cell of which contains carbon.
- The hydrosphere contains dissolved carbon dioxide.
- The atmosphere and pedosphere contain CO_2 gas.

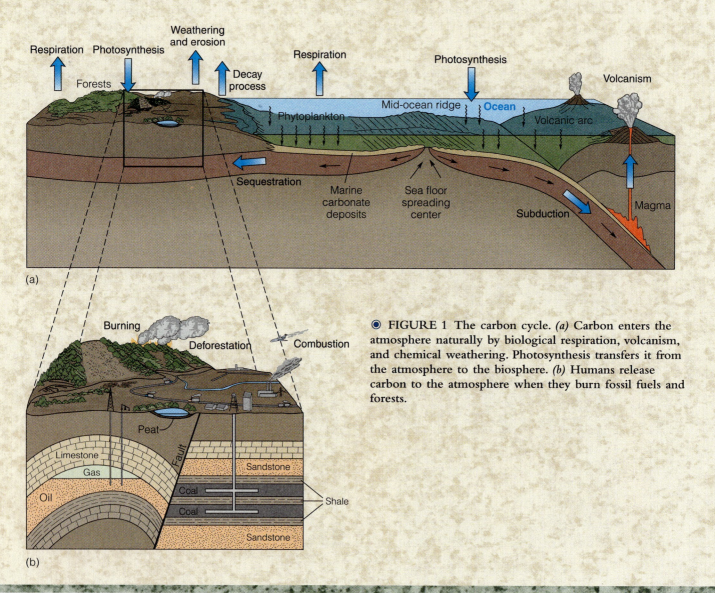

⦿ **FIGURE 1** The carbon cycle. *(a)* Carbon enters the atmosphere naturally by biological respiration, volcanism, and chemical weathering. Photosynthesis transfers it from the atmosphere to the biosphere. *(b)* Humans release carbon to the atmosphere when they burn fossil fuels and forests.

■ Carbon is stored in the lithosphere in deeply buried organic materials, in petroleum and coal, and as calcium carbonate in limestone and marble.

The transfer of carbon among these geospheres is termed the carbon cycle (◉ Figure 1, below left). The carbon cycle is not just the cycling of CO_2; it incorporates all of the earth's carbon compounds, and there are many of them.

The biosphere is the key in the carbon cycle. Plants continuously remove CO_2 from the atmosphere and utilize it, by photosynthesis, to produce their food (sugar), much of which, in turn, is consumed by animals. Plant and animal waste materials contain carbon, which ultimately is carried to the oceans by surface runoff. Dead organisms decay by combining with atmospheric oxygen to form more CO_2, which is then cycled back to the atmosphere. So fast is the cycling of carbon through the biosphere that a complete recycling of atmospheric CO_2 occurs about every four years.

In the ocean, plankton, shellfish, and other marine life use CO_2 and calcium to form shells and skeletons. These materials eventually sink to the sea floor, where their carbon becomes sequestered in calcareous sediment. By the motion of the crustal plates, ocean-floor sediments eventually are cycled into the mantle by subduction. Heat drives the CO_2 out of the sediment, and CO_2 dissolved in magma eventually rises to the surface through volcanism. When a volcano erupts, large amounts of CO_2 are returned to the atmosphere to restart the cycle.

Not all dead organic matter in the biosphere decays to CO_2. Some is sequestered in sedimentary rock, forming reservoirs of carbon in petroleum and coal, the so-called fossil fuels. Some of this material naturally reenters the atmosphere when uplift and erosion expose the rock and its trapped carbon compounds (◉ Figure 2).

Atmospheric CO_2 is the smallest of all the carbon reservoirs, accounting for only about 0.03 percent of the atmosphere. Because of the atmosphere's relatively small carbon content, it is very sensitive to changes in the larger carbon reservoirs with which it exchanges. Of special concern is the recent acceleration of human-induced CO_2 production by the clearing of forests and the burning of fossil fuels. Both of these release CO_2 to the atmosphere from the biosphere and the lithosphere much faster than would occur naturally. The rate at which natural processes remove CO_2 from the atmosphere has not changed, however, and that is why there is a buildup of CO_2 in the atmosphere. Accordingly, this anthropogenic atmospheric-CO_2 imbalance to the greenhouse effect is the focus of the concern about global warming.

Despite the availability of high-speed computers and our ability to do sophisticated computer modeling, our knowledge about the carbon cycle is remarkably scant. Scientists have yet to

◉ FIGURE 2 These mountains in the Montana Rockies are composed largely of ancient carbonates that were formed as sea-floor sediments about 1.5 billion years ago. Such rock bodies constitute reservoirs of carbon that have been isolated from the carbon cycle. Combined uplift, erosion, and weathering processes are now slowly releasing the long-sequestered carbon back into the carbon cycle.

account for about 25 percent of the sequestered carbon. Apparently, it is not in the ocean, but somewhere in the terrestrial biosphere, that is, in land-living plants and animals. Solving the puzzle of the "missing" carbon might be a big step toward solving the bigger global-climate-change problem. Such an effort will require the collaboration of paleoclimatologists, ecologists, ecosystem modelers, geochemists, and others. Such coordinated problem-solving is the key to understanding the carbon cycle and global change.

correlation seems compelling, but a cause-and-effect relationship is still unknown: does the change in atmospheric gas content initiate a glacial stage, or does the global cooling of a glacial stage cause a change in the gas content? That greenhouse gases play a significant role in the glacial-interglacial cycle seems obvious, but it is still uncertain precisely what causes their percentages to change. On a human time scale, the ice reveals that since preindustrial times, the level of CO_2 has increased by about 40 percent, and CH_4 has increased more than 100 percent.

The Vostok core also reveals that the atmosphere was dusty during glacial stages. Winds, lifting fine dust from glacial deposits and from arid regions, caused a hazy sky, which would have scattered incoming solar radiation back into space, causing further cooling of the earth.

Volcanic activity has been suggested as a contributor to global cooling, but it probably could not have caused the ice ages. The summers of 1992 and 1993 were a bit cooler than normal, for example, and the cooling was attributed to the dust and sulfur dioxide (SO_2) gas released in the 1991 eruption of Mount Pinatubo in the Philippines (see Chapter 5). The SO_2 combined with water to form tiny droplets of sulfuric acid, which acted as little mirrors and reflected solar energy back into space. This reduced the amount of solar energy the earth received. Consequently, global temperatures averaged $0.5°C$ ($1°F$) cooler by the summer of 1992, and continued cool in 1993. In 1994 temperatures returned to "normal," and the pattern of global warming resumed. Similar reductions in average global temperatures following eruptions of other volcanoes are recognized, such as after the eruptions of Tambora in Indonesia in 1815 and El Chichón in Mexico in 1982.

Vagaries in ocean currents, too, seem to play a role in plunging the earth into glacial and interglacial climates (see Case Studies 12.6, page 351 and 12.7, page 353).

Global Warming

The earth's surface is warmed by radiant energy from the sun. Part of the sun's radiation is reflected back to space by clouds and the atmosphere so that about 51 percent of the solar energy that could reach the earth actually does. This energy received from the sun is visible light and near-infrared radiation, particular wavelengths of solar radiation that do not "see" the atmosphere through which they pass on their way to earth, where they heat the ground and oceans. This incoming solar radiation is sometimes referred to as *shortwave radiation*. Because all warm objects radiate heat, the warm surface of the earth then radiates energy back to the atmosphere, but at longer wavelengths known as *infrared*, or *longwave, radiation*. The atmosphere is warmed by this longwave radiation due mainly to the absorption and reradiation of heat by atmospheric CO_2, water vapor, and other trace gases. Consequently, the atmosphere acts as a blanket that absorbs most of the longwave radiation and keeps the earth warm. This process, called the **greenhouse**

CONSIDER THIS . . .

Considering what you have learned about the increasing concentration of greenhouse gases in the atmosphere, what changes might you expect over the next 50 years where you live if the global temperature should rise by 2° C?

effect by analogy to the solar-heating method that warms greenhouses, serves to keep the earth's surface temperatures from dropping excessively during winter and at night.

Scientists have been measuring and recording the level of CO_2 in the atmosphere since 1880. Their measurements over the years show that the concentrations of CO_2 and other greenhouse gases as well as average global temperatures have risen steadily, a phenomenon known as **global warming** (Figure 12.35). This increase in atmospheric CO_2 is unprecedented in human history and is believed to be largely due to humankind's excessive use of fossil fuels. Other greenhouse gases—including methane (CH_4), which is expelled as flatulence from the world's cattle population, from rice paddies, and of all things, from termites in the tropical rainforests, and nitrogen oxides from fertilizers and automobile exhausts—combine to increase global warming.

IS GLOBAL CLIMATE CHANGING TODAY?

For every complex problem there is always a simple answer—and it's usually wrong.

H. L. MENCKEN (1880–1956), AMERICAN EDITOR

Information from Glaciers

Deposits of Pleistocene glaciers and ocean-floor sediments are providing key information on major climate changes over the past 2 to 3 million years. Modern glaciers also provide relevant information, because they respond to climate changes by either advancing or retreating. Their behavior serves as an indicator of climate change and also as a "filter," smoothing out the record of seasonal and annual variations in temperature and precipitation.

Because of their small size and ice volume, alpine glaciers are remarkably sensitive indicators of climate change; they respond quickly even to small perturbations in climatic elements. Data on the behavior of alpine glaciers (Table 12.3, page 354) reveal that the number of advancing glaciers has decreased abruptly in Washington State, Italy, Switzerland, and Norway. For example, whereas all eight glaciers on Mount Baker in the North Cascades of Washington were advancing in 1976, by 1990, all eight were

The Oceanic Conveyor Belt

A giant, salty "conveyor-belt" current snakes through the world's oceans (●Figure 1). Originating in the North Atlantic by northward-flowing warm, salty (and thus dense) water because of evaporation, the belt is chilled as it moves northward, which further increases its density. It sinks into the depths flowing southward out of the Atlantic and into the other oceans, much of it rising in the Pacific, where it offsets the evaporation losses there. This conveyor is the main agent of oceanic heat transfer, with one part of it a deep, cold, salty (and thus dense) current (shown in green in the figure), and the other part of it a warm, less salty, surface current (shown in yellow). The northward-flowing surface water averages 8°C warmer than the cold, deep water flowing south.

When the conveyor is operating, the climate of the North Atlantic is mild, like that of today. It is vulnerable to abrupt disruptions and shutdowns, however, which can cause drastic shifts in global climate (see Case Study 12.7) and force the North Atlantic into bitter, ice-age cold.

When excessively large volumes of cold fresh water from melting icebergs and river runoff enter the North Atlantic, the conveyor shuts down and winter temperatures drop by five or more degrees.

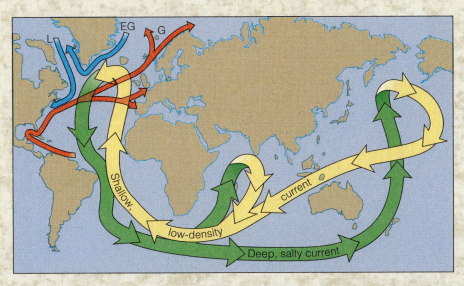

● FIGURE 1 The giant "conveyor belt" current that snakes through the world's oceans. Two surface currents of major importance in determining the modern climates in Greenland, eastern North America, and western Europe are shown. The cold Labrador (L) and East Greenland (EG) currents (both shown in blue) cool the adjacent coasts. The Gulf Stream (G, red) transfers heat northward, influencing Arctic air currents and warming the coasts of Iceland, Greenland, and western Europe. This current was deflected southward toward North Africa during the coldest part of the last maximum glaciation. The importance of these currents, in combination with the atmospheric wind system, is illustrated by comparing the modern equitable temperature of northern Scandinavia, which is essentially unglaciated, with that of Greenland, which, at the same latitude, is covered by an enormous ice sheet.

retreating. Four of 47 alpine glaciers studied in the North Cascades have ceased to exist since 1984. The Palisade Glacier, the largest glacier in California's Sierra Nevada, retreated a third of a mile between 1940 and 1980 and is still retreating. Swiss glaciologists estimate that glaciers in the Alps have lost as much as half their mass since 1850. These glaciers are now responding to the higher average global temperature between 1977 and 1990, which was 0.4°C (0.7°F) above the 1940–1976 mean. No significant change in the behavior of Alaskan alpine glaciers was noted in that time interval, but that is assumed to be due to their much greater size, which results in a greater lag time in their response to climate variation.

Information from the Ozone Layer

In addition to their being greenhouse gases, **chlorofluorocarbons (CFCs)**, the compounds used as coolants in refrig-

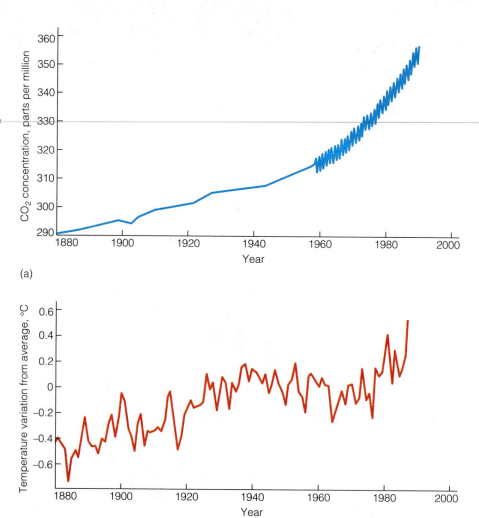

● FIGURE 12.35 Atmospheric CO_2 and temperature changes since 1880. *(a)* Changes in concentration of CO_2 obtained from ice cores (smooth curve) and measured annual oscillations at Mauna Loa, Hawaii. *(b)* Average annual surface-air temperature at Mauna Loa, plotted as a deviation from the average, denoted as zero (0).

erators and air conditioners and formerly as propellants in aerosol-spray cans, break down the upper-atmospheric (the stratospheric) ozone layer 10 to 48 kilometers (6 to 30 mi) above the earth's surface. Ozone, a form of oxygen, serves as a vital shield that protects the earth's organisms from the harmful solar ultraviolet radiation that bombards the earth. The chlorine in one molecule of CFC has the potential to destroy as many as 100,000 ozone molecules before it becomes inert. Any thinning of the ozone shield leads to global change.

Using data from satellites and airborne sensors, scientists discovered in the late 1970s and early 1980s that atmospheric ozone levels were decreasing at an alarming rate. By the spring of 1987, an "ozone hole" the size of the United States and Mexico combined had developed over Antarctica. The hole has continued to enlarge, appearing to document the continuing effects of human activity on earth. The hole is seasonal; it is larger in the Antarctic spring and summer, the seasons when plants and animals face the greatest danger from ultraviolet rays. Ozone thinning is now recognized over the North Pole and the middle latitudes as well. Ozone loss in the stratosphere causes warm-

ing on the earth's surface as well as dangerous cooling in the stratosphere.

A thinning of the atmospheric ozone layer threatens humans with increased risks of skin cancer and cataracts and of weakening the immune system. The increased ultraviolet radiation reaching the earth also threatens to damage crops and phytoplankton, the microscopic plants that are the basis of ocean food chains. Some medical authorities have reported that a person's chance of developing skin cancer in the 1990s is about one in 75, up from one in 1,500 about 40 years earlier. Humans, of course, can protect themselves from ultraviolet radiation, but other animals cannot. For example, ultraviolet-radiation damage to DNA has been shown to account for severe losses of fertilized eggs in at least two species of Cascade Mountain frogs.

Also contributing to the depletion of the ozone layer is the burning of aviation fuel at altitudes above 9 kilometers (5.5 mi). Nitrogen oxides (NO_x) and water-vapor molecules released at altitudes between 9 and 13 kilometers (5.5 and 8 mi)—the usual cruising altitude for commercial jetliners—stay in the atmosphere about 100 times longer than those released near ground level. Although it seems contradictory,

Is There a New Beat to the Rhythm of the Ice Ages?

If you think today's weather (Table 12.4) seems weird, try to imagine weather conditions during the last ice age. Oceanographers' studies of sea-floor sediments have revealed that massive fleets of icebergs have moved across the North Atlantic Ocean every 2,000 to 3,000 years as the climate has flip-flopped back and forth between warm interglacial and glacial periods. The melting icebergs have chilled the North Atlantic and left trails of ground-up rock debris in layers on the ocean floor. After two or three such cycles there would be a larger flood of icebergs, which left a layer with a different composition. These greater floods, occurring every 7,000 to 12,000 years, have become known as *Heinrich events* (for the German oceanographer who discovered them), and the smaller, more frequent events are informally called *flickers*.

It is now believed that the Heinrich events, recognized by layers of carbonate-rock grains in the sea-floor rock debris, resulted when the giant Keewatin ice sheet (Figure 12.14, part a) surged into Canada's Hudson Strait, causing enormous numbers of bergs to calve and drift into the North Atlantic. The more numerous flickers, in contrast, left layers of reddish, iron-coated sand on the ocean floor that has been traced to the present-day St. Lawrence Valley, which also was ice-covered at the time. The significance of the Heinrich and flicker events in the oceanic record is their implication that in the late stages of the Pleistocene ice ages, the Northern Hemisphere experienced numerous rapid climate changes unlike anything seen in historic times (◉ Figure 1).

Meanwhile in Greenland, glaciologists' studies of oxygen isotope ratios in ice cores have revealed wild flip-flops in climate from bitter cold to warm, between 115,000 and 140,000 years ago (an interval known as the *Eemian*), with the same timing as the Heinrich and flicker events recorded in the sea-floor rocks. What does this all mean? One hypothesis is that the St. Lawrence ice sheet accumulated and surged several times as the Keewatin ice sheet was growing, and

that when the Keewatin ice sheet surged, the ice over the St. Lawrence did also. The two ice sheets, each acting independently and having different characteristics, must have surged as a consequence of climate warming. Variable currents in the North Atlantic—perhaps related to abrupt shutdowns in the oceanic conveyor system (Case Study 12.6), brief ones of a century or less—may cause temperature swings every 2,000 to 3,000 years in a cycle that may run all the time, even today.

The debate continues on the explanations for the cycles of sediment deposition, the oceanic conveyor system, rapid swings in climate, and the discrepancies in the Greenland and Antarctic ice-core data. Conflicting data from different sources may drive the scientific community to develop more complicated, yet more

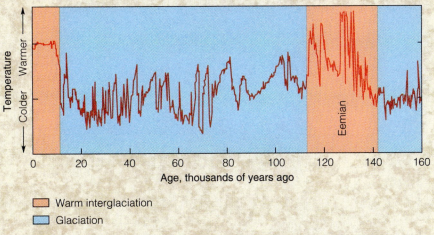

◉ **FIGURE 1** The continuous GRIP temperature record (based on oxygen-18/oxygen-16 ratios) from the Greenland ice core. This record shows that during the Eemian interglacial interval, generally warm conditions suddenly gave way to drastic cold on several occasions. Compare the Eemian of this record with that of the Vostok core from Antarctica (Figure 12.34).

realistic, views of the earth's climate and why it changes. With the earth expected to be bulging with people in 50 to 100 years, we should spare no effort in deciphering the seemingly chaotic behavior of the global climate system.

	NUMBER OF GLACIERS		
	Advancing	*Stationary*	*Retreating*
Switzerland			
1980	70	3	28
1983	46	6	57
1986	42	9	13
1987	35	13	58
1989	19	3	83
1993	6	0	73
North Cascades			
1974	9	0	1
1984	5	11	31
1988	2	9	36
1990	1	5	41
1992	1	2	44
1995	0	0	47
Norway			
1980	5	0	7
1985	1	1	10
Alaska			
1976	15	18	18
1985	16	14	24
1994	15	12	27
Italy			
1981	25	10	10
1985	25	6	14
1988	26	13	92
1989	11	5	106
1990	9	9	123

SOURCES: Mauri Pelto, Director of the North Cascade Glacier Climate Project; Swiss Glacier Commission; Norwegian Water Resources and Electricity Board; U.S.G.S.; and *Ice* 98, no. 1, p. 3.

NO_x emissions can either increase or decrease ozone concentrations in the atmosphere. In the lower atmosphere, where we live, NO_x, together with hydrocarbons, forms ozone (a major component of smog) in the presence of ultraviolet light. In the upper atmosphere, however, NO_x breaks down ozone. CFC climate modeling, which does not take NO_x emissions from air traffic into account, has underestimated the amount of ozone depletion. According to researchers, the amount of ozone loss over the Northern Hemisphere is about twice that calculated using only CFC climate modeling. Regular (subsonic) air traffic is believed to be responsible for a 10 percent increase in the nitric acid concentration at 21 kilometers (13 mi) above sea level, which may be the most critical altitude for ozone holes. Clearly, the chemistry of the atmosphere is complex, and it is the subject of continuing scientific research.

Some people, notably a few radio talk-show hosts and some government policymakers, have expressed doubts about the danger of CFCs and the depletion of the ozone layer. They argue that chlorine from CFCs is trivial in comparison to the great quantities of chlorine released to the atmosphere from natural sources such as volcanic eruptions, which have been contributing chlorine to the atmosphere for many hundreds of millions of years. In spite of this the ozone layer still exists, so, they argue, the concern about CFCs is overstated. Their argument is spurious, however, because natural chlorine is chemically reactive in the lower atmosphere and soluble in water; consequently it gets rained out before it can reach the stratosphere. CFCs, in contrast, are absolutely inert and thus insoluble in the lower atmosphere. This allows them to rise through the lower atmosphere and into the stratosphere, where ultraviolet rays cause the release of chlorine. It was recognition of their chemical inactivity in the lower atmosphere—they react with *nothing*—that made them so appealing as coolants and propellants.

By 1987, the concern of the international scientific community was so great that the Montreal Protocol on Substances that Deplete the Ozone Layer was organized. The group's initial goal was simply to control the production and use of CFCs in refrigeration systems and as aerosol propellants. Upon learning of the acceleration of ozone depletion between 1990 and 1992 and that North Pole ozone also was being depleted, the group asked for a total ban on CFC production by 31 December 2000. As evidence of CFCs' effects on the ozone layer mounted, then-President Bush called for an accelerated phasing-out of CFC manufacture in the United States by the end of 1995. In November 1992, another meeting of environmentalists and scientists from more than 80 nations agreed on a new plan for meeting the year-2000 phaseout deadline.

The chemical industry has conducted vigorous research and development of CFC substitutes. Compounds that have the useful properties of CFCs but that lack chlorine or decompose in the lower atmosphere include the hydrofluorocarbons (HFCs) and the hydrochlorofluorocarbons (HCFCs). The hydrofluorocarbon HFC-134a is now replacing CFC R-12 (known commercially as *Freon*) in refrigerators and automobile air conditioners, and HCFC-22 is being used in air conditioners and as a puffing agent in Styrofoam manufacture. Because HFC-134a reacts with hydrogen and oxygen in the lower atmosphere to form trifluoroacetic acid, which can return to the earth's surface as acid rain, questions have been raised about its desirability. Although the concentrations of the acid do not appear to be immediately harmful, sufficient amounts of the acid could build to harmful levels in tundra or seasonal wetlands, which remain moist for part of the year and then lose their moisture by evaporation.

Freon, widely used for decades as a refrigerant in automobile air conditioners, has been a major contributor to CFC pollution of the upper atmosphere. For some time in the United States, all repairs to automobile air-conditioning systems have been required to be done by certified mechanics, and the Freon that is bled from the systems has been required to be collected and recycled, not vented to the atmosphere as in the past. U.S. production of CFCs ended in December 1995, and HFCs and HCFCs are now being used. Worldwide production of CFCs, by international agreement, will end by 2000. The changeover requires that both refrigeration units and manufacturing facilities be modified to use the new refrigerants. Older refrigerators and air conditioners presently using CFCs are to be recharged only with recycled CFCs, but eventually they will need retrofitting to use the newer refrigerants.

Because of the approaching deadline banning all manufacture of CFCs, a thriving black market in contraband CFC-12 has developed, with Miami as the central point for U.S. entry and distribution. In the ranking of contraband items by estimated street value, CFCs are second to illicit drugs in the United States. To encourage the use of alternatives, a federal tax is charged on legal CFC sales, and by the mid 1990s prices had soared to $15 a pound from $1 in 1989. The smugglers buy the CFCs from Russia and other countries and sell it in the United States for less than the taxed product. By 1995 an estimated 10,000 tons of black-market refrigerant was being smuggled into the United States each year.

Whether or not the use of CFC-12 is harmful to the ozone layer may become academic depending on the action of the Congress. At a hearing in the summer of 1995 the House Science Subcommittee on Energy and the Environment attempted to decide whether scientists have overstated the case for CFC damage to the ozone layer. Subsequently, legislation was introduced in the House of Representatives to repeal the ban on CFCs.

Weather's Puzzling Signals

In August 1995 a record-breaking dry spell in the Northeast brought drought warnings to New York City, Westchester County, and the Catskills. The following winter brought record-breaking snowfall to the entire region. Similar puzzling weather anomalies (● Table 12.4) and their consequences have been episodic from the mid 1980s through the mid 1990s, a decade that included wide swings in the weather from cold winters (except for a peculiarly warm one in 1995) in New England to warm winters in Alaska, a six-year drought in California, a bitter cold snap for the 1994 Winter Olympics in Norway, the record-breaking flooding in the Mississippi River valley in 1993, and summer heat waves in central Europe. Worldwide, the warmest ten years on record occurred in the last 15 years in spite of cooling due to the eruption of Mount Pinatubo.

■ **TABLE 12.4 Weather Extremes of 1994–1996**

LOCATION	EVENT
Northeast New Jersey	100 municipalities imposed water-use restrictions in August 1995, due to prolonged drought
Susquehanna River basin (stretching from Maryland to New York)	below-normal precipitation for 9 of 11 months in 1994–1995 forced water-use restrictions
Long Island, New York	brush fires in August 1995
Chicago, Illinois	severe 8-day heat wave in July 1995, with temperatures of 41° C (106° F) and relative humidities at 88%
Tibet	early snowmelt in spring 1995
British Isles	record-breaking drought in summer 1995
Himalaya Mountains	forest fires in summer 1995
California	severe flooding in January 1995; heaviest snowpack in decades in mountains in 1995
Eastern North America	unseasonally warm December 1994–January 1995; record snowfall in winter of 1996
Tropical Pacific Ocean	3 of the 4 winters between 1992 and 1995, unusually warm near the international date line in the central Pacific
Worldwide	warmest May–December in history, 1994
Oregon	severe flooding, winter 1996
Southern plains and southwestern United States	drought conditions during much of 1996.

Are such weather extremes likely to continue, and do they indicate global warming, or are they just coincidence? Scientifically, it is hard to tell, because climate is fickle. This episode of warming could be part of the natural changes that scientists have traced over the millennia as the earth has cooled, then warmed, then cooled again. (Case Study 12.7). Weather patterns change naturally in the short term, and climate changes naturally in the long term. Thus, any human-induced contributions to climate change are superposed on natural changes.

Living with Uncertainty

Atmospheric scientists are still uncertain about why climate changes, and they lack techniques for distinguishing between long-range climate trends and short-term weather events. "The temptation . . . is often to attribute any of these temporal and sometimes local variations to a wider and more pervasive change in climate," note several atmospheric scientists with the National Oceanic and Atmospheric Administration (NOAA). A respected senior climate scientist has shown that U.S. weather has become more extreme since 1980 and has estimated with 80 percent certainty that the extremes are due to human-induced greenhouse warming. In the case of greenhouse warming, built-in delays in the earth's climate system compound the potential dangers of waiting for 100 percent certainty; decades are required to warm or cool the atmosphere and oceans. If the world's policymakers wish to head off the potential for greater climate changes, they must act before the major effects of greenhouse-gas pollution are apparent. Regrettably, global warming has become political, and communication between scientists and policymakers is sometimes difficult (◉ Figure 12.36). If it *is* occurring, it presents a dilemma. If it is due to human activity, we must take measures now to slow or halt it, and enormous up-

front costs will be required for mediating the impact. But the determination of certainty must be based on the uncertainty of the past record, and any actions taken may prove to be expensive mistakes if humans are not the primary cause for the apparent climate change.

Some climatologists are skeptical about global warming, and some of them doubt that it is occurring. Others feel that warming is indeed happening, but believe that the apocalyptic projections for the future are overblown. Supporting this latter view are satellite observations of the average surface temperature of the earth collected since the late 1970s, which do not indicate any rise in global temperature. The satellite observations measure the temperature of the entire planet, whereas the warming trend frequently reported by the news media is based on weather-station data, which is collected only on continents and islands. It is the land-based data that have suggested that the years since the mid 1980s have been progressively warmer. In addition, the amount of warming reflected in that data is much less than the amount that had been predicted by widely accepted computer-generated climate models.

Factors suggested for the variation of the measured warming from the predicted warming include these:

1. The amount of CO_2 uptake by oceanic organisms is a significant unknown and is not readily measurable.
2. Trees and plants in the Northern Hemisphere's temperate latitudes (30°–60°) were found to absorb and store about half of the global fossil-fuel emissions (as determined by measurements of two isotopes of carbon in the atmosphere during a study in 1992 and 1993). The data suggest that forests and other plant life are a more important terrestrial "sink" for carbon than previously thought.
3. Computer-generated models do not treat cloud cover satisfactorily. The models used for predicting the earth's

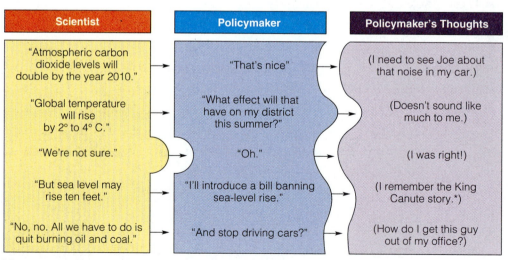

◉ FIGURE 12.36 A hypothetical scientist briefs a hypothetical policymaker on the dangers of the greenhouse effect.

Scientist	Policymaker	Policymaker's Thoughts
"Atmospheric carbon dioxide levels will double by the year 2010."	"That's nice"	(I need to see Joe about that noise in my car.)
"Global temperature will rise by 2° to 4° C."	"What effect will that have on my district this summer?"	(Doesn't sound like much to me.)
"We're not sure."	"Oh."	(I was right!)
"But sea level may rise ten feet."	"I'll introduce a bill banning sea-level rise."	(I remember the King Canute story.*)
"No, no. All we have to do is quit burning oil and coal."	"And stop driving cars?"	(How do I get this guy out of my office?)

*Canute was a king of Denmark, England, and Norway, 1016–1035, who supposedly commanded the oceanic tide to stop rising.

future climate are based on assumptions of increasing concentrations in atmospheric CO_2. Presumably, the increase in CO_2 will initially warm the atmosphere, permitting it to hold more water vapor. The increase in water vapor will increase cloud cover. The cooling effect of additional cloud cover, the types of clouds formed, and so on, are uncertain factors, which computer models are unable to deal with effectively.

4. Computer models are unable to predict polar climate variability.

5. Computer models are unable to predict the potential responses of the oceans to global warming—such things as biological changes, changes in ocean heat content, and others.

For the sake of evaluation let us examine a scenario developed by one computer model. In October 1994 the Forum on Global Change Modeling arrived at the following consensus of anticipated global changes based on the assumption of continuing anthropogenic (human-caused) increases in greenhouse gases. Most of the group's computer modeling experiments are based on a 1 percent per year increase in CO_2.

Very Probable:

■ Globally, the mean surface temperature will increase by the mid twenty-first century by about 0.5° to 2° C (0.9° to 3.6° F). Such warming will be due to the concentrations of greenhouse gases alone, assuming that no significant actions are taken to reduce these gases. If the CO_2 concentrations in the atmosphere should double, there is a range of estimates for the warming that will result. The best estimate is 1.5° to 4.5° C (2.7° to 8° F), and the most probable estimate is 2.5° C (4.5° F).

■ Globally, mean precipitation will increase, but how it will be distributed is uncertain. A warming of the earth's surface temperatures will increase global mean precipitation because of the well-established relationship between evaporation rates and surface temperature.

■ Northern Hemisphere sea ice will be reduced. The projected effects and timing of changes in Southern Hemisphere sea ice are uncertain.

■ Arctic land areas will experience wintertime warming. Paleoclimate studies and computer modeling offer evidence for polar warming and reduction of snow cover on land areas. The magnitude of the warming is uncertain because of the complexity in understanding all of the variables.

■ Globally, sea level will rise at an increasing rate, although the rate of rise may not be significantly greater than at present. The most reasonable estimate for the rate of sea level rise is 5–40 centimeters (2–16 in) by 2050, as compared to a rise of 5–12 centimeters (2–4.7 in) if the rates of the last century continue. Incorporated in the modeling is a component due to seawater expansion, which is closely dependent on the amount of atmospheric warming, and an estimate of water derived from the retreat of mountain glaciers.

The volume of water contributed by the melting of the polar ice sheets is less certain. If global temperatures increase by 3° C by the year 2050, as some climatologists believe, it is estimated that meltwater from mountain glaciers alone will raise sea level by another 20 centimeters (8 in).

Probable:

■ Precipitation at high latitudes will increase, with potential feedback effects related to the influence of additional fresh water to the oceans causing changes in oceanic circulation. There will be increased precipitation on the polar ice caps.

■ Summer dryness will increase in the mid latitudes of the Northern Hemisphere. Evaporation increases strongly with increasing temperatures. Soil moisture in the Northern Hemisphere could decrease by perhaps 40 percent if greenhouse gases continue to increase at predicted rates. This would make "dry" farming impossible or nearly so in much of North America and Eruope.

■ Episodic, explosive volcanic eruptions will cause short-term relative cooling of a few tenths of a degree lasting up to a few years.

Uncertain:

■ Details of the climate changes over the next 25 years are uncertain.

■ Biosphere–climate feedbacks are expected, but whether these feedbacks will modify or amplify climate change is uncertain.

■ Changes in climate variability will occur.

■ Regional-scale climate changes will differ from the global averages. Unfortunately, there is limited capability to estimate how various regions will respond to climate change.

■ Tropical-storm intensity may change. An increase in intensity is likely but uncertain because of the potential changes in poleward heat flow and other conditions.

Conclusion

You have to live with uncertainties. A lot of people don't like that.

BRUCE MURRAY, JPL/CAL TECH PLANETARY SCIENTIST

Let us assume that global warming is occurring and that its cause is largely anthropogenic. Can anything be done to slow or halt it? One effort, in August 1995 at a meeting of representatives of some 150 countries in Geneva, Switzerland, was to debate the latest evidence on global warming and work out details for a United Nations agreement on climate change. Among the many topics considered were setting national CO_2-emission quotas and the United States' ability to meet the agreed-upon goal of reducing greenhouse gases to the 1990 level by the year 2000. At the time of the meeting the United States remained 30 percent short of that goal. It was recognized that such a reduction

Wonders of Glaciation

Some of the world's most-visited scenic areas are popular because of the spectacular landscapes that remain as superlative legacies of the work of Pleistocene glaciers. Figures 12.37 to 12.41 are a collage of photographs exemplifying the beauties of glaciers and glacial landscapes.

◉ FIGURE 12.37 (right) The Vaughan Lewis Glacier, a valley glacier, Alaska.

◉ FIGURE 12.38 A U-shaped glacial valley sculptured by an Ice Age valley glacier; Kern Canyon, Sequoia–Kings Canyon National Park, California.

⦿ FIGURE 12.39 A glacial "erratic" that was transported by the Pleistocene valley glacier that occupied Lee Vining Canyon, California. This erratic boulder's location is more than 25 kilometers (15 mi) from the nearest outcropping of the particular bedrock type that could have yielded it.

⦿ FIGURE 12.40 A cirque, an armchair-shaped depression at the head of a U-shaped valley, viewed from the summit of 4,313-meter (14,150-ft)-high Mount Sneffels, Colorado.

⦿ FIGURE 12.41 Malaspina Glacier, a 2,200-square-kilometer (850-mi²) piedmont glacier, which was formed on Alaska's coastal plain by the discharge of about two dozen valley glaciers. It is the largest glacier in North America. Its contorted pattern is due to its entrained moraine rock debris being subjected to differential flowage when the valley glaciers that nurture the Malaspina have surged (episodic surging).

would have a negative impact on the U.S economy, cause reductions in employment, and negatively influence congressional attitudes toward funding climate and energy research. Many policymakers, as well as some scientists, are cautious and reluctant to support drastic measures because of the large uncertainties in scientific knowledge, such as those considered in the preceding discussions; scientists recognize trends and patterns, but no certainties.

Some researchers believe that most of the proposed global emission-reduction plans will do relatively little to slow greenhouse-gas emissions, because they consist mainly of modest, voluntary policies. Furthermore, the industrialized countries' CO_2 emissions may not be the major problem. The United Nations' economic and science advisors on global climate change predict that by 2025 the developing countries and those that are moving away from centrally planned economies (including the former Soviet-bloc countries) will be responsible for nearly 70 percent of all energy-related CO_2 emissions.

A U.S. trade group representing manufacturers, utilities, mining companies, coal and oil producers, and railroads insisted that reducing industrial emissions by 20 percent or more below the 1990 levels, as proposed at the 1995 Geneva conference, would be disastrous for the U.S. economy. They estimated that cuts of this magnitude could cost an average of 600,000 jobs a year every year until the year 2010.

An innovative alternative to a mandated emissions-reduction program, known as the Joint Implementation (JI) pilot program, was being tested by the U.S. government in cooperation with the private sector. The program creates an international market for emissions permits modeled after the successful U.S. pollution-permit trading program. In this system, a firm reducing greenhouse gas emissions in a foreign country would receive an emissions credit, which could be retained or sold to another company. Regrettably, the House of Representatives eliminated the funding for the JI pilot program for the 1996 fiscal year. Whether the Senate will take steps to restore the funding is uncertain.

Offering hope for mitigating CO_2 buildup is a chemical process that combines the gas with materials such as calcium oxide and magnesium oxide to form stable minerals. The process is viewed as a short-term insurance policy if political pressure resulting from the perceived threat of global warming forces industry to limit CO_2 emissions. The monetary costs of capturing CO_2 on an industrial scale would be very high. For example, it is estimated that removing the gas from coal-fired electric generating plants would double or triple the cost of electricity. Oceanographers have been cautiously exploring the feasibility of dumping finely ground iron particles into the ocean to stimulate the growth of phytoplankton, which, like all plants, metabolize CO_2 by photosynthesis. From initial experiments it is estimated that spreading a half-ton of iron across 100 square kilometers (39 mi^2) of a tropical ocean would stimulate enough plant growth to absorb some 350,000 kilograms (771,800

pounds) of CO_2 from seawater. Performed on a much larger scale, the iron fertilization of seawater could absorb billions of tons of CO_2, offsetting perhaps a third of global CO_2 emissions. The environmental effects of such large-scale "iron fertilization" are difficult to foresee, but the enhanced bloom of plants could enrich the entire oceanic ecosystem.

Despite international conferences and considerable research, we still lack a clear vision of the climatic future. For the present, a widespread conclusion is that scientific theory and evidence show that it is likely that the climate is warming and that it will continue to do so as we add greenhouse gases to the atmosphere. Because of the immense uncertainties, the concern about climate change resolves into a game of chance—risk assessment. Assuming that global warming continues, people, and their policymakers, will ultimately have to decide how much risk they believe they can live with.

Global warming seems to be rapid now, regardless of whether it is natural and anthropogenic or just anthropogenic, and it may become even more rapid due to the warming itself. The means of slowing the process that we have discussed so far are reasonable, but they are moderate. What is really necessary for slowing or stopping it is to halt deforestation, institute a massive program of reforestation, and reduce the global consumption of fossil fuels by 50 percent. Most concern about the effects of potential global warming has focused on the physical consequences it might wreak: the destruction of much of the world's agriculture, widespread death of forests, depletion of surface- and ground-water supplies, and the flooding of low-lying coastal areas. Overall, warming will stress human activities because it will demand rapid adjustment of economic patterns and changes in much of society's infrastructure. There is also a growing indication that the potential effect on human health is no less serious.

In their worst-case scenarios, some global climate models point to the possibility of tens of millions more cases of infectious diseases, as mosquitoes and other pests expand their ranges, and of hundreds of thousands of additional deaths each year due to an increasing incidence of heat waves. Epidemiologists say the most deadly punch of global warming will be delivered to the developing world, the so-called Third World. Their models predict an increase in such diseases as sleeping sickness, malaria, schistosomiasis, and yellow fever. These diseases already afflict more than 600 million people each year, killing more than two million. Some researchers estimate that global warming would increase the populations of the insects and snails that transmit these diseases, as they are cold-blooded insects and invertebrates that respond to subtle changes in temperature. Some scientific speculation links the 1°C increase in Rwanda's average temperature in 1987 with the country's 337 percent increase in the number of malaria cases that year. Medical researchers hope that monitoring vector (disease-carrying organism) populations will enable them to anticipate and combat such potential threats to human life. One global-warming model predicts that the annual mortality rate from heat-related deaths in New York City would increase from the current average of 320 per year (the highest in the U.S.) to 824, and that Montreal, Chicago, Cairo, and many other cities would have similar increases. Air-conditioning might lower the mortality rate in some cities somewhat, but it is unlikely that it would affect the numbers in Karachi or Cairo.

In the short term—that is, on the scale of human lifetimes—it appears that there will be global warming. From a long-term, geological perspective, however, the warming is seen as a brief event and a nonrepeatable one—nonrepeatable because the earth's reserves of fossil fuels will be consumed within the next few hundred years. After they are consumed, atmospheric conditions should eventually return to normal. True, human-induced greenhouse warming may last a thousand years, but ultimately the variations in the earth's orbital geometry will change, and the earth will be thrust into yet another ice age.

 SUMMARY

Wind

DESCRIPTION Movement of air due to unequal heating of the earth's surface.

CAUSES

1. Variations in the heating of the earth at different latitudes because they receive varying amounts of solar heat energy due to the differing angles of the sun's rays.
2. Due to the earth's rotation, the Coriolis effect deflects winds to the right in the Northern Hemisphere and to the left in the Southern Hemisphere.

Air Pressure, Wind, and Climate Belts

POLAR HIGHS Polar regions of high pressure; subsiding air produces variable winds and calms.

WESTERLIES Zones of generally consistent winds lying between 35° and 60° north and south latitudes.

SUBTROPICAL HIGHS Zones of high pressure and subsiding air masses lying between 30° and 35° north and south latitudes; have variable winds and calms and, commonly, clear and sunny skies. Often called the *horse latitudes*, these zones are the regions in which most of the earth's deserts are located.

TRADE WINDS Belts of generally consistent easterly winds lying between 5° and 30° north and south latitudes.

EQUATORIAL LOW AND INTERTROPICAL CONVERGENCE ZONE Zone of variable winds and calms lying between latitudes 5° north and 5° south. Sometimes called *the doldrums*.

Ocean Circulation

DESCRIPTION Each ocean has a characteristic current pattern, the general pattern being clockwise in the Northern Hemisphere and counterclockwise in the Southern Hemisphere. The net effect is the movement of cool water from polar regions toward the equator.

IMPORTANCE Currents moderate coastal climate and weather, and coastal upwelling brings cool water and nutrients that are important for biologic productivity.

Deserts

DESCRIPTION Regions where annual precipitation averages less than 25 centimeters (10 in) and that are so lacking in vegetation as to be incapable of supporting abundant life.

CAUSES

1. High-pressure belts of subsiding, warming air that absorbs water and precludes cloud formation.
2. Isolation from moist maritime air masses by position in deep continental interior.
3. Windward mountain barrier that blocks passage of maritime air.

CLASSIFICATION May be classed into four categories determined by climate belts and a fifth category due to human misuse.

TYPES

1. *Polar deserts,* regions of perpetual cold and low precipitation.
2. *Mid-latitude deserts* are found within the interior of continents in the middle latitudes, remote in distance from the influence of oceans.
3. *Subtropical deserts,* the earth's largest realm of arid regions, lie in and on the equatorial side of subtropical zones of subsiding high-pressure air masses in the western and central portions of continents.
4. *Coastal deserts* occur on the coastal side of a land or mountain barrier in subtropical latitudes. Because they are bordered by ocean, they are cool, humid, and often foggy.
5. *Deserts of infertility,* generally variants of mid-latitude deserts, are not so much due to climate as to human misuse.

Wind as a Geologic Agent

EROSION

1. *Deflation*—earth materials are lifted up and blown away.
2. *Abrasion*—mineral grains are blown against each other and into other objects.

CONTROL OF MIGRATING SAND Accomplished with sand fences, paving, windbreaks, and stabilizing plants.

Glaciers

DESCRIPTION Large masses of ice that form on land where, for a number of years, more snow falls in winter than melts in summer, and which deform and flow due to their own weight because of the force of gravity.

LOCATION High latitudes and high altitudes.

CLASSIFICATION Major types are continental (ice-sheet), ice-field, ice-cap, valley (alpine), and piedmont.

Some Important Glacial Features

END MORAINE Ridge formed at the melting end of a glacier composed of glacially transported rocky debris.

KETTLE LAKES Water-filled, bowl-shaped depressions formed as glaciers retreated.

U-SHAPED CANYON Remains after a valley glacier melts.

FIORD U-shaped canyon drowned by the sea.

Effects of Pleistocene Glaciation

Ice Age glaciers are directly or indirectly responsible for:

1. Some soil conditions, including loess deposits transported by Ice Age winds and some areas of glacial till.
2. Local ground-water conditions.
3. Shore-line configuration due to sea-level changes and isostatic rebound.
4. Human transportation routes.

Ice-Age Climate and Causes of Ice Ages

DESCRIPTION Temperatures 4° to 10°C cooler than at present.

CAUSES A number of interrelated factors contribute to climate changes in ways that are not always clearly understood. They include:

1. Obliquity of earth's axis.
2. Eccentricity of earth's orbit.
3. Precession of earth's axis.
4. Variations in content of atmospheric gases.
5. The "greenhouse effect."
6. Changes in landmass positions due to plate tectonics.
7. Tectonic changes in the elevation of continents.
8. Volume and temperature of the oceans.
9. Changes in ocean currents

Some Model-Based Predictions of Climatic Change

1. Globally, average surface temperatures will increase.
2. Globally, average precipitation will increase, but its distribution is uncertain.
3. Northern Hemisphere ice will decrease; Southern Hemisphere ice may increase.
4. Arctic land areas will experience wintertime warming.
5. Sea level will rise at an increasing rate with drowning of low-level coastal plains.
6. Decreasing soil moisture in the Northern Hemisphere may make farming nearly impossible in much of North America and Europe.
7. Tropical-storm (hurricane) intensity and frequency may increase.

KEY TERMS

ablate	kettle lake
abrasion	mid-latitude desert
calving	moraine
chlorofluorocarbons (CFCs)	obliquity
coastal desert	polar desert
Coriolis effect	polar front
deflation	polar high
desert	precession
desertification	proglacial lake
eccentricity	rain-shadow desert
end moraine	recessional moraine
equatorial low	subpolar low
erratic	subtropical desert
esker	subtropical highs
firn	(horse latitudes)
fiord	surging, glacial
glacial outwash	till, glacial
glacier	upwelling
global warming	ventifact
greenhouse effect	wind
isostatic rebound	zone of flowage

STUDY QUESTIONS

1. What makes the wind blow? Why do winds in the Northern Hemisphere deflect to the right, and winds in the Southern Hemisphere to the left?
2. Why are the world's largest deserts located at or near the horse latitudes?
3. Is it simply drought that is responsible for the desertification and starvation so prevalent in Africa and elsewhere? (*Hint:* Consider carrying capacity, as discussed in Chapter 1.)
4. How do you explain that in some of the driest places in the world—for example, the Devil's Golf Course, at about 75 meters (250 ft) below sea level in Death Valley, and the Qattara Depression, at about 135 meters (443 ft) below sea level in the Sahara Desert— ground water is found only a few centimeters to a meter (3 ft) or so below the salt-encrusted surface?
5. What measures can be taken to halt sand dune migration?
6. Explain the generalization "Glaciers are found at high latitudes and at high altitudes."
7. Glacial ice begins to flow and deform plastically when it reaches a thickness of about 30 meters (100 ft). What is the driving force that causes the ice to flow?
8. Describe the behavior of a glacier in terms of its *budget;* that is, the relationship between accumulation and ablation. How does a glacier behave when it has a balanced budget; a negative budget; a positive budget? What kind of a budget does a surging glacier have?

9. Areas far beyond the limits of the ice sheets were affected by Pleistocene continental glaciation. What are some of the effects? (*Hint:* Consider coastlines, today's arid regions, and modern agricultural regions.)

10. Describe isostatic rebound. What effect has the melting of the great ice sheets had on the crustal regions that were covered by thick masses of ice until 8,000 to 11,000 years ago? What are the effects of this rebound on society?

11. Describe the greenhouse effect. What causes greenhouse warming? What can be done to modify (or even control) greenhouse warming? What are some potential effects of greenhouse warming on human life?

12. What causes ice ages? Will there be more ice ages? Explain.

FURTHER INFORMATION

BOOKS AND PERIODICALS

Andersen, Bjørn G., and Harold W. Borns. 1994. *The Ice Age world*. Oslo: Scandinavian University Press.

Barron, Eric J. 1995. Global researchers assess projections of climate change. *Eos* 76, no. 18: 185–190.

Broecker, Wallace S. 1995. Chaotic climate. *Scientific American* 273, no. 5: 62–68.

Burger, Jack. 1992. New York: Take a geological field trip. *Earth: The Science of our Planet* 1, no. 2; 60–67.

Collier, Richard. 1977. *The war in the desert*. Chicago: Time-Life Books.

Fairbridge, Rhodes. 1960. The changing level of the sea. *Scientific American*. 202, no. 5; 70–79.

Greenland Ice Core Project (GRIP). 1993. Climate instability during the last interglacial period recorded in the GRIP ice core. *Nature*. 364, no. 6434 (15 July): 203–207.

Hambrey, Michael, and Jürg Alean. 1992. *Glaciers*. New York: Cambridge University Press.

Holstrom, David. 1992. Cooling that won't heat up the globe. *The Christian Science monitor*, 9 December, p. 12.

Huber, N. King. 1987. The geologic story of Yosemite valley. *U.S. Geological Survey bulletin* 1595. Washington, D.C.: U.S. Government Printing Office.

Kirk, Ruth. 1983. Of time and ice. *Glacier Bay: Official National Park handbook*. Washington, D.C.: Division of Publications, National Park Service, U.S. Department of the Interior, pp. 22–103.

Mayo, L. 1988. Advance of Hubbard Glacier and closure of Russel Fjord, Alaska: Environmental effects and hazards of the Yakutat area. *Geologic studies in Alaska by the United States Geological Survey during 1987*. U.S. Geological Survey circular 1016, pp. 4–16.

Monroe, James S., and Reed Wicander. 1995. *Physical geology: Exploring the earth*, 2d ed. St. Paul: West Publishing Co.

Motavalli, Jim. 1996. Some like it hot; global warming is not just a scientific prediction—it's also a hot political football. *Environmental Magazine*, 7, no 1: 28–35.

National Research Council. 1989. *Ozone depletion, greenhouse gases, and climate change: Proceedings of a joint symposium by the Board of Atmospheric Sciences and Climate and the Committee on Global Change, National Research Council*. Washington, D.C.: National Academy Press.

Scott, Ralph C. 1992. Climates of the world. pp. 181–220 in *Physical geography*, 2d ed. St. Paul: West Publishing Co.

Trent, D. D. 1983. California's Ice Age lost: The Palisade Glacier: *California geology* 36, no. 12: 264–267.

Van Diver, Bradford B. 1992. *Roadside geology of New York*, 5th ed. Missoula, Mont.: Mountain Press.

Williams, Richard S., Jr. 1983. *Glaciers: Clues to future climate?* Washington, D.C.: U.S. Geological Survey.

ENERGY

I suspect the energy crisis is over, until we have our next energy crisis.

JAMES SCHLESINGER, FORMER SECRETARY, DEPARTMENT OF ENERGY, 1982

F rom the simplest pond scum to the most complex ecosystem, energy is essential to all life. Derived from the Greek word *energia* meaning "in work," *energy* is defined as the capacity to do work. The units of energy are the same as those for work, and the energy of a system is diminished only by the amount of work that is done.

Prosperity and quality of life in an industrialized society such as ours depend in large part on the society's energy resources and its ability to use them productively. We may illustrate this in a semiquantitative fashion with the equation:

OPENING PHOTO
A solar-powered electrical generating facility at Rancho Seco near Sacramento, California. The photovoltaic panels are capable of generating 2 megawatts of electricity. In the background is a fully functional 900 megawatt nuclear-power facility that has been idle since 1989 by a vote of the people.

$$L = \frac{R + E + I}{\text{population}}$$

where L represents quality of life, or "standard of living," R represents the raw materials that are consumed, E represents the energy that is consumed, and I represents an intangible we shall call *ingenuity*. As the equation expresses, when high levels of raw materials, energy, and ingenuity are shared by a small population, a high material quality of life results. If, on the other hand, a large population must share low levels of resources and energy, a low standard of living would be expected. Some highly ingenious societies with few natural resources and little energy can and do enjoy a high quality of life. Japan is a prime example. Some other countries that are selfsufficient in resources and energy—such as Argentina—are having difficult times. Thus we see that ingenuity, which is reflected in a country's political system, technologies, skills, and education, is heavily weighted in the equation and it can cancel out a lack of resources and a large population. ◉ Figure 13.1 shows annual world and U.S. energy consumption between 1925 and 1985.

To a physicist there is no energy shortage, because she or he knows that energy is neither created nor destroyed; it is simply converted from one form to another, such as from nuclear energy to heat energy. Fuels of all kinds are warehouses of energy, which can be tapped by some means and applied in some way to do work. Coal and oil, for example, are fossil fuels that have been storing solar energy within the earth for millions of years.

Some forms of energy are *renewable;* that is, they are replenished at a rate equal to or greater than the rate at which they are used. Examples include solar, water, wood, wind, ocean and lake thermal gradients, geothermal, and tidal energy. The energy in all of these resources except for geothermal and tidal (gravitational) energy was originally derived from the sun. Renewable resources are dependable only if they are consumed at a rate less than or equal to their rate of renewal. If they are overexploited, some period of time will be required to replenish them. If geothermal heat is overexploited, for example, subsurface rocks or magma will cool to temperatures below the temperature at which steam is produced, and they will be reheated only after some time has passed. Peat, a fuel used extensively for space heating and cooking in Ireland and Russia, is estimated to accumulate at a remarkable 3 metric tons per hectare per year (1.3 tons/acre/year). Nonetheless, the conversion from plant litter to peat may take a hundred years. Wood energy may renew in a matter of a few decades, and water and wind are renewed continuously.

Nonrenewable resources are not replenished as fast as they are utilized, and once consumed, they are gone forever. Crude oil, oil shales, tar sands, coal, and fissionable elements are nonrenewable energy resources. Their quantities are finite. Supplies of crude oil, for example, are within a few human generations of exhaustion. The problem is that, owing to the use of gasoline-powered engines for trans-

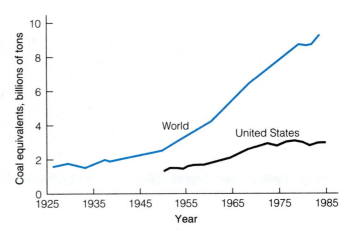

◉ **FIGURE 13.1** World energy consumption, 1925–1985. Consumption increased more than 500 percent between 1925 and 1985, and increased 300 percent between 1950 and 1985. The increase beginning about 1950 was mostly in developing countries. The United States is the world's largest consumer of energy, and its rate of consumption has increased very little since the early 1970s.

portation, demand far exceeds replenishment. Oil underground was discovered almost fifty years before the first automobile was operational. Prior to then, gasoline was a minor byproduct of refining oil for kerosene lamps. In 1885 Germans Gottlieb Daimler and Karl Benz independently developed gasoline engines, and in 1893 Massachusetts bicycle makers Charles and Frank Duryea built the first successful U.S. gasoline-powered automobile. By the turn of the century, production automobiles were hitting the roads, horses were being put out to pasture, and refineries were stepping up production to satisfy an increasing demand for gasoline. The transportation revolution to automobiles spawned the largest private enterprise on earth: the exploration and production of petroleum. Today this industry employs half the world's geologists and hundreds of thousands of engineers, technicians, and managers. More than half the world's energy needs are met by oil and gas (◉ Figure 13.2). This has led to major geopolitical and economic problems, because the nations that consume the most do not produce in comparable amounts.

PETROLEUM

Although the carbon content of the earth's crust is less than 0.1 percent by weight, carbon is one of the most important elements to the earth's people. It is indispensable to life, and it is the principal source of energy and the principal raw material of many manufactured products. Crude oil, or petroleum (Latin *petra*, "rock," and *oleum*, "oil"), is composed of many **hydrocarbon compounds,** simple and complex combinations of hydrogen and carbon (■ Table 13.1). Petroleum occurs beneath the earth's surface in liquid and

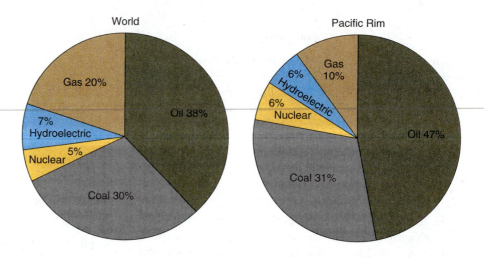

● FIGURE 13.2 World and Pacific Rim energy consumption by type in 1990. Almost 60 percent of the world's energy needs are met by oil and gas.

gaseous forms and at the surface as oil seeps, tar sands, solid bitumen (gilsonite), and oil shales. In addition to its use as a fuel, hydrocarbon compounds derived from petroleum are used in producing paints, plastics, fertilizers, insecticides, soaps, synthetic fibers (nylon and acrylics, for example), and synthetic rubber. Carbon combines chemically with itself and with hydrogen in an infinite variety of bonding schemes; some two million hydrocarbon molecules have been identified to date. The manufacturing process of separating crude oil into its various components is known as *refining,* or *cracking.*

Origin and Accumulation of Hydrocarbon Deposits

Carbon and hydrogen did not combine directly to form petroleum; they were chemical components of living organisms before their transformation to complex hydrocarbons in crude oil. Porphyrin compounds found in petroleum are derived either from chlorophyll, the green coloring in plants, or from hemin, the red coloring matter in blood, and their presence is solid evidence for an organic origin for crude oil. The fact that large quantities of oil are not found in igneous or metamorphic rocks also rules out an inorganic source for oil.

Four conditions are necessary for the formation and accumulation of an exploitable petroleum deposit in nature:

- a source rock for oil,
- a reservoir rock in which it can be stored,
- a caprock to confine it, and
- a geologic structure or favorable strata to "trap" the oil.

Even where these geologic conditions are met, the human elements of exploration, location, and discovery still remain. Oil companies utilize the skills of geologists to interpret surface and subsurface geology, to locate the potential oil-bearing structure or stratum, and to specify the optimal location for a discovery well. Until a well is drilled, the geologists' interpretation remains in doubt, much like a medical doctor's diagnosis of an ailment subject to surgery; the diagnosis is tentative until the patient is opened up.

A **source rock** is any volume of rock that is capable of generating and expelling commercial quantities of oil or gas. Source rocks are sedimentary rocks, mostly shales or limestones, usually of marine (ocean) origin and sometimes of *lacustrine* (lake) origin. The biological productivity (biomass) of surface waters must have been high enough to generate a settling "rain" of dead organisms, and the bottom waters must have been low enough in oxygen to prevent the deposited organic matter from being oxidized or consumed by scavengers. Almost all source rocks are dark-colored, indicative of high organic-matter content, and some of them carry a fetid or rotten-egg odor.

Favorable marine environments are rich in microscopic single-celled plants known as *diatoms* (phytoplankton), which form the largest biomass in the sea. Where there are diatoms, we also find zooplankton—tiny protozoans and larvae of large animals. These together with diatoms provide the molecules that make up crude oil. As the rocks are buried, they are heated. The conversion from organic matter to petroleum takes place mostly between 50° C and 200° C. Thus a proper thermal history, ideally between 100° C and 120° C, is necessary to form liquid petroleum—too cool, and oil does not form; too hot, and the hydrocarbons "boil" away.

After petroleum has formed in a source bed, it is squeezed out and migrates through a simple or complex plumbing system into a **reservoir rock.** This migration is a critical element in the formation of an economically exploitable accumulation of oil. Reservoir rocks are porous and permeable (see Chapter 10). Commonly they are sandstones, porous limestones, or in some cases, fractured shales. Reservoir porosities range from 20 to 50 percent, meaning that for each cubic foot of reservoir rock we will find 1 to 4 gallons of oil. The unit of oil volume is the barrel (equal to 42 gallons), and a so-called giant field, such as the north

MOLECULAR TYPE		HYDROCARBON COMPOUND		PERCENTAGE OF WEIGHT IN MEDIUM-GRADE CRUDE OIL
Name	*General Formula*			
Paraffins	C_nH_{2n+2}	methane, CH_4 ethane, C_2H_6 propane, C_3H_8 butane, C_4H_{10}* pentane, C_5H_{12}**	*lighter* ↕	25
Aromatics	C_nH_{2n-6}***	benzene, C_6H_6		17
Naphthenes	C_nH_{2n}	asphalt		50
Asphaltenes	(solid hydrocarbons)	gilsonite	*heavier*	8
				100

* Butane gives gasoline quick-starting capability.

** Pentane gives smooth engine warm-up.

*** Aromatics improve mileage and "knock" resistance.

slope of Alaska, will yield a billion barrels of oil in its lifetime. Using standard recovery techniques, however, as much as 40 to 80 percent of the oil may be left in pore spaces and as films on mineral grains.

An impermeable **caprock** prevents oil from seeping upward to form tar pits at the surface or dissipate into other rocks. Such seals are analogous to aquicludes in groundwater systems and are most commonly clay shales or limestones of low permeability. Our discussion of the fourth requirement, a suitable geologic structure for trapping the oil, requires a separate subsection.

Geologic Traps—Oil and Gas Stop Here

STRUCTURAL TRAPS. An **anticline** is an ideal structure for trapping gas and oil. It is a convex-upward fold in stratified rock (● Figure 13.3, part a). Analogous to a teacup inverted in a pan of water that traps a layer of air inside it, an anticline holds a reservoir of gas and oil. This occurs because crude oil floats on water and natural gas rises to the top of the reservoir. The anticlinal theory of oil accumulation was not developed until 1900, forty-one years after oil was discovered. Strata dip away from the central axis of an anticline at the ground surface, and most anticlines with surface geologic expression have been drilled. Today the search for oil is more difficult, because less obvious and geologically more complex traps need to be discovered. Many times faults form impermeable barriers to hydrocarbon migration, and oil becomes trapped against them (see Figure 13.3, part a). Faulted and folded stratified rocks in a single oil field may thus contain many isolated oil reservoirs.

Salt domes are of much interest to geologists (Figure 13.3, part b). Not only do they create oil traps, but they are valuable sources of salt and sulfur and may be potential underground storage sites for petroleum and hazardous waste. More than 500 salt domes have been located along the U.S. Gulf Coast, both offshore and on land (● Figure 13.4). They rise as flowing fingers of salt, literally puncturing their way through the overlying rocks and buckling the overlying shallow strata into a dome (● Figure 13.5). Some fingers rise as high as 13 kilometers (8 mi) above the "mother" salt bed, the Louann salt, and would be taller than Mount Everest if they were at the surface of the earth. Because the salt is considerably less dense than the overlying rock, buoyancy forces drive the salt upward—much as a blob of oil will rise through water. Brittle solids, such as salt or ice, will flow over long periods of time, and many salt domes are still rising measurably. The "mother" salt was deposited in Jurassic time by evaporation of seawater when the embryonic Gulf Coast was connected to the open ocean by a shallow opening. The opening allowed seawater to enter the basin but did not allow the denser salt brines at the bottom to escape. Thus the brine became concentrated to the point of saturation, and the great thickness of salt that now underlies the entire Gulf Coast was deposited. Oil accumulates against the salt column and also in the dome overlying the salt. "Old Spindletop," a salt dome near Beaumont, Texas, that was discovered in 1901, is perhaps the most famous U.S. oil field. Soon after its discovery it was producing more oil than the rest of the world combined, and the price of oil plummeted to two cents a barrel.

STRATIGRAPHIC TRAPS. Any change in sedimentary rock lithology (its physical character) that causes oil to accumulate is known as a **stratigraphic trap** (in contrast to anti-

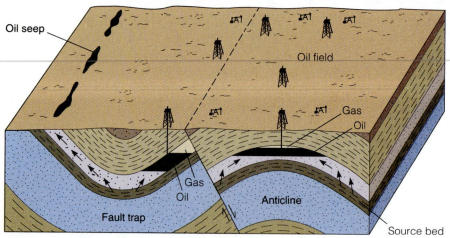

Oil seep

Oil field

Gas
Oil

Gas
Oil

Anticline

Fault trap

Source bed

(a) Anticline and fault trap

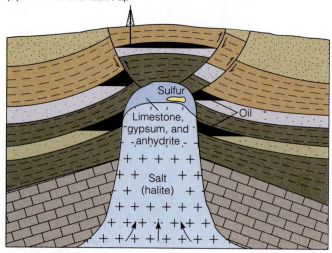

Sulfur

Limestone,
gypsum, and
anhydrite

Oil

Salt
(halite)

(b) Salt dome

Gas
Oil

Coral reef

Source bed

(c) Stratigraphic traps

⊙ **FIGURE 13.3** (*a* and *b*) Common structural oil traps. *(c)* Stratigraphic oil traps.

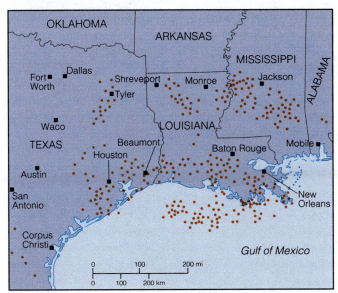

● FIGURE 13.4 Locations of salt domes of the U.S. Gulf Coast. More than 500 domes have been discovered on land and in the shallow parts of the Gulf of Mexico, and more are known to be in deep water offshore.

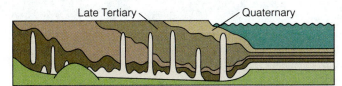

● FIGURE 13.5 Ancient salt deposits are buried deeply, and because salt is less dense than the overlying sediments, it rises buoyantly as pillars of salt, creating salt domes.

clines and salt domes, which are structural traps). Thus, if a stratum changes laterally from a permeable sandstone to an impermeable shale or mudstone, oil may be trapped in the stratum (Figure 13.3, part c).

Ancient coral reefs are ideal reservoirs, because they are porous and were biologically productive when they were living. Oil may be trapped in the porous, permeable debris on the flanks of the reef, and production from such fields can be measured in thousands of barrels per day (bbl/d). Oil fields of the Middle East are of this type, and their production potential is tremendous. A comparison of reef production to that of sandstone reservoirs such as those of California or Texas, which typically yield only a few hundred barrels per day, explains why the Middle East can control oil production and therefore price.

Oil Production

The first successful oil well was drilled in Titusville, Pennsylvania, in 1859 (Case Study 13.1). In modern jargon, this well would be called a **wildcat well,** because it was the discovery well of a new field. There is 1 chance in 50 of a

CONSIDER THIS . . .

An acquaintance has offered you a share in a sure-shot wildcat oil-drilling venture in the Sierra Nevada. How should you repond to the offer?

wildcat well being successful, less favorable odds than those of winning at roulette. The probability improves to 1 in 10 when we consider all wells drilled, including those in known oil fields. Independent oil entrepreneurs take considerable risks; it is not a business for the faint of heart. The potential rewards are also great, however, and much of our country's wealth depends on the business of finding and producing oil.

Most successful wells require pumping. If gas and water pressure are sufficient, however, oil may simply flow to the ground surface. High reservoir pressures develop from water pressing upward (buoyancy) beneath the oil and gas pressure pushing downward on the oil (● Figure 13.6). In some cases dissolved gases "drag" the oil along with them as they spew forth, as though from a bottle of champagne. If high pressures are not controlled, rocks, gas, oil, and even drill pipe may shoot into the air as a "gusher." High-pressure wells—749 in all—were set afire by Iraqi soldiers in Kuwait during Operation Desert Storm (● Figure 13.7). Professional oil-fire-fighting firms from all over the world were summoned to squelch the fires, among them the well-known Red Adair Company and others with such interesting names as Boots

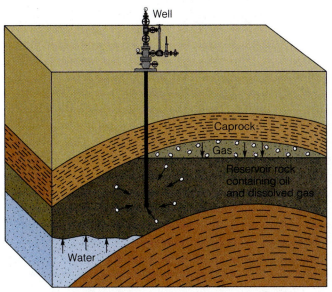

● FIGURE 13.6 Oil in an anticline is driven by gas pressure from above and by buoyant water pressure from below.

Rock Oil and the Colonel

The history of the petroleum industry is full of interesting characters, and probably none was more charismatic than the self-styled Colonel of Titusville, Pennsylvania, Edwin L. Drake (⊙ Figure 1). In the 1840s the demand for whale oil and lard had outstripped the supply and was being supplemented by "rock oil" from seeps and tar pits. The oil on northwest Pennsylvania's Oil Creek, so-named because oil bubbled to its surface, had long been skimmed by the Seneca Indians and other Native Americans and used to waterproof their homes and boats. Later, Europeans skimmed it for use as lamp oil, paraffin, liniments, and lubrication.

Drake was a railway conductor in New Haven, Connecticut, who had an interest in petroleum exploration but no experience. There are several versions of why Drake went to Titusville. The most credible one is that he had bought some shares in the Seneca Oil Company in New Haven and was sent by one of its founders, a friend of his, to stimulate the seepage at Oil Creek, which the company had leased from the Pennsylvania Rock Oil Company. Drake had a "Lincolnesque" appearance and he frequently wore a knee-length topcoat and stovepipe hat. The company bestowed upon him the honorary title of *Colonel* to enhance his credibility in Pennsylvania.

Colonel Drake tried digging trenches to stimulate the flow, and he installed a series of skimmers to collect the oil. The process was so inefficient that he persuaded the company to lease the land and attempt drilling. Drake enlisted the help of local "salt borers," who drilled for brine from which salt was extracted. Often these drillers had struck oil instead of water and they had been run out of town because of the terrible mess it created. The Colonel finally found an experienced driller named Uncle Billy Smith, who used percussion tools featuring a chisel on a cable to chip the rock and a bailer to remove the cuttings. Trouble with cave-ins ensued, and Drake ingeniously drove a large-diameter pipe into the caving zone and drilled the hole through it. On Friday, 27 August 1859, they struck oil, although they didn't know it until they checked the well on Sunday. A black goo had filled the 21-meter (69-ft)-deep hole, and the world had its first successful oil well.

Drake's well produced 10–35 barrels per day, almost doubling the previous world output of rock oil. The oil price jumped to $20/bbl, and a thousand oil pits sprang up around Titusville. Within a year the price had dropped to $0.20/bbl, and the Seneca Oil Company sold the Drake Well. Nearby towns such as Pithole City flourished for a few years and then became ghost towns. At Oil City a few miles south of Titusville, a Cleveland hay and grain dealer named John D. Rockefeller built the first commercial oil refinery in 1863. Seven years later he established the Standard Oil Company.

⊙ **FIGURE 1** Colonel Edwin L. Drake *(right)* in front of his well in 1861 with his friend, Titusville druggist Peter Wilson. The men in the background are probably drillers.

Because Edwin Drake had filed no patents and established no lease claims, he did not benefit from his vision and labor. He later lost his money speculating on oil stocks and became impoverished. He was forced to live on the generosity of friends. Learning of his dire need in 1873, Titusville citizens took up a collection for him. Three years later the Pennsylvania legislature voted to give him an annual income of $1,500. Drake was buried in Bethlehem, Pennsylvania, in 1880, but his body was reinterred in Titusville in 1901, ironically, the same year that a gusher in Texas, "Old Spindletop," drastically changed the oil supply of the world.

Today, Oil Creek between Titusville and Oil City is the site of the Oil Creek State Park. The ghost town of Pithole is just east of the park, and Drake Well Memorial Park is at the park's northern boundary.

and Coots and Joe Bowden's Wild Well. With the help of Kuwaiti *roughnecks* (the nickname applied to drilling-rig workers), the fires were extinguished in nine months, three months earlier than the most optimistic estimate.

It is possible to drill a well so that the drill hole slants from the vertical to penetrate reservoir rocks far from the drilling site (● Figure 13.8). This is desirable where the oil structure is offshore and must be "slant-drilled" from land or from a drilling platform. It is also employed for tapping reservoirs beneath developed land as in Beverly Hills, where a large number of oil wells are slant-drilled. **Slant drilling** is called *whip-stocking* after the wedge-shaped tool that was placed in the hole to deflect the drill bit at the desired angle. Modern methods use "smart" drill bits that are remotely controlled. Producers can thus withdraw oil from below a large area more economically and at the same time minimize the visual blight of drilling towers.

Oil-filled fractures

Horizontal hole (many fractures intercepted)

Slant hole (more fractures intercepted)

Vertical hole (few fractures intercepted)

● FIGURE 13.8 Comparison of vertical, slant, and horizontal drilling. Reservoirs at considerable horizontal distance from the well can be trapped by slant-drilling, and horizontal drilling allows reservoirs beneath an even larger area to be intercepted.

The ultimate technology of the 1990s is "horizontal" drilling, a technique by which the drill pipe and bit follow the plane of a flat-lying reservoir, rather than cutting across it (Figure 13.8). By this means a single borehole can access a much larger volume of oil-bearing strata than would a vertical hole. The method is being used in Texas to extract oil from fractures and voids in the Austin Chalk, a very productive limestone.

SECONDARY RECOVERY. Secondary recovery methods extract oil that remains in the reservoir rock after normal withdrawal methods have ceased to be productive. As much as 75 percent of the total oil may remain. Secondary recovery methods can be grouped into three categories: thermal, chemical, and fluid-mixing (miscible) methods (● Figure 13.9). All of these methods require *injection* wells

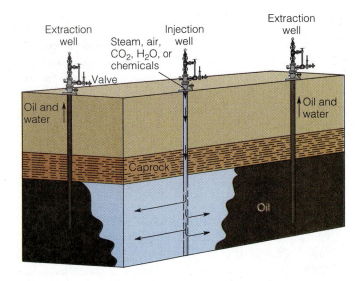

Extraction well

Injection well

Extraction well

Steam, air, CO_2, H_2O, or chemicals

Valve

Oil and water

Oil and water

Caprock

Oil

● FIGURE 13.9 Secondary recovery. Steam, air, carbon dioxide, or chemicals dissolved in water are injected into a sluggishly producing formation in order to stimulate the flow of oil to extraction wells.

for injecting a fluid or gas and *extraction* wells for removing the remobilized oil.

Thermal methods include steam injection, which makes the adhering oil less viscous and thus more free to flow, and fire flooding—in which air is injected into the reservoir in order to set fire to the oil and thus produce gases and heat that will increase the flow of oil. Water injection is a chemical method that utilizes large-molecule compounds, which when added to water, thicken it and increase its ability to wash or sweep the adhering oil films and globules toward an extraction well. Fluids that mix with oil, that are **miscible,** are very effective in removing "stuck" oil from the reservoir. Miscible recovery methods use mixtures of water with propane or ethane extracted from natural gas or mixtures of CO_2 and water. Even after the use of these secondary recovery methods, as much as a fifth of the total oil may remain in the reservoir.

Quality and Price

The price of a barrel of crude (unrefined) oil varies with its grade—its quality—and with market demand at the time. Light oils bring higher prices, because they contain large proportions of paraffins and aromatic hydrocarbons, which are desirable for gasoline and bottled gases. Heavy crudes contain lower proportions of those components and greater proportions of the heavy, less valuable asphalts and tars (see Table 13.1). A scale of crude-oil quality has been established by the American Petroleum Institute based upon its weight. Light crude oil is very fluid and yields a high percentage of gasoline and diesel fuel. Heavy crude, on the other hand, is about the consistency of molasses. The percentages of fuels and lubricating oils yielded by a barrel of medium-weight crude oil are shown in ■ Table 13.2.

Future Energy Estimates—Who Cares?

Our way of life is possible only if we have abundant energy, even if our only use of it is for household electricity and gas for our car. For this reason, we all should care about future energy supplies. In the case of nonrenewable energy forms, these supplies are referred to as **reserves,** the amount of an identified resource that can be extracted *economically.* A gen-

■ TABLE 13.2 Typical Composition of an A.P.I. Medium-Grade Crude Oil

COMMON NAME	NUMBER OF CARBON ATOMS★	PERCENTAGE OF CRUDE-OIL WEIGHT
Gasoline	5–10	27
Kerosene	11–13	13
Diesel fuel	14–18	12
Heavy gas oil	19–25	10
Lubricating oil	26–40	20
Heavy fractions	>40	18
		100

★ Volatility decreases as the number of carbon atoms increases, C_3 to C_{40}.

CONSIDER THIS . . .

Historically, Pennsylvania crude oil (such brands as Quaker State and Pennzoil) has been known for its superior lubricating properties compared to, say, California or Texas crude oils. Why do you suppose this is so? (Table 13.1 can help you deduce a reasonable answer.)

eral awareness of this began in 1980, when the U.S. Geological Survey (USGS) predicted that the world's known oil reserves would be exhausted by the year 2006 if oil continued to be consumed at the 1980 consumption rate. In 1989, the USGS advanced the projected year of "the last drop of oil on earth" to 2049 and projected a "lifetime" of 120 years for natural gas reserves at 1988 consumption rates. ● Figure 13.10 summarizes the amounts of oil reserves by country in 1990. Oil and gas exploration and discovery are dynamic enterprises, however, and much can happen in a short period of time. In 1995 the amount of technically recoverable U.S. oil—that is, the sum of proven reserves, inferred reserves in existing fields, and undiscovered accumulations—had grown to 110 billion barrels (BB), an increase of almost 41 percent over the similarly derived 1989 estimate of 78 BB. World reserves estimates were increased slightly in 1995 from those of 1989 to about 1.1 trillion barrels, yielding a lifetime of slightly more than 50 years. However, in an astounding 1995 report, USGS geologists estimated the world's "ultimate resources" of conventional oil at double that amount, 2.1 trillion barrels, of which 700 BB had been used by 1995. "The last drop" is now forecast to occur at the end of the twenty-first century (● Figure 13.11). These changes illustrate the dynamic nature of the energy business.

The earth's richest oil regions are in an equatorial belt (loosely defined by the 30° latitudes) that includes the Middle East, the Gulf of Mexico, North Africa, and Southeast Asia. Altogether, these areas account for 68 percent of the world's oil and gas. Geologically, this climatic zone favors the formation of limestone, an excellent reservoir rock, and evaporite (salt) caprock seals. Only about four percent of the world's known oil has been found in areas that made up the southern part of Gondwana, and this is due to their position before and after the breakup of Pangaea in late Mesozoic time (see Chapter 3). The significant Paleozoic oil deposits in North America and other northern regions were formed when the northern part of Gondwana was in

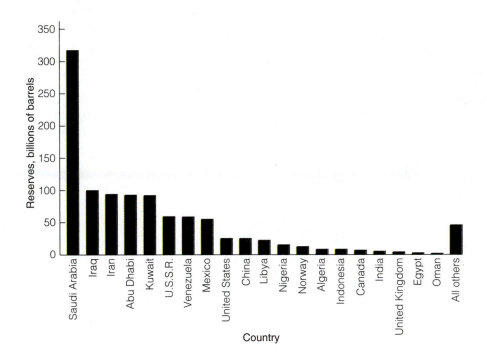

the equatorial belt. The southern part of Gondwana, on the other hand, was near the south polar area, a region not geologically favorable for oil accumulation.

The Organization of Petroleum Exporting Countries (OPEC) includes all the major oil-producing countries of the equatorial climatic zone with the exception of Mexico. Other significant nonmembers are countries of the former Soviet Union. OPEC has the ability to control oil prices, as the oil "shock" of the 1970s revealed (● Figure 13.12). The United States and other nations responded to OPEC's price increases by instituting conservation measures and accelerating exploration—actions that eventually drove prices down from a high of $38/bbl to a low of $12/bbl. Persian Gulf nations have more than half of the world's known reserves, so it is likely they will dominate oil production and marketing well into the next century.

In 1986, the United States was the most energy-deficient of the 20 largest oil-producing countries, with only 9 years of reserves at the 1986 production rate of 8.4 million bbl/d

CONSIDER THIS . . .

It has been forecast that the last drop of oil will not be pumped until some time beyond your life expectancy. In light of this, do you think we should go on consuming energy as usual, or should the world's peoples and their governments do something to delay the day when that last drop will be used?

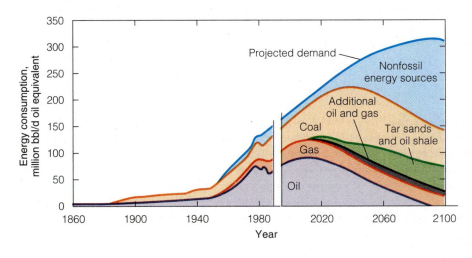

● FIGURE 13.11 Consumption of world energy resources by type, 1860–2100, in million barrels of oil per day equivalents. Fossil-fuels are projected to provide only half the demand in the year 2100. The consumption of coal is expected to remain steady and the use of nonconventional petroleum (oil shale and tar sands) to increase significantly in the twenty-first century. The break in the curves represents 1989.

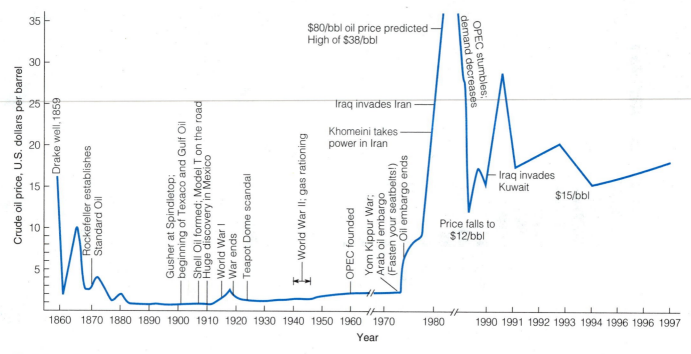

● FIGURE 13.12 Petroleum industry milestones and oil prices, since 1859. Note that since the founding of OPEC in 1960, prices have risen and have fluctuated with political stability and harmony in the Middle East.

(× 365 = 3 billion bbl/y), followed closely by Egypt, Canada, the United Kingdom, and the countries of the former Soviet Union. The difference between the U.S. consumption of 16 million bbl/d and its production of 8.4 million bbl/d represents its dependency on imported oil. This dependency has a major impact on the U.S. trade deficit and foreign policy. Another consequence of having to import oil is the risk of environmental damage due to oil spills. Tanker-transported oil is always subject to accidental spillage or catastrophic spills due to collisions or groundings (Case Study 13.2, page 376).

The future is not so bleak when we factor U.S. ingenuity into the energy equation. Much oil remains to be discovered using current exploration techniques, although a crude-oil price of at least $25/barrel is required to make the expensive searches feasible. In addition, there are nonconventional sources of petroleum such as oil shale, tar sands, and coal from which liquid fuel can be derived. Coal is also a source of fuel-grade natural gas at present.

COAL

Coal is the carbonaceous residue of plant matter that has been preserved and altered by heat and pressure. Next to oil and oil shales, coal is the earth's most abundant reservoir of stored energy. Deposits are known from every geologic period since Devonian time and the appearance of widespread terrestrial plant life some 390 million years ago. Permian coal is found in Antarctica, Australia, and India—

pre-continental drift Gondwanaland (Chapter 3). The large fields of North America, England, and Europe were deposited during the Carboniferous Period, so-named because of the extensive coal deposits in rocks of that age. In the United States the Carboniferous Period is divided into the Mississippian and Pennsylvanian Periods, named after the rocks found in those states. Tertiary coals are found in such diverse locations as Spitsbergen Island in the Arctic Ocean, the western United States, Japan, India, Germany, and Russia.

Coalification and Rank

The first stage in the process of *coalification* is the accumulation of large amounts of plant debris under conditions that will preserve it. This requires high plant production in a low-oxygen depositional environment, the conditions usually found in nonmarine brackish-water swamps. The accumulated plant matter must then be buried to a depth sufficient that heat and pressure expel water and volatile matter. The degree of metamorphism (conversion) of plant material to coal is denoted by its **rank**. From lowest to highest rank, the metamorphism of coal follows the sequence *peat*, to *lignite*, to *subbituminous*, to *bituminous*, to *anthracite*. The sequence is accompanied by increasing amounts of fixed carbon and heat (Btu) content, and a decreasing amount of quickly burned volatile material (■Table 13.3). Once ignited, it is the carbon that burns (oxidizes) and gives off heat, just as wood charcoal does in a barbecue pit.

■ TABLE 13.3 Evolution of Coal and Its Properties

PEAT →	LIGNITE →	SUBBITUMINOUS →	BITUMINOUS →	ANTHRACITE
dried moss	*crumbly*	*intermediate*	*soft coal*	*hard coal*

Increasing heat and pressure ———————————————————→
Increasing fixed-carbon content ———————————————————→
Increasing heat content (Btu) ———————————————————→
Decreasing volatiles, ash, and water contents ———————————————————→

Bituminous, or soft coal, usually occurs in flat-lying beds at shallow depths that are amenable to surface mining techniques. Anthracite, the highest rank of coal, is formed when coal-bearing rocks are subjected to intense heat and pressure—a situation that sometimes occurs in areas of plate convergence. In the Appalachian coal basin, high-volatile bituminous coals occur in the western part of the basin and increase in rank to low-volatile bituminous coals to the east. Anthracite formed close to the ancient plate-collision zone of North America and Africa (pre-Pangaea), where the coals were more intensely folded and subject to higher levels of heat. Traveling east from the Appalachian Plateau to the folded Ridge and Valley Province, one goes from flat-lying bituminous terrain to folded anthracite terrain. West Virginia, Kentucky, and Pennsylvania are the leading coal producers of the eastern states. The interior coal basins of the Midwest are characterized by high-sulfur bituminous coal, whereas the younger Western deposits are low-sulfur lignite and subbituminous coal. ■ Table 13.4 projects the development of coal-mining activity in the United States to the year 2000. The expansion of mining activity will occur mainly in the Western coal basins because the coal seams there are very thick and easily mined and the coals are low in sulfur. For environmental reasons, as we shall see, the desirability of different coals depends largely on how many tons of sulfur dioxide (SO_2) they generate per million Btu produced. Wyoming, with its low-sulfur coal, was the leading coal producer in 1993 (190 million short tons), followed in order by West Virginia, Kentucky, Pennsylvania, and Illinois.

Reserves and Production

The future of coal-derived energy in the United States is far more optimistic than the outlook for petroleum. Coal constitutes 80 percent of U.S. energy stores but only 18 percent of present usage. In 1989, electric utilities accounted for 86 percent of the coal consumed, with residential and industrial use accounting for the remainder. It is estimated that there are 84 trillion (84×10^{12}) recoverable tons of coal in the world, the equivalent of 34 trillion barrels of oil. For various reasons, much of the world's coal cannot be mined. For example, some deposits are too thin to be mined by conventional methods; others are located in politically or ecologically sensitive areas. Thus it is necessary to distinguish between total reserves and "recoverable" reserves. The United States has 283 billion tons of recoverable reserves (◉ Figure 13.13, page 378), which could last 200–300 years at the current rates of production and use. These reserves could meet only 50 years of total U.S. energy demand, however.

By the year 2000 it is estimated that energy demand will be between 100 and 150 **quads** (a quad is a quadrillion British thermal units, 10^{15} Btu), compared to 73 quads in 1973 and about 80 quads in 1980 (see ■ Table 13.5 and ◉ Figure 13.14, page 378). Coal production will double to meet the demand, and this increase will be mostly in the Western coal basins, where the relatively inexpensive surface-mining techniques can be used. Underground mining costs more, requires a greater capital investment, and takes

■ TABLE 13.4 Active U.S. Coal Mines by Region, 1977 and 2000 (Projected)

REGION	SURFACE MINES 1977	SURFACE MINES 2000	UNDERGROUND MINES 1977	UNDERGROUND MINES 2000	TOTAL 1977	TOTAL 2000
Appalachia★	185	130–175	205	380–500	390	510–675
Midwest	91	95–135	54	120–180	145	215–315
West	141	700–1,005	13	80–110	154	780–1,115

★ Encompasses parts of Pennsylvania, eastern Kentucky, West Virginia, Virginia, Maryland, Tennessee, and Ohio.
SOURCE: U.S. Geological Survey.

CASE STUDY 13.2

The Tragedy in Prince William Sound

Late in the evening of 24 March 1989, the supertanker *Exxon Valdez* left Valdez Fiord—in the area called the "Switzerland" of Alaska—and sailed into the designated outbound shipping lane on a southwest heading. The seas were calm and the winds light, and visibility was 10 miles. Everything was normal as the pilot left the ship outside the entrance to the fiord.

A short time later the officer in command radioed the Coast Guard, requesting permission to move over to the inbound lane so as to avoid a small iceberg that had calved off the nearby Columbia Glacier (see Chapter 12). The request was granted, and the ship altered its course 25° to port to change lanes. The ship continued on this more southerly heading for 30 minutes, sailing *past* the inbound lanes and into the shallow waters near Bligh Island (named by James Cook for Captain William Bligh of *Mutiny on the Bounty* fame). Realizing his error, the officer on the bridge gave urgent commands to turn starboard in order to return to the shipping lanes, but it was too late. At 4 minutes past midnight, the *Exxon Valdez* drove up onto Bligh Reef, tearing a gash in her hull and coming to a stop balanced on a pinnacle of rock (●Figure 1).

Within the next two days, 10.1 million gallons of oil spilled into a bay teeming with marine mammals, fish, waterfowl, and eagles. Exxon and Alyeska Pipeline Service Company, which jointly operate the terminal at Valdez, responded to the emergency, but their efforts to mobilize containment booms were slowed because the booms were buried under snow. By the time the booms could be put into place to keep the oil from spread-

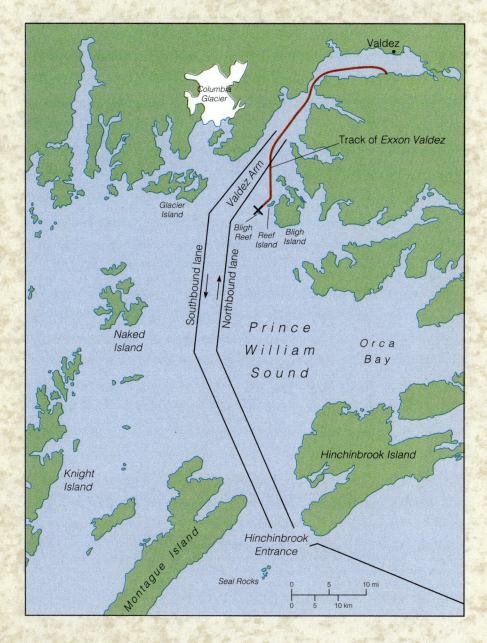

● FIGURE 1 The Prince William Sound tanker channel and the track of the *Exxon Valdez* from the loading docks in Valdez Fiord to its grounding on the reef off Bligh Island.

● **FIGURE 2** An oil-soaked bird is rescued and prepared for clean-up after the spill. Thousands of other marine birds were not so lucky.

ing and skimmers could begin removing the oil from the sea surface, much damage had already been done to the shoreline and its inhabitants (● Figure 2).

It was estimated that 1,600 otters and 37,000 marine birds died from the oil. Some biologists believe this bird count represents only 10–30 percent of the actual number killed, that the losses were 100,000–350,000 birds. About 150 bald eagles died after scavenging oil-covered corpses, and more than 80 percent of the area's nesting pairs failed to produce young that year, further affecting the eagle population. Compared to other oil-tanker spills, such as those of the *Amoco Cadiz* off Spain that killed 20,000 birds and the *Torrey Canyon* off England that killed 30,000 birds, the *Valdez* spill was a much larger catastrophe (● Figure 3).

Fish kill is very difficult to estimate. After a few weeks on the sea surface, the oil thickened and sank to the bottom, where it affected such fish as snapper and halibut. The impact on invertebrate marine populations and the ecological damage cannot be quantified. There is some evidence that the shoreline cleanup involving steam cleaning did as much damage to the environment as the oil spill did. It was learned in the California oil spill of 1969 that oil-eating bacteria do an efficient job removing oil from bottom sediments and rocks. Steam cleaning the shores of Prince William Sound killed the oil-consuming bacteria in the tidal zone, worsening the problem.

The lesson learned there is that oil spills of this magnitude are difficult if not impossible to contain. "Throwing money at it" will not change the results if weather and currents hamper the cleanup. As they say, "It doesn't do any good to close the barn

● **FIGURE 3** Comparison of the area of the 24 March 1989 oil spill with the Eastern Seaboard.

door after the horse is out." In these days of satellite positioning systems, it would seem that restricted areas could easily be programmed into the navigation systems, and that audible warnings could be sounded when a vessel strays from the approved shipping lane. The double hulls on newer supertankers offer assurance that at least some groundings will not result in oil spills. The final cost to Exxon of this "mistake" will probably be in excess of $4 billion ($1.025 billion of that in fines). As long as humans are at the controls, however, incidents such as this one can be expected.

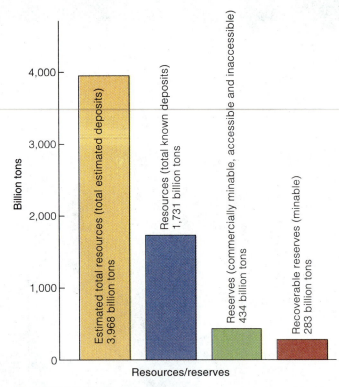

● **FIGURE 13.13** U.S. coal resources and reserves based upon 57 percent recoverability in underground mines and 80 percent recoverability in surface mines.

■ **TABLE 13.5 Energy (Power) Units**

UNIT	EXPLANATION
British thermal unit (Btu)	The amount of heat required to raise the temperature of 1 pound of water 1° F; ≈ energy released by a burning match
Quad (quadrillion)	10^{15} Btu = 172 million bbl/oil
Barrel of oil	5.8 million Btu = 42 gallons
Bituminous coal (average)	25 million Btu/ton
Natural gas	Variable Btu content, measured in cubic feet (cf)
Megawatt (MW)	1,000 kilowatts (kw), or a million watts
Gigawatt (GW)	1,000 megawatts, or a billion watts

more time to get into production. Also, miners who work underground face greater risks than do those who work at the surface.

The economic future of coal lies in its use as a substitute fuel for oil and gas. The technology is in place to make methane from coal—the product is called *synthesis gas*—and it has been used on a large scale to manufacture petro-chemicals, including methanol (methyl alcohol). Methanol can be converted to gasoline using a mineral catalyst. This process is currently used in New Zealand, where synthesis methanol is being converted to gasoline at a rate of 14,500 barrels per day. Methanol is an interesting fuel: its combustion generates little nitric oxide (NO) and ozone, and it has a high octane number—about 110. Its high octane has made it attractive for years as a fuel for race cars and high-compression engines. Although it has a lower heat (Btu) content than gasoline, it burns more efficiently, which evens things out. On the negative side, methanol releases greater emissions of carbon dioxide. **Biomass** alcohol (ethanol) made from sugar cane has been used in Brazil for

● **FIGURE 13.14** Distribution of U.S. energy consumption to 1992.

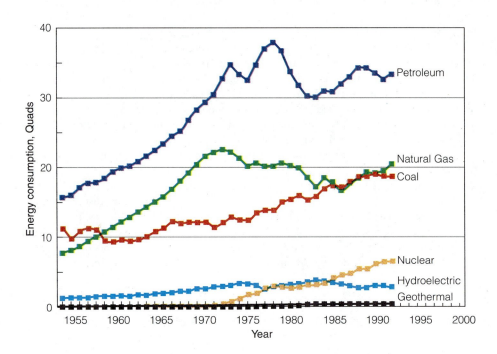

● FIGURE 13.15 Mining Athabasca tar sand, Alberta, Canada. After extraction by huge machinery, the sands are washed with hot water to extract the oil adhering to the grains. Cleaned sand is then returned to the open pit, and the land is restored to its original contours.

years as an auto fuel, and at one time almost 90 percent of the cars there burned it. Ford Motor Company has been developing an engine that can burn gasoline, methanol, or mixtures thereof. As the use of lead in gasoline is phased out, methanol can be added as an octane enhancer.

Direct liquefaction of coal shows great promise. Using coal- or petroleum-derived solvents, coal is dissolved and hydrogenated, resulting in about 75 percent gasoline and the remainder propane and butane. About 5.5 barrels of liquid are derived from a ton of coal at the approximate cost of the most expensive oil during the 1970s energy crunch, $35–$40/bbl. As long as we continue to rely on hydrocarbon-fueled engines, a major facet of our energy future lies in converting coal to liquid or gaseous **synthetic fuels,** often called simply *synfuels*.

NONCONVENTIONAL FOSSIL FUELS

TAR SANDS. Sands containing heavy oil that is too thick and viscous to flow at normal temperatures have been found in Canada, Russia, Venezuela, Madagascar, and the United States. Where the sands occur at shallow depths, such as at Athabasca, Alberta, Canada, they may be mined. The Athabasca Field is the largest oil field in the world, estimated to contain more than 10^{15} barrels of oil, of which 300 billion barrels are recoverable. The sand is extracted by open-pit mining techniques, and the oil is separated from the sand using steam or other thermal methods (● Figure 13.15). About 300,000 bbl/day were being produced profitably in 1992. The cleaned sand is returned to the pit, which is then leveled and landscaped. Tar sands are found at relatively shallow depths in Utah, California, and Texas. The cost of tar-sands oil is equivalent to that of high-cost offshore oil or oil from remote areas. When oil prices rise sufficiently, some of these sands will be exploited.

The Cold Lake area of Alberta is underlain by tar sands that are too deep for surface exploitation and that are estimated to contain 1.7 trillion barrels of heavy oil. Recovery methods are being developed for this site that include steam injection, horizontal drilling, and borehole mining. The borehole mining technique was developed for the uranium and potash industries. Water is pumped under high pressure through a drill pipe to create a cavern in the sand. The caved sand is continuously pumped to the surface through the space between the drill pipe and a larger-diameter casing. The oil is then removed from the sand, and the clean sand is returned to the cavern.

OIL SHALES. Oil shales are sedimentary rocks that yield petroleum when they are heated. Found on all continents, including Antarctica, they were originally deposited in lakes, marshes, or the ocean. The original rock oil used in kerosene lamps was produced from black oil shales. The most extensive U.S. deposits, the Green River shales, were formed during Eocene time in huge fresh-water lakes in the present-day states of Wyoming, Colorado, and Utah (● Figure 13.16). They are not oily like tar sands, but when heated to 500° C (900° F) they yield oil. The source of oil in the rock is **kerogen,** a solid bituminous substance, and a ton of oil shale may yield 10–150 gallons of good-quality oil. Potential deposits of

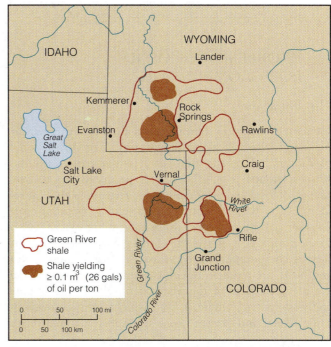

● FIGURE 13.16 Green River oil shale deposits.

● FIGURE 13.17 Green River oil shale exposed in Book Cliffs near Grand Junction, Colorado.

economic interest yield 25–50 gallons per ton, and the U.S. Geological Survey includes 3 trillion barrels of oil from shale in their estimates of reserves. Kerogen-rich lake deposits are also found in China, Yugoslavia, and Brazil.

Water is the limiting factor in oil-shale utilization, because processing the shale requires water volumes 3–4 times greater than that of the oil extracted. Furthermore, the extraction process destroys the natural landscape, and some of the richest shales are in scenic wilderness areas (● Figure 13.17). Because oil shales expand like popcorn when they are heated, disposal of the waste rock is a problem. At this time the oil shale industry cannot compete economically because of the high cost of transporting unprocessed oil shale to consumers. The future for the trillions of barrels of oil in shales rests in developing *in situ* ("in place") extraction methods that will not disrupt the surface environment or create waste-rock disposal problems.

● PROBLEMS OF FOSSIL FUEL COMBUSTION

Air Pollution

Obtaining energy by burning fossil fuels creates environmental problems of global proportions. It produces oxides of carbon, sulfur, and nitrogen and fine-particulate ash. Carbon monoxide (CO), an oxide produced by combustion of all fossil and plant fuels, is converted to carbon dioxide (CO_2), which contributes to global warming (see Chapter 12). Burning coal also releases sulfur oxides (SO_x) to the atmosphere, where they form environmentally deleterious compounds. Nitrogen oxides (NO_x), mostly NO and NO_2, are products of combustion in auto engines and are the precursors of the photochemical oxidants, ozone and peroxyacetyl nitrate (PAN), which we associate with smog. Water and oxygen in the atmosphere combine with SO_2 and NO_2 to form sulfuric acid (H_2SO_4) and nitric acid (HNO_3), the main components of **acid rain**—rain with increased acidity due to environmental factors such as atmospheric pollutants (Case Study 13.3).

The Clean Air Act of 1963 as amended in 1970 and 1990 specifies standards for pollutant oxides and hydrocarbon emissions. By 1980 new cars were 90 percent cleaner than their 1970 counterparts. The Clean Air Act Amendments of 1990 (CAAA) set goals and timetables that are affecting how and what we drive. Beginning in 1992, fuel suppliers were required to sell only reformulated gasoline in 39 areas where winter air quality is a problem and to sell only reformulated gasolines in the high-ozone cities of Baltimore, Chicago, Hartford, Houston, Los Angeles, Milwaukee, New York, Philadelphia, and San Diego. Reformulated gasolines that enhance burning and reduce emissions will be the required auto fuel in the late 1990s. Reformulations using methyl tertiary butyl ether (MTBE) add as much as 2.5 percent oxygen to the fuel, causing it to burn cleaner and create less ozone, at a cost of about 10 cents more per gallon.

The CAAA mandated that automakers significantly reduce emissions leading to ozone formation in all new vehicles sold in the United States beginning in 1996 and that fleet owners phase in clean-fueled vehicles that use only reformulated or diesel fuels. California took this a step farther by requiring manufacturers to offer a mix of low-emission vehicles including ZEVs (zero-emission vehicles), which are essentially electric cars. By 1998, 2 percent of all new cars sold in California must be ZEVs, and the requirement rises to a whopping 10 percent by 2003.

SULFUR EMISSIONS, ACID RAIN, AND HEALTH. Sulfur occurs in coal as tiny particles of iron sulfide, most of which is the mineral pyrite, or "fool's gold" (● Figure 13.18). Coal also contains organic sulfur originally con-

CASE STUDY 13.3

Baking Soda, Vinegar, and Acid Rain

The term *acid rain* refers to the atmospheric deposition of acidic substances including rain, snow, fog, dew, particles, and certain gases. Although volcanic activity (see Chapter 5) is the greatest source of materials that may form acids—sulfur, carbon dioxide, and chlorine—sulfur and nitrogen oxides introduced by human activities are approaching the amount of nature's contributions. Combustion of fossil fuels and the refining of sulfide ores are the major human sources of these contaminants. Acids form when these gases come in contact with water in the atmosphere or on the ground. Whereas carbon dioxide forms carbonic acid, a weak acid (see Chapter 6), chloride ion and the oxides of sulfur and nitrogen form the strong hydrochloric, sulfuric, and nitric acids, respectively.

The acidity or basicity (the chemical opposite of acidity) of an aqueous solution is referenced to the pH scale (● Figure 1), a measure of the solutions' hydrogen-ion activity. A solution containing no acid and no base will have a pH value of 7.0. The more hydrogen ions floating around in the solution, the more acidic it is, and the lower its pH will be. Note that the scale is reversed, so the speak; a solution with a pH below 7.0 is acidic, and one with a pH above 7.0 is basic, or "alkaline." Acid-rain events with acidities below pH 2.8, the pH of vinegar, have been reported in large cities and heavily industrialized areas. In contrast, neutral waters may become quite basic. For example, lakes without outlets, such as Lake Natron in Africa, may become extremely alkaline and have pHs greater than 11.0—well in excess of the pH of a concentrated solution of baking soda and water.

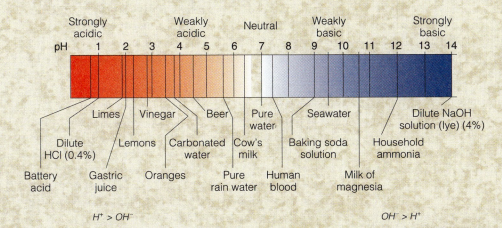

● **FIGURE 1** The pH scale. A neutral solution has a pH of 7.0. A solution whose pH is less than 7.0 is acidic, and a solution whose pH is above 7.0 is basic, or alkaline.

tained in the coal-forming vegetation. Some coals are mostly organic sulfur, having very little or no pyrite. Upon combustion, mostly in coal-burning electrical generating plants, the sulfide is oxidized to sulfur dioxide (SO_2) and carried out through smokestacks into the environment. Emissions of SO_2 are known to have a negative impact on human health (■ Table 13.6). The residence time for sulfur oxides in the atmosphere is only a few days; if all the sources were shut down, the air would be rid of human-generated sulfur compounds within 3 to 5 days. Although most coal-burning in the United States occurs in the Midwest and the East, significant volumes of SO_x can travel 1,000 kilometers (600 mi) or more from their source. In 1980, 51 power plants in six Eastern states emitted 25 percent of the total U.S. sulfur dioxide. ● Figure 13.19 shows typical U.S. SO_2 emission densities, which reflect the pattern of coal usage very closely.

Mobile sources using petroleum fuels emit less than 4 percent of the total. Nevertheless, these emissions amounted to almost a million tons of SO_2 in 1980 (■ Table 13.7).

Sulfur dioxide emitted from a smokestack may be deposited dry near the facility, where it will damage plant life and pose a health threat to animals. If the gas plume from the smokestack is transported away from the facility, its SO_2 will react with the atmosphere and eventually fall to the earth surface as acid rain. Acid rain can damage crops; acidify soils; corrode rocks, buildings, and monuments; and contaminate streams, lakes, and drinking water. Scientists and government agencies disagree on the impact of acid rain on forests. The 1990 report of the National Acid Precipitation Assessment Program shows the distribution of acid rain to include southern and midwestern states and minimizes acid rain's impact on forests. The Environmental

● FIGURE 13.18 Small pyrite masses in a bituminous coal matrix (gray). The light-gray material is the cell walls of the original plant material, and the darker gray is the solid material that was contained within the cell cavities. Earth pressures compacted the plant material, squeezed out any water, and collapsed the cavities. That pyrite formed within the coal before completion of compaction is evidenced by plant materials bent around solid pyrite masses. The pyrite "blebs" are about 50 microns across.

Protection Agency, on the other hand, believes it has evidence that acid rain does extensive forest damage.

Great Britain has acknowledged that its power plants contribute to acid rain in Scandinavia, and it is known that U.S. sources are responsible for much of the acid rain that has fallen throughout eastern Canada. U.S. "clean-air" laws have proven to be ineffective against acid rain, and no international law prohibits polluted air from invading a neighboring country. The Canadian government estimates that acid rain deposition from the United States threatens industries that account for 8 percent of its gross national product. An acid-rain agreement between the United States and Canada is needed, but political progress is slow because of pressure from the coal and power industries and because there is no cheap method of removing sulfur from coal. The Clean Air Amendment of 1989 mandates that power plants reduce their sulfur-oxide emissions by 10 million tons from their 1980 emission levels by the year 2000. Meanwhile, it is estimated that annual U.S. emissions of sulfur gases will increase almost 30 percent between 1990 and 2010.

NITROGEN OXIDES, SMOG, AND HEALTH. Nitrogen oxide emissions (NO_x) are significant from both stationary (coal-burning) and mobile (vehicular) sources (Table 13.7). Mobile sources are the most difficult to control, and their contribution to total NO_x is increasing as the number of automobiles increases. NO_x forms when engine combustion causes nitrogen in air to oxidize to NO and NO_2, which then react with oxygen, free radicals (incomplete molecules that are highly reactive), and unburned hydrocarbons to produce **photochemical smog**. Photochemical smog requires sunlight for its formation. Although the word *smog* is a contraction of the words *smoke* and *fog,* the air pollution in Los Angeles, Mexico City, and Denver has little direct relationship to either smoke or fog (Case Study 13.4).

Both NO_2 and ozone (O_3) are reactive oxidants and cause respiratory problems. A level of only 6 parts per million (ppm) of ozone can kill laboratory animals by pulmonary edema (water in the lungs) and hemorrhage within four hours. On 17 March 1992, Mexico City, the largest city in the world, had such high ozone concentrations (four times the normal amount) that it was necessary to close schools, restrict traffic, and have factories cut back production 50 percent. In humans, alcohol consumption, exercise, and high temperatures have been found to increase adverse reaction to ozone, whereas vitamin C and previous exposure seem to lessen the reaction. One investigator noted that the person most likely to succumb to ozone would be "an alcoholic with a 'snootful' arriving at LAX for the first time on a hot, smoggy day and jogging all the way to Beverly Hills." Such a scenario may not be all that unlikely in Southern California. A frightening aspect of ozone is that it is so reactive with human cells that when it is inhaled, only a fourth of it gets past the nasal and bronchial passages to the lungs, and that appears to be more than enough to cause lung damage. NO_2 has the same effect on humans as ozone, except at dosages that are 15 times greater.

Clean Air

Sulfur oxides can be removed from combustion gases in smokestacks by using a device called a *scrubber* that selectively reacts with SO_2 and absorbs or neutralizes it. Many coal-burning facilities are switching to a process known as fluidized-bed combustion. Coal and pulverized limestone

■ TABLE 13.6 Health Effects of Atmospheric Sulfates

SULFATE EXPOSURE

Level	Micrograms per m^3	HEALTH EFFECT
Short events, high concentration levels	500 ↓	Increased difficulty breathing in animals
	100 ↓	Decrease in lung-clearing mechanisms in healthy adults
Frequent 24-hour levels	25 ↓	Aggravation of heart and lung disease in the elderly; increased mortality with 24-hour exposure
Rural levels	5	

★ For northeastern United States.
SOURCE: U.S. Congress, Office of Technology Assessment.

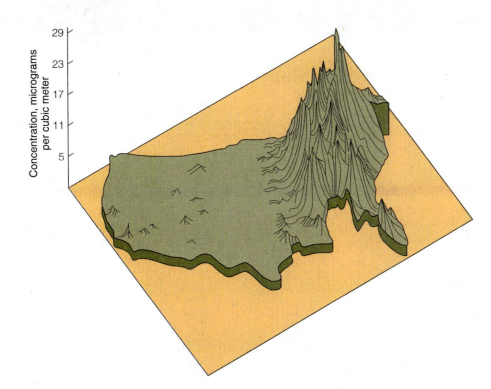

Concentration, micrograms per cubic meter

29
23
17
11
5

● **FIGURE 13.19** Atmospheric concentration of sulfate for a 26-quad U.S. coal-use scenario in 1990. Note that the heaviest concentrations are in the Northeast, where most coal is burned.

are burned together, and the heat of combustion converts the limestone into calcium oxide (CaO) and CO_2. CO_2 is emitted, and the CaO combines with sulfur and oxygen to form calcium sulfate ($CaSO_4$, a gypsumlike substance) as a solid waste product. Sulfur emissions are reduced, but CO_2 emissions increase. It is thus a tradeoff between the health benefits of reduced atmospheric sulfur and the environmental effects of increased contributions to global warming.

Low-sulfur coals are most desirable, but air-quality standards are such that even these coals must be cleaned before combustion. Physical coal cleaning employs differences in density to separate pyrite (5.0 g/cc) from coal (1.4 g/cc); in

a dense fluid, the pyrite sinks and the coal floats. Companies favor this method, because it is relatively cheap, requires minimal power-plant modification, and may improve the combustion process. The standards require that sulfur content be reduced to less that 1 percent, which would still release 20 pounds of sulfur to the atmosphere for every ton of coal burned.

Energy-efficient cars and new fuels have gone a long way to decreasing NO_x emissions. Exhaust-system catalytic conversion of nitrogen oxides, carbon monoxide, and hydrocarbons into carbon dioxide, nitrogen, and water are mandatory automotive equipment on new cars in many states.

 ALTERNATIVE FORMS OF ENERGY

Although fossil fuels are still the energy of choice for most people, air pollution, global warming, oil spills, and geopolitics are making alternative forms more attractive. Currently, greatest interest is focused on solar energy and the technologies that harness it for producing electricity and fuels, largely because it is nonpolluting and renewable. Converting solar heat to electricity is a *direct* use of the sun's energy. The sun also provides energy *indirectly* through plants (biomass), the hydrologic cycle, wind, and ocean waves. In short, the sun is the source of almost all known renewable, nonpolluting alternative energy resources.

Hydropower is the most utilized indirect solar energy. It provides about 30 percent of the electricity in developing nations. Significantly, these projects tap only 10 percent of the potential hydropower in those countries. Hydrogen gas

■ **TABLE 13.7 U.S. Sources of Sulfur and Nitrogen Oxide Emissions, 1980**

SOURCE	PERCENTAGE OF TOTAL EMISSIONS	
	Sulfur Oxides★	*Nitrogen Oxides*
Mobile sources★★	3.8	44.0
Stationary sources★★★	80.0	51.0
Industrial processes	16.0	2.9
Other (agriculture, etc.)	0.2	2.1

★ SO_2 emissions (tons/year) decreased 33% between 1975 and 1990.
★★ All transportation including recreational vehicles.
★★★ Utilities, industrial, institutional, and residential sources.
SOURCE: Council on Environmental Quality, *Twelfth Annual Report*, 1981, and various other sources.

Make Smog while the Sun Shines

Hell is a city much like London,
A populous and smoky city.

SHELLEY, *PETER BELL THE THIRD*

Air pollution is the blight that results from burning fossil fuels. It was probably first recognized as a serious problem in Elizabethan London, filthy with smoke from thousands of domestic and workshop coal fires. In 1661 John Evelyn, one of the founders of the Royal Society and a noted do-gooder, published by royal decree a paper entitled "Fumifugium, or the Inconvenience of Aer and Smoke of London Dissapated; together with Some Remedies Humbly Proposed." It is said that Londoners' lungs have been black since that time, and that for many years when the city's medical students saw pink lungs in the anatomy laboratory, they suspected the deceased was from Ireland.

Many catastrophic air-pollution incidents have occurred since the industrial revolution. The few listed here give an idea of the magnitude of the problem.

YEAR	PLACE	FATALITIES
1872	London	thousands (estimated)
1930	Meuse Valley, France	60
1948	Donora, Pennsylvania	20 (and many ill)
1950	Poza Rica, Mexico	22
1952	London	4,000 (estimated)

The word *smog,* a contraction of the words *smoke* and *fog,* was probably coined in the course of explaining the smoke-fog deaths that occurred in Scotland in 1909.

All health-endangering smog events begin with an atmospheric temperature inversion, which "puts a lid on" polluted air, holding it close to the earth surface. Temperature normally decreases with altitude, about 2° C (3.6° F) per 1,000 feet, which causes air that has been warmed and expanded at the surface to rise and denser cold air from above to sink. When an inversion forms, a layer of warm air at, say, 1,000 feet above ground level prevents vertical circulation and keeps the bad vapors at the surface (● Figure 1).

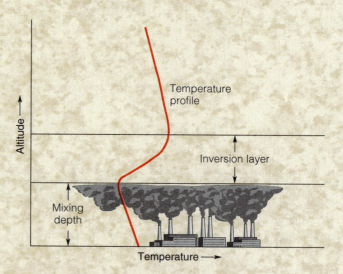

● FIGURE 1 Temperature and altitude factors of an inversion layer, which traps polluted air. If the base of the inversion layer lowers, the mixing depth decreases and air quality worsens.

offers promise as a fuel for aircraft and automobiles and it can be produced in quantity from water using solar energy. Hydrogen is available and totally nonpolluting, forming only water when it is burned. Solar technologies are advancing, and solar-thermal, wind, and fuel from biomass is becoming economically competitive. Large-scale use of photovoltaic cells, which convert radiation directly into electricity, and production of liquid fuels from biomass will probably not be realized until the twenty-first century.

Geothermal energy is being utilized many places in the world. At present about 1 percent of the world's energy needs can be satisfied by earth heat, but many untapped sources remain within reach of drilling capability. Nuclear energy offers vast potential, but that potential is clouded by issues of safety, radioactive-waste disposal, and nuclear-weapon proliferation.

Direct Solar Energy

Although the amount of solar energy that reaches the earth far exceeds all human energy needs, it is very diffuse. For example, the amount that strikes the atmosphere above the British Isles is 80 times Great Britain's energy needs; it averages about 1 kilowatt per square meter. Cloud cover reduces the amount by 80 percent, but there is still more than enough. So why aren't there solar-electrical plants in every city and hamlet? The reason is cost. The current technology requires a great deal of space, which makes installations expensive. There are essentially two ways to put the sun to work for us:

1. Solar-thermal methods utilize panels or "collectors" to warm a fluid, which is then used for heating or for generating electricity.

The worst air-pollution disaster in history was the infamous "black fog" event in London in December 1952. The fog was so thick that the racing dogs at a greyhound track lost sight of the rabbit and became disoriented, and a large duck flying blind crashed through the glass roof of a train station. In two days the fog turned black when coal-derived soot mixed with sulfur oxides, and people, most of them elderly, began dying. Many stricken people could not get to hospitals, and 50 bodies were recovered from a small park in the city.

Photochemical smog (as opposed to true smog) is produced when automobile emissions are activated by ultraviolet radiation to produce ozone. In simplest terms, the process is this:

1. The high temperatures and pressures involved in fuel–air combustion in the engine form nitric oxide (NO), which, with unburned hydrocarbons, is emitted as exhaust.
2. NO reacts with the air to form nitrogen dioxide (NO_2).
3. Under ultraviolet radiation, NO_2 reacts reversibly with oxygen to yield ozone (O_3) and nitric oxide.

In chemical notation, steps 2 and 3 look like this:

$$\underset{\substack{\text{nitrogen}\\\text{dioxide}}}{NO_2} + \underset{\text{oxygen}}{O_2} \underset{\textit{sunlight}}{\rightleftarrows} \underset{\text{ozone}}{O_3} + \underset{\substack{\text{nitric}\\\text{oxide}}}{NO}$$

A "tossed salad" of reactions occur between the bits and pieces of unburned hydrocarbons that remain. One of the reactions forms PAN (peroxyacetyl nitrate), which makes your eyes tear and which itself decomposes to form more NO_2 (a feedback mechanism), which then forms more ozone. So once the smog begins,

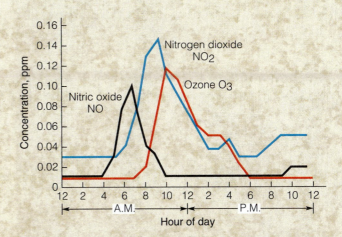

● **FIGURE 2** Typical daily variation of smog components in Los Angeles.

it is self-perpetuating as long as there is ultraviolet radiation to drive the reactions.

We can see that photochemical smog is a daytime blight; it abates at night, only to build up the next morning (● Figure 2). It's not uncommon for the Los Angeles Basin to be clear at 5:00 A.M. on a summer day and for visibility to be restricted to a few miles five hours later. Topography is an influence, and smog is chronic in cities such as Denver, Log Angeles, and Mexico City that have mountain barriers. However, any city with a plethora of cars and a temperature inversion can have air pollution problems.

2. Photovoltaic cells use the second-most common element on earth, silicon, to convert light directly into electricity.

In 1991, the electricity produced from photovoltaic cells cost 14.5 cents per kilowatt hour (kwh), about 5–6 times that produced by conventional means (■ Table 13.8). However, it is estimated that photovoltaic electricity will soon cost less than 10 cents per kilowatt hour, and will compete with conventional electrical generation by the next century.

Solar-thermal facilities called *power towers* illustrate a creative use of solar energy. The "Solar One" station in the Mojave Desert near Barstow, California, generated electricity between 1982 and 1988 for the Southern California Edison Company. It employed 1,818 suntracking mirrors focused on a liquid-filled receiver atop a tower where temperatures reached 900° C (● Figure 13.20). The superheated liquid was transferred to a plant where it flashed to steam and drove turbines that operated an electrical generator. Solar One occupied 40 hectares (100 acres) and its receiver was visible for miles, looking like a miniature sun. Power towers are practical only in areas that offer more than 300 sunny days per year and inexpensive land. Because of the success of Solar One, a 10-megawatt facility was built on the same site and began operating in mid-1996. Solar Two uses molten nitrate salt as the heat transfer medium. Molten salt has excellent heat storing properties; the heat can be used immediately to generate steam, or it can be stored and used later, during cloudy periods or after sundown.

In 1981 the *Solar Challenger* aircraft, its wings covered with 16,000 photovoltaic (PV) cells, crossed the English

■ **TABLE 13.8 Cost of Electricity by Energy Source**

ENERGY SOURCE	COST, CENTS/KWH
Sun	15.8★
Geothermal	13.3★
Biomass	13.0★
Wind	11.5★
Oil	4.4†
Gas	2.9†
Coal	1.2†
Nuclear	0.93†

★ The rates the Southern California Edison company was paying for purchased and manufactured energy in December 1995.
† The rates the Southern California Edison Company was paying for purchased and manufactured energy in December 1991 (more recent data not available).

◉ **FIGURE 13.20** Solar One, a direct solar-energy power conversion facility in the Mojave Desert near Barstow, California. The heliostat (reflector) field directs sunlight to the central tower. Solar Two now exists on this site using a new molten salt heat-transfer technology.

Channel in just over 5 hours. Impressive though this accomplishment is, the cost of the venture could have sent four conventional aircraft around the world. If clouds had appeared, the power delivered to the motor would have been diminished, if not cut off completely, requiring an abrupt change in flight plan. In PV cells, photons—pulses of light energy—are absorbed by semiconductor materials, most commonly silicon or gallium compounds, producing an electrical current. It is now possible to apply thin films of these materials, about the thickness of a human hair, cutting the cost and making PV cells a viable energy alternative in remote locations. Several U.S. utility companies have installed PV systems to supply small users in out-of-the-way localities without building costly power-line extensions. Most of Western Europe, Japan, and India are involved in PV electrical production. In India, which lacks an extensive electrical distribution network, PV power is more economical than diesel generators in remote locations. India has 4–5 million diesel water pumps that could be powered by PV during daylight hours.

Indirect Solar Energy

WIND ENERGY. Wind is solar energy that has been converted to motion, mechanical energy. It has been used to do work for several thousand years. In fact, windmills were once the only large machines capable of doing the work of many people. It is estimated that more than a million old-fashioned windmills are still being used in the United States and Australia to pump underground water to the surface. Electrical generation by windmills is straightforward, and areas with strong winds are favored. This is because the wind's ability to do work increases as *the cube of the increase in wind velocity;* with a doubling of wind velocity 8 times (2^3) as much power is available (◉ Figure 13.21). Of the 1,660 megawatts (MW; a million watts) of wind-generated electricity produced globally, 85 percent is produced in California. It is not that California is windier than other areas; rather, development of the technology was encouraged by a tax-shelter offered during the energy "crunch" of the 1970s, and taxpayers eagerly jumped on the bandwagon. In Altamont Pass (Spanish "high mountain") in central California there are 7,500 windmills, and in Banning Pass near Palm Springs there are more than 1,000 (◉ Figure 13.22). At both localities narrow passes through mountain ranges separate hot valleys from areas with maritime climates. As hot air rises in the valleys, cool coastal air is drawn through the topographically narrow passes to replace it. The original wind generators were

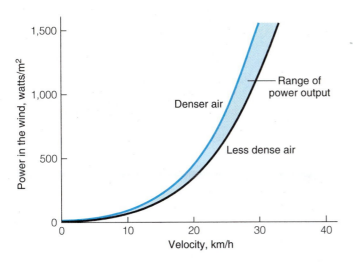

FIGURE 13.21 Wind's ability to do work increases exponentially with velocity. For this reason, wind "farms" are located in areas where winds are strong.

plagued by design flaws and maintenance problems, but with modern composite materials and programmed "smart" turbines, more-reliable 300-kilowatt windmills are now mass-produced. Costs have been reduced to the point that wind-generated electricity is now price-competitive (Table 13.8). ⊙ Figure 13.23 shows the wind-electric potential of the 48 conterminous states expressed as a percentage of each state's total energy need. Note in the figure that practically all of the electricity needed in the United States could be provided by winds over 12 states. Wind can provide electricity during peak demand periods, which typically occur late in the afternoon on hot summer days and early in the evening in the winter. Peak-demand power is most costly to the consumer, and an immediate gain can be realized where wind electricity is available to satisfy peak demand. Of course, windmills can be "becalmed." In areas where electricity is critical, they are used in conjunction with diesel electricity or other conventional generators.

Environmental problems that accompany wind generation of electricity include noise, interference with TV reception by steel blades, land acquisition difficulties, and most importantly, visual blight. Modern generators are quiet, and their wood or composite blades do not interrupt television signals. Although land disruption is minimal with windmills, they are hard to ignore, and many people feel they destroy the visual esthetics of a region. On the other hand, ranchers at Altamont love the windmills; they can still graze their cattle on the land and they receive land-use royalties that have increased the value of their land—to five times its prewindmill value.

HYDROELECTRIC ENERGY. Falling water, our largest renewable resource next to wood, has been used as an energy source for thousands of years. First used to generate electricity about 100 years ago on the Fox River near Appleton, Wisconsin, today it provides a fourth of the world's electricity. In Norway, 99 percent of the country's electricity and 50 percent of its total energy is produced by falling water. The principle is relatively simple: impound water with a dam and then cause the water to fall through a system of turbines and generators to produce electricity. Because of this simplicity, electricity coming from existing facilities is also the cheapest source of this power. Aitupu Dam on the Parana River between Paraguay and Brazil, completed in 1982, is the largest-producing electrical complex in the world; its potential is 12,600 megawatts. It provides much more electricity than the present demand in this part of South America.

Because of land costs and environmental considerations, it is doubtful that any more large dams will be built in the United States. Therefore, dams that were built for other

⊙ FIGURE 13.22 Wind "farm" on the Mojave Desert side of Banning Pass, near Palm Springs, California. The San Bernardino Mountains are in the background.

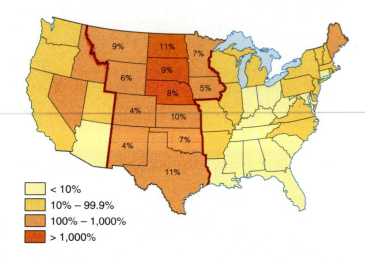

Legend:
- < 10%
- 10% – 99.9%
- 100% – 1,000%
- > 1,000%

⦿ **FIGURE 13.23** Wind-electric potential of the 48 contiguous states as a percentage of each state's energy need. Twelve states in the central part of the country could provide about 90 percent of the total U.S. need.

purposes have become attractive for retrofitting with generators. Hydroelectrical facilities are particularly useful for providing power during times of peak demand in areas where coal, oil, or nuclear generation provides the base load. Hydroelectricity can be turned on or off at will to provide peak power, and many utilities have built pumped-water-storage facilities for just that reason. During off-peak hours, when plenty of power is available, water is pumped from an aqueduct or other source to a reservoir at a higher level. Then during peak demand the water is allowed to fall to its original level, fulfilling the temporary need for added electricity.

Although hydroelectric energy is clean, dams and reservoirs change natural ecological systems into ones that require extensive management. One of the social consequences of dam building is displaced persons; for example, 80,000 people were forced to move when Lake Nasser was created after construction of the Aswan High Dam. This displacement is unacceptable in most societies today. Also unacceptable are recurring attempts to dam areas of great aesthetic value, such as the Grand Canyon. Finally, dam failures—most often geological failures of dams' foundations—result in floods that can take enormous human and economic tolls.

Geothermal Energy

The interior of the earth is an enormous reservoir of heat produced by the decay of small amounts of naturally radioactive elements that occur in all rocks. When the deep heat rises to shallower depths, this **geothermal energy** can be tapped for human use (*geo-*, "earth"; *-therm*, "heat"). The heat loss from the earth's interior would more than satisfy the power needs of all the nations of the world if it could be harnessed. Furthermore, this natural heat is

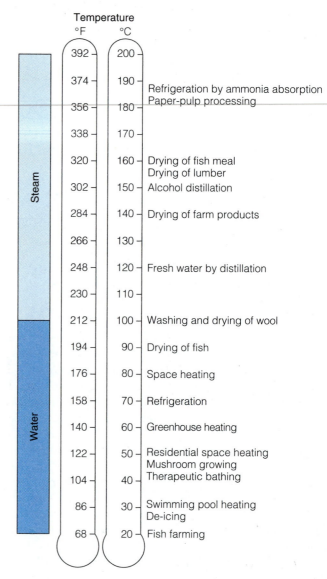

⦿ **FIGURE 13.24** Some established commercial uses of geothermal energy.

cleaner than many other energy sources, and there are fewer negative environmental consequences of using it. Not surprisingly, the prospects for using geothermal energy are best at or near plate boundaries where active volcanoes and high heat flow are found. The Pacific Rim (Ring of Fire), Iceland on the Mid-Atlantic Ridge, and the Mediterranean belt offer the most promise. The first facility to utilize geothermal energy was built in 1905 at Larderello, Italy. Today, the United States, Japan, New Zealand, Mexico, and countries of the former USSR are utilizing energy from earth heat. Iceland uses geothermal energy directly for space heating (one geothermal well there actually began spewing lava), and many other countries have geothermal potential. The versatility of earth heat ranges from using it to grow mushrooms to driving steam-turbine generators with it (⦿ Figure 13.24).

Geothermal energy fields may be found and exploited where magma exists at shallow depths and where there is sufficient underground water to form steam. Geothermal fields are classified as either *steam-dominated* systems or *hot-water-dominated* systems. Whereas deep, insulated reservoirs may produce live steam, shallower reservoirs contain hot water, some of which flashes to steam at the surface. Yellowstone National Park and Wairaki, New Zealand, are examples of hot-water reservoirs (⦿ Figure 13.25). The two kinds of fields require different power-plant designs to generate electricity, as illustrated in ⦿ Figure 13.26. A third kind of geothermal electrical generation utilizes hot water at temperatures below 100° C. The water is used to heat a low-boiling-point liquid, typically isopropane, and under pressure, the "steam" that is produced drives the generators. These *binary* plants are so named because they use two operating liquids.

Larderello, Italy, and The Geysers near Napa Valley in northern California (⦿ Figure 13.27) are examples of steam-dominated fields. Such fields are the rarest and most efficient energy producers. They occur where water temperatures are high and discharge is low so that steam forms. They are much like an oil reservoir in that porous rocks that hold steam or very hot water are overlain by an imperme-

⦿ FIGURE 13.25 Wairaki, one of several large volcanic centers on the North Island of New Zealand. Significant amounts of geothermal energy are tapped here in a country that is not energy-rich.

able layer that prevents the upward escape of the steam or water. ■ Table 13.9 shows the electrical generating and direct-use capacities of selected U.S. geothermal fields in 1990. The Geysers is the largest producer of geothermal

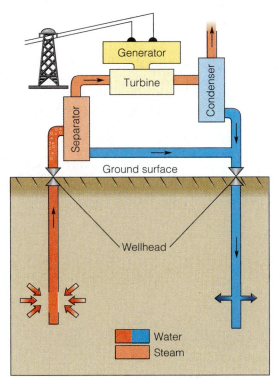

(a) Hot-water system

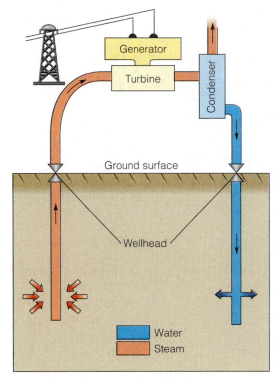

(b) Steam system

⦿ FIGURE 13.26 *(a)* Generation of electricity with a hot-water geothermal system. Steam used to drive the generator is diverted at the separator, and wastewater from the separator and condenser are reinjected underground to keep the system going. *(b)* A steam-dominated electric-power generation system. Steam from the well is injected directly onto the turbine generator. Wastewater from the condenser is reinjected underground to help extend the life of the system.

● FIGURE 13.27 The Geysers geothermal field in Sonoma County, California, is steam-dominated. It is the largest geothermal-energy producer in the world.

tricity are produced at Mammoth Lakes, Coso Springs, and the Salton Sea in California, all of which are underlain by cooling magma.

Geothermal energy is clean. Coal-burning plants typically generate 5.5 metric tons of CO_2 for each megawatt of electricity produced, and relatively clean-burning natural gas (methane) releases 3 tons of CO_2 per megawatt. Geothermal energy, in contrast, releases from 0.5 metric ton to as little as 10 kilograms of CO_2 per megawatt. Similar contrasts exist for emissions of sulfur and sulfur oxides.

The problems associated with geothermal energy production are related to water withdrawal and water quality. Some geothermal waters contain toxic elements such as arsenic and selenium and heavy metals such as silver, gold, and copper. Technology is in place to transfer the heat of the toxic brine deep within the wells to a clean, working fluid that can be brought safely to the surface. This eliminates the easier but environmentally less desirable process of bringing salty water to the surface, using its heat to make electricity, and then disposing of it there. In some geothermal areas removal of underground water can cause surface subsidence and perhaps even earthquakes. At The Geysers, subsidence of 13 centimeters (5 in) has been measured without noticeable impact. Reinjecting the cooled water back into the geothermal reservoir decreases the subsidence risk and helps to ensure a continued supply of steam to the electrical generators. All geothermal fields in California are

power in the world; its production is sufficient to supply electricity to a U.S. city of 1.3 million people. More than 600 wells had been drilled there by 1994, some of them as deep as 3.2 kilometers (10,400 ft). To illustrate geothermal energy's potential, consider this: whereas a nuclear power plant generates about 1,000 MW, The Geysers produced 1,300 MW in 1992. Lesser but important amounts of elec-

■ TABLE 13.9 Electrical and Direct-Use Capacities of Selected U.S. Geothermal Fields, 1990

STATE	FIELD	ELECTRICAL CAPACITY, MW	DIRECT-USE CAPACITY, MW (THERMAL)
California	The Geysers★	1,970[1]	—
	Coso	256[2]	—
	Salton Sea area★	214[2]	—
	East Mesa	10[3]	—
	Long Valley (Mammoth Lakes)	43[3]	—
Colorado	Pagosa Springs	—	14
Hawaii	Puna	25[2]	—
Idaho	Boise	—	57
Nevada	Dixie Valley	50[2]	—
	Steamboat Springs	18[2,3]	—
New Mexico	Animas	—	8
New York	Auburn	—	1.2
Oregon	Klamath Falls	—	37.8
Utah	Cove Fort	13[2,3]	—
Wyoming	Thermopolis	—	9.0

★ 1993 data.
[1] Steam is piped directly to a power plant.
[2] Flash wells produce a mixture of steam and water that must be separated.
[3] Binary wells produce hot water that is piped to a power plant, where heat is transferred to another fluid.
SOURCE: U.S. Geological Survey circular 1125, 1994.

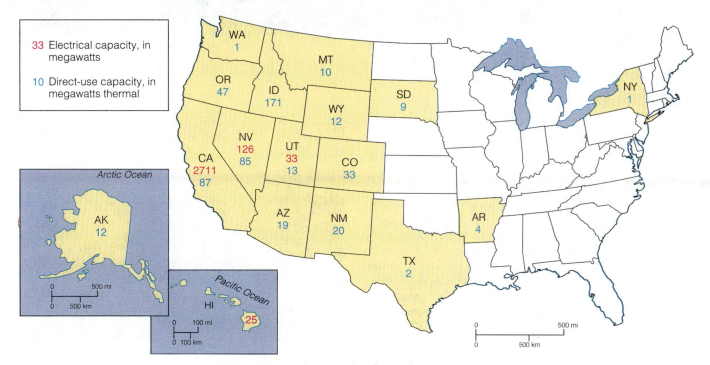

Electrical capacity, in
33 megawatts

Direct-use capacity, in
10 megawatts thermal

◉ FIGURE 13.28 Total capacities of electrical and nonelectrical geothermal developments in the United States, 1990.

being monitored for possible seismic activity related to steam production. Small quakes have been reported from The Geysers, but none of significant magnitude.

The future for geothermal energy looks good. About 20 geothermal fields in the United States are generating electricity, and direct-use hydrothermal systems are being developed at more than 30 sites (◉ Figure 13.28 and Case Study 13.5). In spite of some technical and environmental problems, it is anticipated that geothermal energy will supply 1 percent of U.S. energy needs by the early twenty-first century.

NUCLEAR ENERGY

Nuclear energy provides 16 percent of the world's electricity and 19 percent in the United States. At one time it was considered *the* solution to the world's future energy needs—being clean, limitless, and "too cheap to meter." More than 400 nuclear-power plants were built worldwide, 111 of them in the United States, and most of them are currently operating (◉ Figure 13.29). As with many new technologies, however, some segments of society viewed power derived from nuclear reactors with fear and apprehension. Natural gas, electricity, and even the automobile were met with similar doubts and suspicions, but most people consider these things necessities today (even through 4,000 people per year are accidentally electrocuted, and California alone records one automobile death every 2 hours). Heightened by the accidents at Three Mile Island and Chernobyl, public opposition became such that no new orders for nuclear reactors have been placed in the United

States since 1978. In the early 1990s Sweden passed a referendum calling for the dismantling of its 12 reactors by the year 2010, and other Western European countries have *de facto* moratoriums on new reactors.

Controlled nuclear fission was developed in 1942 by a U.S. effort led by Enrico Fermi at the University of Chicago. Fermi found that by stacking graphite blocks con-

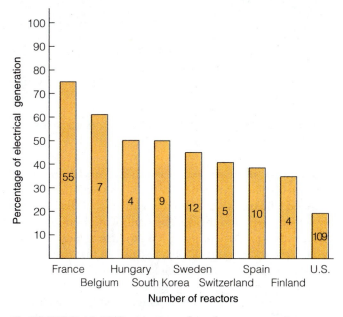

◉ FIGURE 13.29 Percentage of total energy need supplied by nuclear energy and the number of operating reactors for selected countries, 1994.

It's Hot In Iceland

Legend has it that in A.D. 874 Norwegian Ingólfur Arnarson, the first settler in Iceland, threw his shipboard throne overboard and followed it, relying on the gods to direct him where to land. The throne washed ashore where smoke rose out of the ground, and he and Mrs. Arnarson named the site *Reykjavík,* "smoky bay." The smoke was actually steam rising above a hot-water spring, and today geothermal water is the island-country's most important natural resource. Icelanders have utilized naturally occurring hot water for space heating since 1928, when a school building was plumbed to a well yielding 87° C water (189°F). Today, 50 wells in the Reykjavík region produce water at temperatures of 62°–132° C (144°–270° F). This water keeps 24,000 houses warm during bitter winters and provides inexpensive energy (the cost being about half that of oil) for 145,000

◉ FIGURE 1 Reykjavík, Iceland. A hot-water well is in the brown shed in the distance, and wells under development are seen at the far right and in the foreground.

◉ FIGURE 2 The Blue Lagoon, a great hot-water pool 40 kilometers (24 mi) from Reykjavík. It uses geothermal power-plant outlet water, which is salty and grows algae, giving it its blue color. Open every day, the lagoon attracts about 100,000 visitors annually.

people, well over half of Iceland's population (◉ Figure 1). Geothermal waters are also used in public swimming pools (◉ Figure 2) and to heat the country's many greenhouses (35,800 acres of them), which produce fresh vegetables and flowers year round.

Because Iceland is split by a divergent plate boundary and consists of young volcanic rock, warm water is encountered in almost every hole drilled on the island. Wells close to active volcanism (the actual boundary) provide steam for producing electricity. A little more than 45 percent of the country's primary energy is derived from geothermal sources. Hydroelectric power provides 17 percent, and coal and oil account for the rest. Iceland serves as an excellent example of human utilization of a natural resource with positive results and few negative environmental consequences.

taining lumps of uranium, a self-sustained nuclear reaction could be obtained that generated heat. This experiment was in essence the first atomic reactor. The work at the University of Chicago led to the development of the atomic bomb, a device that produces energy at a prodigious rate. Atomic reactors produce energy at such a slow rate that they could not possibly give rise to an atomic explosion.

In 1979 the United States sustained its first and only nuclear power plant accident at Three Mile Island in Pennsylvania. Although media coverage was extensive, there were no deaths or injuries. This is not to minimize the

health hazards of handling radioactive materials, but electrical generation using nuclear energy is cleaner and, so far, safer than hydroelectric or fossil-fuel facilities.

Nuclear reactors and fossil-fuel plants generate electricity by similar processes. Both operate on the principle of heating a fluid, which then directly or indirectly makes steam used to spin turbine blades that drive an electrical generator. Uranium is the fuel of atomic reactors, because its nuclei are so packed with protons and neutrons that they are capable of sustained nuclear reactions. All uranium nuclei have 92 protons and between 142 and 146 neutrons.

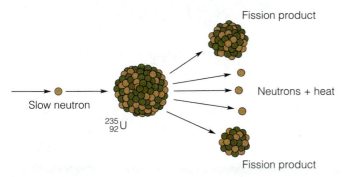

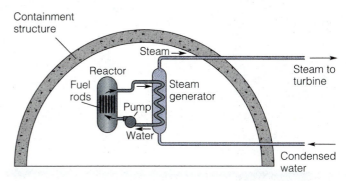

● FIGURE 13.30 Nuclear fission. A slow neutron penetrates the nucleus of a U-235 atom, creating fission products (also called *daughter isotopes*), 2 or 3 fast neutrons, and heat.

● FIGURE 13.31 Main components of a light-water reactor. Note that the reactor is surrounded by a containment structure and that the steam supplied to the turbine is isolated from the reactor core.

The common isotope uranium-238 (^{238}U) has 146 neutrons and is barely stable. The next-most common isotope, ^{235}U, is so unstable that a stray so-called slow neutron penetrating its nucleus can cause it to split apart completely, a process known as **fission**—a term particle physicists borrowed from the biological sciences. Nuclei that are split easily, such as those of ^{235}U, ^{233}U, and plutonium-239 (^{239}Pu), are called **fissile isotopes.** Fission occurs when a fissile nucleus absorbs a neutron and splits into lighter elements, called *fission products,* while at the same time emitting several "fast" neutrons and energy, 90 percent of which is heat (● Figure 13.30). The lighter fission products are elements that recoil from the split nucleus at high speeds. This energy of motion and subsequent collisions with other atoms and molecules creates heat, which raises the temperature of the surrounding medium.

Meanwhile, the stray fast neutrons collide with other nuclei, repeating the process in what is called a **chain reaction.** A controlled chain reaction occurs when one free neutron on average from each fission event goes on to split another nucleus. If more than one nucleus is split by each emitted neutron (on average), the rate of reaction increases rapidly, and the reaction eventually goes out of control. Because ^{235}U makes up only 0.7 percent of all naturally occurring uranium, the rest being ^{238}U, it must be "enriched" for use in commercial nuclear reactors.

In 1994, 109 commercial nuclear power plants were operating in the United States. Reactors that use slow neutrons to create controlled fission have a moderating material surrounding the "hot" fuel rods for slowing down the emitted fast neutrons and increasing the probability of nuclear collisions. Most U.S. reactors use water as the moderator and are known as *light-water reactors (LWR).* The components of an LWR reactor are shown in ● Figure 13.31. Note that a thick containment dome surrounds the fuel rods in the reactor core, the surrounding plumbing for cooling the core, and the plumbing for heat transfer to the turbines. This is to prevent the release of radioactivity to the environment in the event of an accident (see Case Study

13.6). If the reactor fuel vessel (the core) is not cooled properly, it will overheat, and it may reach the temperature where the fuel rods and the vessel melt. This is known as a **central-core meltdown.** When it occurs the molten material may eat its way through the floor of the containment structure and release radioactivity to the atmosphere.

Applications of probability risk analysis to modern reactors place a very low probability of a core meltdown. The major risk associated with nuclear energy is the risk of contamination during processing, transportation, and disposal of ^{235}U and high-level nuclear waste products. The locations of these risks in the nuclear fuel cycle are apparent in ● Figure 13.32, page 396.

GEOLOGICAL CONSIDERATIONS. One of the many criteria in the siting of nuclear reactors is geologic stability. The selected site must be free from landsliding, tsunami, volcanic activity, flooding, and the like. The Nuclear Regulatory Commission (NRC), the federal government body that licenses nuclear reactors for public utilities, requires that all active and potentially active faults within a distance of 320 kilometers (200 mi) of a nuclear power plant be located and described. Reactors must be built a distance from an active fault determined by the fault's earthquake-generating potential. Further, the NRC defines an active fault as one that has moved once within the last 30,000 years or twice in 500,000 years. This is a very strict definition, and sites that meet these criteria are difficult to find along active continental margins. To provide additional safety and allow for geological uncertainty, reactors are programmed to shut down immediately at a seismic acceleration of 0.05 *g*. This is conservative indeed, and the shutdown would prevent overheating or meltdown should the earthquake be a damaging one.

 NUCLEAR WASTE DISPOSAL

There is probably no more sensitive and emotional issue involved in geology today than the need to provide for safe disposal of the tons upon tons of radioactive materials that

Three Mile Island and Chernobyl—To Err Is Human; To Forgive Is Difficult

It started on 28 March 1979 with a valve that stuck and allowed core coolant to escape from the reactor vessel. No problem; the emergency core-cooling system (ECCS) automatically goes into action, adding "makeup" water to the core and keeping it cool. Enter the human factor. The operators misinterpreted the information fed to them by their instruments and *shut off* the ECCS; that is, they literally overrode the safety system that was designed and installed to mitigate just such an event. This human error at Three Mile Island Nuclear Station in Pennsylvania initiated the worst accident in the history of U.S. commercial power reactors.

With the ECCS shut down, the heat that built up in the core boiled off the cooling water surrounding the fuel rods. Before operators realized they had made a mistake, about a third of the fuel had melted and dropped into the cooling water in the bottom of the vessel. The vessel did not break, and the water quenched the molten rods, but the damage had been done. The material covering the fuel rods had deteriorated, and fission products had escaped into the steam to be carried into the containment building. Essentially all of the radioactivity remained in the building, but another human error allowed overflow water with low-level radioactivity to be pumped out of the leakproof containment structure to an adjacent building that was not leakproof.

Although "Three Mile Island" was a frightening incident, no adverse health effects have yet been detected. It is estimated that the gases that leaked from the containment structure gave the local residents a total radiation dosage equal to that of 4 chest X rays.

On 26 April 1986, Reactor Number 4 at Chernobyl, a nuclear facility 100 kilometers (60 mi) north of Kiev in the Ukraine, exploded (●Figure 1). The cause was similar to that at Three Mile Island, only here an experiment was being conducted in which the graphite-moderated reactor was running while the emergency core-cooling system was turned off. Miscalculation by operators allowed neutron buildup in part of the core, and the chain reaction went out of control—became *supercritical,* in the jargon of the nuclear scientist. Two explosions are known to have occurred, one of which blew the top off the reactor building. This latter explosion is believed to have been caused by a buildup of hydrogen when steam reacted with graphite, but both may have been gigantic steam explosions. With the core exposed, the graphite ignited and burned, furiously spewing radioactive materials directly into the atmosphere. Ten days were required to put out the fire, and when it was out, the reactor core was buried in concrete. Two workers at the plant were killed instantly, and 29 others died later due to radiation exposure. In all, 500 people were hospitalized.

●FIGURE 1 Chernobyl Reactor Number 4 after the explosion.

FIGURE 2 The Chernobyl fallout area.

Area of elevated radiation

LATVIA

LITHUANIA

RUSSIA

POLAND

BELARUS

Mensk

Gomel

Chernobyl Atomic Energy Plant

Kiev

UKRAINE

0   100 mi
0   100 km

⊙ **FIGURE 3** A colt born in the Ukraine after the Chernobyl accident. Genetic damage is indicated by the colt's deformed legs and extra legs and hooves.

The significance of Chernobyl is that large quantities of radioactive materials were released directly into the atmosphere. In fact, scientists who were measuring atmospheric radiation in Sweden were the first to tell the world that an accident had occurred somewhere in the Soviet Union. Sadly, it took 36 hours to evacuate people within a 30-kilometer (18 mi) radius of the plant, and it is expected that within this group, there will be 500 more cancer-related deaths than normal in the next 70 years. Seventy-five million people lived within the fallout area in the western USSR and Eastern Europe, and the Chernobyl disaster may have increased the number of future cancer deaths in this group by as much as 24,000. All of the water, soil, and vegetation within the 30-kilometer fallout zone had to be decontaminated. Many areas are still "hot" and probably will never be cleaned up (⊙ Figure 2). Gomel, 150 kilometers (90 mi) north of Chernobyl, has a child thyroid-cancer rate 80 times the world average. A reasonable medical explanation is that the Gomel residents' thyroid glands have absorbed and concentrated the radioactive iodine-131 that was released from the burning reactor. The prognosis is indeed grim for the long-term cancer toll from radiation exposure in the fallout area (⊙ Figure 3).

The accidents at Three Mile Island and Chernobyl demonstrated differences in reactor design. The Russian reactor would not have been licensed in the United States, because it did not meet U.S. design and safety standards. Had Chernobyl Reactor Number 4 been enclosed by a containment building like the one at Three Mile Island, the escaping radioactivity would have been trapped and the disaster would have been averted, or at least minimized. Today, Number 4 is buried beneath concrete, but the other reactors at Chernobyl are back in operation. Six plant officials were convicted of negligence and are serving prison terms.

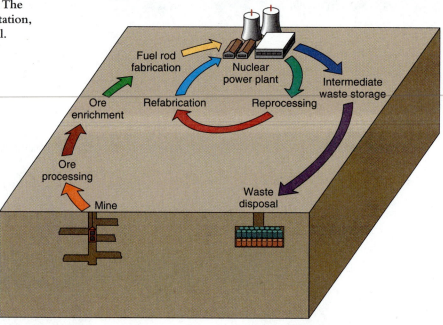

● **FIGURE 13.32** The nuclear fuel cycle. The steps include mining, enrichment, transportation, power generation, reclamation, and disposal.

have accumulated over many decades. These wastes are the largest deterrent to further development of nuclear energy, as they present a legitimate hazard to humans, their offspring, and generations yet to come. Nuclear-waste products must be isolated, because they emit high-energy radiation that kills cells, causes cancer and genetic mutations, and causes death if individuals are exposed to large doses (as at Chernobyl; see Case Study 13.6). The need to develop the means for safe, permanent disposal led Congress to pass the Nuclear Waste Policy Act in 1982 (NWPA). This law gave the Department of Energy (DOE) responsibility for locating a fail-safe underground disposal facility, a **geological repository,** for high-level nuclear waste and designated nine

locations in six states as potential sites (● Figure 13.33). Based upon preliminary studies, President Reagan approved three of the sites for intensive scientific study (known as *site characterization*): Hanford, Washington; Deaf Smith County, Texas; and Yucca Mountain, Nevada. In 1987 Congress passed the Nuclear Waste Policy Amendments Act, which directed DOE to study only the Yucca Mountain site. This site's suitability is discussed later in this section.

Radioactive wastes are differentiated by the intensity of their radiation and by their physical form; that is, whether they are a solid, liquid, or gas. Their physical form determines how they need to be handled and their potential as a health and environmental hazard.

● **FIGURE 13.33** The nine potential repository sites for high-level nuclear waste. A repository has yet to be built, and the Yucca Mountain location was the only one being studied in 1996. The Texas, Louisiana, and Mississippi sites are all salt domes or bedded salt deposits.

Types of Nuclear Waste

HIGH-LEVEL WASTES. High-level wastes, by-products of nuclear power generation and military uses, are the most intensely radioactive and dangerous. About once a year, a third of a reactor's fuel rods are removed and replaced with fresh rods. Replacement is necessary because, as the fissionable uranium in a reactor is consumed, the fission products capture more neutrons than the remaining uranium produces. This causes the chain reaction to slow, and eventually it would stop. The radioactive rods, called *spent fuel,* are the major form of high-level nuclear waste. Currently they are stored in water-filled pools at the individual reactor sites, supposedly a temporary measure until a final grave is approved and prepared for them. More than 20,000 metric tons of spent fuel is stored across the country, and at many of the reactors, pool storage capacity is nearly filled (⊙ Figure 13.34). It is expected that 40,000 metric tons of spent fuel will be in "temporary" storage by the year 2000. As on-site pools become filled with spent fuel, the rods must either be packed more closely or transported to sites with more storage space. Two major concerns about pool storage are that an unintended nuclear reaction might start in the pool and that the rods might deteriorate and release fuel pellets. It should be noted that although the fuel removed from a reactor every year weighs about 31,000 kilograms (65,000 lbs), because of its high density, it is only about the size of an automobile.

Alternatively, the rods are chopped up and *reprocessed* to recover unused uranium-235 and plutonium-239. However, the liquid waste remaining after reprocessing contains more than 50 radioactive isotopes, such as strontium (^{90}Sr) and iodine (^{131}I), that pose significant health hazards. In 1992, the Department of Energy (DOE) decided to phase out reprocessing of spent fuel for the purpose of recovering chemical stocks for nuclear weapons. Unfortunately, the volume of existing reprocessed waste in temporary storage

would cover a football field to a depth of 60 meters (200 ft). These wastes are currently stored in metal tanks at four U.S. sites. The oldest site, the one at Hanford, Washington, originally stored wastes in single-walled steel tanks. In 1956 a tank leak was detected, and since then, 750,000 gallons of high-level waste have leaked underground from 60 of 149 storage tanks. It is said by some that "radioactive waste was invented here." Cleanup at Hanford may be the costliest environmental cleanup in world history; the cost has been estimated as at least $57 billion.

TRANSURANIC WASTES. Most (96 percent) of the nuclear fuel in a reactor is composed of uranium-238, which is not involved in the chain reaction. Instead, ^{238}U absorbs neutrons and becomes a **transuranic element**—one whose atomic number is higher than that of uranium. Thus the ^{238}U in a fuel rod changes into one of the 11 unstable transuranic elements *(TRU),* of which plutonium, fermium, nobelium, einsteinium, and berkelium are examples. Transuranic waste decays slowly and requires long-term isolation from humans and the environment. The wastes typically include tools, gloves, rags, protective clothing, and debris contaminated with plutonium during facility operations. The DOE Waste Isolation Pilot Plant (WIPP) has been designed to store transuranic waste in vast salt deposits 660 meters (2,150 ft) beneath the desert surface 44 kilometers (26 mi) east of Carlsbad, New Mexico. These salt deposits are impermeable, easily mined, and capable of eventually self-sealing around voids, such as the waste-storage rooms that are created by mining. Part of the DOE's long-term planning for radioactive-waste disposal, the WIPP site is scheduled to begin accepting waste in April 1998.

LOW-LEVEL WASTES. Low-level wastes have neither the radioactivity of high-level wastes nor the long decay times of transuranic elements. They are created in research

Pearl Harbor

⊙ **FIGURE 13.34** Location of temporarily stored spent nuclear fuel and other high-level nuclear waste awaiting permanent disposal, 1994. The symbols do not reflect precise locations.

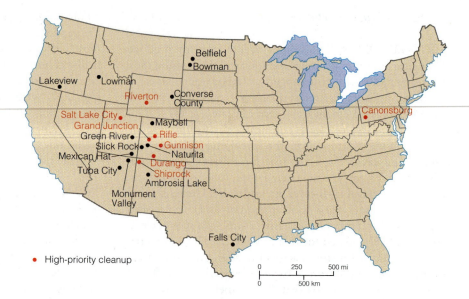

Belfield
Bowman
Lakeview • Lowman
Riverton • Converse County
Canonsburg
Salt Lake City
Grand Junction • Maybell
Green River • Rifle
Slick Rock • Gunnison
Mexican Hat • Naturita
Tuba City • Durango
Shiprock
Ambrosia Lake
Monument Valley
Falls City

• High-priority cleanup

0   250   500 mi
0   500 km

labs, hospitals, and industry as well as in reactors. Ocean dumping of these wastes in 55-gallon drums was a common practice between 1946 and 1970. Since then, shallow land burial has been the preferred disposal method.

MILL TAILINGS. Uranium mill tailings, the finely ground residue that remains after uranium ore is processed, are typically found in mountainous piles outside the mill facility. Large volumes of tailings exist that still contain low concentrations of radioactive materials, including thorium (^{230}Th) and radium (^{226}Ra), both of which produce radioactive radon gas (^{222}Rn), the so-called hidden killer (Case Study 13.7). Most of the abandoned tailings piles are in sparsely settled areas of the West (◉ Figure 13.35.). A former Atomic Energy Commission policy permitted the use of uranium-mill tailings in the manufacture of building materials such as concrete blocks and cement. For almost twenty years, 30,000 citizens of Grand Junction, Colorado, lived in homes whose radon levels were up to seven times the maximum allowed for uranium miners. Uranium tailings were used in building materials in Durango, Rifle, and Riverton, Colorado; Lowman, Idaho; Shiprock, New Mexico; and Salt Lake City, Utah.

The Uranium Mill Tailings Radiation Control Act (1978) makes the DOE responsible for 24 inactive sites in 10 states left from uranium operations. The act specifies that the federal government is to pay 90 percent of the cleanup cost, with the respective state to pay the remainder. Without some form of management, uranium-mill tailings could become a serious problem, because they contain the largest volume of radioactive waste in the country.

Isolation of Nuclear Waste

Scientists agree that nuclear waste must be isolated from contact with all biological systems for at least 250,000 years in order for radioactive materials—particularly plutonium-239, whose half-life is 24,000 years—to decay to harmless levels. Should the ban on plutonium recovery be rescinded, the conditions for high-level-waste disposal may change. Several methods of storing wastes have been proposed, but any method must meet these conditions:

- Safe isolation for at least 250,000 years,
- Safe from sabotage or accidental entry,
- Safe from natural disasters (hurricanes, landslides, floods, etc.),
- Noncontaminating to any nearby natural resources,
- Geologically stable site (no volcanic activity, earthquakes, etc.), and
- Fail-safe handling and transport mechanisms.

A number of disposal methods have been proposed. Six of the more serious ones are discussed here.

1. *Rocket the waste into the sun or deep space.* Rocket launchings are prohibitively expensive, and an explosion during launch could contaminate the atmosphere, ocean, and land.
2. *Continue the present tank storage.* This is not a permanent solution, because the tanks have limited lifetimes and leakage has already occurred.
3. *Dispose of waste in a convergent plate boundary (subduction zone).* The feasibility of this very imaginative proposal has not been demonstrated, because the rate of plate movement is very slow.
4. *Place containerized waste on the ice sheets of Antarctica or Greenland.* Radioactive heat would cause the containers to melt downward to the ice-bedrock contact surface, which could place the radioactive waste two miles beneath ice in either Greenland or Antarctica. This proposal assumes that long-term global warming will not occur in the 250,000-year period. This is unsound, because the last ice sheet of the Pleistocene melted back to Greenland in less than 10,000 years.

5. *Geologic repository: Salt mines or domes.* Salt is dry, flows readily, and is self-healing. Because it is a good conductor of heat, temperatures in the repository rocks would not become excessive. Salt is subject to underground solution, however, and some hydrated minerals in salt beds give up their water when heated. Salt beds and salt domes occur in geologically stable regions such as in Texas, Louisiana, and Mississippi.

6. *Geologic repository: Deep chambers in granite, volcanic rocks, or some other very competent, relatively dry rock formation.* This would require detailed site analysis to ensure that all geologic criteria are met.

YUCCA MOUNTAIN GEOLOGIC REPOSITORY.

The DOE's Civilian Radioactive Waste Management Program has been conducting a site characterization of Yucca Mountain, Nevada, as a high-level-nuclear-waste repository since the early 1980s (◉ Figure 13.36). The 7-square-kilometer (2.7-mi²) site is in welded ashflow tuffs (see Chapter 5), and the waste would be placed more than 200 meters (650 ft) below the surface (◉ Figure 13.37, page 402). A favorable geologic condition at Yucca Mountain is the depth to the water table, 224–366 meters (800–1,200 ft) below the level of the proposed storage chambers. Important geologic questions in the site investigation include:

- What is the depth, thickness, and extent of the host rocks?
- How much water is in the zone of aeration, and how does it move through the rocks?
- What effect might nearby volcanic activity have on the repository?
- How might an earthquake affect the repository and the water table below it?

All geologic studies indicate that Yucca Mountain could be an acceptable geologic repository, but until it receives final certification, radioactive wastes are piling up in temporary storage facilities at an alarming rate.

Completion of the Yucca Mountain repository was originally scheduled for 1998, but the year has been extended twice to date, first to 2003 and then to 2010. The legal stumbling blocks placed in the way of the project by the State of Nevada are classic examples of NIMBY (see Case Study 15.1). When and if the repository is ever built and filled, the DOE will be required to monitor it for leakage at least 50 years from the opening date. Eventually the site will be returned to its original condition.

The problem of radioactive-waste disposal is common to all countries that use nuclear power. Debate is heated over a repository in Wordsworth's backyard in England's world-famous Lake District, and the Swiss, who get 40 percent of their power from reactors, are looking at a tunnel into Mount Wellenberg for storing their wastes.

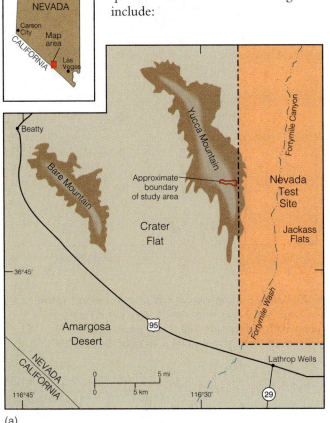

(a)

(b)

◉ **FIGURE 13.36** *(a)* Location of the proposed Yucca Mountain nuclear repository. *(b)* Aerial view of the Yucca Mountain study area. Structures are at the site of the proposed repository.

CASE STUDY 13.7

Radon and Indoor Air Pollution

R adon (^{222}Rn) is a radioactive gas formed by the natural disintegration of uranium as it transmutes step-by-radioactive-
step to form stable lead. The immediate parent element of radon
is radium (^{226}Ra), which emits an alpha particle (^{4}He) to form
^{222}Rn. Radon-222 has a half-life of only 3.8 days, but it in turn
breaks down into other elements that emit dangerous radionu-
clides. *Emanation* is sometimes used to describe the behavior of
this element, and concern has been growing about the health
hazard posed by radon emanating from rocks and soils and seeping into homes. Radon was first suggested as a danger to miners
by Georgius Agricola (1494–1555). Known as "the father of
mineralogy," Agricola proposed that the lung disease contracted
by miners might have something to do with emanations inside
the mines. The EPA estimates that between 5,000 and 20,000
people die every year of lung cancer because they have inhaled
radon and its decay products.

Radon is only a problem in areas underlain by rocks whose
uranium concentrations are greater than 10 ppm—three times
the average amount found in granite. For comparison, uranium
ore contains more than 1,000 ppm of uranium. As radon disintegrates in the top few meters of rock and soil, it seeps upward
into the atmosphere, or into homes through cracks in concrete
slabs or around openings in pipes. The measure of radon activity
is *picocuries per liter (pCi/L)*—the number of nuclear decays per
minute per liter of air. One pCi per liter represents 2.2 potentially cell-damaging disintegrations per minute. Inasmuch as the
average human inhales between 7,000 and 12,000 liters
(2,000–3,000 gallons) of air per day, high radon concentrations in
household air can be a significant health hazard. In addition,
radon can access the human body through well water if it is
injested within a few weeks of withdrawal from underground.
Average uranium concentrations of some common rocks are

black shale	8.0 ppm
granite	3.0 ppm
schist	2.5 ppm
limestone	2.5 ppm
basalt	1.0 ppm

Bear in mind that these figures do not indicate the *range* of uranium content in the rocks, which for black shales and granites
can be quite great.

Cancer due to bombardment of lung tissue by breathing radon-
222 or its decay products is a major concern of health scientists.

pCi/L	Comparable Exposure Level	Comparable Risk *
200	1,000 times average outdoor level	More than 75 times nonsmoker risk of dying from lung cancer
100	100 times average indoor level	4-pack/day smoker / 10,000 chest X rays per year
40		30 times nonsmoker risk of dying from lung cancer
20	100 times average outdoor level	2-pack/day smoker
10	10 times average indoor level	1-pack/day smoker
4		3 times nonsmoker risk of dying from lung cancer
2	10 times average outdoor level	200 chest X rays per year
1	Average indoor level	Nonsmoker risk of dying from lung cancer
0.2	Average outdoor level	

* Based on lifetime exposure

● **FIGURE 1** Lung cancer risk due to radon exposure,
smoking, and X-ray exposure. A few houses have been
found with radon radiation levels greater than 100 pCi/L,
and one in Pennsylvania had a level of 2,000 pCi/L.

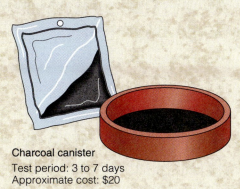

● FIGURE 2
Canister and plastic-
film radon detectors.

Charcoal canister
Test period: 3 to 7 days
Approximate cost: $20

Alpha-track detector
Minimum test period: 2 to 4 weeks
Approximate cost: $26 for one
detector; discounts for multiple detectors

The EPA has established 4 pCi/L as the maximum allowable indoor radon level. This is equivalent to smoking a half-pack of cigarettes per day or having 300 chest X rays in a year. ● Figure 1, opposite, vividly illustrates the dangers of inhaling indoor radon (and of smoking). Such cancer is rarely apparent before 5–7 years of exposure, and its incidence is rare in individuals under the age of 40.

Widespread interest in radon pollution first appeared in the 1980s with the discovery of high radon levels in houses in Pennsylvania, New Jersey, and New York. The discovery was made accidentally when a nuclear-plant worker set off the radiation monitoring alarm when he arrived at work one day. Since his radioactivity was at a safe level, he did not worry about it. Returning to work after a weekend at home, he once again triggered the alarm. An investigation of his home revealed very high levels of radon. This was the first time radon was recognized as a public health hazard.

There are currently two types of radon "detectors." One is a charcoal canister that is placed in the household air for a week and then returned to the manufacturer for analysis. The other is a plastic-film alpha-track detector that is placed in the home for 2–4 weeks (● Figure 2). It is recommended that the less-expensive canister type be used first, and that if a high radon level is indicated, a follow-up measurement with a plastic-film detector be made. Any reading above 4 pCi/L requires follow-up measurements. If the reading is between 20 and 200 pCi/L, one should consider taking measures to reduce radon levels in the home (● Figure 3). Common methods are increasing the natural ventilation below the house in the basement or crawl-space, using forced air to circulate subfloor gases, and intercept-

ing radon before it enters the house by installing gravel-packed plastic pipes below the floor slabs. For houses with marginal radon pollution, sealing all openings from beneath the house that go through the floor or the walls is usually effective.

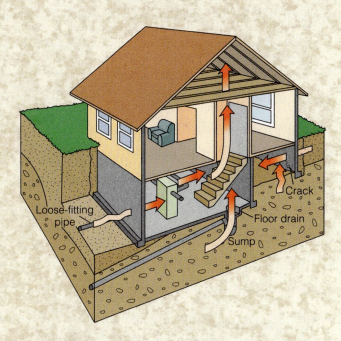

Loose-fitting pipe

Crack

Floor drain

Sump

● FIGURE 3 Routes of radon entry into a home. Plugging radon leaks is the most common method of reducing residential radon accumulations.

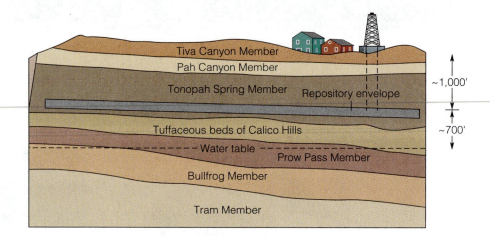

● FIGURE 13.37 Paintbrush Tuff Formation at Yucca Mountain. Rock layers called *members* are distinguishable beds within the formation. The proposed position of the repository is shown.

Health and Safety Standards for Nuclear Waste

There is general agreement that any nuclear-waste repository will leak radiation over the course of the next 10,000 years. Nuclear waste emits three types of radiation:

1. Alpha (α) particles, the most energetic but least penetrating type of radiation (they can be stopped by a piece of paper). If α emitters are inhaled or ingested, alpha radiation can attack lung and body tissue.
2. Beta (β^-) radiation, penetrating but most harmful when a β^- emitter is inhaled or ingested. For example, strontium-90 is a β^- emitter that behaves exactly like calcium (builds bone) in the food chain.
3. Gamma (γ) rays, nuclear X rays that can penetrate deeply into tissue and damage human organs (● Figure 13.38).

Nuclear waste policy limits the radioactivity that may be emitted from a repository, as measured by cancer deaths directly attributable to leaking radiation. The mortality limit is 10 deaths per 100 years, and for this reason the maximum allowable radiation dose per year per individual in the proximity of a repository is set at 25 millirems. The *rem* is the unit used for measuring the amounts of ionizing radiation that affect human tissue, and a **millirem** (mrem) is 0.001 rem. Doses on the order of 10 rems can cause weak-

ness, redden the skin, and reduce blood-cell counts. A dose of 500 rems would kill half of the people exposed. Rem units take into account the type of radiation—apha, beta, or gamma—and the amount of radioactivity deposited in body tissue. ● Figure 13.39 shows how the average annual dose of 208 mrems is partitioned for U.S. citizens and common sources of radiation. (Case Study 13.7 explains the hazards of high levels of radon gas in the home and methods for testing and mitigating them.) It is estimated that a person living within 5 kilometers (3 mi) of the Yucca Mountain repository would receive less than 1 mrem per year as a result of nuclear waste being stored there (currently no one lives that close to the site). The 25-mrem standard for maximum annual dosage is equal to the radiation a person would receive in 2 or 3 medical X rays.

Is there a Future for Nuclear Energy?

Nuclear reactors are very complex; a typical U.S. reactor may have 40,000 valves, whereas a coal-fired plant of the same size has only 4,000! This fact alone makes reactors relatively unforgiving of errors in operation, construction, or design. A newer generation of "standard" reactors are now under consideration that use natural forces such as gravity and convection instead of the network of pumps and valves currently required. For example, if the core of one of these new reactors should overheat, a deluge of water stored in tanks above the reactor would inundate the core. In addition, design would be standardized, which would greatly simplify construction, maintenance, and repair. Standardization would also facilitate better operator training and would eliminate the possibility of some human errors.

The ultimate, clean nuclear-energy source is released by **fusion,** whereby two nuclei merge to form a heavier nucleus and in the process give off tremendous amounts of heat. Deuterium ($_1^2$H) is a heavy isotope of hydrogen that at very high temperatures fuses to form helium ($_2^4$He). This is the very same nuclear process that energizes our sun. The fuel is inexhaustible, the reaction cannot run out of control, and it is environmentally nonpolluting. A not insignificant

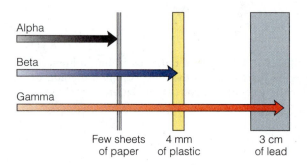

Alpha

Beta

Gamma

Few sheets of paper   4 mm of plastic   3 cm of lead

● FIGURE 13.38 Penetrating abilities of alpha, beta, and gamma radiation, the three types of radiant energy emitted from nuclear wastes.

FIGURE 13.39 The typical American receives 208 millirems of radiation per year from the various sources. Some medical, research, and industrial workers' exposures may exceed the 208-millirem average.

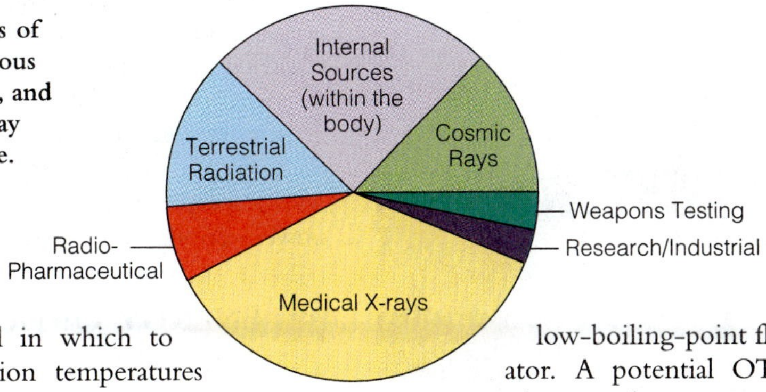

Internal Sources (within the body)

Cosmic Rays

Weapons Testing

Research/Industrial

Medical X-rays

Radio-Pharmaceutical

Terrestrial Radiation

problem is designing a vessel in which to "make a small star" at ignition temperatures between 100 and 250 million degrees—several times the temperature of the sun's interior. The high temperature is required to strip the electron from the deuterium nucleus, to overcome electric repulsion, and to maintain the high velocity necessary for overcoming any repulsion between the atoms' nuclei. Energy from nuclear fusion is a long way off, but it could well become the energy source of the twenty-first century.

Plutonium-239 is fissionable and theoretically can be produced faster than it is consumed when fast neutrons collide with U-238 in a reactor. A reactor based upon this principle is called a **breeder reactor,** because plutonium is "bred" in the process. This is like getting something for nothing, not likely in this world, and it is said that "breeder reactors are in the future and always will be." No work is being done on breeders in the United States because of the fear that the bred plutonium may fall into the hands of belligerents who will use it to make a bomb.

 ## ENERGY FROM THE SEA

Ocean Thermal Energy Conversion

Ocean thermal energy conversion (OTEC) utilizes the temperature difference between warm surface water and colder deep waters, at least 22° C (72° F), to vaporize a low-boiling-point fluid and operate a generator. A potential OTEC site must be in a coastal area where water depth increases rapidly enough to obtain the needed temperature differential and close enough to shore for power transmission. The Island of Oahu in Hawaii has the population, need, and oceanographic conditions for this form of energy conversion. A mini-OTEC of about 50 kilowatts was operated in Hawaii in the 1980s. The electricity produced was not transmitted to the shore, but a continuously burning light bulb on a float near the village of Mokapu testified to OTEC capability. At present, large-scale utilization of the ocean thermal gradient cannot compete with other sources, but it may be the energy solution for some island communities.

Wave Energy

Capturing wave energy and converting it to electricity is not a new idea, and its practicality has improved very little since 1900. Wave energy is so diffuse—spread over such a large area—that concentrating and utilizing it is difficult. In order to convert a worthwhile amount of energy from wave motion, the energy must be focused by "horns," or funnels, extending seaward. In a 1971 experiment, a wave-powered generator hanging from a pier was used to power a string of lights of Pacifica, California. The Japanese currently market a wave-power machine for providing energy to lighthouses and buoys, and they have been working on wave-powered electrical-generating barges for offshore installation. So far, no large facility has proved feasible.

Ocean Currents

Ocean currents have potential for electrical generation. The kinetic energy of the Florida Current, between Cuba and the tip of Florida, for example, is equivalent to 25 1,000-MW generating plants, and the Kuroshio Current along the east coast of Japan is swift (about 6 knots) and close to users. Unfortunately, oceanic-current energy also is diffuse, and currents meander and change course unpredictably. Thus, very large installations are required.

Tidal Currents

Tidal currents have been used by coastal dwellers to power mills for at least a thousand years. Restored, working tidal

CONSIDER THIS . . .

Many utilities are getting out of the nuclear energy business, as it has become too expensive due to nuclear-waste management and storage problems. Yucca Flats in Nevada is widely regarded as the preferred geologic repository, except in Nevada. The State of Nevada has been using every legal means available to halt its completion. In view of the 20,000 tons of high-level waste sitting in pools around the country, do you think a state should have the right to stop this project that will benefit all citizens of the United States?

● FIGURE 13.40 The Rance River tidal-power electrical generating facility near St. Malo, province of Brittany, France. Because tidal difference at sea level is relatively small, 24 small, 10-MW generators are required to produce a significant amount of electricity.

mills in New England and Europe are popular tourist attractions. Both flood (landward-moving) and ebb (seaward-moving) tidal currents are used. Modern designs utilize basins that fill with seawater at high tide and empty at low tide, generating electricity in the process. A potential site must have a large tidal range—the height difference between high and low tides—and adequate area for storing the elevated water. It is estimated that there are more than 100 suitable sites in the world that, together, are capable of generating 5 percent of the world's present generating capacity.

The first and best-known large-scale tidal power plant is on the Rance River near St. Malo on the English Channel in northwest France (● Figure 13.40). The tidal range here varies from 9 to 14 meters (30–46 ft), and the power output is a reversible operation; that is, the plant generates power at both flood and ebb tides. Because the hydrostatic head (water drop) is only a few meters, 24 small 10-MW generators with a total capacity of 240 MW are used. The estuary of the Rance River serves as a reservoir for high water, and locks through the dam allow navigation upstream. The Rance project is nonpolluting, and it has been a boon to the tourist-dependent province of Brittany. On balance, the operation has been successful.

In North America interest has focused on the Bay of Fundy in the Atlantic Ocean's Gulf of Maine. This bay has the greatest tidal range in the world, about 16 meters (52 ft), and a storage area in excess of 350 square kilometers (135 mi²). The bay is made up of 9 smaller bays with a total generation potential of 30 gigawatts (1 GW = 1,000 MW), equivalent to 30 conventional power plants. Under consideration by Canada and the United States for more than a half-century, the proposed project is known in both countries as the *Passamaquoddy Site*. Through a series of dams

and locks, high-tide water would be stored in Passamaquoddy Bay and then drained at low tide to produce electricity (● Figure 13.41). It is a very ambitious and costly project, but as population and demand for nonpolluting energy increase, the project is likely to be built in the next century. With an anticipated generating capacity of 65–345 MW, its capacity as well as its size would dwarf the Rance power station.

⊙ ENERGY OPTIONS FOR THE TWENTY-FIRST CENTURY

The ability of the United States, as well as every other country in the world, to meet the energy demands of the next century depends on its ability to slow and eventually to end population growth. As emphasized in Chapter 1, this is imperative if we are to have any future at all. All scenarios for energy sufficiency in the United States assume that early in the twenty-first century energy demand will stabilize at about 20–25 quads above the 1980 consumption rate. (Reminder: a quad is a quadrillion (10^{15}) Btu.) In 1990, nine to ten percent of our total energy needs, or about 6.0–7.5 quads, were being supplied by renewable resources. Because fossil fuels and fissionable elements are not inexhaustible, some economists and energy planners argue that renewable energy could and should supply the entire additional need for the twenty-first century goal of energy self-sufficiency.

■ Table 13.10 presents two scenarios for the twenty-first century: one for sources of energy that will supply an increased demand of 25 quads over 1980 levels, and the other for meeting an increased demand of 50 quads. Much of the additional energy supplied in the 50-quad scenario occurs in the geothermal, wind, photovoltaic, and biomass sectors—alternative energy sources that are technologically ready for exploitation. Whether these energy sources will supply the additional demand in the twenty-first century

■ TABLE 13.10 Two Scenarios for Meeting Increased U.S. Energy Demand in the 21st Century

ADDITIONAL ENERGY SOURCE	25-QUAD SCENARIO	50-QUAD SCENARIO
Hydroelectricity	5	6
Geothermal	1	4
Wind	1	6
Photovoltaic	1	6
Biomass	9	18
Direct solar	7	9
Other	1	1
TOTAL	25 quads	50 quads

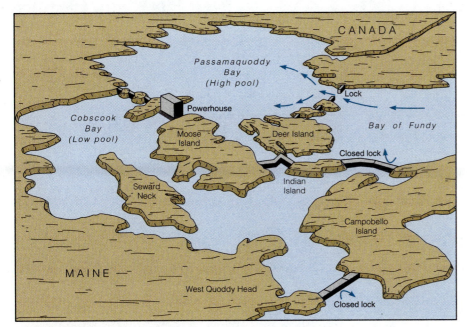

(a) High tide

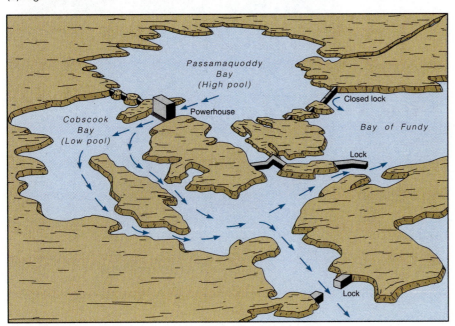

(b) Low tide

● FIGURE 13.41 Proposed Passamaquoddy site tidal-power project. *(a)* Desired flow of water into Passamaquoddy Bay at high tide and *(b)* from Passamaquoddy Bay into Cobscook Bay and the Bay of Fundy at lowtide. A power plant at the narrows between Moose Island and the mainland would generate electricity as water drains from the high pool to the low pool.

depends a great deal upon government leadership and economic incentives for private enterprise to exploit their potential. The last drop of oil and perhaps even the last lumps of coal and uranium will be used within the next century. Ultimately, total U.S. energy demands must be supplied by renewable energy resources. It is only a matter of time—better now, but later for sure.

Some economists and energy planners believe that renewable clean energy sources could and should supply the entire additional need. Note that the big increases in energy

CONSIDER THIS . . .

What nonfossil-fuel energy sources are likely to be expanded in the United States in the coming century? In other words, if you wanted to make a long-term investment in one of them, which one would you choose?

Energy Is Where You Find It

H umans devote a great deal of mental and physical energy to harnessing the earth's energy. Energy, like fresh water, is invaluable, and when it is in short supply a family, a tribe, or a country will do almost anything to obtain it. The photos in this gallery, ⊙ Figures 13.42 to 13.45, illustrate some of the earth's nonconventional energy sources and some relics of fossil energy. Keep in mind that all these energy resources derive from our sun.

⊙ FIGURE 13.43 Fossil scale-tree stumps in their living positions; Victoria Park, Glasgow, Scotland. These trees flourished in the coal forests and swamps of the Carboniferous Period of geologic time. After the stumps rotted out, sand spilled into the hollows, making perfect sandstone casts of the trees.

⊙ FIGURE 13.42 A community gathers around its solar-cell-powered television set; the Republic of Niger. More than likely they are watching a cricket or soccer match, not "Wheel of Fortune."

production from the 25-quad to the 50-quad scenario occur in the geothermal, wind, photovoltaic, and biomass sectors. You might just tuck that away in your memory bank for future economic potential.

Energy itself is neither created nor destroyed—only changed in its form. Human ingenuity will play a major role in the way that the earth's energy potential will be developed and utilized. It seems clear, however, that renewable resources, along with nuclear fission and perhaps fusion, will be *the* energy resources of the twenty-first century.

SUMMARY

Energy

DEFINED   Capacity or ability to do work. A society's quality of life depends in large part on the availability of energy.

TYPES

1. Renewable energy is replaced at least as fast as it is consumed; e.g., solar, wind, and hydrologic energy.

● FIGURE 13.44 Peat mining in Northern Ireland. Cut into blocks and dried, the peat is used for heating homes in some parts of the British Isles.

● FIGURE 13.45 A new windmill design, the Darrieus rotor, is utilized at Altamont Pass, California. The vertical-axis machine is nicknamed the "eggbeater" because of its appearance. The design is a major departure from horizontal-axis machines used to drive electrical generators. Darrieus rotors are more efficient and less expensive than propeller-driven types, and they do not need to face the wind, so they do not require pitch and pivoting mechanisms. Their principal drawback is their inability to self-start; a generator is needed to start the blades spinning.

2. Nonrenewable energy is replenished much more slowly than it is utilized; e.g., energy derived from coal, oil, and natural gas (fossil fuels).

Petroleum

DEFINED Volatile hydrocarbons composed mostly of hydrogen and carbon. Crude oil contains many different hydrocarbon compounds.

ORIGIN AND ACCUMULATION
The necessary conditions for an oil field:

1. Source rock—origin as organic matter;
2. Reservoir rock—porous and permeable;
3. Caprock—impermeable overlying stratum;
4. Geologic trap—such as an anticline, faulted anticline, reef, or stratigraphic trap.

PRODUCTION Drilled wells are either pumped or they flow if under gas and water pressure. Even with the best drilling and producing techniques, primary production may recover only half the oil.

SECONDARY RECOVERY Oil remaining in pore spaces

underground is stimulated to flow to a recovery well by injecting water, chemicals, steam, or CO_2. Much of the "stuck" oil can be stimulated to flow to the extraction well by these methods.

VALUE Varies with weight. Light crude oil is the most valuable.

RESERVES At the 1988 rate of consumption, known world reserves of crude oil will last until 2050 and natural gas until 2110. The latest forecast is that world oil reserves will last until 2100.

Coal

DEFINED Carbonaceous residue of plants that has been preserved and altered by heat and pressure.

RANK As coal matures, it increases in heat content and decreases in the amount of water and volatile matter it contains. The sequence is plant matter to peat, to lignite, to subbituminous, to bituminous, to anthracite.

RESERVES The U.S. has enough coal to last 200–300 years.

SYNFUELS The manufacture of combustible gas, alcohol, and gasoline from coal has a promising future.

Other Fossil Fuels

TAR SANDS Sands containing oil that is too thick to flow and that can be surface-mined. Oil is then washed from the sand. One place the sands are mined is the Athabasca Field in Alberta, Canada.

OIL SHALE Eocene lake deposits containing light kerogen oil that can be removed by heating. Deposits are found in Wyoming, Utah, and Colorado.

Problems with Fossil-Fuel Combustion

AIR POLLUTION Products of combustion contribute to global warming, smog, and acid rain. Acid rain is a hazard in the U.S. East and Southeast and in eastern Canada, which receives some of the U.S. emissions. This is a major international air-pollution problem.

MITIGATION Sulfur content must be reduced below 1 percent. Methods include fluidized-bed combustion, neutralizing sulfur dioxide in smokestacks with limestone scrubbers, cleaning coals by gravity separation, and chemical leaching methods.

Alternative Energy Sources

DIRECT SOLAR
1. Solar heat collectors warm a fluid, which is then put to work.
2. Photovoltaic cells convert light directly into electricity.

INDIRECT SOLAR
1. Wind is produced by unequal heating of the earth's surface.

GEOTHERMAL Facilities for generating electricity with geothermal energy currently are in place and may supply as much as 2 percent of U.S. electrical energy needs. Geothermal heat is also used worldwide for heating interior spaces, keeping sidewalks free of winter snow and ice, etc.

HYDROELECTRIC Hydroelectric dams harness the power of water distributed over the earth by the hydrologic cycle.

NUCLEAR ENERGY Provides 16 percent of world and 19 percent of U.S. energy needs. Heat generated by fission of uranium isotopes and their daughter-isotopes such as plutonium is converted to electricity. There are 109 commercial nuclear power plants operating in the United States.

1. *Problems*—Mining, transportation, and disposal of high-level nuclear materials are the most dangerous aspects of the nuclear energy cycle. If the chain reaction runs out of control, the heat generated may lead to a central-core meltdown such as occurred at Chernobyl.
2. *Geological considerations*—Very careful geological site studies and investigations are performed to assess hazards from earthquakes, mass-wasting, subsidence, and coastal and riverine floods.
3. *Nuclear waste considerations*—High-level nuclear waste is a major problem. Tons of these wastes sit in water-filled pools at reactor sites waiting for permanent storage. Yucca Mountain in Nevada is the only geological repository site being studied, and due to political and environmental delays, the earliest it can be ready to accept waste is 2010.

OCEANS The energy in waves, currents, and tides can be converted to do work for humans. Ocean thermal energy conversion (OTEC) shows some promise, but little has been done with waves and currents. Tidal power is being used successfully only at the Rance River, France, at present.

 KEY TERMS

acid rain	millirem
anticline	miscible
biomass fuel	photochemical smog
breeder reactor	quad
British thermal unit (Btu)	radon
caprock	rank (coal)
central-core meltdown	reserves
chain reaction	reservoir rock
fission	salt dome
fusion	slant drilling
geological repository	(whipstocking)
geothermal energy	source rock
high-level nuclear wastes	stratigraphic trap
hydrocarbon compound	synthetic fuels (synfuels)
kerogen	transuranic element
low-level nuclear wastes	wildcat well

STUDY QUESTIONS

1. What four conditions are required for an oil deposit to form?
2. What oceanographic conditions favor the formation of a good source rock for oil?
3. Draw a geologic cross section that shows an anticline and a faulted anticline. Assume your cross section is looking north. What are the directions of the dips and strikes of the two limbs of the anticline? Dip and strike determination is explained in Appendix 3.
4. What nonrenewable energy resources exist in the United States, and which one is most abundant?
5. Describe atomic fission. How does it create heat for electrical generation?
6. Explain the difference between nuclear fission and nuclear fusion. What would be the advantage of producing nuclear energy from fusion, rather than controlled fission?
7. What are the environmental consequences of burning fossil fuels? What will be the impact upon the global environment of continued dependence upon fossil fuels?
8. How can solar energy be collected and used for heating, generating electricity, and producing fuels?
9. What sector of the U.S. economy uses the most fossil fuel, and what international problem is connected with this use?
10. Explain how photochemical smog is formed. (Show the chemical steps and the required conditions.)
11. Explain the meaning of *renewable energy*. Which renewable energy sources have been developed, and which undeveloped renewable sources show the most promise?
12. List the pros and cons of living near a nuclear electrical-generating facility and of living near one powered by fossil fuel. Which one would you rather have as a neighbor?

FURTHER INFORMATION

BOOKS AND PERIODICALS

Abelson, Philip H. J. 1987. Energy futures. *American scientist* 75 (Nov./Dec.), pp. 584–593.

Blackburn, John O. 1987. *The renewable energy alternative.* Durham, N. C.: Duke University Press.

Brinkworth, Malcom. 1985. *Energy.* London: British Broadcasting Corp.

California, State of, Office of Appropriate Technology. 1983. *Common-sense wind energy.* Andover, Mass.: Brick House Pub. Co.

Carr, Donald. 1965. *The breath of life: The problem of poisoned air.* New York: W. W. Norton and Co.

Carter, L. M. H., ed. 1995. *Energy and the environment: Application of geosciences to decision-making.* U.S. Geological Survey circular 1108, 134 pp.

Charlier, Roger. 1982. *Tidal energy.* New York: Van Nostrand Reinhold Co.

Chiles, James. 1987. Spindletop. *Invention and technology,* Summer.

Deutsch, Robert W. 1987. *Nuclear power: A rational approach.* 4th ed. Columbia, Md.: G. P. Courseware.

Duffield, W. A.; J. H. Sass; and M. Sorrey. 1994. *Tapping the earth's natural heat.* U.S. Geological Survey circular 1125, 63 pp.

Goodger, E. M. 1975. *Hydrocarbon fuels.* New York: John Wiley and Sons.

Hafele, Wolf. 1990. Energy from nuclear power. *Scientific American* 263, no. 3.

Hirsch, Robert L. 1987. Impending United States energy crisis. *Science* 235, pp. 1467–1473.

Hoyle, Fred, and Geoffrey Hoyle. 1980. *Common sense in nuclear energy.* New York: W. H. Freeman and Co.

McCabe, Peter J., and others. 1993. *The future of energy gases.* U.S. Geological Survey circular 1115, Public Issues in Earth Science, 57 pp. series.

Magoon, Leslie B., ed. 1988. *Petroleum systems of the United States.* Washington, D.C.: U.S. Geological Survey professional paper 1870.

Mossop, Grant D. 1980. Geology of the Athabasca oil sands. *Science* 207, p. 145.

National Research Council. 1986. *Acid deposition: Long-term trends.* Washington, D.C.: National Academy Press.

Office of Technology Assessment (OTA), U.S. Congress. 1979. *The direct use of coal.* Washington: Superintendent of Documents.

Rolph, Elizabeth. 1979. *Nuclear power and the public safety.* Lexington, Mass.: Lexington Books, D.C. Heath and Co.

Ross, David. 1987. *Energy from waves,* 2d ed. New York: Pergamon Press.

Schmandt, Jurgen; J. Clarkson; and Hilliard Roderick, eds. 1988. *Acid rain and friendly neighbors: The policy dispute between Canada and the United States.* Durham, N.C.: Duke University Press.

Scientific American. 1990. Energy for planet earth. *Scientific American* special issue, September.

Scollon, Reed T., and E. E. Sheridan. 1978. Utilization of lignite in the United States. *Geological Society of America special paper 179,* pp. 85–90.

Stern, A. C., ed. 1976. *Air pollution,* 3d ed., vol. I. New York: Academic Press.

Sweet, William. 1988. *The nuclear age,* 2d ed. Washington: Congressional Quarterly, Inc.

U.S. Geological Survey. 1995. *National assessment of the United States oil and gas resources.* Circular 1118, to accompany CD-ROM, USGS Digital Data Series 30, for Macintosh-and Windows-based PC systems. Available on Internet for downloading.

Vernon, Bill. 1988. Probing earth's history. *Earth science,* Spring.

Young, Robert, and Jan Young. 1971. *Gusher: The search for oil in America.* New York: Julian Messner, a Division of Simon and Schuster.

MINERAL RESOURCES AND SOCIETY

14

Our entire society rests upon—and is dependent upon—our water, our land, our forests, and our minerals. How we use these resources influences our health, security, economy, and well-being.

JOHN F. KENNEDY, 23 FEBRUARY 1961

E very material used in modern industrial society is derived from the earth's natural mineral resources. These mineral resources are usually classified as either metallic or nonmetallic, as shown in ▪Table 14.1. Production and distribution of the thousands of manufactured products and the food we eat is dependent upon the utilization of metallic and nonmetallic mineral resources. The per capita U.S. consumption of mineral resources used directly or indirectly in providing shelter, transportation, energy, and clothing is enormous (◉Figure 14.1). The availability and cost of mineral and rock products influences our nation's standard of living, gross national product, and position in the world.

OPENING PHOTO
The Castle Mountain Mine near the Nevada border is an open-pit mine in California's Mojave Desert. This operation is typical of numerous currently-active mines in the western United States.

■ TABLE 14.1 Classification of Mineral Resources

METALLIC MINERAL RESOURCES	NONMETALLIC MINERAL RESOURCES
Abundant Metals	*Minerals for Industrial and Agricultural Use*
iron, aluminum, manganese, magnesium, titanium	phosphates, nitrates, carbonates, sodium chloride, fluorite, sulfur, borax
Scarce Metals	*Construction Materials*
copper, lead, zinc, tin, gold, silver, platinum-group metals, molybdenum, uranium, mercury, tungsten, bismuth, chromium, nickel, cobalt, columbium	sand, gravel, clay, gypsum, building stone, shale, and lime-stone (for cement)
	Ceramics and Abrasives
	feldspar, quartz, clay, corundum, garnet, pumice, diamond

SOURCE: James R. Craig, David L. Vaughan, and Brian J. Skinner, *Resources of the Earth* (Englewood Cliffs, N.J.: Prentice Hall, 1988).

With the value of nonfuel minerals produced in the United States running $30–35 billion per year and the annual value of materials processed from minerals running several hundred billion dollars (about five percent of the gross national product), it is amazing that the general public has relatively little knowledge of where these minerals naturally occur, the methods by which they are mined and processed, and the extent to which we are dependent on them. It is important that the public recognize that exploitable mineral resources occur only in particular places, having formed there due to unique geologic conditions, and that all deposits are exhaustible. Furthermore, development and exploitation of mineral resources is capital-intensive, requiring substantial long-term investment, and has environmental effects.

Understanding the origins, economics, and methods of mining and processing of mineral resources enables an individual to understand many of the difficult resource-related issues that are currently facing all levels of government. Increased land-use competition among the interests of mining, housing, wilderness preservation, agriculture, logging, recreation, wetlands preservation, and industrial development is forcing our government officials to make decisions

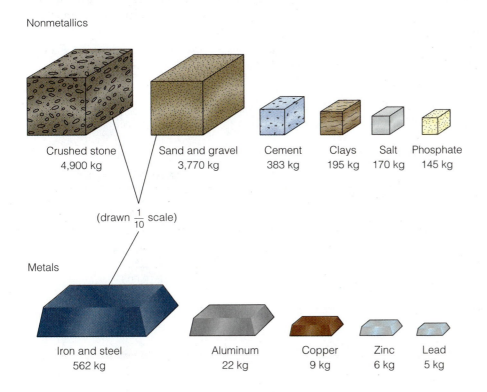

Nonmetallics

Crushed stone
4,900 kg

Sand and gravel
3,770 kg

Cement
383 kg

Clays
195 kg

Salt
170 kg

Phosphate
145 kg

(drawn 1/10 scale)

Metals

Iron and steel
562 kg

Aluminum
22 kg

Copper
9 kg

Zinc
6 kg

Lead
5 kg

● FIGURE 14.1 Approximate annual per capita consumption of nonfuel mineral resources in the United States. Nonmetallic resources constitute about 94 percent of the mineral use by weight. (A kilogram equals 2.2 pounds.)

that affect mineral-resource development and management and, thus, all of our lives.

MINERAL ABUNDANCES AND DISTRIBUTION

Economic Mineral Concentration

The elements forming the minerals that make up the earth's crust exist in many forms in a great variety of rocks. The higher the abundance, the richer the deposit and the more economically feasible it is to extract and process the desired element. Locally rich concentrations of minerals are called **mineral deposits.** If they are sufficiently enriched, they may be **mineral reserves,** deposits of earth materials from which useful commodities are economically and legally recoverable with existing technology. Reserves of metallic minerals are called **ores,** and the minerals comprising the deposits are referred to as **ore minerals.**

A particular ore deposit may be described in terms of its **concentration factor,** its enrichment expressed as the ratio of the element's abundance in the deposit to its average continental-crustal abundance. ■ Table 14.2 illustrates that the minimum concentration factors for profitable mining vary widely for eight metallic elements. Note in the table that aluminum, one of the most common elements in the earth's crust, has an average crustal abundance of 8 percent. A "good" aluminum-ore deposit is defined at this time as one that contains about 35 percent aluminum; thus, the present concentration factor for profitable mining of aluminum ore is about 4. In contrast, deposits of some rare elements such as uranium, lead, gold, and mercury must have concentration factors in the thousands in order to be considered mineral *resources.*

Some elements are extremely common. In fact, 97 percent of the earth's crust is comprised of only eight elements: oxygen, silicon, aluminum, iron, sodium, calcium, potas-sium, and magnesium. For most elements, the **average crustal abundance,** the amount of a particular element present in the continental crust, is only a fraction of a percent (Table 14.2). Copper, for example, is rather rare; its average crustal abundance is only 0.0063 percent. In some localities, however, natural geochemical processes have concentrated it in mineral deposits that are 2 to 4 percent copper. A deposit with a high concentration of a desired element is called a **high-grade deposit.** A deposit in which the mineral content is minimal but still exploitable is called a **low-grade deposit.** For either grade of deposit, localities where desirable elements are concentrated sufficiently for economic extraction are relatively few.

Factors that Change Reserves

Reserves are not static, because they are defined by the current economics and technology as well as by the amount of a mineral that exists. Reserves fluctuate due to several factors: changing demand, discoveries of new deposits, and changing technology. In a free economy, a mineral deposit will not be developed at an economic loss, and prices will rise with demand. Consequently, low-grade deposits that are marginal or submarginal in today's economic climate may, if demand and prices rise, eventually become profitable to exploit. Advancements in technology also may increase reserves by lowering the cost of development or processing.

For example, the exploitable reserves of gold increased dramatically in the United States in the late 1960s due to a combination of changes in government policies and advances in technology. In 1968, when the price of gold was $35 per troy ounce,★ the Treasury Department suspended gold purchases to back the dollar and began allow-

---

★A troy ounce equals 31.103 grams, whereas the avoirdupois ounce of U.S. daily life equals 28.330 grams.

■ TABLE 14.2 Concentration Factors for Profitable Mining of Selected Metals

| METAL | PERCENTAGE ABUNDANCE | | CONCENTRATION FACTOR |
	Average in Earth's Crust	*In Ore Deposit*	
Aluminum	8	35	4
Copper	0.0063	0.4–0.8	80–160
Gold	0.0000004	0.001	2,500
Iron	5	20–69	4–14
Lead	0.0015	4	2,500
Mercury	0.00001	0.1	10,000
Silicon	28.2	46.7	2
Titanium	0.57	32–60	56–105

SOURCE: U.S. Geological Survey Professional Paper 820, 1973.

ing the metal to be traded on the open market. Then in the 1970s the federal government removed restrictions prohibiting private ownership of gold bullion. In 1980 the gold price rose to more than $800 an ounce, and by 1996 it had settled at about $380 an ounce. Thus in a period of about 20 years the price of gold increased by 900 percent, meaning that many submarginal mining claims became economically profitable mines in that time. Furthermore, gold exploitation became more cost-efficient in the 1980s with the development of the new technology of cyanide heap-leaching, by which disseminated gold is dissolved from low-grade deposits and then recovered from the solution. A combination of changing economics and the new heap-leach technology initiated a new gold rush in the West beginning in the 1980s (see Case Study 14.1).

But there are also factors that may reduce reserves, because extracting the earth's riches almost always requires the trade-off of aesthetic and environmental consequences (■ Table 14.3 provides some examples). For example, large low-grade mineral deposits can be mined only by open-pit methods. Not only do these methods devastate the landscape, their excavations may reach nearly to the water table, which can lead to ground-water contamination, and they produce enormous amounts of waste rock and tailings that can pollute surface-water runoff. **Smelters,** large industrial plants that process ore concentrates and extract the desired elements, produce more air pollution, in the form of flue dust, than any other single industrial activity. Smelters also produce slag, a solid residue that can contain thousands of times the natural levels of lead, zinc, arsenic, copper, and cadmium. The real and perceived negative environmental impacts that serve to rule out exploitation and development of known deposits also reduce reserves. In recent years there have been several widely publicized instances of wildlife and wilderness values being judged more important than exploitation of mineral or petroleum deposits. An example is the 1992 decision not to allow drilling for oil in the Arctic National Wildlife Area in Alaska, which will be argued for years to come. A mid-1990s conflict between mining interests and wilderness values in Yellowstone National Park is the subject of Case Study 14.2 (page 416). We can expect that controversies over land use will be increasingly common in the future as more and more lower-grade deposits will need to be exploited when higher-grade deposits become exhausted.

Political factors also may further limit or reduce resources. The United States imports substantial amounts of the mineral resources it needs from other countries. (This is discussed in detail later in the chapter.) As an illustration, consider that the Gulf War of 1991 was basically a resource war, and that the energy crisis of the 1970s was due to geopolitical events. Furthermore, resource availability may affect diplomatic policy. In spite of the general trade sanctions the United States enacted in the 1980s against South Africa to protest its apartheid policy, ten minerals the

United States imports from South Africa were exempted because of their importance to the United States.

Distribution of Mineral Resources

The global distribution of mineral resources has no relationship to the locations of political boundaries or technological capability. Some mineral deposits, like those of iron, lead, zinc, and copper, originate by such a variety of igneous, sedimentary, and metamorphic processes that they are widely distributed, though in varying abundance. On the other hand, bauxite, essentially the only ore mineral of aluminum (one of the most common elements in the earth's crust), is concentrated by only one geochemical process: deep chemical weathering in a humid tropical climate. For this reason, economically exploitable deposits of bauxite and other such minerals are unequally distributed and restricted to a few, highly localized geologic and geographic sites. Such localization of mineral deposits means that desired mineral resources do not occur in all countries. When this unequal supply pattern results in the scarcity of particular mineral resources that are crucial to a nation's economy, those minerals become critical or even strategic in the event of a national emergency. What constitutes "critical" minerals changes from time to time, mainly due to technological advances in mining or processing technologies.

Throughout history, nations' domestic mineral-resource bases and their needs for minerals unavailable within their borders have dictated international relations. Empires and kingdoms have risen and fallen due to the availability or scarcity of critical mineral resources. Two classic examples are those of the Greek and Roman empires. Silver mined by the Greeks near Athens financed the building of their naval fleet that defeated the Persians at the Bay of Salamis in 480 B.C. A century later, Greek gold supported the ambitious conquests of Alexander the Great. The Romans helped to finance their vast empire by mining tin in Britain, mercury in Spain, and copper in Cyprus. In the twentieth century, the efforts of Japan and Germany to secure mineral and fuel resources lacking within their borders were major factors leading to World War II, and the Soviet Union's annexation of the Karelian highlands of Finland following World War II provided the Soviets with new reserves of copper and nickel. More recently, Morocco acquired valuable phosphate deposits when it occupied part of the Spanish Sahara in 1975.

ORIGINS OF MINERAL DEPOSITS

Specific geological processes must occur in order to concentrate minerals or native elements in a mineral deposit. A simple classification of mineral deposits is based on their concentrating processes (■ Table 14.4, page 418). The geologic processes responsible for many mineral deposits may be

Nevada's Modern Gold Rush

The biggest bonanza for Nevada mining interests since the 1860s boom days of the Comstock Lode was the discovery of the Carlin Trend, a 60-kilometer (37-mi) line of predominantly low-grade but very large gold deposits in the north-central part of the state (● Figure 1). It is the largest gold district discovered in North America in the twentieth century. Typically the gold is submicron-sized, "invisible" discrete grains incorporated within sulfide minerals. So small are the grains that no colors can be obtained by the tried-and-true method of gold panning, even of high-grade ores. About a dozen major mines and several smaller ones, most of them open-pit operations, were mining the Trend by the mid 1990s.

The Gold Quarry Mine, the largest of Newmont Gold Company's three major open-pit mines, is representative of mines on the Carlin Trend (● Figure 2). Newmont began operations in Carlin in 1965, and its annual production between 1989 and 1995 was exceeding 1.5 million ounces. Combined production at the company's three open-pit mines in 1994 was 1.56 million ounces. In 1995, Newmont was developing several underground mines for extracting higher-grade ores. Most of their operation is on private land, for which the company pays royalties ranging from 7 to 16.5 percent of net profit. It pays no royalty on federal lands.

Several processing techniques are used to extract the gold. These include milling of the high-grade oxide ores, refractory treatment of the sulfide ores (which entails roasting to burn off the sulfides and carbon), conventional cyanide heap-leaching, and bioleaching (using bacteria to loosen microscopic amounts of gold from problematic sulfide ores that would otherwise end up as mine waste). The cost of recovering gold from Newmont's open-pit mines runs about $240 per ounce.

● FIGURE 2 (right) Newmont's Gold Quarry Mine. The open pit is at the top of the photo, and the tailings pond at the upper left. Waste-rock disposal sites and the processing plant occupy the remainder of the view.

● FIGURE 1 (left) Location of the Carlin Trend.

From the outset, Newmont's reclamation staff has worked closely with the mine-design staff to coordinate the planning and design of mining and reclamation. Reclamation bond money is set aside continuously, according to the conditions of the state permit, and it amounts to about a dollar for each ounce of gold that is produced. Topsoil for future reclamation is stockpiled and seeded with grasses to protect it from erosion by rain and wind until it is needed. Conditions of the mining permit require that "the land be returned to a reasonable state" when the mining is completed. Specifically, this requires that the pit be bermed and fenced and that waste-rock dumps be encapsulated with a four-foot cover of topsoil and planted with native plants. A network of dispersion drainage ditches will be installed to collect and divert surface flow from the waste piles. Because Carlin has an arid climate (annual rainfall is about ten inches), water penetration into the waste should be negligible. Thus, in spite of the ore body being rich in sulfides, acid drainage is not perceived as a potential environmental problem.

The operating permit calls for a 3½-year review following completion of mining and obligates Newmont to work on reclamation for 15 years. The final closure date of the company's open-pit mines is projected as 2030.

■ TABLE 14.3 Some Negative Environmental Consequences of Mineral Extraction and Processing Operations

LOCATION	OPERATION	ENVIRONMENTAL CONSEQUENCES
Sudbury, Ontario, Canada	nickel–copper mining and smelting	This was one of the world's best-known environmental "dead zones." Little vegetation survived in 10,400-hectare (40-mi^2) area around the smelter. Acid fallout destroyed fish populations in lakes within 65 km (40 mi). Conditions improved after completion of $530 million SO_2 smelter-abatement project and construction of the world's tallest smokestack; the city has a successful landscape reclamation program, but there is still significant SO_2 damage downwind of Sudbury.
Pará State, Brazil	Grande Carajas iron ore project	Wood requirements for smelting ore will require cutting 50,000 hectares (193 mi^2) of tropical forest annually during the 250-year life of the project.
Amazon basin, Brazil	gold mining	The region has been invaded by hundreds of thousands of miners digging for gold, clogging rivers with sediment, and releasing some 100 tons of mercury into the ecosystem annually.
Ilo-Locumba area, Peru	copper mining and smelting	Each year, the Ilo smelter emits 600,000 tons of sulfur compounds, and nearly 40 million cubic meters (523 million yd^3) of tailings containing lead, zinc, copper, aluminum, and traces of cyanide are dumped into the ocean, poisoning marine life in a 20,000-hectare (77-mi^2) area. Nearly 800,000 tons of slag are dumped in the sea yearly.
Panguna mine, Bougainville, Papua New Guinea	copper sulfide ore mining	Before the mine closed in 1989, the operation had dumped 600 million tons of tailings into the Kawerong River. The wastes cover 1,800 hectares (7 mi^2) of the Kawerong River system, including a 700-hectare (2.7-mi^2) delta. No aquatic life survives in the river. Local anger generated by the destruction was a major cause of a civil war.
Nauru, South Pacific	phosphate mining	When mining is completed in 1998–2008, 80% of the 2,100-hectare (8-mi^2) island will be uninhabitable. The people of Nauru have initiated legal action to control the mining.
Summitville, San Juan Mountains, Colorado	gold mining	Leaking leach pad, contaminated with cyanide and heavy metals, poisoned 27 km (17 mi) of the Alamosa River, on which region's agriculture depends. When cleanup costs exceeded the posted bond, the Canadian mining company and its parent declared bankruptcy. Cleanup may cost U.S. taxpayers over $100 million.
Guyana, South America	gold mining	Cyanide spill from the Omai mine—backed by the same investors as Colorado's Summitville mine—released 3.2 million m^3 (113 million ft^3) of poisonous runoff into one of Guyana's largest rivers.

Source: *Mining the Earth, State of the World* (Washington: World Watch Institute, 1992).

understood in terms of plate tectonics, as illustrated in ◉ Figure 14.2, page 418, with mineral concentrations occurring in distinct tectonic regimes. For example, the great porphyry copper deposits of the Western Hemisphere were formed by igneous processes at convergent plate margins at the sites of former volcanic island arcs. On the other hand, deposits of

Does Mining Endanger Yellowstone National Park?

People return to Cooke City every summer to work claims. Many of these faithful pilgrims, both men and women, believe that the days of big production will come.

BILLINGS GAZETTE, 21 DECEMBER 1952

I cannot think of an area more sensitive than that being proposed by the New World Mine. . . . I am not willing to gamble with a national treasure for short-term economic gain.

U.S. SENATOR MAX BACUS (D-MONTANA), 1995

Most of the land surrounding Yellowstone National Park is undeveloped wilderness, but not all of it. Just four kilometers (2½ mi) from the park's northern boundary is Henderson Mountain, where miners have been digging out gold for more than a century. Crown Butte Mines, part of Noranda, Incorporated, Canada's largest natural-resources conglomerate, has found a mother lode within the mountain that eluded the early miners. Crown Butte proposes developing the New World Mine, a massive high-tech gold, silver, and copper mining complex in this alpine terrain on the edge of the nation's, and the world's, first national park. (⊙ Figures 1 and 2).

In 1992, the company began the lengthy process required by federal law for evaluating the potential environmental impacts of mining $750 million worth of gold ore from the mountain. The process created a mountain of its own, a mountain of documents about seven feet tall. Twenty or more federal and state agencies must review the massive environmental impact statement before the decision of whether to allow the mining operation to proceed can be made.

The National Forest Wilderness Area surrounding the mine site is a popular winter and summer recreation site, an important area along the threatened grizzly bear's migration route, and habitat for moose, bighorn sheep, and other wildlife. The site, at an elevation of about 2,750 meters (9,000 ft), is near the headwaters of the Yellowstone River, but all of the proposed mining activities and facilities would be located in the valley of Fisher Creek, a tributary of the Clarks Fork of the Yellowstone River—which, despite some publicity to the contrary, does not flow into or toward Yellowstone National Park. Headwaters of two other drainages, which contain remnants of former mining, that feed directly into Yellowstone's famous blue-ribbon trout waters are near the site, but the New World operations would not disturb these areas except for the excavation of three small ventilation shafts from underground. No surface mining and no cyanide processing is proposed.

Crown Butte intends to use about half of the tailings resulting from the mining to backfill the underground operation and to place the remainder in an impoundment area on the surface. This surface impoundment is a major concern of conservationists, because it is seen as a potential source of acid mine drainage. Tailings disposal will be behind an enormous rock-walled, clay- and plastic-lined tailings pond that will displace 56 acres of high-elevation wetland in Fisher Creek valley. An impoundment dam as high as a ten-story building and a 72-acre tailings pond (the size of 70 football fields) will be needed for dumping as much as 5.5 million tons of chemical-laced waste rock and tailings. Con-

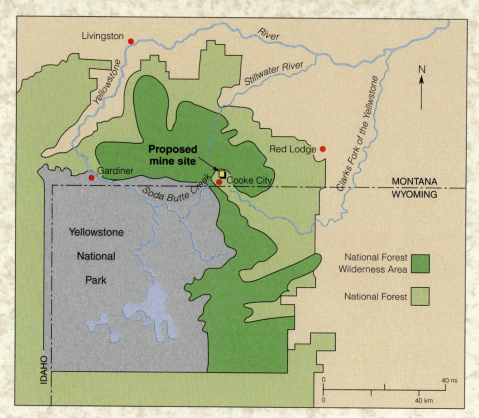

⊙ FIGURE 1 Location of the proposed New World Mine.

servationists are concerned that in this area of harsh winter climate, avalanches, and earthquakes (a swarm of more than 2,000 microearthquakes centered about 80 kilometers [50 mi] from the site was recorded in July 1995), the dam will fail, or certainly erode, and release its burden of toxic waste into the headwaters of the Clarks Fork, Wyoming's only federally designated Wild and Scenic River. Alternative sites for the disposal site are seen as prohibitively expensive.

A spokesperson for Crown Butte Mines argues that the dam's state-of-the-art design will allow it to flex in an earthquake, maintaining its integrity, and that the disposal system's ability to withstand extreme weather has been tested in Sweden, Norway, and Canada, where annual temperatures range from 40 degrees below zero to the mid-20s Celsius (the minus 40s to the mid-80s Fahrenheit). The proposed method for tailings containment purportedly will inhibit the formation of acid by keeping the tailings beneath the water table (see Chapter 10)

● FIGURE 2 The proposed New World Mine site. The peaks of Yellowstone National Park appear in the distance.

and away from oxygen. Besides, the spokesperson continues, the law requires the company to post millions of dollars in bonds as guarantees for remediation and revegetation and to monitor the site for up to 40 years. Crown Butte also points out that building the dam would benefit the park, because Crown Butte would clean up the old mine wastes that have been leaking acid into the Yellowstone drainage for years.

Conservationists feel that the value of the area's wilderness and Yellowstone National Park far exceeds the value of the gold. The only addition to UNESCO's threatened "red list" of the world's top natural or cultural heritage sites in 1995 was Yellowstone; it was seen as being in imminent danger of environmental decay. In response, the mining company countered that the area already has a long mining history and that its mineral potential is the very reason that Congress specifically excluded it from inclusion in the Wilderness system; that because the site was excluded from Wilderness designation, the government has essentially said, "you

can mine here." So the company came, spent $35 million doing exploratory work, and found the ore body. Now the issue is being raised that the company should not be allowed to mine here.

In pleading their case, conservationists have also raised questions about Noranda's environmental integrity, pointing to its previous operations in Canada and the United States. It has been fined for dumping contaminated tailings into streams and lakes, has walked away from a mine and a tailings pond when the price of metal fell, has been fined for water-quality violations, and has paid more than $1.9 million in fines for occupational-safety and environmental violations. At Noranda's abandoned Grey Eagle Mine in California, about a ton of acid mine drainage seeps from the tailings pond each day, which the company treats and hauls away. A state official has described the Grey Eagle containment structure as a "failed containment design."

The U.S. Forest Service, a branch of the Department of Agriculture, is obligated to give the proposed mine a fair review and, along with nearly two dozen other federal and state agencies, it has a mountain of documents to study. On the already-patented private-land claims purchased for the New World mine, the Forest Service has no voice; the only regulators for such land are state and local governments. Although the 1872 mining law guarantees mining companies a vested interest in the use of public land, it does not instruct public agencies to determine if the mining is truly in the public's interest. Eventually, one of these agencies will have to decide if there should be a mine on the doorstep of Yellowstone National Park.

As this text went to press, the Clinton Administration announced that it had reached a land-swap agreement with Crown Butte Mines, under which Crown Butte will cancel its plans to open the New World Mine. In exchange, Crown Butte will receive other lands to be determined through further bargaining.

■ TABLE 14.4 Genetic Classification of Mineral Deposits

GENETIC TYPE	EXAMPLE OF MINERAL(S) FORMED	LOCATION
Igneous Processes		
Pegmatite	tourmaline	Oxford County, Maine; San Diego County, California
	beryl	Minas Gerais, Brazil
Crystal settling	chromite–magnetite–platinum	Bushveld complex, South Africa
Disseminated	copper–molybdenum porphyries	western North and South America
	diamonds	kimberlite pipes in South Africa, Canada and Australia
Hydrothermal	gold–quartz veins	Mother Lode belt, California
	lead–zinc deposits in carbonate rocks	Mississippi River valley
Volcanogenic	massive copper sulfide deposits	Wrangell Mountains, Alaska
Sedimentary Processes		
Chemical precipitates in shallow marine basins	banded-iron formation	Lake Superior region
Marine evaporites	potassium salts	Carlsbad, New Mexico
Nonmarine evaporites	carbonates and borax minerals	Searles Lake, Trona, California
Deep-ocean precipitates	iron–manganese nodules	ocean floor, worldwide
Placer deposits	gold	Yukon Territory, Canada
Weathering Processes		
Lateritic weathering	bauxite	Weipa, Australia; Jamaica
Secondary enrichment	copper minerals	Miami, Arizona
Metamorphic Processes		
Contact-metamorphic	tungsten and molybdenum minerals	eastern Sierra Nevada, California
Regional metamorphic	asbestos	Quebec, Canada

tungsten, tin, some iron ore, gold, silver, and molybdenum originated by hydrothermal activity or by contact metamorphism accompanying the emplacement of granitic plutons (batholiths) within continental lithospheric plates near convergent boundaries.

Igneous Processes

INTRUSIVE DEPOSITS. Pegmatites are small, tabular-shaped, very coarse grained, intrusive igneous bodies that may be important sources of mica, quartz, feldspar, beryl–

● **FIGURE 14.2** Relationships between metallic ore deposits and tectonic processes.

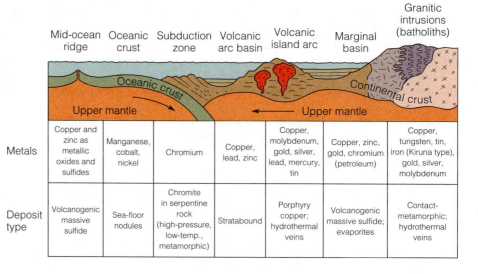

	Mid-ocean ridge	Oceanic crust	Subduction zone	Volcanic arc basin	Volcanic island arc	Marginal basin	Granitic intrusions (batholiths)
Metals	Copper and zinc as metallic oxides and sulfides	Manganese, cobalt, nickel	Chromium	Copper, lead, zinc	Copper, molybdenum, gold, silver, lead, mercury, tin	Copper, zinc, gold, chromium (petroleum)	Copper, tungsten, tin, iron (Kiruna type), gold, silver, molybdenum
Deposit type	Volcanogenic massive sulfide	Sea-floor nodules	Chromite in serpentine rock (high-pressure, low-temp., metamorphic)	Stratabound	Porphyry copper; hydrothermal veins	Volcanogenic massive sulfide; evaporites	Contact-metamorphic; hydrothermal veins

● FIGURE 14.3 Granitic pegmatite, a very coarse grained igneous rock. The grains in this sample have lengths up to 3 centimeters.

● FIGURE 14.4 Tourmaline crystals. Tourmaline is commonly found as an accessory mineral in granitic pegmatites.

lium, lithium, and gemstones. Reserves of pegmatites in the Western Hemisphere occur in the Black Hills of South Dakota; at Dunton, Maine; at Minas Gerais, Brazil; and in the tin-spodumene deposits of North Carolina (● Figures 14.3 and 14.4).

Crystal settling within a cooling magma chamber of ultramafic composition appears to be responsible for forming layers according to density, with the densest and earliest formed on the bottom and the less-dense and later-to-crystallize above or toward the top (● Figure 14.5), although alternate ideas for the layering have been proposed. An example is the layered deposit of the Bushveld Complex in the Republic of South Africa, the largest single igneous mass on earth (roughly the size of Maine). Chromium, nickel, vanadium, and platinum are obtained from this source. Some layered nickel-copper-cobalt deposits also are important sources of platinum-group elements.

DISSEMINATED DEPOSITS. Disseminated (scattered) **deposits** occur when a multitude of mineralized veinlets develop near or at the top of a large igneous intrusion (see Chapter 2). Probably the most common of these are the **porphyry copper** deposits (● Figure 14.6), in which the ore minerals are widely distributed throughout a large volume of granitic rocks that have been emplaced in regions of current or past plate convergence. The world's largest open-pit mine, at Bingham Canyon, Utah, has extracted more than $6 billion worth of copper and associated minerals from a porphyry copper body (● Figure 14.7).

HYDROTHERMAL DEPOSITS. Hydrothermal (hot water) **deposits** originate from hot, mineral-rich fluids that are squeezed from cooling magma bodies during crystallization. Solidifying crystals force the liquids into cracks, fissures, and pores in both the magmatic rocks and the adjacent, "country" rocks (Figure 14.6). Hydrothermal fluids also may be of metamorphic origin, or they may form from subsurface waters that are heated when they circulate near a cooling magma. Regardless of their origin, the heated waters may dissolve, concentrate, and remove valuable minerals and redeposit them elsewhere. When the fluids cool, minerals crystallize and create hydrothermal **vein deposits** which miners often refer to as **lodes**. The specific minerals formed will vary with the composition and temperature of the particular hydrothermal fluids, but they commonly include such suites (associations) of elements as gold-quartz and lead-zinc-silver (● Figure 14.8).

VOLCANOGENIC DEPOSITS. When volcanic activity vents fluids to the surface, sometimes associated with ocean-floor hot-spring **black smoker** activity, **volcanogenic deposits** are formed (● Figure 14.9). These deposits are so-named because they occur in marine sedimentary rocks that are associated with basalt flows or other volcanic rocks, and the ore bodies they produce are called **massive sulfides**. The rich copper deposits of the island of

● FIGURE 14.5 Layered mineral deposit of the Bushveld complex, Republic of South Africa. The lighter layers are plagioclase feldspar; the darker layers, magnetite.

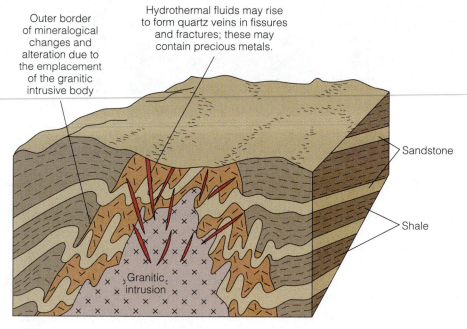

● FIGURE 14.6 A disseminated porphyry copper deposit.

Outer border of mineralogical changes and alteration due to the emplacement of the granitic intrusive body

Hydrothermal fluids may rise to form quartz veins in fissures and fractures; these may contain precious metals.

Sandstone

Shale

Granitic intrusion

Disseminated deposits of copper form when mineralized solutions invade permeable zones and small cracks.

● FIGURE 14.7 Kennecott Corporation's Bingham Canyon mine south of Salt Lake City, Utah, is the world's largest open-pit mine and largest human-made excavation. In the mid-1990s, the pit was more than 800 meters (½ mi) deep and nearly 4 kilometers (2½ miles) wide.

⊙ FIGURE 14.8 Specimen of a gold-quartz vein from the Sunnyside Mine, Silverton, Colorado. Sample is 9 cm (3.5 in) long.

Cyprus, which have supplied all or part of the world's needs of copper for more than 3,000 years, are of this type. (The word *copper* is derived from Latin *Cyprium,* "Cyprian metal.") Cyprus's copper sulfide ore formed millions of years ago adjacent to hydrothermal vents near a sea-floor spreading center. Warping of the copper-rich sea floor due to convergence of the European and African plates brought the deposits to the surface when the island formed.

Sedimentary Processes

SURFICIAL PRECIPITATION. Very large, rich mineral deposits may result from evaporation and direct **precipitation** (the process of a solid forming from a solution) of salts in ocean water, usually in shallow marine basins. Minerals formed this way are called **evaporites.** Evaporites may be grouped into two types: marine evaporites, which are primarily salts of sodium and potassium, gypsum, anhydrite, and bedded phosphates; and nonmarine evaporites, which

⊙ FIGURE 14.9 A "black smoker" hydrothermal vent on the Endeavor segment of the Juan de Fuca mid-oceanic ridge photographed during a dive of the submersible *Alvin.* The "smoke" is hot water saturated with dissolved metallic-sulfide minerals, which precipitate into black particles upon contact with the cold seawater.

are mainly calcium and sodium carbonate, nitrate, sulfate, and borate compounds. Other important sedimentary mineral deposits are the **banded-iron formation** ores (⊙ Figure 14.10). These deposits formed in Precambrian time, more than a billion years ago, when the earth's atmosphere lacked free oxygen. Without free oxygen, the iron that dissolved in surface water could be carried in solution by rivers from the continents to the oceans, where it precipitated with silica to form immense deposits of red chert and iron ore. Banded-iron deposits are found in the Great Lakes region, northwestern Australia, Brazil, and elsewhere, and they are enormous; they provide two hundred years or more of reserves even without substantial conservation measures.

DEEP-OCEAN PRECIPITATION. Manganese nodules (⊙ Figure 14.11) are formed by precipitation on the deep

⊙ FIGURE 14.10 Banded-iron formation specimen. The red bands are chert, and the gray bands are iron oxide minerals that were deposited as sedimentary layers in a shallow marine basin bordering a deeply weathered landmass.

⊙ FIGURE 14.11 Manganese nodules, an example of deep-ocean precipitation. The nodules consist mainly of manganese and iron, with lesser amounts of nickel, copper, cobalt and zinc.

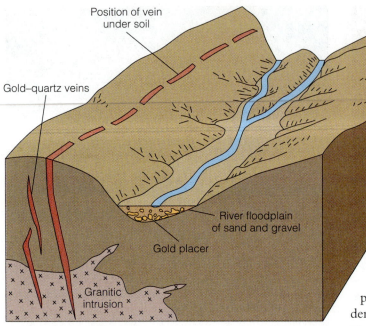

Position of vein under soil

Gold–quartz veins

River floodplain of sand and gravel

Gold placer

Granitic intrusion

(a) Ancient river pattern

Former positions of rivers

Outline of former topography

Position of vein under soil

Bench or terrace paystreak from former stream channel

Modern placer paystreak

(b) Modern river pattern

◉ **FIGURE 14.12 The origin of gold placer deposits.** *(a)* An ancient landscape with an eroding gold-quartz vein shedding small amounts of gold and other mineral grains that eventually become stream sediments. The gold particles, being heavier, settle to the bottom of the sediments in the channel. *(b)* The same region in modern time. Streams have eroded and changed the landscape and now follow new courses. The original placer deposits of the ancient stream channels are now elevated above the modern river valley as "bench" placers.

ocean floor. The nodules are mixtures of manganese and iron oxides and hydroxides, with small amounts of cobalt, copper, nickel, and zinc, that grow in onionlike concentric layers by direct precipitation from ocean waters. Any commercial recovery of this resource appears to be many decades away because of technological, economic, international political, and environmental limitations, but eventually it may be necessary to exploit the deposits.

PLACER DEPOSITS. Mineral deposits concentrated by moving water are called **placer deposits** (Spanish *placer,* "reef"; pronounced "plasser"). Dense, erosion-resistant minerals such as gold, platinum, diamonds, and tin are readily concentrated in placers by the washing action of moving water. The less-dense grains of sand and clay are carried away, leaving gold or other heavy minerals concentrated at the bottom of the stream channel. Such deposits formed by the action of rivers are referred to as **alluvial placers;** ancient river deposits that are now elevated as stream terraces above the modern channels are called **bench placers** (◉Figure 14.12 and Case Study 14.3). Considerable uranium is mined from Precambrian placer deposits in southern Africa. The deposits formed under the unique conditions existing on the earth's surface before free oxygen was present in the atmosphere. Some placer deposits occur on beaches, the classic example being the gold beach placers at Nome, Alaska.

Sand and gravel concentrated by rivers into alluvial deposits are important sources of aggregate for concrete. River deposits formed by **glacial outwash**—that is, resulting from the runoff of glacial meltwater—are another type of placer deposit.

Weathering Processes

The deep chemical weathering of rock in hot, humid, tropical climates promotes mineral enrichment, because the solution and removal of more soluble materials leaves a residual soil of less-soluble minerals. Because iron and aluminum are relatively insoluble under these conditions, they tend to remain behind in *laterite,* a highly weathered, red subsoil or material that is rich in oxides of iron and aluminum and lacking in silicates. When the iron content of the parent rock is low or absent, however, this lateritic weathering produces rich deposits of **bauxite,** the principal ore mineral of aluminum (see Chapter 6, Figure 6.15).

Ground water moving downward through a disseminated sulfide deposit may dissolve the dispersed metals from above the water table to produce an enriched deposit below the water table by **secondary enrichment** (◉ Figure 14.13). At Miami, Arizona, for example, the primary disseminated copper-ore body is of marginal to submarginal grade, containing less than one percent copper, but secondary enrichment has improved the grade to more than three percent.

Metamorphic Processes

The high temperatures, high pressures, and ion-rich fluids that accompany the emplacement of intrusive igneous rocks produce a distinct metamorphic halo around the intrusive body. The result, in concert with the accompanying hydrothermal mineralization, is known as a **contact-metamorphic deposit.** If, for example, granite intrudes limestone, a diverse and colorful group of contact-metamorphic minerals may be produced, such as a tungsten-molybdenum deposit (◉ Figure 14.14). Asbestos and talc originate by **regional metamorphism,** metamorphism that affects an entire region.

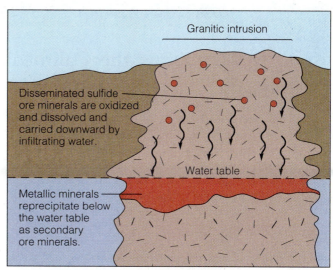

◉ **FIGURE 14.13** The development of ores by secondary enrichment. Descending ground water oxidizes and dissolves soluble sulfide minerals, carrying them downward and leaving a brightly colored residue of limonite and hematite. Below the water table new metallic minerals are precipitated as secondary ore minerals.

Labels on figure:
Granitic intrusion
Disseminated sulfide ore minerals are oxidized and dissolved and carried downward by infiltrating water.
Water table
Metallic minerals reprecipitate below the water table as secondary ore minerals.

◉ MINERAL RESERVES
Metallic Mineral Reserves

The metallic minerals are often grouped on the basis of their relative crustal abundance.

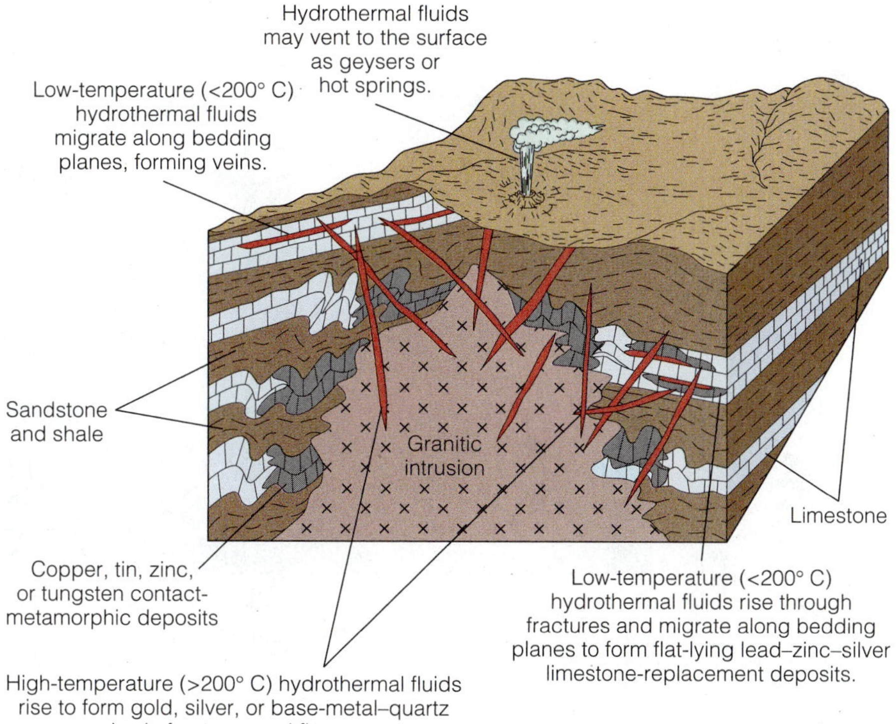

◉ **FIGURE 14.14** The relationship of limestone and a granitic intrusion in the origin of contact-metamorphic mineral deposits.

Labels on lower figure:
Hydrothermal fluids may vent to the surface as geysers or hot springs.
Low-temperature (<200° C) hydrothermal fluids migrate along bedding planes, forming veins.
Sandstone and shale
Granitic intrusion
Limestone
Copper, tin, zinc, or tungsten contact-metamorphic deposits
High-temperature (>200° C) hydrothermal fluids rise to form gold, silver, or base-metal–quartz veins in fractures and fissures.
Low-temperature (<200° C) hydrothermal fluids rise through fractures and migrate along bedding planes to form flat-lying lead–zinc–silver limestone-replacement deposits.

Mining the California Mother Lode

Although rumors of gold in Spanish-held California were circulating as early as 1816, the first rush for California's gold did not begin until 1842, following ranchero Francisco Lopez's discovery of gold in what is now Placerita Canyon on Rancho San Fernando near Pueblo de Los Angeles. The lack of water at the site resulted in mining by only the crudest of methods until the arrival of experienced miners from Sonora, Mexico, who introduced the method of dry washing. The spurt of interest quickly waned, however, because the small amount of gold available was soon extracted, and the lack of water for washing the gravels was discouraging to all but the most experienced miners. In 1848, however, James Marshall's recognition of gold at Sutter's Mill on the American River electrified the world and triggered the greatest gold rush in history, ultimately altering the development of western North America and perhaps of the United States (⊙ Figure 1).

The crude early mining methods by which knives and spoons were used to extract gold from placer deposits soon gave way to the gold pan and "rocker," an open-topped wooden box mounted on curved boards. Using a rocker was a two-person operation: one miner shoveled sand and gravel and poured water into the box, while the second rocked the box back and forth. The dirt would wash through the box and the gold would be caught behind "riffles," slats placed across the bottom of the box perpendicular to the flow of water. Rockers in turn led to the development of the elongated rocker, and the wooden sluice box, with similar transverse riffles (⊙ Figure 2). The sluice was positioned in a stream so that water would flow down the length of the box. Miners shoveled in goldbearing sand and gravel, and the gold was caught by the riffles. Cocoa matting was sometimes used to catch very small particles of gold, and oftentimes a copper plate coated with mercury was used to recover the very fine particles more efficiently by forming an *amalgam,* an alloy of mercury and gold—or silver; hence the reason that mercury is called *quicksilver.* When "cleaning up," the miners would heat the amalgam in a vessel to separate the dissolved gold from the amalgam by vaporizing and condensing the mercury, which could then be reused. The miners were careful to avoid breathing the fumes in recovering the vaporized mercury, because they knew of its toxicity.

The easily obtainable gold was soon exhausted, and by 1853 the method of hydraulic mining—extracting gold by blasting a jet of water through a nozzle, called a *monitor,* against hillsides of ancient alluvial deposits—was developed (⊙ Figure 3). Once the

⊙ FIGURE 1 The Mother Lode district, California's nineteenth-century gold-rush belt.

⊙ FIGURE 2 American and Chinese gold miners working a sluice box at Auburn, California, in 1852.

◉ **FIGURE 4** Interior of a large stamp mill, like those used in the Mother Lode gold belt in the nineteenth century. Mills like this crushed ore 24 hours a day six days a week with a noise that must have been deafening. It is said that one could travel the length of the Mother Lode in the 1880s and never be out of earshot of the pounding stamp mills.

◉ **FIGURE 3** Hydraulicking a bench placer deposit.

necessary ditches, reservoirs, penstocks (vertical pipes), and pipelines had been installed, it was possible to wash thousands of cubic yards of goldbearing gravels without hand labor. An attempt that same year to mine alluvial gold by dredging failed; it was not until 1898 that the first real success was achieved with a bucket-conveyor dredge (text Figure 14.23).

Experienced miners immediately began searching for the *mother lode*—the bedrock source veins from which the placer deposits originated. The first gold-quartz lode vein was discovered in 1849 on Colonel John C. Fremont's grant near Mariposa. Discoveries of more gold-quartz veins of the lode system soon followed. In contrast to the simplicity of mining placer deposits, lode mining requires extensive underground workings (which require drilling, blasting, and tunneling) and a complex surface plant for hoisting men and ore from underground, for pumping water out of the mine, and for milling the raw ore. A few banks of old stamp mills (◉ Figure 4), batteries of heavy weights that were alternately lifted and dropped to pulverize the ore, remain today in the gold country. The gold was recovered by gravity separation from a slurry of powdered rock and water as it flowed over riffles and by amalgamation with quicksilver. In the 1890s, cyanide extraction was introduced into the milling process to make recovery even more efficient. It had been learned that gold dissolves readily in a dilute solution of sodium or potassium cyanide. Hence the new method mixed crushed ore with cyanide and removed the gold later by chemical methods.

The influx of thousands of forty-niners, as the fortune-seekers were known, to the Mother Lode district led to widespread prospecting and to the rapid and complete exploration of the previously little-known, sparsely populated region. Within a year of James Marshall's gold discovery at Sutter's Mill, California had attained statehood and become, almost overnight, the most important Western state. California's gold rush also served as the impetus for widespread prospecting rushes in the 1850s that led to important gold strikes in Nevada, Colorado, Montana, Australia, and Canada. A conservative estimate placed California's annual gold production at $750,000,000 by 1865. The major world impacts of the California gold rush were the widespread redistribution of population, with its resulting problems and challenges, and the general increase of money in circulation.

THE ABUNDANT METALS. The geochemically abundant metals—aluminum, iron, magnesium, manganese, and titanium—have abundances in excess of 0.1 percent by weight of the average continental crust. Economically valuable ore bodies of the abundant metals—such as iron and aluminum, for example—need only comparatively small concentration factors for profitable mining (Table 14.2) and are recovered mainly from ore minerals that are oxides and hydroxides. Even though they are abundant, these metals require large amounts of energy for production. Because of this, it is not surprising that the world's industrialized nations are the greatest consumers of these metals.

THE SCARCE METALS. The scarce metals are those whose crustal abundances are less than 0.1 percent by weight of average continental crust. Most of these metals are concentrated in sulfide deposits, but the rarest occur more commonly, or even solely, in rare rock types. Included within this category are copper, lead, zinc, and nickel—which are widely known and used, but nevertheless scarce—along with the more obvious gold and platinum-group minerals. U.S. deposits of chromium, manganese, nickel, tin, and platinum-group metals are scarce or of submarginal grade. Important, less-well-known scarce metals include cobalt, columbium, tantalum, and cadmium, which are important in space-age ferroalloys; antimony, which is used as a pigment and as a fire retardant; and gallium, which is essential in the manufacture of solid-state electronic components. The United States has all of these critical metals in the National Defense Stockpile, a supply of about 100 critical mineral materials that could supply the nation's needs for three years in a national emergency.

Nonmetallic Mineral Reserves

The yearly per capita consumption of nonfuel mineral resources in North America exceeds 10 tons, 94 percent of which is nonmetallic, with the major bulk being construction materials (Figure 14.1). The nonmetallic minerals are subdivided for convenience into three groups based on their use in industry, agriculture, and construction. The number and diversity of nonmetallic mineral resources is so great that it is impossible to treat the entire subject in a book of this sort. Thus, this subsection discusses only those nonmetallic resources generally considered most important.

INDUSTRIAL MINERALS. Industrial minerals include those that contain specific elements or compounds that are used in the chemical industry—sulfur and halite, for example—and those that have important physical properties—such as the materials used in ceramics manufacture and the abrasives, such as diamond, corundum, and garnet. Sulfur, one of the most widely used and most important industrial chemicals, is obtained as a by-product of petroleum refining (◉ Figure 14.15) and from the tops of salt domes in the central Gulf Coast states, where it is associated with anhydrite ($CaSO_4$) or gypsum ($CaSO_4 \cdot 2H_2O$). Salt domes originate from evaporite beds that are compressed by overlying sedimentary rock layers, causing the salt to be mobilized into a dome, which rises, piercing the overlying rocks and concentrating sulfur, gypsum, and anhydrite at the top. Superheated water is pumped into the sulfur-bearing caprock to melt the sulfur, which is then piped to a processing plant. More than 80 percent of domestic U.S. sulfur is used to make sulfuric acid, the most important use for

◉ **FIGURE 14.15** Sulfur awaiting shipment to Asia at Vancouver, British Columbia. The sulfur originated as a contaminant in natural gas produced from gas fields in Alberta. Processing plants have separated the sulfur from the natural gas in order to produce a gas whose combustion products meet environmental air-quality standards. The sulfur is a useful by-product of the process.

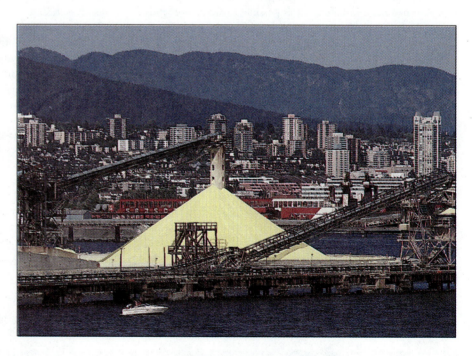

which is the manufacture of phosphate fertilizers. Salt domes are also important sources of nearly pure halite (NaCl), common salt, and many domes have oil fields on their flanks (see Chapter 13).

In addition to being mined from salt domes, halite is mined from flat-lying evaporite beds in New Mexico, Kansas, and Michigan. The most important use of halite is for manufacturing sodium hydroxide and chlorine gas; about 15 percent of halite production is used for de-icing highways, and significant amounts are used in food products and in the manufacture of steel and aluminum.

Diamond is by far the most important industrial abrasive, although corundum, emery (an impure variety of corundum), and garnet also are used. Traditionally, natural diamonds have been important both as gemstones and as industrial abrasives, but artificial diamonds manufactured from graphite now hold about 70 percent of the market for diamond-grit abrasives. The major source of natural diamonds is kimberlite (named for Kimberley, South Africa), a rare and unique ultramafic rock that forms in pipelike bodies. Not only are kimberlite pipes rare, but most of them lack diamonds. The important diamond-bearing kimberlites are in South Africa, Ghana, the Republic of Zaire, Botswana, Australia, Belarus (formerly Belorussia of the USSR), and Yakutia (the largest republic of the Russian Federation).

The metamorphic minerals corundum and emery, natural abrasives for which U.S. reserves are scarce or nonexistent, have largely been replaced by synthetic abrasives such as carbides, manufactured alumina, and nitrides. The United States still imports limited amounts of corundum from South Africa. Garnet, which is found in igneous and metamorphic rocks, is widely used in the United States for optical grinding, in sandblasting, and as a grit on finishing paper for wood. The main source of domestic industrial garnet is metamorphic rocks in the Adirondack Mountains in New York.

AGRICULTURAL MINERALS. With the world's human population doubling every 40 years and the global population expected to reach over 6.25 billion by the turn of the century (see Chapter 1), it seems obvious that food production and the corresponding need for fertilizers and agricultural chemicals also will continue to expand. Therefore, nitrate, potassium, and phosphate compounds will continue to be in great demand. Whereas nearly all agricultural nitrate is derived from the atmosphere, phosphate and potassium come only from the crust of the earth. Phosphate reserves occur in many parts of the world, but the major ones are in the United States, Morocco, Turkmenistan, and Buryat (the latter two are republics of the former Soviet Union). The primary sources of phosphate in the United States are marine sedimentary rocks in North Carolina and Florida (see Figure 14.27, later in the chapter), and there are other valuable deposits in Idaho, Montana, Wyoming, and Utah. The main U.S. supply of potassium comes from

widespread nonmarine evaporite beds beneath New Mexico, Oklahoma, Kansas, and Texas, with the richest beds being in New Mexico. Canada also has large reserves of potassium salts.

CONSTRUCTION MATERIALS. Nonmetallic minerals used as construction materials include *aggregate,* the sand, gravel, and crushed stone used in concrete and for making roadbeds and asphalt road surfaces; limestone and shale, which are used in making cement; clay, used for tile and bricks; and gypsum, the primary component of plaster and wallboard. It is the mining of aggregate that is probably the most familiar to urban dwellers, because quarries customarily are sited near cities, the major market for the product, in order to minimize transportation costs. The quantity of aggregate needed to build a 1,500-square-foot house is impressive: 67 cubic yards are required, each cubic yard consisting of one ton of rock and gravel and 0.7 ton of sand. This amounts to 114 tons of rock, sand, and gravel per dwelling, including the garage, sidewalks, curbings, and gutters. Thus, it is no wonder that aggregate has the greatest commercial value of all mineral products mined in most states, ranking second only in those states that produce natural gas and petroleum. The annual per capita production of sand, gravel, and crushed stone in the United States amounts to about six tons, the total value of which exceeded $15 billion in 1995. The principal sources of aggregate are open-pit quarries in modern and ancient floodplains, river channels, and alluvial fans. In areas of former glaciation, aggregate is mined from glacial outwash and other deposits of sand and gravel that remained after the retreat of the great Pleistocene ice sheets in the northern United States, Canada, and Northern Europe (see Chapter 12).

Concrete, in addition to aggregate, contains cement. The major raw materials for manufacturing cement are large amounts of limestone or marble, from which is produced lime (CaO), and lesser amounts of gypsum or anhydrite, shale, clay or sand, and iron-bearing minerals such as hematite (Fe_2O_3) or magnetite ($FeO \cdot Fe_2O_3$). The principal raw materials are obtained by surface mining, and sufficient deposits of these materials occur worldwide.

THE FUTURE OF MINERAL RESOURCES

Just as the world's fossil-fuel reserves are limited (Chapter 13), so are its mineral reserves. The earth has been well-explored, and fairly reliable estimates exist on the reserves of most important mineral resources. By dividing the known reserves of a resource by its projected rate of use, it is possible to calculate the number of years remaining until the resource is depleted. It was determined in the 1970s that reserves of many important elements would be exhausted by sometime early in the twenty-first century. Although more recent studies have found that the situation is not that

■ TABLE 14.5 U.S. Consumption and Production of Selected Metallic Minerals, 1995

MINERAL	CONSUMPTION, thousands of metric tons	PRIMARY PRODUCTION, thousands of metric tons*	PRIMARY PRODUCTION AS PERCENTAGE OF CONSUMPTION**
Aluminum	7,300	3,300	45.2
Chromium	387	0	0
Cobalt	7.2	0	0
Copper	2,800	1,840	65.7
Iron ore	69,300	57,000	82.3
Lead	1,500	365	24.3
Manganese	695	0	0
Nickel	131	0	0
Tin	46	negligible	0
Zinc	1,220	540	44.3
Platinum-group metals, metric tons	127	8.3	16.3

*Primary production is production from new ore.
**Primary production divided by consumption and rounded to the appropriate number of significant figures.
SOURCE: *Mineral Commodity Summaries*, U.S. Bureau of Mines, 1995.

serious, a few experts still view the exhaustion of some mineral deposits in the near future as likely.

U.S. consumption and production data for selected metallic and nonmetallic mineral resources appear in ■ Tables 14.5 and 14.6, respectively. Notice in Table 14.5 that we consume greater amounts of many metallic mineral commodities than we produce. In the cases of chromium, cobalt, manganese, nickel, tin, and platinum-group metals, all or virtually all of those materials must be imported. The extent of U.S. import reliance for these and other mineral resources is shown in ● Figure 14.16. Perhaps more humbling is ■ Table 14.7, the projected U.S. reserves of selected minerals, assuming 1994 rates of consumption, production, and imports without recycling. Thus, the United States is far from self-sustaining. If nothing else, the data constitute

a clear case against isolationism if we hope to continue enjoying our present life-style.

A brighter forecast is shown by the data in Table 14.6. Note in the table that the United States was self-sustaining in its consumption of domestic nonmetallic and rock resources as of 1995 and that it was producing a surplus of one mineral commodity. There are some apparent disagreements between these tables and Figure 14.16; the low need for imported lead reported in Figure 14.16, for example, does not seem to agree with the value shown in Table 14.5. The discrepancy results from extensive recycling of lead in the United States. Further, the high domestic production of aluminum reported in Table 14.5 results from a high reliance on imported bauxite (reported in Figure 14.16), because domestic sources of bauxite are insignificant.

■ TABLE 14.6 U.S. Consumption and Production of Selected Nonmetallic Minerals, 1995

MINERAL	CONSUMPTION, thousands of metric tons	PRIMARY PRODUCTION, thousands of metric tons*	PRIMARY PRODUCTION AS PERCENTAGE OF CONSUMPTION**
Clay	37,900	42,300	112
Gypsum	26,000	17,300	66.5
Potash	5,390	1,425	26.4
Salt	47,400	39,500	83.3
Sulfur	13,450	11,300	84
Sand and gravel	922,000	922,000	100
Crushed stone	1,198,000	1,195,000	95.7

*Primary production is production from new ore.
**Primary production divided by consumption and rounded to the appropriate number of significant figures.
SOURCE: *Mineral Commodity Summaries*, U.S. Bureau of Mines, 1995.

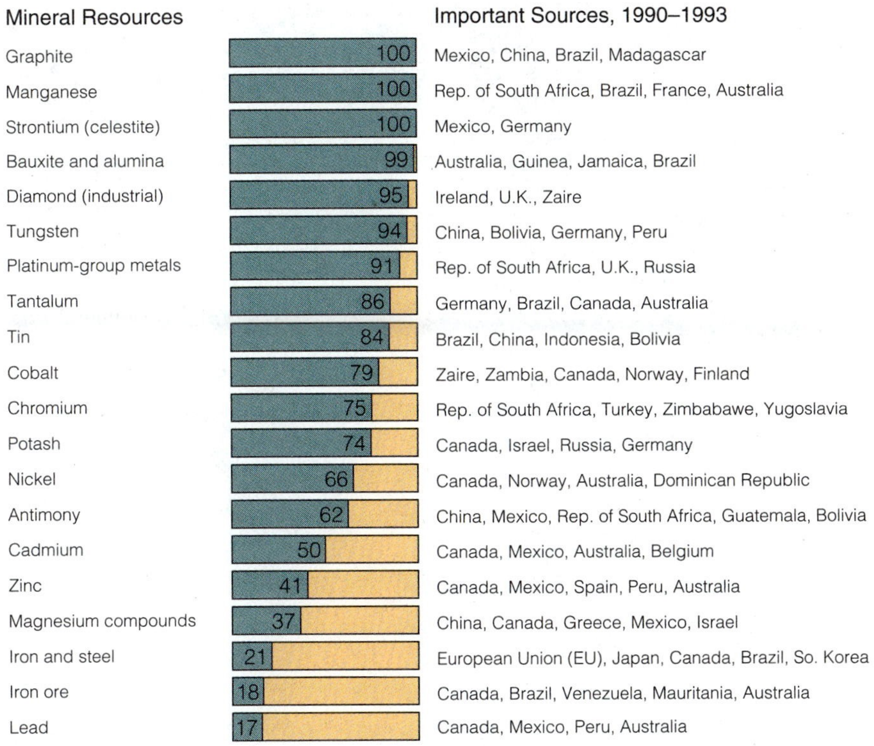

Mineral Resources

Mineral Resources	Net import reliance (%)	Important Sources, 1990–1993
Graphite	100	Mexico, China, Brazil, Madagascar
Manganese	100	Rep. of South Africa, Brazil, France, Australia
Strontium (celestite)	100	Mexico, Germany
Bauxite and alumina	99	Australia, Guinea, Jamaica, Brazil
Diamond (industrial)	95	Ireland, U.K., Zaire
Tungsten	94	China, Bolivia, Germany, Peru
Platinum-group metals	91	Rep. of South Africa, U.K., Russia
Tantalum	86	Germany, Brazil, Canada, Australia
Tin	84	Brazil, China, Indonesia, Bolivia
Cobalt	79	Zaire, Zambia, Canada, Norway, Finland
Chromium	75	Rep. of South Africa, Turkey, Zimbabawe, Yugoslavia
Potash	74	Canada, Israel, Russia, Germany
Nickel	66	Canada, Norway, Australia, Dominican Republic
Antimony	62	China, Mexico, Rep. of South Africa, Guatemala, Bolivia
Cadmium	50	Canada, Mexico, Australia, Belgium
Zinc	41	Canada, Mexico, Spain, Peru, Australia
Magnesium compounds	37	China, Canada, Greece, Mexico, Israel
Iron and steel	21	European Union (EU), Japan, Canada, Brazil, So. Korea
Iron ore	18	Canada, Brazil, Venezuela, Mauritania, Australia
Lead	17	Canada, Mexico, Peru, Australia

● FIGURE 14.16 Net U.S. import reliance for selected mineral resources as a percentage of apparent consumption, 1994. *Net import reliance = imports − exports + adjustments for government and industry stock changes. Apparent consumption = U.S. primary + secondary production + net import reliance.*

■ TABLE 14.7 Estimated Lifetimes of U.S. Reserves of Selected Minerals Assuming 1994 Rates of Consumption, Primary Production, and Imports

MINERAL	PRODUCTION, thousands of metric tons	RESERVE BASE, thousands of metric tons*	ESTIMATED LIFETIME, years**
Bauxite	3,300	40,000	12
Chromium	0	0	0
Cobalt	0	860	negligible
Copper	1,840	90,000	49
Gold, metric tons	330	5,500	17
Gypsum	17,300	>700,000	>40
Iron ore	57,000	25,000,000	439
Lead	365	22,000	60
Manganese	0	0	0
Nickel	0	2,500	negligible
Phosphate	41,000	4,440,000	110
Platinum-group metals, metric tons	8.3	780	94
Potash	1,425	>6,000,000	>4,210
Silver, metric tons	1,400	72,000	51
Sulfur	11,300	230,000	20
Tin	negligible	40	negligible
Zinc	560	50,000	89

*The reserve base consists of the in-place demonstrated resources including presently economic mineral reserves, marginal reserves, and identified subeconomic resources.

**Estimated lifetimes were calculated by dividing the reserve base by the 1994 U.S. production (data from *Mineral Commodity Summaries,* U.S. Bureau of Mines, 1995) and rounding to the appropriate number of significant figures.

SOURCE: Data from *Mineral Commodity Summaries,* U.S. Bureau of Mines, 1995.

■ TABLE 14.8 Estimated Lifetimes of Global Reserves of Selected Minerals Assuming 1994 Rates of Consumption and No Recycling

MINERAL	PRODUCTION, thousands of metric tons	RESERVE BASE, thousands of metric tons*	ESTIMATED LIFETIME, years**
Bauxite	110,000	55,000,000 to 75,000,000	500 to 682
Chromium	9,400	11,000,000	1,170
Cobalt	22	8,800	400
Copper	9,300	1,600,000 (land), 700,000 (sea floor)	172 (land), 75 (sea floor)
Gold, metric tons	2,300	75,000	33
Gypsum	110,000	>2,400,000	>22
Iron ore	1,000,000	>800,000,000	>800
Lead	2,900	1,500,000	520
Manganese	7,100	4,900,000	690
Nickel	850	130,000	153
Phosphate	130,000	34,000,000	262
Platinum-group metals, metric tons	250	100,000	400
Potash	22,000	250,000,000	1,140
Silver, metric tons	14,000	420,000	30
Sulfur	52,000	180,000,000	350
Tin	180	10,000	56
Zinc	6,700	1,800,000	270

*The reserve base consists of the in-place demonstrated resources including presently economic mineral reserves, marginal reserves, and identified subeconomic resources.

**Estimated lifetimes were calculated by dividing the reserve base by the 1994 production (data from *Mineral Commodity Summaries,* U.S. Bureau of Mines, 1995) and rounding to the appropriate number of significant figures.

SOURCE: Data from *Mineral Commodity Summaries,* U.S. Bureau of Mines, 1995.

As resources diminish in the coming decades, we can expect shortages of some minerals that presently are common. In many cases technological advances will develop substitute materials that will suffice, but in other cases, we will have to do without. Some minerals will become sufficiently scarce that deposits now classed as "submarginal" will be upgraded to "low-grade" and will be economically exploitable due to market forces. Projections of worldwide mineral reserves are shown in ■ Table 14.8.

The need for mineral resources may require exploration and exploitation in more remote regions and more hostile environments. Coal has been mined for more than a half-century well north of the Arctic Circle on Norway's island of Spitsbergen (Svalbard), for example, and sources of fuels and minerals are currently being sought and exploited in the harsh climates of Canada's Arctic islands and Greenland. The Red Dog zinc-lead-silver mine on the lands of the Inupiat Eskimos north of the Arctic Circle in Alaska went into full production in 1991. Even environmentally sensitive Antarctica, which is known to have extensive deposits of coal and iron, may be the target of future exploitation as more is learned about that continent's mineral wealth (see Case Study 14.4). Deeper parts of the continents and the sea floor may also be targeted for exploration and exploitation as geophysical innovations and technological advances open up these presently inaccessible regions.

In the future we will experience more recycling of critical metals, such as has been done with gold for thousands of years. The gold of the crown on your tooth, for example, may have been in jewelry during the age of Pericles in Ancient Greece. Such metals as lead, aluminum, iron, and copper have been recycled for years. In many cities the large-scale recycling of metals, especially lead and aluminum, as well as of paper, glass, and plastics, is being conducted successfully as part of municipal trash collection (see Chapter 15). Significant amounts of several metals consumed in the United States are now being obtained from recycled scrap. In 1994 these amounts were

METAL	AMOUNT OBTAINED BY RECYCLING
aluminum	37%
copper	40%
iron	59%
lead	65%
cobalt	22%
tin	27%
nickel	34%

Penguins, Treaties, and Antarctica's Potential Mineral Wealth

Fifty years of geological study reveal that Antarctica was connected to Africa, Australia, and South America before the fragmentation of Gondwanaland. Thus, plate tectonics serves as a basis for speculating about Antarctica's possible mineral wealth by extrapolating from known reserves in the other Southern Hemisphere continents (see Chapter 3).

Limited exploration in the Antarctic has already shown that there are substantial deposits of coal and iron; and traces of copper, tin, chromium, nickel, cobalt, gold, uranium, and other metals also have been found (● Figure 1). A layered intrusion covering about 50,000 square kilometers (19,500 mi²) of the Pensacola Mountains (about midway between the South Pole and the Weddell Sea) is notably similar to the Bushveld complex of South Africa, one of the richest mineral sites in the world. Considering that 98 percent of the continent is ice-covered and that 90 percent of that area is unmapped, there is a strong likelihood that substantial mineral deposits ultimately may be discovered.

The continent's thick ice cap, harsh climatic conditions, and environmental sensitivity would seem to preclude mineral exploitation in the near future, but for some countries that have already made territorial claims in Antarctica, political and strategic motivation might override these obstacles. In 1978 the Antarctic Treaty members began debating whether any exploitation should be allowed and, if so, under what conditions. After ten years of sometimes tense negotiations, the 67 articles of the *Convention on Regulation of Antarctic Mineral Resource Activities* were agreed upon. Since 1988, an additional 12 articles, the "Annex for an Arbital Tribunal," have been adopted. The spirit of the *Convention* appears to be the desire to protect the integrity of Antarctica's ecosystems while exploring and exploiting its mineral resources, but such goals seem to be contradictory. Would it be *possible* to prospect and mine mineral resources and yet adhere to the *Convention's* Article IV, which states that these activities must avoid ". . . significant adverse effects on air and water quality; significant changes in atmospheric, terrestrial, and marine environments; and significant changes in the distribution, abundance, or productivity of populations of species of flora and fauna"?

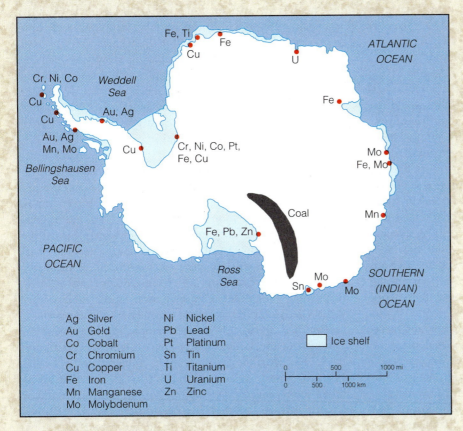

● **FIGURE 1 Known mineral deposits of Antarctica.**

Collectively, the value of recycled metal and mineral scrap amounted to about 30 percent of the nation's mineral production.

 ## MINING AND ITS ENVIRONMENTAL IMPACTS

The Mining Law of 1872 was important legislation that helped establish the mining industry as a fundamental element in the U.S. economy. At the same time, the law was the ultimate in *laissez-faire* regulation, as it allowed miners to exploit, to take profits at public expense, to pay little in return for the privilege of mining on federal land, and to walk away from the scars and waste materials remaining when they abandoned their mines. The original nineteenth-century law allowed miners to stake a mining claim on potentially profitable public land and, should the claim prove to contain valuable minerals, to obtain a patent (legal title) and reap the profits. This law helped "win" the West by enticing thousands of prospectors to seek their fortunes in gold, silver, and other valuable mineral commodities.

The 1872 law gave no consideration whatever to the environmental consequences of mining. The law is staunchly defended today by some public officials and the mining industry, who claim that it does a good job of providing the country with valuable minerals, employment, and economic development. The same law exists today, but it has been much modified over the years by more than 50 amendments, so that today's body of mining law fills six volumes. An example of the changes is the Mineral Lands Leasing Act of 1920, by which the government retains title to all federal lands possessing energy resources (oil, gas, and coal) and assesses royalties on the profits from the lands leased by developers. In addition, environmental restrictions added in the 1970s and 1980s dictate that modern mining must be conducted within a different policy framework than the simply drafted law of 1872.

Although interest in environmental protection and reclamation in coal mining districts began to develop in the 1930s (see Case Study 14.5), serious efforts at regulation and reclamation did not begin until the late 1960s due to the increasing amounts of unreclaimed land, and water pollution and problems arising from the lack of uniform standards among state mining programs. Consequently, the Surface Mining Control and Reclamation Act (**SMCRA**) was passed in 1977. This act established coordination of federal and state efforts to regulate the coal industry in order to prevent the abuses that had prevailed in the past. SMCRA regulations apply to all lands—private, state, and federal. SMCRA established a special fee that is charged on coal production and generates funds for reclaiming abandoned coal-mine lands. SMCRA also provides for state primacy once a state's coal mine reclamation laws and policies have been approved by the Secretary of the Interior. This gives state regulators primary SMCRA enforcement jurisdiction as long as the state programs meet federal standards.

Federal reclamation standards and other safeguards established by SMCRA for the coal-mining industry have no counterpart for noncoal underground and surface mineral mining on state or private lands. Although there are federal regulations for noncoal mining on federal lands, it is left to the individual state governments to codify safeguards and enforce noncoal mine reclamation requirements.

Mining on federal lands must comply with the regulations of the *National Environmental Policy Act* (NEPA) of 1969 (see Chapter 11). A mining company must file an operating plan and an *Environmental Impact Report* (EIR), and conduct public hearings before the administrating agency can approve an operating permit. Even though an applicant may spend several thousand dollars in complying with NEPA permit application requirements, there is no assurance that an application will be approved (see Case Study 14.2).

There is no specific fund for cleaning up abandoned noncoal mines on state and private lands, nor are there uniform regulatory requirements for these lands. Action is left to the states and the mining companies. However, once any state with coal production has implemented safeguards and standards for coal mining, it may use the SMCRA abandoned mine fund to address reclamation of abandoned mines that exploited other commodities. For example, Utah's abandoned mine reclamation program uses SMCRA funds to reclaim metals mines.

Environmental pollution abatement has become a major concern of domestic gold producers in some states, especially in Alaska, where there are many small placer mines. Planning for reclamation of mined land has become an integral part of an increasing number of gold mine plans. Abandoned mine sites, as well as other contaminated industrial sites that are deemed to pose serious threats to health or safety, may be cleaned up under the Comprehensive Environmental Response, Compensation and Liability Act (**CERCLA**) of 1980, which is overseen by the *Environmental Protection Agency* (EPA). For example, the Forest Service, a branch of the United States Department of Agriculture, has CERCLA authority to clean up contaminated mines on National Forest System lands. It was CERCLA that established the so-called *Superfund* to clean up the worst toxic or hazardous waste sites that have been assigned to the National Priorities List (see Case Study 14.7 and Chapter 15).

Some states have enacted *State Environmental Policy Acts* (SEPAs) that are patterned after NEPA, while other states

⦿ FIGURE 14.17 A large underground mining operation; the Homestake Mine near Lead, South Dakota. The shaft headframe and the hoist building are the large structures on the ridge crest. The mills for processing the ore are the large "stair-stepped" buildings at the center left and the lower right of the photograph. This mine began operations in 1876 and is still producing gold.

may have little regulation. The differences in the various states' permit procedures, their mining and reclamation standards, and their enforcement of underground and surface mining laws can be illustrated by comparing the laws of three states, each noted for over a century of mining activity: Colorado, Arizona, and Michigan. Colorado has strong regulatory provisions relating to mine operation, environmental protection, reclamation bonding, inspections, emergency response by the state, and even the authority over mine operation. In contrast, Arizona, the only state in the country without a mine-reclamation law until 1994, allows companies to self-bond their reclamation fund, to wait as long as 17 years after mining has stopped to begin reclamation, and to set their own reclamation standards. Furthermore, there is an escape clause that may allow companies a variance from any of the law's provisions. Michigan, which in 1994 ranked sixth among all states in the production value of nonfuel minerals, has no comprehensive mining law. No financial bond is required to ensure that a mine site will be properly closed and reclaimed if the company abandons the project or declares bankruptcy. No permit is needed to open a mine except a local zoning permit and a federal water-discharge permit—permits that are required of *every* industrial operation. Reportedly, there is no enforcement of the state's voluntary reclamation law.

As we consider the environmental impacts of mining in the United States in this section, it is important to distinguish between (1) the legacy of careless exploitation that is obvious at long-abandoned mines with their residual scars and toxic wastes and (2) the current mining scene, in which, depending upon the particular state and whether the site is on land administered by the U.S. Forest Service or the Bureau of Land Management, more than 30 different permits may by needed in order to develop and operate a mine

legally. Many mine owners consider the current permitting requirements to be excessive, and they question whether they will be able to continue to operate in the United States at all if the regulations become more restrictive.

Mining in the past was conducted largely by underground methods with a surface plant for milling and processing and for hoisting workers, ore, and equipment (⦿ Figure 14.17). Surface mining methods, typified by vast open-pit excavations such as the Bingham Canyon mine (see Figure 14.7 and Case Study 14.1), have now largely replaced underground methods. The surface plant remains much the same, however, although it usually operates on a much larger scale than in the past. Both surface and underground mining create significant environmental impacts on the land and air and on biological and water resources. In addition, the needs for housing and services in mining areas have social impacts.

Impacts of Coal Mining

Much coal now mined in the United States is extracted by surface operations, which involve removing rock and soil that overlie the coal beds. Compared to underground mining (⦿ Figure 14.18, page 436), surface mining is generally less expensive, safer for miners, and facilitates more complete recovery of coal. Surface mining, however, causes more extensive disturbances to the land surface and has the potential for serious environmental consequences unless the land is carefully reclaimed. The three major methods of surface mining of coal are contour mining, area mining, and mountain-top mining. Each of these is explained here.

Contour mining is typical in the hilly areas of the eastern United States, where coal beds occur in outcrops along hillsides. Mining is accomplished by cutting into the hillside to expose the coal and then following the coal seam around

Reclamation of the Copper Basin

A well-known area of severe, long-lasting industrial degradation is the Copper Basin, near Ducktown, Tennessee, a region straddling the borders of Tennessee, North Carolina, and Georgia (● Figure 1). The area is also remembered as the site of the 1996 Summer Olympics white-water canoe and kayak events, which were held on the Ocoee River from south of Ducktown to Parksville Lake. Little did the contestants know of the area's history of degradation, which began in the 1850s with small-scale copper mining and the region's first smelter. Removal of trees to fuel the ore roasters and smelters quickly stripped the basin of its vegetation. Heap-roasting (open-pit smelting) of the copper ore yielded high concentrations of sulfur dioxide (SO_2) gas, which drifted over neighboring lands and extended into the neighboring states of Georgia and North Carolina, devastating the Ducktown area and causing severe plant and crop damage in neighboring areas. The basin's natural environment contributed to the problem. Its naturally acidic soils were highly leached and deficient in calcium; its meteorologic conditions (warm temperatures, high humidity, frequent fog, usually stagnant air, abundant and sometimes intense rainfall)

in conjunction with its topographic setting produced atmospheric inversions that concentrated the acid and limited the lateral movement of the air. Consequently, the humid, unrefreshed, contaminated air lingered over the basin for extended periods of time.

Large-scale mining and smelting operations began in 1891. The fuelwood cutting for the smelters, mining activities, SO_2 emissions, and related disturbances killed all vegetation, eventually affecting about 130 square kilometers (50 mi^2) of the mountain enclave. Subsequently, intense erosion of topsoil and subsoil occurred on about 9,290 hectares (23,000 acres). By 1905 the effects of the smelters' acid smoke and the acres of roasting ore piles on vegetation and agriculture were so devastating that the state of Georgia hauled Ducktown Sulphur Company and Tennessee Copper Company into court in an effort to get them to stop their polluting.

The mining companies began extending their smokestacks, one of them to 99 meters (325 ft) tall. High stacks have often served to satisfy complaints about industrial pollution, especially at copper smelters; the common solution for pollution being dilution. The tall stacks successfully removed fumes from ground level, but due to their higher release points, the fumes traveled even further and damaged vegetation over an even larger region. Court action dragged on for a decade, eventually reaching the U.S. Supreme Court. The Court ruled that the operations must "prevent the diffusing of sulphurous fumes over the Georgia border." In response to these problems, the mining companies constructed sulfuric acid plants to recover the SO_2 from the smelters, thereby converting an injurious polluting gas into a useful by-product needed by the fertilizer industry. So successful was the sulfuric acid production that by 1911 southeastern Tennessee had the two largest sulfuric acid plants in the world, and the sulfuric acid recovered from Ducktown's smelters accounted for 20 percent of U.S. production. There was a slow but encouraging improvement in the air quality, enough to satisfy Georgia and the Supreme Court, but the land in the Copper Basin by this time had become so devastated that it was a biological desert (● Figure 2). The barren landscape of deeply gullied red soil drew tourists out of their way to see it.

Erosion in the Copper Basin has now decreased the flood-storage capacity of the Tennessee Valley Authority's (TVA's) three Ocoee River reservoirs and increased TVA's hydroelectric production costs due to sediment damage to turbines and power-generation facilities (● Figure 3). Heavy-metal pollution and siltation to surface waters

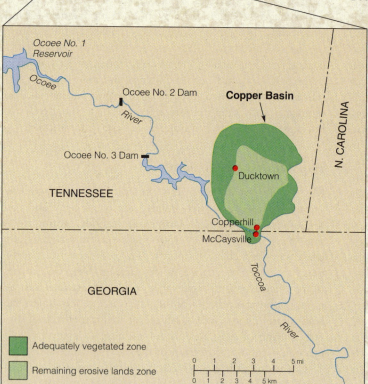

● FIGURE 1 The Copper Basin.

● FIGURE 2 The Tennessee Copper Company's smelter and sulfuric acid plant at Copperhill, Tennessee, 1912. The barren, denuded ground had been sterilized by acid fallout.

● FIGURE 3 Sediment deposits behind Ocoee Dam No. 2, a diversion dam for a hydroelectric generator. Sediment from the Copper Basin adversely impacts the Ocoee River's drainage, decreases dams' flood-storage capacities, and increases the cost of hydroelectric energy production.

draining the basin has become a high-priority water-resource problem in the Tennessee Valley. Measured annual erosion rates from the basin's denuded lands are as much as 195 tons per acre.

Efforts by the mining companies, the TVA, and other government agencies to revegetate the devastated 9,290 hectares began in the 1930s. The Tennessee Copper Company (TCC), in cooperation with the TVA, began research on restoration of the lands and instituted an extensive tree-planting program in 1939. The work was continued by the company and its successors, Cities Service, and BIT Manufacturing. By the late 1970s, aerial applications of fertilizer were enhancing tree growth and survival, and the increased vegetative cover had contributed to a marked slowing of erosion.

By 1984, driven by the need to improve the Ocoee's water quality, reduce the rate of sedimentation behind the Ocoee River dams (Dam No. 3 had lost nearly 90% of its water-storage capacity), and reduce the adverse impact of sediment loads on

hydropower operations, new techniques were implemented by the TVA, TCC, and others. These included chemically treating more than 1,200 hectares (3,000 acres) of land and planting more trees. By 1990, more than 16 million trees, most of them genetically improved strains of loblolly pine that can tolerate the poor soil, had been planted (● Figure 4). Additional projects planned for the 1990s and beyond included containing contaminated surface waters in critical drainage basins, operating wastewater treatment plants in critical areas, installing ground-water-monitoring wells in six towns, cleaning up the aniline-chemical and petroleum contamination on the property of a chemical plant, and continuing the reforestation of the TCC property. By combining these projects with the ongoing reclamation effort, major environmental improvements can be expected in the Copper Basin.

(a)

(b)

● FIGURE 4 The Copper Basin (a) before reclamation, March 1986, and (b) after reclamation, July 1991.

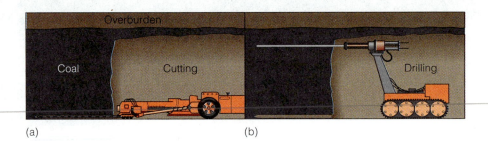

● FIGURE 14.18 The steps in conventional large-scale underground bituminous-coal mining.

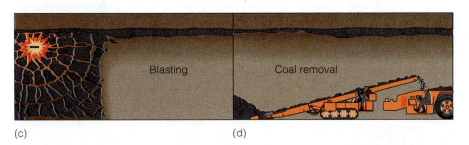

the perimeter of the hill. Successive, roughly horizontal strips are cut, enlarging each strip around the hillside until the thickness of the overburden is so great that further exposure of the coal bed would not be cost-effective. At each level of mining is a *highwall,* a clifflike, excavated face of exposed overburden and coal that remains after mining is completed (●Figure 14.19). Augers (giant drill bits) are used to extract coal from the areas beneath the highwall. The SMCRA requires that the sites must be reclaimed, that is, returned to the original contour and highwalls must be

● FIGURE 14.19 Abandoned coal mine in an early stage of reclamation, 2.6 kilometers (1.25 mi) long and more than 122 meters (400 ft) deep and less than 100 meters (330 ft) from a residential area. One death resulted at the site when a boy fell off the edge of a highwall. The highwall in the right foreground has already been eliminated and revegetated, while in the distance the work continues. The site is in Pennsylvania.

covered and stabilized after contour mining is completed. This is accomplished by backfilling **spoil,** the broken fragments of waste rock removed in order to mine the coal, against the highwall, and spreading and compacting as necessary to stabilize the reclaimed hillside. *Hydroseeding,* spraying seeds mixed with water, mulch, fertilizer, and lime onto regraded soil, is used on steep reclaimed slopes to aid in establishing vegetation that will help prevent soil erosion (Case Study 14.6).

Area mining is commonly used to mine coal in flat and gently rolling terrain, principally in the Midwestern and Western states (●Figure 14.20). Because the pits of active area mines may be several kilometers long, enormous equipment is used to remove the overburden, mine the coal, and reclaim the land. Topsoil is stockpiled in special areas and put back in place when mining is completed. After replacing, it is tilled with traditional farming methods to reestablish it as crop or pasture land. Commonly, the land is more productive after reclamation than it was before mining.

Mountain-top mining is used predominantly in the eastern United States to recover coal that underlies the tops of mountains. After the coal is removed, the mined area is returned to its approximate original shape or left as relatively flat terrain. Mountain-top mining allows for nearly complete removal of the coal bed. In many instances, removal of the mountain top results in terrain that is suitable for agriculture, grazing, or residential development after mining ceases.

Impacts of Underground Mining

The impacts of underground mining range from moderate to critically damaging, but they are often less severe than those of surface mining. Underground workings usually follow the ore body closely, which minimizes the amount of barren rock that must be removed and piled on the sur-

Steps in Surface Mine Reclamation

There are five steps in reclaiming abandoned coal mine lands: (1) drainage control to eliminate acid mine drainage, (2) stabilization of landforms (◉ Figure 1), (3) revegetation (◉ Figures 2 and 3), (4) ongoing monitoring, and (5) returning the land to use (◉ Figure 4). The reclamation work pictured in Figures 1, 2, and 3 was carried out near Beckley, West Virginia. The reclaimed site in Figure 4 is near Harding, West Virginia.

◉ FIGURE 2 Immediate seeding, here done with a hydroseeder, follows emplacement of topsoil in order to establish vegetation quickly. A biodegradable green dye is added to the seed to help the operator determine the coverage.

◉ FIGURE 1 Drainage control and diversion at the disturbed area. Spreading and compacting of waste rock, subsoil, and soil against the highwall stabilizes slopes and buries reactive sulfide minerals.

◉ FIGURE 3 A fast-growing, temporary cover crop such as rye grass prevents soil erosion and adds organic matter to the soil. The cover crop is selected carefully in order to avoid using a plant that may be undesirably dominant or persistent.

◉ FIGURE 4 After the cover crop has died back, permanent legumes and grasses take over. They eventually restore the disturbed land to meadow or pasture so that it can be used for wildlife habitat, recreation, or livestock grazing. Trees eventually may return the site to forest. Reclamation of this site eliminated more than a mile of highwall and sealed four hazardous abandoned mine openings. An underdrain was placed along the length of the highwall to collect drainage from the site. Today the site bears no resemblance to its appearance before reclamation.

FIGURE 14.20 Area (open-pit) coal mining. As the ground is opened up to expose more of the coal and the pit enlarges, bulldozers backfill the mined-out area to prepare it for reclamation.

face near the mine entrance. Some waste materials at currently active underground mines, such as the Homestake Mine in South Dakota (Figure 14.17), are used to backfill abandoned tunnels. This reduces the costs of energy for hoisting as well as minimizing unsightly surface waste-rock piles.

The most far-reaching effects of underground mining are ground subsidence (Chapter 8), the collapse of the overburden into mined-out areas (⦿ Figure 14.21), and **acid mine drainage (AMD)**, the drainage of acidic water from mine sites. AMD is generally considered to be the most serious

⦿ FIGURE 14.21 Store building subsiding into an abandoned underground coal mine south of Pittsburgh, Pennsylvania.

environmental problem facing the mining industry today, as some acid drainage may continue for hundreds of years. AMD can result when high-sulfur coal and metallic-sulfide ore bodies are mined. In both cases, pyrite (FeS_2) and other metallic-sulfide minerals are prevalent in the walls of underground mines and open pits; in the **tailings,** finely ground, sand-sized waste material from the milling process that remains after the desired minerals have been extracted; and in other mine waste. The reaction of pyrite or other sulfide minerals with oxygen-rich water produces sulfur dioxide (SO_2), which, facilitated by bacterial decomposition, reacts with water to form sulfuric acid. As the reaction proceeds, heat is generated and acidity increases, which results in an increased rate of reaction. This not only acidifies surface and underground waters, but it also expedites the leaching, release, and dispersal of iron, zinc, copper, and other toxic metals into the environment. Such substances kill aquatic life and erode human-made structures such as concrete drains and bridge piers, sewer pipes, and well casings. (Zinc concentrations as low as 0.06 mg/L and copper concentrations as low as 0.0015 mg/L are lethal to some species of fish.) Estimates are that between 8,000 and 16,000 kilometers (5,000 and 10,000 mi) of U.S. streams have been ruined by acid drainage. The oxidized iron in AMD colors some polluted streams rust-red (⦿ Figure 14.22). Where polluted water is used for livestock or irrigation water, it may diminish the productive value of the affected land. Technologies for preventing and controlling AMD are still being developed.

⊙ FIGURE 14.22 Rust-red sludge of acid mine drainage from West Virginia coal mines that were abandoned in the 1960s. The acidic water has eaten away at the Portland cement–based concrete retaining wall, and the bridge supports will soon share that fate. Problems such as this are rarely caused by modern surface mines, because current mining and reclamation practices eliminate or minimize acid mine drainage.

⊙ FIGURE 14.23 Placer mining with a bucket-conveyer dredge near Platinum, Alaska, 1958. The dredge operated from the early 1930s until the late 1970s, during which time it was the major producer of platinum-group metals in North America.

Impacts of Surface Mining

A variety of surface-mining operations contribute to environmental problems. Hydraulic mining is largely an activity of the past, but **dredging**—scooping up earth material below a body of water from a barge or raft equipped to process or transport materials—continues in the 1990s (⊙ Figure 14.23). Much of the current U.S. production of sand and gravel is accomplished by dredging rivers, and large dredges are used to mine placer tin deposits in Southeast Asia. Dredging causes significant disruption to the landscape; it washes away soil, leaves a trail of boulders, and severely damages biological systems. Scarification remains from former dredging in river bottoms in many Western states and elsewhere in the world (⊙ Figure 14.24). Few gold-dredge areas have been reclaimed, but an outstanding example of what can be accomplished appears in ⊙ Figure 14.25. Another example is a 73-square-kilometer (28-mi²) dredge-scarified area near Folsom, California. The area was reclaimed by reshaping the waste piles, covering the surface with topsoil, and then building a subdivision of homes on it. Landscaped with trees and grasses, no evidence remains of the once scarified surface.

In **hydraulic mining,** or *hydraulicking,* a high-pressure jet of water is blasted through a nozzle, called a *monitor,* against hillsides of ancient alluvial deposits (Case Study 14.3). Hydraulicking requires the construction of ditches, reservoirs, a *penstock* (vertical pipe), and pipelines. Once constructed, it is possible to wash thousands of cubic yards of gold-bearing gravel per day. The gold is recovered in sluice boxes, where mercury (the liquid metal that dissolves gold and silver to form an amalgam; it's explained in Case Study 14.3) usually is added to aid recovery at a low cost. Hydraulicking is efficient but highly destructive to the land (⊙ Figure 14.26). Because nineteenth-century hydraulick-

⊙ FIGURE 14.24 Placer gold mining with a bucket-conveyor dredge on the Yuba River, California, in 1993. Surrounding the dredge operation is the scarified landscape of waste-rock piles.

● FIGURE 14.25 Reclaimed dredge-mined site, a former placer gold mine; Fox Creek, Fairbanks District, Alaska. The State of Alaska awarded a certificate of commendation for the reclamation work.

ing was found to create river sediment that increased downstream flooding, clogged irrigation systems, and ruined farmlands, court injunctions stopped most hydraulic mining in the United States before 1900 (although it still continues in Alaska on a limited scale). Hydraulicking is still being used in Russia's Baltic region for mining amber, and in 1979 thousands of gold seekers began invading remote areas of Brazil's Amazon basin and using it to extract placer gold. The danger of using mercury is that when it escapes from the mining activity, it can accumulate in the food chain; consumption of mercury-tainted food causes birth defects and neurological problems in humans and animals. Its current use in gold recovery in Brazil as well as in Indonesia and the Philippines is causing extensive contamination. It has been estimated that 100 tons of mercury is working its way into the ecosystem of the Amazon basin each year.

Strip-mining is used most commonly when the resource lies parallel and close to the surface. The phosphate deposits

● FIGURE 14.26 A monument to the destructive force of hydraulic mining of bench placer deposits; Malakoff Diggins State Park, Nevada County, California. When the hydraulic mining operations here ended in 1884, Malakoff Diggins was the largest and richest hydraulic gold mine in the world with a total output of about $3.5 million. More than 41 million cubic yards of earth had been excavated to obtain gold. The site is now marked by colorful, eroded cliffs along the sides of an open pit that is some 7,000 feet long, 300 feet wide, and as much as 600 feet deep.

● FIGURE 14.27 A rich phosphate bed in Florida is strip-mined by a mammoth drag-line dredge. The bucket, swinging from a 100-meter boom (that's longer than a football field), can scoop up nearly 100 tons of phosphate at a time. The phosphate is then piped as a slurry to a plant for processing. Fine-grained gypsum, a by-product of processing, is pumped to a settling pond (*upper left*), where it becomes concentrated by evaporation.

of North Carolina and Florida are strip-mined by excavating the shallow, horizontal beds to a depth of about 8 meters (26 ft) (● Figure 14.27). After the phosphate beds are removed, the excavated area is backfilled to return the surface to its original form.

Open-pit mining is the only practical way to extract many minerals when they occur in a very large low-grade-ore body near the surface. The process requires processing enormous amounts of material and is devastating to the landscape. The epitome of open-pit mining is the Bingham Canyon copper mine in Utah, where about 3.3 billion tons of material—seven times the volume moved in constructing the Panama Canal—have been removed since 1906 (Figure 14.7). Now a half-mile deep, the pit is the largest human excavation in the world, and the waste-rock piles literally form mountains.

The environmental consequences of open-pit mining are several. The mine itself disrupts the landscape, and the increased surface area of the broken and crushed rocks from mining and milling sets the stage for erosion and the leaching of toxic metals to the environment. This is especially true of sulfide-ore bodies. They produce ADM, because the waste rocks and tailings are highly susceptible to chemical weathering.

Impacts of Mineral Processing

Except for some industrial minerals, excavating and removing raw ore are only the first steps in producing a marketable product. Once metallic ores are removed from the ground, they are processed at a mill to produce an enriched ore, referred to as a *concentrate.* The concentrate is then sent to a smelter for refining into a valuable commodity.

Concentration and smelting are complex processes, and a thorough discussion of them is well beyond the scope of this book. Briefly, the concentration process requires:

1. crushing the ore to a fine powder,
2. classifying the crushed materials by particle size by passing them through various mechanical devices and passing on those particles of a certain size to the next step, and
3. separating the desired mineral components from the noneconomic minerals by a flotation, gravity, or chemical method.

These means of separation are discussed individually here.

FLOTATION. The **flotation** separation process is widely used, especially for recovering sulfide-ore minerals such as lead, zinc, and copper sulfides from host rock. The process is based on the principles of wettability of mineral particles and surface tension of fluids. After crushing and concentration, the wettability of the *undesired* mineral particles is increased by treating the crushed ore chemically—usually with liquid hydrocarbons—to ensure that the undesired minerals will sink. Air is then bubbled into the slurry of crushed ore and water, forming a froth that collects the *desired* mineral particles of low wettability. The froth, with the attached desirable mineral particles, is skimmed off the top of the flotation tank and dried; this is the concentrate. The undesired mineral particles, the tailings, sink to the bottom of the flotation tank. They are drawn off and piped to the tailings pond. Although they are usually environmentally undesirable, tailings are an unavoidable waste product of mining.

GRAVITY SEPARATION. Gravity separation methods are used in recovering high-density ore minerals such as gold, platinum-group metals, tungsten, and tin. By this

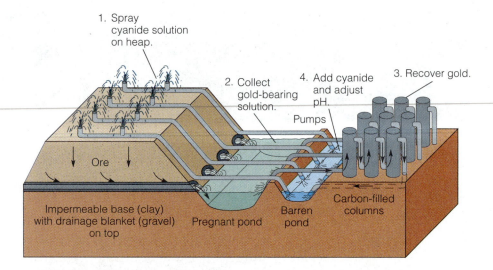

● **FIGURE 14.28** The major components of cyanide heap-leach gold recovery.

1. Spray cyanide solution on heap.

2. Collect gold-bearing solution.

4. Add cyanide and adjust pH.

3. Recover gold.

Pumps

Ore

Impermeable base (clay) with drainage blanket (gravel) on top

Pregnant pond

Barren pond

Carbon-filled columns

process, mineral particles mixed with water are caused to flow across a series of riffles placed in a trough. The riffles trap the desired high-density particles, and water carries away the undesired low-density minerals, the tailings. The rockers, sluice boxes, and dredges used in gold placer min-

● **FIGURE 14.29** Cyanide drip lines on a leach heap. The cyanide concentration is 200 parts per million (one five-thousandth of one percent), and the solution is kept highly alkaline to prohibit cyanide gas from forming.

ing, explained in Case Study 14.3, are all examples of gravity separation methods.

CHEMICAL METHODS. For minerals whose physical properties make them unsuitable for separation by flotation or gravity methods, chemical processes are used, the major ones being leaching and cyanidation. *Leaching* is often used in treating copper-oxide ores. Sulfuric acid is added to crushed ore to dissolve the copper and produce a solution of copper sulfate. The dissolved copper is then recovered by placing scrap iron in the copper sulfate solution; the copper plates out onto the iron. The acidic waste materials are chemically neutralized by treating them with lime.

CYANIDATION. *Cyanidation,* used to recover gold and silver since 1890, makes use of the special property of cyanide to dissolve gold and silver. An innovation of cyanide recovery, **cyanide heap-leaching,** began to be widely used in the United States in the 1980s. During the 1990s there have been about 150 heap-leach operations active in Alaska, Montana, Idaho, Colorado, Nevada, California, Washington, Utah, South Dakota, and South Carolina. Although heap-leaching is efficient, it is controversial in environmental circles because the open-pit mining, waste-rock dumps, and tailings piles are destructive to the landscape. Furthermore, cyanide is perceived as a hazard to wildlife and a contaminant to ground and surface waters. The process is illustrated in ● Figure 14.28. Ore from an open-pit excavation is pulverized, spread out in piles over an impervious clay or plastic liner, and sprayed with a dilute cyanide solution, commonly about 200 parts per million (0.02 %) (● Figure 14.29). The solution dissolves gold and silver (and several other metals) present in small amounts in the ore as it works its way through the heap to the pregnant pond (● Figure 14.30). The gold and silver are recovered from the resulting "pregnant" solution by adsorption on activated charcoal, and the barren cyanide solution is recycled to the leach heap. The precious metals are removed from the charcoal by chemical and electrical techniques,

● FIGURE 14.30 A typical "pregnant pond" for collecting the cyanide solution containing dissolved gold from a leach heap. The face of the leach heap is at the upper right of the photograph. The pipes on the far side of the pond drain the "pregnant" solution from beneath the leach heap. The pond is lined with two layers of high-density polyethylene underlain by a layer of sand and a leakage detection system of perforated PVC pipe. In the rare chance that leakage should occur, it is trapped in the pregnant pond, and appropriate repairs are made. The pond is covered with fine-mesh netting to keep birds from the toxic solution.

melted in a furnace, and poured into molds to form ingots.

Monitoring wells are placed downslope from leach pads and ponds containing cyanide for detecting possible leakage. Maintaining the low concentrations of cyanide required by state regulations requires regular monitoring of the solution. The cyanide solution is kept highly alkaline by additions of sodium hydroxide, a strong base, in order to inhibit the formation of lethal cyanide gas. As of 1994, no measurable cyanide gas had been detected above leach heaps where instrumental testing for escaping gas is carried out. Unforeseen peculiarities of the ore chemistry and unusual weather have caused unexpected increases in cyanide values at some mines. This has required an occasional shutdown of operations until the chemistry of the

solution could be corrected to the established standards. In a few cases, stiff fines have been assessed where cyanide levels exceeded the limits of the operator's permit. Upon abandonment of a leach-extraction operation, federal and state regulations require flushing and detoxification of any residual cyanide from the leach pile. Regular sampling and testing at ground-water-monitoring wells also may be required for several years following abandonment.

Securing the open cyanide ponds to keep wildlife from drinking the lethal poison is an environmental concern at several operations. After 900 birds died on the cyanide-tainted tailings pond at a mine in Nevada in 1989–1990, the company pleaded guilty to misdemeanor charges, paid a $250,000 fine, and contributed an additional $250,000 to the Nature Conservancy for preservation of a migratory bird habitat. Reported wildlife mortalities at cyanide-extraction gold mines in Arizona, California, and Nevada from 1980 to 1989 appear in ■ Table 14.9. Of major concern are a number of endangered, threatened, or rare mammal species whose geographic ranges include cyanide-extraction mines. About 34 percent of the mammal mortalities reported in Table 14.9 are bat species that are on the federal government's list of rare and endangered species.

Various techniques for discouraging wildlife from visiting cyanide-extraction operations have been tried: stringing lines of flags across leach ponds in an effort to scare birds away, covering pregnant ponds with plastic sheeting, and blasting recorded heavy-metal rock music from loudspeakers to frighten wildlife away. The current standard practice is to enclose the operations with chain-link fencing to keep out larger animals and to stretch netting completely across the ponds to protect birds (Figure 14.30). The fences and netting have been successful; California's 12 to 15 active heap-leach operations, for example, have reduced bird losses to 50 to 60 a year in the cyanide ponds. This number is less than trivial when compared to the thousands of birds killed each year in the state when they fly into roof-mounted television antennas and the paths of automobiles.

SMELTING. Historically, smelters have had bad reputations for causing extensive damage to the environment (see Case Study 14.5). Sulfurous fumes emitted as by-products of

■ TABLE 14.9 Mammal, Bird, Reptile, and Amphibian Mortalities Reported at Cyanide-Extraction Gold Mines in Arizona, California, and Nevada, 1980–1989

STATE	NO. OF MINES	MAMMALS	BIRDS	REPTILES	AMPHIBIANS	TOTAL
Arizona	1	52 (12.7%)	357 (87.3%)	0	0	409
California	11	34 (5.2%)	606 (92.7%)	14 (2.1%)	0	654
Nevada	63	433 (6.6%)	6,034 (92.2%)	24 (0.4%)	55 (0.8%)	6,546
Totals	75	519 (6.8%)	6,997 (92.0%)	38 (0.5%)	55 (0.7%)	7,609

SOURCE: D. R. Clark, and R. L. Hothem, "Mammal Mortality at Arizona, California, and Nevada Gold Mines Using Cyanide Extraction," *California Fish and Game* 77 (1991), pp. 66–67.

smelting processes have polluted the air, and toxic substances from smelting operations have contaminated soils and destroyed vegetation. Because water was necessary for operating early-day smelters, they were located near streams. The accepted practice at the time was to discharge mill wastes and tailings into streams or settling ponds, which commonly spilled over into streams. Improved smelting technologies are now eliminating these problems. For example, Kennecott's former smelter at the Bingham Canyon, Utah, copper mine released 2,136 kilograms (4,700 pounds) of sulfur dioxide (SO_2) per hour to the atmosphere. Kennecott's new smelter, designed to meet or exceed all existing and anticipated future federal and state emission standards, was put into service in 1995. SO_2 emissions have been cut by 96 percent to only 91 kilograms (200 pounds) per hour, less than the rate for the world's cleanest smelters now operating in Japan.

MINE LAND RECLAMATION

The volume of coal-mine spoil and hardrock-mining waste rock is usually greater than the volume occupied by the rock before mining. SMCRA requires that coal-mine spoil be disposed of in fills, usually in the upper reaches of valleys near the mine site (● Figure 14.31). Because some settlement of this material can be expected over time, construction on these reclaimed sites may be risky. Waste rock from hardrock mining is usually deposited in great mesa-like piles near the mines. In recent years, many of these piles have

● FIGURE 14.31 A 16-hectare (40-acre) coal-mine-waste disposal site in Belmont County, Ohio, (*a*) before and (*b*) after reclamation. The mining company spread 103,200 cubic meters (129,000 yd³) of topsoil on the refuse and revegetated the site.

(a)

(b)

been contoured to break up their unnatural, flat-topped appearance. In both cases the features are engineered and structured for stability with terraces and diversion ditches for controlling surface-water flow and preventing erosion, and then they are landscaped. Ground water within the fills is commonly channeled through subsurface drains.

Surface- and Ground-Water Protection

High-quality, reliable water is critical for domestic, industrial, and agricultural activities, and it is especially precious in arid and semiarid regions. SMCRA requires that a surface coal-mining operation be conducted in a way that will maintain hydrologic balance and ensure the availability of an adequate water supply for postmining use. By 1992, nearly 1,500 abandoned U.S. coal mines had been identified as having water problems. Of these, about 500 have been reclaimed or have been funded to begin reclamation.

During surface coal mining all of the runoff water that collects in the pit is required to be collected and treated. Because surface mining destroys the original plant cover and exposes the soil to erosion, mitigation of erosion and sediment loss is needed during mining and reclamation. Watershed protection begins at the onset of development with planning for the proper handling and disposal of spoil or other materials that could cause AMD. AMD problems have arisen from coal mining in Pennsylvania, West Virginia, Maryland, and Ohio and from the hardrock mining of sulfide ores in the Western states and in many areas of Canada. More than 100,000 abandoned hardrock mines in the West may pose AMD problems.

SMCRA requires coal mine operators to treat water from their mines before releasing it into streams and rivers. Control measures for AMD include:

- Holding mine drainage water in entrapment ponds and neutralizing it there by adding alkaline materials from other industries such as kiln dust, slag, alkaline fly ash, and limestone before releasing it from the mine-permit area.
- Grading and covering acid-forming materials to promote surface-water runoff and inhibit infiltration.
- Backfilling underground mines with alkaline fly ash, which absorbs water and turns into weak concrete, in order to minimize the flow of water from the mine and reduce the amount of oxygen in the mine. Filling is done by pumping a slurry down several boreholes. This process has been carried out by the U.S. Bureau of Mines to reduce AMD and stabilize the ground surface of many communities built over mined-out areas.
- Using wetlands to treat AMD issuing from both operating and abandoned coal mines. Cattails and other wetlands plants have been found to be effective in removing toxic metals and other substances from water. One such Pennsylvania wetland has saved an electric power company $50,000 in water treatment costs each year since it was built in 1985.

● FIGURE 14.32 Settling ponds for trapping sediment downstream from a mine; Granite Creek, Iditarod District, Alaska. The area along the west bank (*right*) is a reclaimed mining area.

Wastewater from hardrock mines and from processed ore tailings is another source of contamination. These waters are usually treated by chemical or physical processes before disposal. When those processes proved unsuitable at the Homestake Gold Mine at Lead, South Dakota (Figure 14.17), the company developed a biological treatment. The bacterial bioxidation process breaks down residual cyanide from the mill's wastewater into harmless components that are environmentally compatible with the disposal creek's ecosystem. The creek now supports a viable trout fishery for the first time in a century. A secondary benefit of the bacterial treatment is that precious metals are also recovered by the process.

Another problem at some unprotected mine properties is siltation that clogs streams and increases the threat of flooding. This problem may be mitigated by constructing settling ponds for trapping the sediment downstream from the source (● Figure 14.32).

Mill and Smelter Waste Contamination

As ore is milled to produce concentrate, about 90 percent of the mined material is discarded as tailings. Smelters that process metallic sulfide concentrates release waste products that can contain thousands of times the natural levels of heavy metals such as lead, arsenic, zinc, cadmium, and beryllium. These contaminants remain in smelter slag (rendered inert by the smelting process so that it is not bioavailable) and in flue dust that commonly remains around former smelter sites long after smelting has ended. The primary concerns with such contamination are the environmental impact of stormwater runoff into nearby streams or lakes and the risks to human health that result when people unwittingly inhale or ingest the heavy metals. If the site is on the EPA's Superfund list, the cleanup will take one of two directions—either removal or in-place remediation such as enclosing or encapsulating the heavy-metal sources—determined by which one is considered appropriate for preventing human contact (see Case Study 14.7). At

Reclamation in Big Sky Country

Butte, Montana, sitting on "the richest hill on earth," was the copper capital of the West at the turn of the century (● Figure 1). With its bars, brothels, and copper smelters running wide open day and night, fortunes were being made. Unfortunately, the smelters were belching out toxic fumes and arsenic- and lead-laced, yellow smoke that were poisoning the ground and killing the vegetation. By the 1980s, the mines had nearly played out. Industry had found substitutes for copper, and the world's copper price had dropped, as had Butte's population. The local economy was devastated.

Until the 1990s the city lived with a toxic legacy of slag heaps, waste-rock piles, and smelter emissions scattered among its abandoned mine headframes and residential backyards. The toxic contaminants—arsenic, cadmium, copper, lead, and zinc—were picked up as dust by the wind, tracked into houses on shoes, and inhaled and ingested by the residents. The community has largely been cleaned up (● Figure 2), but the lake at the abandoned

Berkeley Pit Mine pit between the Uptown Historic District and Interstate 15 to the east remains. With a pH about like lemon juice, it too is laced with toxic metals (● Figure 3). The pit waters do not yet threaten the ground-water quality, but researchers are working to solve that eventual problem.

Butte has the distinction of being at the heart of one of the nation's largest complex of toxic-waste cleanup projects, a part of the federal government's ambitious Superfund effort. Butte is at the southeast end of a trail of mining-residue sites that extends 140 miles down the Clark Fork River almost to Missoula. By 1995, Atlantic Richfield Corporation (ACRO) had spent and committed more than $250 million to Superfund cleanups around Butte. The ongoing remediation entails removing earth and waste at numerous sites (see Figure 14.33), rebuilding catchment ponds, reshaping waste-rock piles, encapsulating waste with soil, and revegetation (Figure 2). The most serious hazards have been cleaned up, but the cleanup efforts will continue for many

● **FIGURE 1** Location of Butte and Anaconda, Montana.

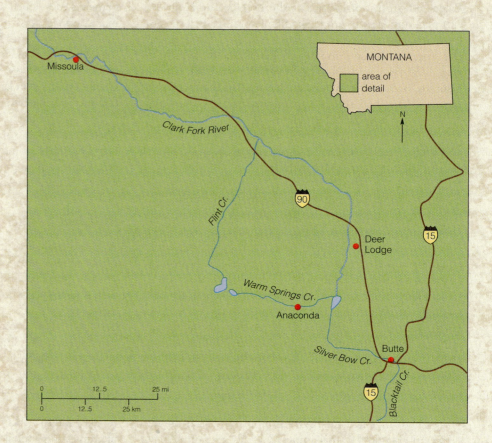

● **FIGURE 2** A reclaimed mine site at Butte, Montana, 1994. The site had been recontoured and seeded with grass. Headframes are left at some reclaimed sites as symbols of the area's heritage.

● **FIGURE 3** Acidic mine water remains in the abandoned pit at Berkeley Pit Mine, Butte, Montana. The headframe and hoist works of an abandoned underground mine appear at lower right.

years. Although ARCO is a latter-day arrival in Butte, having acquired the Anaconda Copper Mining Company properties in 1977, it was tagged by the EPA as the Potentially Responsible Party and, therefore, must pay for the remediation. The scope of the EPA's Superfund liability assessment is retroactive and nearly unlimited. With ownership of land, even if only recently purchased, comes the responsibility of cleaning up a contaminated site, even if the dumping occurred 100 years ago. Thus, ARCO's eventual cleanup costs at Butte may exceed $300 million.

The cleanup effort has attracted a variety of high-tech companies that specialize in treating mining wastes, and Butte is becoming a center of mine-waste-treatment research and development. The city's Montana College of Mineral Science and Technology provides an excellent setting for academic research and support for companies that specialize in mine-waste remediation. Research is being conducted on new remediation techniques such as methods to remove valuable metals from mine waste and neutralize it, bioremediation techniques using bacteria that consume toxic materials, and techniques for economically recovering iron and copper from contaminated water.

Mine-waste cleanup has the potential of becoming a major industry. In the Western states alone some 100,000 abandoned mines need remediation. The U.S. mining industry currently produces an estimated two billion tons of new mine waste a year, most of which requires ongoing remediation from the time operations begin in order to meet the operating permits' conditions. Butte is hoping not only to clean up its own act, but to become the nation's mine-cleanup capital as well.

Anaconda, part of the Superfund complex about 20 miles west of Butte, is the site of a smelter that ceased operations in 1980, leaving behind veritable mesas of slag and waste. ARCO, the present owner of this site as well, has developed a 606-hectare (1,500-acre) world-class golf course as part of its cleanup plan. Designed by golf pro and golf-course designer Jack Nicklaus, the 18-hole course is scheduled to open in 1997. The new course encapsulates the scarred landscape of smelter waste with 5 centimeters (2 in) of crushed limestone overlain by 41 centimeters (16 in) of topsoil on which grasses and other vegetation have been planted. To ensure that the course is not overwatered, which could leach out contaminants, irrigation water applied to the course is monitored. A network of subsurface piping collects excess water and carries it to an evaporation pond. ARCO, the EPA, and the Montana Department of Health and Environmental Sciences (MDHES) agree that the golf course area is now safe for both human health and the environment. EPA and MDHES will monitor the reclaimed course indefinitely. It is hoped that the course will help to attract tourists to the region, as their "clean" money is desired for revitalizing the economy.

● FIGURE 14.33 Removal of mill tailings at a Butte, Montana, Superfund site. Immense amounts of material must be moved at some Superfund sites.

● FIGURE 14.34 Preparation of a repository site for contaminated smelter-flue dust near Anaconda, Montana. The pit will have a heavy polyethelene liner.

one Superfund site the discovery of elevated levels of arsenic in the hair and blood of children in a small community built on contaminated soil resulted in the EPA's moving the entire community of about 30 families and destroying the houses. If a contaminated Superfund area is populated, the EPA may require an ongoing blood-testing program to monitor residents' blood levels of heavy metals. Contaminated stormwater runoff may require construction of a stormwater management system with engineered drainage channels, ponds, and wetlands.

The EPA procedure for remediating contaminated smelter slag, flue dust, and tailings that pose a threat to surface or ground water is to remove the material and deposit it in specially constructed repository sites (● Figure 14.33). Commonly these are pits underlain with heavy plastic liners (● Figure 14.34) into which the contaminated waste is placed and covered with crushed limestone, sealed with topsoil, and revegetated. To eliminate leaching of the material by rainwater or snowmelt, a subsoil drainage system collects excess water and disposes of it in such a way that it poses no threat to human health or the environment. (Disposal of toxic wastes is discussed thoroughly in Chapter 15.)

Revegetation and Wildlife Restoration

SMCRA regulations for coal mining and most state reclamation regulations for hardrock mining require healthy vegetation to be reestablished once mining has ended. Permanent vegetation is the principal means of minimizing

CONSIDER THIS ...

Why does mining that exposes limestone or marble as it removes ore not have AMD problems?

erosion and reducing stream siltation. The types of vegetation that are to be used in reclamation are stipulated in the original mine permit, based on premining vegetation and intended postmining uses. Commonly, a straw mulch or chemical soil stabilizer is applied after seeding to inhibit erosion and retain moisture. Mine operators are responsible for maintaining the new plant cover until it is successfully reestablished; the SMCRA minimum is five years in the East and Midwest and ten years in the semiarid West.

Wildlife habitat is one of the most common uses of postmined land. Among the techniques the mining industry uses to meet the EPA, SMCRA, and state regulations for attracting and supporting desirable species of wildlife are contouring the land, introducing plant species that will support browsing and foraging, creating wetland habitats, and stocking ponds for sport fishing (● Figure 14.35).

THE FUTURE OF MINING

Concern about the future reserves of mineral resources has led to the obvious question, Are we running out? Numerous studies have been conducted over the years to determine reserves, and the answer to the question seems to be "Not yet." The question of availability is probably not the important one, however. Increasing needs for resources will require mining lower and lower grades of minerals. This will require improved technologies and larger, more powerful machines, which in turn will produce greater quantities of waste material. This will impose more stress on environmental systems. Because of these factors, the question that *should* be asked is, Can we afford the environmental and human costs required to satisfy our increasing need for minerals?

Today, we residents of industrialized countries enjoy living in comfortable homes, traveling by automobile, and having labor-saving appliances, all of which provide us with life-styles that only 100 years ago would have been considered luxurious. These life-styles are possible only because of

the availability of inexpensive raw materials for manufacturing the products to which we have grown accustomed. Few of the earth's people realize that today's prices of mineral commodities typically reflect only the short-term, tangible costs of wages, equipment, fuel, financing, and transportation—just as they have since mining began in ancient times. Much of the *real cost* of exploiting mineral and fossil-fuel resources is intangible, and it has been externalized. Historically, the environment has borne a great deal of the cost of extracting raw materials; the consumer has paid only part of the real cost. The low prices of raw materials today do not cover the costs of polluted surface waters, dammed rivers, squalid mining towns, and devastated landscapes. Even in the United States, where regulations designed to protect the environment are imposed, mining and processing still cause substantial damage.

Worldwide poor mining and mineral-processing practices contribute significantly to soil erosion, water contamination, air pollution, and deforestation, especially in developing nations, as illustrated by the examples in Table 14.3. Fortunately, the negative environmental impacts of mining and mineral processing are gaining more attention in the developing nations. For example, Mexico's new administration adopted new mining and environmental laws in 1992 before encouraging mining by privatizing that industry. And in 1994, Tanzania and Guinea began requiring environmental impact assessments to be submitted along with applications for mining licenses.

For over a century, the world's indigenous peoples unfortunate enough to live on mineral deposits have been pushed aside as miners removed the riches. This also is changing. A 1992 decision by Australia's High Court overturned the doctrine of *terra nullis,* the principle that the country's land belonged to no one when the first European settlers arrived. This decision is expected to result in mining companies having to pay compensation fees to the Aborigines and being prevented from extracting minerals in locations established as traditional Aboriginal lands or sacred sites. In this country, the Crandon Mine Project in Wisconsin, which could become one of the world's largest zinc-copper mines, would border some of the state's most productive wild rice areas in the waters of the Wolf River. The Native Americans on the Menominee Reservation downstream from the mine site and other Native Americans have joined with the local Chippewa Band in opposing the project, because it threatens the region's Native Americans who depend on the river's annual wild rice harvest for much of their income.

Mining generates twice as much solid waste as all other industries and cities combined, and most mining wastes in the United States are currently unregulated by federal law. In addition, the mineral industry is one of the greatest consumers of energy and a contributor to air pollution and to global warming. It is estimated, for example, that the processing of bauxite into aluminum alone consumes about one percent of the world's total energy budget. Miners admit that mines leave holes in the ground, but they go on to point out that when they abandon their currently active mines in the United States, they will be restoring the site to as near a natural appearance as is reasonable, except for the open pits.

In evaluating mining and its impact on the environment, we must recognize that the only way not to disturb the landscape is *not to mine.* This would mean not having the raw materials to build automobiles, houses, farm machinery, airplanes, radios, television sets—virtually all of the objects we view as essential to a modern society. The question thus becomes, What must be done to minimize the impacts?

The Legacy of Mines and Mining

The collage of pictures in this Gallery illustrates the good, the bad, and the ugly of past and present mining in the western United States and Germany (⊙ Figures 14.36–14.41).

⊙ FIGURE 14.36 Toxic legacies left for perpetuity by turn-of-the-century Colorado silver mining; waste-rock piles from abandoned mines above Silver Plume, Colorado. Although the mines have been shut down for decades, the barren piles still lack vegetation because of the accumulation of acid salts (indicated by the white coloration) due to evaporation of acidic seepage water that has percolated through the material. There are thousands of such abandoned mines in Colorado and perhaps hundreds of thousands in the Western states.

⊙ FIGURE 14.37 A modern rotary-ball mill; Colosseum Mine, San Bernardino County, California. Ball mills have replaced the antiquated stamp mills. Two ball mills are usually required for grinding ore to a consistency finer than that of talcum powder. The powdered ore is mixed with water to form a *slurry,* which is then processed chemically and mechanically to remove the metal. The residual slurry becomes *tailings,* which are disposed of at the tailings pond.

⊙ FIGURE 14.38 The legendary "Mother Lode" gold-quartz vein is exposed in the wall of an open-pit mine; Jamestown Mine, Jamestown, California.

◉ FIGURE 14.39 Creek near an open-pit copper mine at Morenci, Arizona. The water's distinct blue color reveals a high concentration of dissolved copper leached from mine wastes.

◉ FIGURE 14.40 The *entire* landscape visible in this photo is productive reclaimed land; near Cologne, Germany. Extensive surface coal mining was carried out here in the 1960s and 1970s.

◉ FIGURE 14.41 Plattberg Hill, an accumulation of about 40 years' spoil from underground coal mining, near Kamp-Lintford, Westphalia, Germany. Landscaped hills of mine spoil such as this provide welcome relief to the otherwise flat terrane of the lower Rhine Valley. They provide refuges for wildlife and year-round recreation areas for humans. Mine headworks are at the lower right.

U.S. Mining Laws

In the 1970s some genuinely important environmental concerns related to mining and processing mineral materials became obvious in the United States and many other countries, and laws began being written to address these concerns. Consequently, in the United States, the EPA now administers a number of acts regulating water quality and toxic wastes, and the Fish and Wildlife Service requires noninterference with rare and endangered species. Further constraints are imposed by numerous state and local laws requiring that mining operations meet certain air-pollution standards. Many states will not issue a permit for an operation to begin without prior approval of an adequately funded reclamation plan specifying that the waste and tailings piles will be restored to resemble the surrounding topography to the extent that it is possible, that the disturbed ground and the tailings will be replanted with native species, that all buildings and equipment will be removed, and that the open pits will be fenced and posted with warning signs.

It is claimed by some that enforcement of federal regulations pertaining to mining has been weak. Initially, the EPA did little to regulate mining wastes. Further, Congress specifically exempted hardrock-mining wastes and tailings from regulation as hazardous wastes in the Resource Conservation and Recovery Act (1976). Complicating the issues is the fact that instead of the federal government, it is the individual states that play the role of regulator on state and private lands.

Further reform of the Mining Law of 1872 may occur. Beginning in the 1980s public concern grew over the problems of protecting nonmineral values on public lands, the lack of meaningful federal reclamation standards, limited environmental protection, and the lack of royalty collection for exploitation of public land—issues of public-resource management that are not addressed in the current mining law. These issues were once again addressed in Congress in 1995, and a mining law reform bill that had passed in the House of Representatives was killed in the Senate in the final days of the 103rd Congress. A point of contention has been the issue of imposing royalties on mine income. This royalty income could be used to finance the reclamation of some 100 thousand abandoned hardrock mines in the Western states, much as is already done by SMCRA with abandoned coal mines using the Abandoned Mine Fund. Some critics of mining law reform believe that charging royalties for hardrock operations on federal lands will close down U.S. mining and drive more mineral production to foreign countries. Nevertheless, efforts at reform are expected to continue.

The future may also bring some changes in the basic attitudes and assumptions that underlie our capitalistic society. The economic assumption that prosperity is synonymous with mineral production is now being questioned. Environmental deterioration from today's unprecedented rate of mineral production will, if continued, eventually overwhelm the benefits gained from increased mineral supplies. As the world's developing nations strive to achieve the economies and life-styles currently enjoyed by the developed countries, there will be severe demands on, and competition for, the world's remaining mineral and fuel resources. By the middle of the next century when world population has reached 10–15 billion (see Chapter 1), new technologies and economic strategies must be in place. The goals of protecting and managing the environment while at the same time exploiting and expanding the mineral resource base in an environmentally responsible manner are not necessarily mutually exclusive. But how can both goals be achieved? Society will need to replace the current materials-intensive, high-volume, planned-obsolescent manufacturing processes—and consumption patterns—with those that use raw materials and fuels more efficiently, that generate little or no waste, and that recycle most of the waste that is generated. Whether technology can modernize fast enough to conserve natural resources at the same time that the world's wealth is increasing is the big question. This is the formidable challenge that must be faced jointly by industry, governments, and society.

 SUMMARY

Mineral Abundances, Definitions

MINERAL DEPOSIT   Locally rich concentration of minerals.

ORE   Metallic mineral resources that can be economically and legally extracted at the time.

MINERAL RESERVES   Known deposits of earth materials from which useful commodities are economically and legally recoverable with existing technology.

Factors that Change Reserves

TECHNOLOGY CHANGES   that affect the costs of extraction and processing.

ECONOMIC CHANGES that affect the price of the commodity and the cost of extraction.

POLITICAL CHANGES that cut off or open up sources of important mineral materials.

AESTHETIC AND ENVIRONMENTAL FACTORS may reduce reserves.

Distribution of Mineral Resources

DESCRIPTION Mineral deposits are highly localized; they are neither uniformly nor randomly distributed.

CAUSE Concentrations of valuable mineral deposits are due to special, sometimes unique, geochemical processes.

Origins of Mineral Deposits

IGNEOUS DEPOSITS

1. *Intrusive*—occur (a) as a pegmatite, an exceptionally coarse-grained igneous rock with interlocking crystals, usually found as irregular dikes, lenses, or veins, especially at the margins of batholiths; and (b) by crystal settling, the sinking of crystals in a magma due to their greater density, sometimes aided by magmatic convection.
2. *Disseminated*—a mineral deposit, especially of a metal, in which the minerals occur as scattered particles in the rock but in sufficient quantity to make the deposit a worthwhile ore.
3. *Hydrothermal*—a mineral deposit precipitated from a hot aqueous solution with or without evidence of igneous processes.
4. *Volcanogenic*—of volcanic origin, e.g., volcanogenic sediments.

SEDIMENTARY DEPOSITS Mineral deposits resulting from the accumulation or precipitation of sediment. These include

1. Surficial marine and nonmarine precipitation.
2. Deep-ocean precipitation.
3. Placer deposits.

WEATHERING DEPOSITS

1. *Lateritic weathering*—concentrations of minerals due to the gradual chemical and physical breakdown of rocks in response to exposure at or near the earth's surface.
2. *Secondary enrichment*—mineral development that occurred later than that of the enclosing rock, usually at the expense of earlier primary minerals by chemical weathering.

METAMORPHIC DEPOSITS Concentrations of minerals, such as gem or ore minerals, in metamorphic rocks or from metamorphic processes.

Categories of Mineral Reserves

ABUNDANT METALS Average continental crustal abundances are in excess of 0.1 percent.

SCARCE METALS Average continental crustal abundances are less than 0.1 percent.

NONMETALLIC MINERALS Minerals that do not have metallic properties; include industrial minerals, agricultural minerals, and construction materials.

Environmental Impacts of Mining

1. Surface coal mining and land reclamation is regulated by federal law, the Surface Mining Control and Reclamation Act (SMCRA). No similar federal law regulates hardrock mining and land reclamation.
2. SMCRA requires surface coal mine pits to be backfilled with the mine waste (spoil) and then covered with topsoil and landscaped. Some states in which hardrock mining is conducted require similar handling of waste rock and tailings at currently operating mines.
3. Ground subsidence due to underground coal mining is less common than in the past because of the required backfilling. Surface collapse from hardrock (underground) mining is not common, and stabilization may be expensive.
4. Acid mine drainage (AMD) may be a problem at active and abandoned coal and sulfide-ore mines. Pit water in surface coal mines is required to be held and treated before it is released to streams or rivers. Sealing abandoned mines, covering waste rock with soil and landscaping, and installing drainage courses to direct surface drainage off tailings and waste-rock piles are some methods of controlling AMD formation. Wells are installed below waste-rock and tailings dumps to monitor water quality.
5. Scarified ground, despoiled landscapes, and disrupted drainage that remain after dredging and hydraulic mining can be reclaimed by reshaping, adding topsoil, and landscaping.
6. Abandoned open pits with oversteepened side walls despoil the landscape and may pollute hydrologic systems. They are difficult or impossible to reclaim. The public is protected by surrounding the pits with chain-link fencing and posting warning signs. The pit floor may be partially backfilled and sealed with impermeable clay to protect ground water from pollution.
7. Cyanide heap-leach gold-extraction methods normally cause minimal environmental problems because they are strictly regulated by state and federal agencies. Wildlife losses have been mitigated by placing nets over the lethal ponds. There is no demonstrated hazard from lethal fumes. Surface and ground waters are monitored by sampling and monitoring wells, respectively.
8. The main concerns associated with heavy-metal contaminated mill and smelter waste are the environmental impacts resulting from stormwater runoff into nearby surface waters and the human health problems caused by inhaling or ingesting the heavy metals in dust. Superfund cleanups require either removal or

in-place remediation by encapsulating the heavy-metal sources. Wetlands or ponds with appropriate plants may be constructed to clean toxins from surface runoff.

The Future of Mining

WORLDWIDE Mining and processing of mineral commodities are contributing to soil erosion, water contamination, air pollution, and deforestation.

THE UNITED STATES Mining and processing of mineral commodities must meet standards of air and water quality set by the EPA and other government agencies. The Mining Law of 1872, although much amended, is being criticized and evaluated in light of current economics, domestic needs, and conflicting land uses.

KEY TERMS

acid mine drainage (AMD)
alluvial placer
area mining
average crustal abundance
banded-iron formation
bauxite
bench placer
black smoker
CERCLA
concentration factor
contact-metamorphic deposit
contour mining
crystal settling
cyanide heap-leaching
disseminated deposit
dredging
evaporite
flotation
glacial outwash
gravity separation
high-grade deposit
hydraulic mining
hydrothermal deposit

lode
low-grade deposit
manganese nodule
massive sulfide deposit
mineral deposit
mineral reserves
mountain-top mining
open-pit mining
ore
ore mineral
pegmatite
placer deposit
porphyry copper
precipitation
regional metamorphism
secondary enrichment
SMCRA
smelter
spoil
strip-mining
tailings
vein deposit
volcanogenic deposit

STUDY QUESTIONS

1. Distinguish between mineral deposits, ores, and reserves.
2. Why do some metals, such as gold, require concentration factors in the thousands, while others, such as iron, require only single-digit concentration factors?
3. Of which metals does the United States have ample reserves? How do we obtain the scarce mineral resources we need but lack within our borders?
4. How do you expect domestic and foreign supplies of critical mineral resources to change in the next 25 years?

5. What would cause currently marginal or submarginal mineral deposits to become important mineral reserves?
6. Where are the major deposits of porphyry copper and porphyry copper-molybdenum deposits located? How do these deposits relate to plate tectonic theory?
7. What steps are involved in the origin of a placer gold deposit? Of a bench placer deposit?
8. What minerals may eventually be harvested from the deep-ocean floor? What might be the environmental consequences of such mining?
9. Explain why you wouldn't go to the Hawaiian Islands to prospect for gold.
10. What is hydrothermal activity? What metallic mineral deposits are commonly associated with hydrothermal action?
11. Contrast the gold mining and extraction techniques of the nineteenth century with those of today. What are the environmental legacies of the nineteenth-century mining methods? What environmental concerns accompany modern gold-extraction technology?
12. Contrast the environmental hazards and mitigation procedures of underground, open-pit, and surface mining.
13. Why are mercury and cyanide, very dangerous poisons if they are mishandled, used in gold extraction?
14. Briefly describe the cyanide heap-leach gold-extraction process.
15. It has been suggested that as society exhausts the earth's reserves of critical minerals, lower-grade mineral deposits will be mined and supply-and-demand economics will dictate prices and costs. This rationale could be extended to a scenario in which average rock would eventually be mined for critical materials. What is the fallacy of such a rationale?

FURTHER INFORMATION

BOOKS AND PERIODICALS

Baum, Dan, and Margaret L. Knox. 1992. In Butte, Montana, A is for arsenic, Z is for zinc. *Smithsonian* 23, no. 8: 46–56.

Craig, James R., David J. Vaughan; and Brian J. Skinner. 1988. *Resources of the earth.* Englewood Cliffs, N.J.: Prentice-Hall.

Crown Butte Mines, Inc. 1995. *New World project.* Crown Butte Mines, Inc., 2501 Catlin, Suite 201, Missoula, MT 59801.

Cutter, D. C. 1948. The discovery of gold in California. In Jenkins, O. P., ed. Geologic guidebook along Highway 49: Sierran gold belt, the Mother Lode country. *California Division of Mines bulletin* 141, (centennial ed.): 13–17.

Henning, Robert A., ed. 1982. Alaska's oil/gas and minerals industry. *Alaska geographic* 9, no. 4: 216.

King, Trude V. V., ed. 1995. *Environmental considerations of active and abandoned mine lands: Lessons from Summitville, Colorado.* U.S. Geological Survey bulletin 2220. Washington, D.C.: Government Printing Office.

Logan, C. A. 1948. History of mining and milling methods in California. In Jenkins, O. P., ed. Geologic guidebook along

Highway 49: Sierran gold belt, the Mother Lode country. *California Division of Mines bulletin* 141 (centennial ed.): 31–34.

Maxwell, Jessica. 1995. New World blues. *Audubon* 97, no. 5 (September–October): 82–89.

Muncy, Jack A. 1991. *A plan for cooperatively completing revegetation of Tennessee's Copper Basin by the year 2000.* Technical note TVA/LR/NRM-91/3. Norris, Tenn.: Tennessee Valley Authority.

Newmont Gold Company. 1995. *Newmont.* Newmont Gold Company, P. O. Box 669, Carlin, NV 89822-0669.

Office of Surface Mining Reclamation and Enforcement, Department of the Interior. 1992. *Surface coal mining reclamation: 15 years of progress, 1977–1992, Part 1.* Washington, D.C.: Government Printing Office.

Quinn, M. L. 1991. The Appalachian Mountain's Copper Basin and the concept of environmental susceptibility. *Environmental management* 15, no. 2: 179–194.

Silva, Michael A. 1988. Cyanide heap leaching in California. *California geology* 41, no. 7: 147–156.

Stewart, K. C., and R. C. Severson, eds. 1994. *Guidebook on the geology, history, and surface-water contamination and remediation in the area from Denver to Idaho Springs, Colorado.* U.S. Geological Survey circular 1097. Washington, D.C.: Government Printing Office.

VIDEOS

Focus: Endangered park? 1995. Five-minute report on the Crown Butte project, Montana. MacNeil/Lehrer Newshour, 14 September. PBS Video, 1320 Braddock Place, Alexandria, VA 22314; 800-328-PBS1.

Gifts from the earth. 1986. The Planet Earth series, Annenberg/ CPB project. Films Incorporated, 5547 North Ravenswood, Chicago, IL 60640-1199; 312-878-2600, ext. 43.

The magic of Homestake gold from the discovery of gold to mining today. Homestake Mining Tours, Lead, SD 57754

Mining: Discoveries for progress. American Mining Congress, Communications and Education Department, 1920 N Street, NW, Suite 300, Washington, DC 20036-1662.

Poison in the Rockies. 1992. NOVA. WGBH, P. O. Box 2284, South Burlington, VT 05407-2284; 800-255-9424.

Public lands, private profits. 1994. Overview of the controversy surrounding the Mining Law of 1872. Frontline, 24 May. PBS Video, 1320 Braddock Place, Alexandria, VA 22314; 800-328-PBS1.

WASTE MANAGEMENT AND GEOLOGY

15

Garbage isn't generic gunk; it's specific elements of our behavior all thrown together.

WILLIAM RATHJE, PROFESSOR OF ANTHROPOLOGY, UNIVERSITY OF ARIZONA

Getting rid of trash is a major environmental problem in industrialized nations. Growing populations of consumers have created an explosion of solid, liquid, and hazardous wastes, much of which requires special handling. At risk due to careless waste disposal are ground- and surface-water purity, air quality, public health, and less threatening but still important scenic and land-surface degradation. Municipal solid waste generated in the United States in 1990 amounted to 178 million metric tons (196 million tons), the equivalent of 1.9 kilograms (4.3 lbs) per person per day. Municipal solid waste includes yard cuttings, garbage, construction materials, paper products, metal cans, plastics, and glass (◉ Figure 15.1). It does not include wastes from agriculture, industry, utilities, or mining, which account for more

OPENING PHOTO
Greenpeace garbage barge looking for a home in the harbor of New York City. Disposal of solid waste is a worldwide problem in large urban areas.

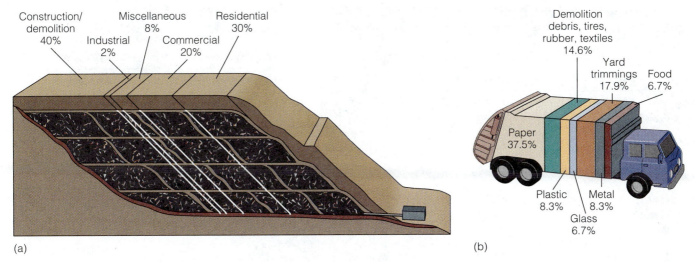

(a)

(b)

⦿ FIGURE 15.1 Components of urban solid waste *(a)* by sector of the economy and *(b)* by material. Note that paper products—packaging and printed paper—comprise the largest volume of waste and that residential trash accounts for only about a third of the total solid waste.

than 96 percent of the solid waste generated in the United States (⦿ Figure 15.2). After recovery for recycling and composting, 1.6 kilograms (3.6 lbs) per person per day went to a landfill or was incinerated. Just one day's total U.S. waste would cover 15 square kilometers (almost 6 mi²) to a depth of 3 meters (10 ft). If it were loaded into ten-ton trucks lined up bumper-to-bumper, the trucks would stretch around the world 20 times. Those trucks containing only the day's *municipal solid waste* would circle the earth almost 3 times. The spectrum of solid-waste disposal problems ranges from the need to isolate highly dangerous nuclear waste to the challenge of dismantling and scrapping an astounding number of cars every year, estimated at 15 million in the United States alone. (One year 60,000 cars were abandoned on the streets of New York City.)

The amount of trash a country or governmental entity generates per unit of land area is just as important as the total amount it generates. If small countries with limited disposal sites generate large quantities of trash, major problems are created. Although the United States generates the highest total amount of waste in the world, the tiny city-state of Singapore generates the most trash per unit area of land, almost 2,500 tons per square kilometer per year in 1985, compared to less than 2 tons per square kilometer for Canada and 20 tons per square kilometer for the United States. Poland generates the most industrial waste per unit area, followed by Japan, and Hungary produces the most toxic waste per unit area, followed by the United States (▪ Table 15.1).

The impetus to develop environmentally safe trash disposal sites, known as *landfills,* was the passage of two important laws, the National Environmental Policy Act of 1969 **(NEPA)** and the Resource Conservation and Recovery Act of 1976 **(RCRA)**. NEPA requires an in-depth field study

and issuance of an Environmental Impact Statement on the consequences of all projects on federal land. The RCRA mandates federal regulation of waste products and encourages solid-waste planning by the states. The RCRA addresses the problems of so-called hazardous waste and its impact upon surface- and ground-water quality, and it created the framework for regulating hazardous wastes. These laws sent the message that the then-prevailing "out of sight, out of mind" approach to disposing of wastes of all kinds, hazardous and benign, was no longer acceptable. The laws mandated "cradle to the grave" systems for impounding wastes and monitoring them to ensure their low potential for migrating into fresh-water supplies. The Environmental

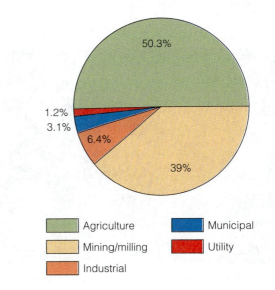

⦿ FIGURE 15.2 Estimated relative contributions to total U.S. solid waste, 1992.

| COUNTRY | URBAN WASTE | | INDUSTRIAL WASTE | HAZARDOUS WASTE |
	kg/person	tons/km²	tons/km²	tons/km²
U.S.	762★	19.4	68.5	28.9
Canada	630	1.7	6.6	0.4
Korea	396	159.5	71.2	1.8
Japan	344	110.3	828.6	1.8
Singapore	585	2,456.0	—	—
Belgium	313	93.9	244.0	27.9
France	272	27.3	90.9	3.6
Hungary	657	75.8	229.0	76.7
Poland	212	25.9	902.8	11.4

★ The highest amount and the next-highest amount in each column are in boldface.
SOURCE: World Resources Institute, *World Resources, 1990–1991* (New York: Oxford University Press, 1991).

Protection Agency (**EPA**) is charged with monitoring and policing the provisions of these and subsequent laws and amendments intended to prevent pollution of natural systems (Case Study 15.1).

Disposing of solid and liquid wastes, regardless of how they are generated, falls into two main categories:

- **isolation,** encapsulating, burying, or in some other way removing waste from the environment; and
- **attenuation,** diluting, incinerating, or spreading trash or a pollutant so thinly that it has little impact.

Attenuation (dilution) is effective with sewage, some waste chemicals, and certain gaseous pollutants. About 10 percent of our waste, including sewage sludge, is incinerated in large, modern incinerators, usually found where land-disposal sites are not available. Historically, most of our solid waste has been isolated in dumps. Open dumps, the local "garbage

dumps" of small towns, are no longer acceptable because of their attendant insect, vermin, odor, and air-pollution problems (◉ Figure 15.3). In the mid 1990s, more than 70 percent of U.S. trash was being disposed of in 5,000 landfills, down from 7,900 landfills in 1988. In 1993 alone almost 1,000 landfills were closed, most because they had reached their capacity, but many because they could not meet the stringent EPA standards for protecting ground water. Deep wells are used successfully for isolating hazardous liquid wastes, although the tremendous volume of used motor oil remains a problem. Today, good **waste management** consists not only of waste disposal, but also of reducing the amount of waste at its sources and promoting waste recovery and recycling programs.

◉ MUNICIPAL WASTE DISPOSAL
Municipal Waste Disposal Methods

SANITARY LANDFILLS. In 1912 the first solid-waste sanitary landfill was established in Great Britain. Known as "controlled tipping," this method of isolating wastes expanded rapidly and had been adopted by thousands of municipalities by the 1940s. The city of Fresno, California, claims to be the first city in the United States to employ a sanitary landfill for waste disposal. A **sanitary landfill** differs from an open dump in that each day's trash is covered with a layer of soil to isolate it from the rest of the environment. Although this sounds simple, the method involves spreading trash in thin layers, compacting it to the smallest practical volume with heavy machinery, and then covering the day's accumulation with at least 15 centimeters (6 in) of soil (◉ Figure 15.4). When finished, a landfill is sealed by 50 centimeters (20 in) of compacted soil and is graded so that water will drain off the finished surface. This prevents water

◉ **FIGURE 15.3 The open town dump is no longer an acceptable way to dispose of waste; Terrytown, New York.**

The Sociology of Waste Disposal

The EPA is charged with the Herculean tasks of monitoring existing land and ocean dump sites, approving new hazardous-waste disposal sites, and supervising the cleanup of historic toxic sites and accidental spills and leaks. In carrying out these responsibilities, EPA personnel have encountered the spectrum of political and public attitudes about waste disposal. A wall poster at EPA headquarters in Washington, D.C., provides the following acronyms for those attitudes:*

NIMBY	Not In My Back Yard
NIMFYE	Not In My Front Yard Either
PIITBY	Put It In Their Back Yard
NIMEY	Not In My Election Year
NIMTOO	Not In My Term Of Office
LULU	Locally Unavailable Land Use
NOPE	Not On Planet Earth

---

* Author's addition: **NITL,** Not In This Lifetime.

---

infiltration and the production of potentially toxic fluids within the fill (◉ Figure 15.5). **Leachate** refers to the water that filters down through a landfill, acquiring (leaching out) dissolved chemical compounds and/or fine-grained solid and microbial contaminants as it goes. Protecting the local environment from leachate is a major consideration in designing and operating waste-disposal sites. The dump-and-cover procedure carried out at sanitary landfills is known as the "cell" method, because it isolates waste from the environment in rhomb-shaped compartments, or cells. A well-operated sanitary landfill is relatively free of odors, blowing dust, and debris (Case Study 15.2).

Landfills and individual disposal sites within a given landfill are classified on the basis of their underlying geology and the potential impact that leachate could have upon local ground or surface water. Each of the three classes is certified to accept specific wastes of differing degrees of reactivity (see ◼ Table 15.2 and ◉ Figure 15.6, page 462). Some landfill sites encompass all three classifications within their boundaries (◉ Figure 15.7, page 463).

INCINERATION. Burning is the only proven way to significantly reduce the volume of garbage, and it can reduce the volume by as much as 90 percent. In 1993 about 10 percent of municipal solid waste was being burned at 162 modern municipal incinerators, up about a fifth from 136 in 1988. Use of the process is not expanding rapidly because of the high initial facility cost and the continuing costs of maintaining air-quality standards. In New York City, for example, burning costs are about $200 per ton, higher than the cost of landfilling waste in the mid-Atlantic states, the most expensive waste-disposal area in the United

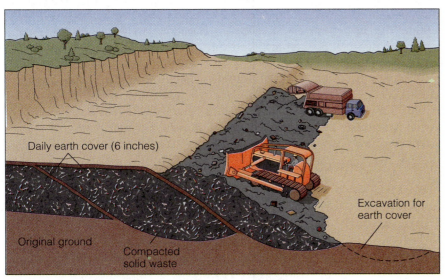

(a)

(b)

◉ **FIGURE 15.4** *(a)* Method of dump and cover employed in a sanitary landfill. *(b)* An urban sanitary landfill in action; Palos Verdes, California.

● FIGURE 15.5 *(a)* Local-government landfill specifications for cells, final cover, and leachate-collection system. Note the barrier for reducing the visual blight of the landfill. *(b)* A full, graded landfill before planting; Puente Hills, Los Angeles County, California. Final grades and slope benches for intercepting runoff are visible in the photo.

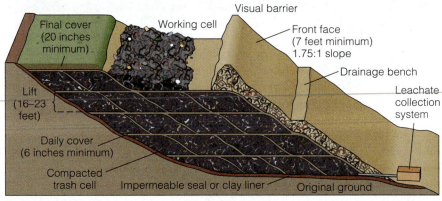

(a)

(b)

States. Heat energy is produced in the process, some of which is converted to electrical energy that can be sold to the local utility or used internally to run the plant. The use of incineration is highest in areas with high population density, little open land, or high water tables. Massachusetts, New Jersey, Connecticut, Maine, Delaware, and Maryland, for instance, burn more than 20 percent of their waste.

New York City is a leader in trash incineration, burning about 720,000 tons per year, or 11 percent of all its waste. Wastes must be burned at very high temperatures, and incinerator exhausts are fitted with sophisticated "scrubbers" that remove dioxins and other toxic air pollutants. Incinerator ash presents another problem, because it is in itself a hazardous waste, containing high ratios of heavy metals that are chemically active (see Case Study 15.2).

The City of Philadelphia operates trash incinerators but lacks landfills for isolating the ash. In desperation, in the mid 1980s 14,500 metric tons (16,000 tons) of the ash from two of its incinerators was loaded onto the steamship *Khian Sea* and sent to sea. Like the sailor in Coleridge's "Rime of the Ancient Mariner," the *Khian Sea* sailed "alone, alone on

a wide, wide sea" for two years, seeking a place to offload its unwanted cargo, with port after port refusing. In 1988 a miracle must have occurred, as the ship sailed into Singapore empty. No explanation has ever been given.

OCEAN DUMPING. Greek mythology relates how Herakles (Latin *Hercules*) was given the task of cleaning King Augeas's stables, which contained 30 years' accumulation of filth from 3,000 head of cattle. Inasmuch as one cow pro-

CASE STUDY 15.2

The Highest Point between Maine and Florida

Fresh Kills landfill on Staten Island, New York, is the world's largest landfill. It contains 25 times the volume of one of the Seven Wonders of the World, the Great Pyramid of Giza, Egypt (●Figure 1). Fresh Kills (Dutch *kil,* "stream") serves New York City and its suburbs and parts of New Jersey. It accepts about 13,000 tons of trash per day, and by 1998 its surface will be 165 meters (500 ft) above sea level—the highest elevation on the eastern shore of the United States. The fill was established in 1948 on a salt marsh with no provisions for constraining leachate. For years, more than 4.2 million liters (a million gallons) of leachate leaked into the marsh and its nearby waters each day. Recent cleanup has rectified many of the pollution problems, but Fresh Kills remains a great mountain of an eyesore to Staten Islanders.

What makes Fresh Kills unique, aside from being built upon marshland, is that trash is transported to the fill by city-owned barges, each with a capacity of 600–700 tons. The barges are loaded from garbage trucks at eight stations located throughout the city and towed to Staten Island by tugboats. The tugboats are followed by specially built "skimmer" boats that pick up trash that falls from the barges or drops into the water at the unloading facilities. At the docks the refuse is unloaded by huge cranes and placed into large, custom-made vehicles that carry the trash to a fill site. After dumping, the refuse is sprayed with a deodorant, compacted by bulldozers, covered by clean earth, and eventually landscaped to "reclaim" the land.

New York City also has three active municipal trash incinerators. In the past the ash residue from these facilities was transported offshore and dumped into the ocean. Now the ash is barged to Fresh Kills and deposited along with raw garbage. Because the ash is loaded with toxic metals, construction of specially designed ash-only fill areas is planned. The ash-only fill area has a double liner and a double leachate-collection system to keep toxic substances within the fill. In addition, a clay slurry "wall" has been excavated around the fill to further confine leachate should there be leaks. Also planned is a waste-to-energy incinerator, and ash from this incinerator also will be dumped in Fresh Kills (●Figure 2).

Major efforts are underway to extend the life of the landfill by recovering resources contained in the trash. Methane is extracted and used for heating and cooking in nearby homes, construction debris is recycled, and yard waste is composted. Grass and leaves from Staten Island and the Borough of Queens are composted and mixed with soil to promote healthy plant growth on the finished fill. Two more recycling facilities with debris-crushing and screening plants are planned, as is an intermediate processing center that will sort and process recyclable glass, cans, paper, and plastics. At the present rate of delivery, Fresh Kills landfill will be filled to capacity by the year 2000.

Great Pyramid of Giza   Fresh Kills landfill

(a)

(b)

● **FIGURE 1** *(a)* Comparison of the volumes of the Great Pyramid of Giza and the Fresh Kills landfill. *(b)* The Great Pyramid *(right)* built by King Cheops at Giza on the west bank of the Nile River in the third century B.C. The length of each side at the base averages 756 feet, and its present height is 451 feet (original height, 481 feet). The smaller pyramid was built by Chephren, a later king. The automobiles give an idea of the immensity of these structures.

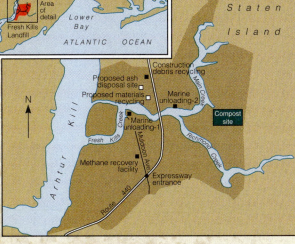

● **FIGURE 2** Fresh Kills landfill on Staten Island.

DISPOSAL SITES

Class I	No possibility of discharge of leachate to usable waters. Inundation and washout must not occur. The underlying lining material, whether soil or synthetic, must be essentially impermeable; that is, have a permeability less than 0.3 cm/year. All waste groups may be received (Figure 15.6, part a).
Class II	Site overlies or is adjacent to usable ground water. Artificial barriers may be used for both vertical and lateral leachate migration. Geologic formation or artificially constructed liners or barriers should have a permeability of less than 30 cm/year. Groups 2 and 3 waste may be accepted (Figure 15.6, part b).
Class III	Inadequate protection of underground- or surface-water quality. Includes filling of areas that contain water, such as marshy areas, pits, and quarries. Only inert Group 3 wastes can be accepted (Figure 15.6, part c).

WASTE GROUPS

Group 1	Consists of but not limited to toxic substances that could impair water quality. Examples are saline fluids, toxic chemicals, toilet wastes, brines from food processing, pesticides, chemical fertilizers, toxic compounds of arsenic, and chemical-warfare agents.
Group 2	Household and commercial garbage, tin cans, metals, paper products, glass, cloth, wood, yard clippings, small dead animals, and hair, hide, and bones.
Group 3	Non-water-soluble, nondecomposable inert solids such as concrete, asphalt, plasterboard, rubber products, steel-mill slag, clay products, glass, and asbestos shingles.

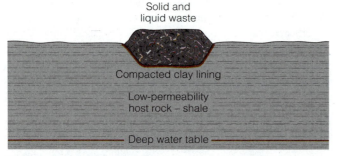

(a) Class I landfill

(b) Class II landfill

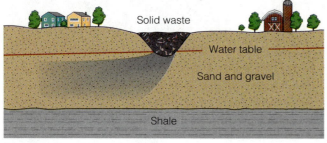

(c) Class III landfill

duces roughly 18 wet tons of manure per year, Herakles was up to his ears in 1.6 million tons of you-know-what, about the amount of sewage sludge that New York City generates in four months. Because Herakles was to perform the cleanup in a day, he ingeniously diverted the courses of two rivers to make them flow through the stables and wash the filth into an estuary. Had an environmental impact statement been required, it would have noted that a huge mass of solid sludge would cover a large area of the wetlands and bury bottom-dwelling organisms. In addition, phosphates and nitrates in the effluent would promote explosive algal blooms at the expense of other organisms, and dissolved oxygen in the water would decrease to the point of mass mortality of swimming and bottom-dwelling organisms.

Historically, all coastal countries have used the sea for waste disposal. This is not surprising, since the sea is convenient, about three times more ocean area exists than land area, and as waste matter disperses, sinks to the bottom, or is diluted, the ocean gives the visual impression that it is able to accommodate an unlimited amount of bad things. This practice is based on the *assimilative capacity* approach to waste disposal; that is, the assumption that a body of seawater can hold a certain amount of material without adverse biological impact. This attitude is still held in many coastal areas, particularly in Third World countries, where floatables in trash present a highly unsightly appearance. No U.S.

⦿ FIGURE 15.6 (left) Geology of the three classes of landfills. Note that a Class I fill offers maximum protection to underground water and that a Class III fill offers no protection.

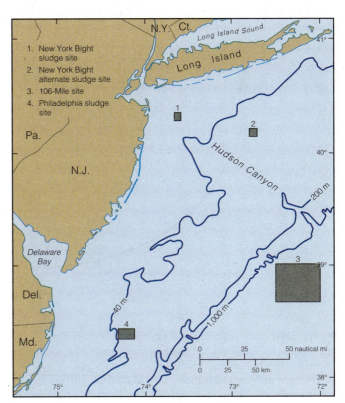

FIGURE 15.7 Oil-field brine is dumped in a Class I landfill. Other sections of this landfill have lower classifications because their bedrock is permeable.

- they are consumed by marine life as a food source,
- they introduce heavy-metal and chlorinated organic compounds when they are attached to the particles, and
- they inhibit light in the water column.

In the late 1970s the EPA announced its intention to ban the dumping of all sludge that degrades the environment by 1981. However, the city of New York challenged enactment of the law and won. Because the city dumps so much sludge (about 5 million wet tons per year), the EPA required the city to move its dumpsite from 12 miles offshore to what is known as *Site 106*, 106 miles off the New Jersey coast in water 2.4 kilometers (1½ mi) deep (Figure 15.8).

Ideally, a sewage outfall that is sited in deep water below the *thermocline*—the zone where water temperature decreases and density increases rapidly with depth—could accept untreated raw sewage. The thermocline acts as a floor for warm surface water and as a ceiling for cool, dense bottom water and prevents the two water layers from mixing. Most outfalls, however, are in water depths *above* the thermocline, and thus sewage effluent mixes with surface waters, which may then pollute shallow-water ecosystems and beaches. According to the EPA, the New York City sludge was supposed to disperse totally during its descent so that it would leave no measurable impact on the bottom fauna. In 1989 a research team from Woods Hole Oceanographic Institution visited Site 106 in the research sub-

municipality has dumped garbage in the sea since New York City stopped the practice in 1934.

Ocean pollution is governed by the Marine Protection Reserve and Sanctuary Act of 1972, better known as the *Ocean Dumping Act*. This act requires anyone dumping waste into the ocean to have an EPA permit and to provide proof that the dumped material will not degrade the marine environment or endanger human health. Hazardous substances entering the sea from the land (Table 15.3) are contained in sewage, sewage sludge (the solid part of sewage), dredged materials, industrial effluents, and natural runoff. Sludge and other particulate matter can create a biological imbalance in the ocean in three ways:

FIGURE 15.8 Dumping sites for sewage sludge off the east coast of the United States. Depth is indicated in meters (m).

■ TABLE 15.3 Examples of Recognized Seawater Pollutants* from Human and Natural Sources

ORGANIC POLLUTANTS	
PCBs	polychlorinated biphenyls
PCDFs	polychlorinated dibenzofurans
PCDDs	polychlorinated dioxins
DDT	chlorinated hydrocarbon pesticide

INORGANIC POLLUTANTS	
Cd	cadmium
Pb	lead
Co	cobalt
Hg	mercury
Zn	zinc

* Pollutants originating in sludge, materials dredged from harbors or estuaries, and sewage-treatment and industrial wastewater that are known to be carcinogenic or to cause mutations, birth defects, nervous disorders, or diseases of the liver, kidney, or lungs.
SOURCE: EPA (adapted).

mersible *DSV Alvin*. Samples taken by the team contained trace metals and bacteria indicative of human sewage. Ongoing studies will provide better understandings of how sewage is transported in deep water and how deep-sea animals are influenced by inputs of sludge.

During the summer of 1987, many beaches were quarantined after high bacteria counts were found in coastal waters and medical debris (including hypodermic needles) washed ashore on East Coast beaches. These findings also led to the Ocean Dumping Ban of 1988, which required cessation of ocean dumping by 1991, subject to stiff financial penalties. New York and other cities now dewater sludge, compost it, and use it as a soil additive or deposit it in landfills.

Problems of Municipal Waste Isolation

Organic refuse in landfills ultimately decomposes. As it does, the ground surface settles; methane, carbon dioxide, and other gases are generated; and noxious or even toxic leachate from it may seep into the water table or bordering streams. Thus, a landfill's *pollution potential* depends upon the change in volume that accompanies decomposition, the waste's reactivity (which determines the chemistry of the leachate), the local geology, and the climate. Organic matter decomposes more slowly in cold, dry climates than in warm, wet ones.

STABILIZATION. Landfills continue to settle as decomposition goes on for many decades after burial. Typically, settlement is rapid at first and diminishes to a much slower

rate after about ten years (● Figure 15.9). The rate of subsidence depends upon many factors, including the climate, the kind of waste, and the amount of compaction during filling. Thirty-ton tractors left on a fill over a weekend have been known to settle as much as a meter into the trash, and total settlement of as much as 30 percent has been measured in landfills (Figure 15.9, part b). Variation is great. A 6-meter-thick landfill in Seattle settled 23 percent in its first year, whereas a 23-meter-thick fill in Los Angeles settled only 1 percent per year in its first three years. Settlement is a problem, because it results in cracks and fissures, ponds, sags, and infestation by vermin. It is monitored to assure that water does not pond on the fill cover and percolate downward, creating a high volume of polluting leachate.

As microorganisms break down organic material *(biodegradation),* they release heat, methane (natural gas), and other gases including hydrogen sulfide, carbon dioxide, and oxygen. Research at the University of Arizona and elsewhere has shown that biodegradation can be surprisingly slow, as illustrated by exhumed garbage containing virtually unblemished 10–20-year-old newspapers (● Figure 15.10), a 14-year-old mound of recognizable guacamole, and positively ancient hot dogs (testimony to the effectiveness of sausage preservatives). The most efficient decomposers are the voracious oxygen-using aerobic bacteria, such as those that quickly turn grass cuttings into compost for our gardens. Most landfills are sealed off from the atmosphere, however, giving rise to anaerobic (no oxygen) microorganisms that manufacture methane and hydrogen sulfide in the slow process of decomposing cellulose. Pumping air into a fill, a procedure known as **composting**, stimulates aerobes,

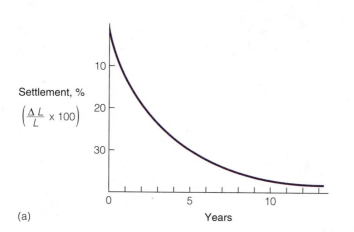

(a)

(b)

● FIGURE 15.9 *(a)* Graph of landfill settlement over time. The percentage of settlement is calculated by dividing the amount of settlement *(ΔL)* by the original thickness of the landfill *(L)* and multiplying by 100. The settlement values here are for illustration only and do not depict typical values. *(b)* Landfill weigh station built on trash-filled land. The station eventually collapsed, requiring that a new structure and scale be established at a different location.

● FIGURE 15.10 Underground newspaper—still readable after at least ten years' burial in a Pomona, California, landfill.

(a)

(b)

● FIGURE 15.11 (a) Mission Canyon landfill near Santa Monica, California, as it was being finished. (b) The reclaimed landfill is now a golf course and upscale residential neighborhood.

● FIGURE 15.12 Methane-extraction well and drill rig. The extracted gas is conveyed to a central location for conversion to electricity.

accelerates decomposition and settlement, and creates more space for trash. The process generates carbon dioxide, sulfur dioxide, and heat up to 70°C (150°F), which drives off other easily volatilized materials in the fill.

GAS GENERATION. Methane (CH_4), a gaseous hydrocarbon, is the principal component of natural gas. Also known as *marsh gas,* it bubbles forth from stagnant ponds and swamps, the result of complex chemical fermentation of plant material by bacteria. It is the same gas that causes the explosions in coal mines that have taken innumerable lives. Methane is explosive when present in air at concentrations between 5 and 15 percent. Although enormous volumes of methane are generated by anaerobic microorganisms in a landfill, there is little danger of the fill exploding because no oxygen is present. Being lighter than air, methane migrates upward in a fill. Methane leaking through the cover of a landfill converted to a golf course near Santa Monica, California, was known to "pop" from ignition when cigarettes were dropped on its greens (● Figure 15.11). Methane can also migrate laterally, and it did so several hundred yards into the wall spaces of structures near a landfill south of Los Angeles. Fortunately, the potentially explosive situation was discovered, and wells were drilled to intercept and extract the gas (● Figure 15.12). One strategy for handling the large volumes of methane generated at large landfills is to extract it through perforated plastic pipes, remove the impurities (mostly CO_2), and use it as fuel—the so-called refuse-derived fuel, *RDF.* This can be accomplished either by piping the gas away via pipelines or by using it on the site to generate electricity. Because the gas may be as much as 50 percent CO_2, it is more expedient to use the fuel directly to generate electricity than to clean out the impurities and put it into pipelines. The Los Angeles County (California) Sanitation District extracts landfill methane and generates sufficient electricity to supply up to 45,000 homes.

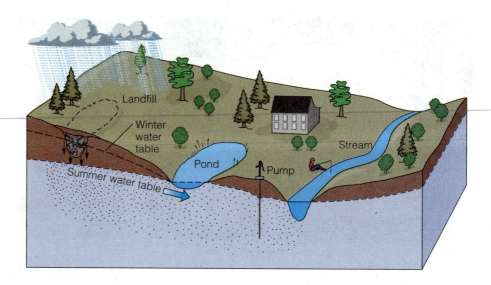

◉ **FIGURE 15.13** The flow of leachate from a landfill can contaminate underground water, ponds, and streams. See also Table 15.4.

LEACHATE. The compositions of leachates in and leaving landfills are highly variable, but they must be considered dangerous to the environment until established otherwise (◉ Figure 15.13). A leachate may be such that a receiving body of water can assimilate it without any impairment in water quality. Some purification of a leachate occurs as it filters through clayey soils, and some purification also occurs when contained organic pollutants are oxidized. Ground-water-quality measurements taken in the vicinity of a landfill, at the fill, and downstream from the fill are shown in ▣ Table 15.4. Note that the leachate was causing deterioration of ground-water quality in the vicinity of the monitoring well. To correct such deterioration, the leachate must be intercepted by wells or a barrier must be installed between the leachate and the water table. Under existing regulations, a landfill that is in a geologic setting where leachate threatens local water bodies may accept only inert (Group 3) wastes.

Resource Recovery from Waste

Resource recovery has become a component of waste management. It is the removal of certain materials from the waste "stream" for the purpose of recycling or composting them. It has been estimated that each year Americans throw out enough wood and paper to heat a billion homes for a year, and that Sunday newspapers alone—many of which are thrown away after 30 minutes—require cutting down 220 million trees a year. We throw away enough aluminum cans in a year to quadruple the size of our airline fleet (Case Study 15.3). Much of the material we toss out is recycled, but we must recycle more. The industrial nations are the "Saudi Arabias" of trash, and all have the same problems—dwindling resources and dwindling space for solid waste.

Until recently, **recycling** held little appeal; that is, it cost more to separate, sort, and recycle trash than to landfill it. It is now mandated in many states, however, because existing

▣ **TABLE 15.4 Ground-Water Quality Measured Near One Landfill**

WATER CHARACTERISTICS	MEASUREMENT SITE		
	Local Ground Water	*Fill Leachate*★	*Monitoring Well*★★
Total dissolved solids, ppm	636	6,712	1,506
pH★★★	7.2	6.7	7.3
BOD, mg/L★★★★	20	1,863	71
Hardness, ppm	570	4,960	820
Sodium content, ppm	30	806	316
Chloride content, ppm	18	1,710	248

★ A saturated fill.

★★ A monitoring well about 150 feet downstream from the landfill at a depth of 11 feet in sandy material.

★★★ A solution with a pH of 7 is neutral, less than 7 is acidic, and greater than 7 is basic (Case Study 13.3).

★★★★ Biological oxygen demand: the amount of oxygen per unit volume of water required for total aerobic decomposition of organic matter by microorganisms. This is discussed later in the chapter in the section on wastewater treatment.

SOURCE: D. R. Brunner and D. J. Keller, *Sanitary Landfill Design and Operation* (Washington, D.C.: U.S. Environmental Protection Agency, 1992).

CASE STUDY 15.3

Don't Mess With Texas

Cleaning up highway litter cost the Texas Department of Highways and Public Transportation $24 million in 1985. To reduce this unnecessary expenditure, Department snoops followed drivers to study their littering habits. The trash spies found that nearly 50 percent of the litterbugs were 18–34-year-old good old boys in pickups. Personal interviews disclosed that the slogans Texas had been using, such as "Don't Be A Litterbug" and "Pitch In," impressed these offenders about as much as drinking a Diet Pepsi on Friday nights.

A new litter-reduction campaign called "Don't Mess With Texas" was implemented in 1986 (◉ Figure 1). The thinking behind the slogan was that no-nonsense language would be remembered and make drivers think before discarding. It worked. A year later almost three quarters of all Texans were familiar with the slogan, and five years later highway litter had been reduced by 64 percent. The slogan now appears on bumper stickers, T-shirts, and trash cans, and Willie Nelson sings "Don't Mess with Texas" jingles on television and radio. Congratulations to the citizens of the Lone Star State!

◉ **FIGURE 1** Anti-litter campaign features this sign on a Texas highway.

landfills are filling and acceptable sites for new ones are difficult to find. Much of the consuming public is "thinking green" these days, and manufacturers are realizing favorable marketing results by using recycled materials in packaging their products. Although an average U.S. family of four generates 2,900 kilograms (6,400 lbs, 3.2 tons) of trash each year, it recycles only 500 kilograms (1,100 lbs) of paper, aluminum cans, glass, and plastic, 17 percent by weight. Much greater resource recovery is possible. For example, the Center for the Biology of Natural Systems recruited 100 families in East Hampton, New York, to recycle as much of their solid waste as possible. With the center's help, the volunteers recycled about 84 percent of their waste. The key to this success was the volunteers' willingness to separate wastes according to their category and grade—glass by color, paper by grade, cans by metal, and so on. In 1993 President Clinton mandated that federal government use more recycled paper, which helped to boost waste-paper prices. Because of such mandates and a growing general awareness that mineral resources are finite, trash recovery and recycling increased from 10 percent to 19 percent between 1985 and 1993.

Because recovery processes generate some wastes, the net landfill savings yielded by recycling and reuse is less than 100 percent. Recycled paper, for instance, must be de-inked and bleached. This process leaves a toxic sludge that must be discarded in a landfill. ◉ Figure 15.14 summarizes

resource recovery at landfills by category. It is believed that as much as 90 percent of municipal solid waste could be recovered by intensive recycling. Curbside recycling programs grew enormously between 1988 and 1996, with more than 100 million people participating in the programs in 1993—almost 40 percent of the U.S. population. Because waste paper reached a high of $200 per ton in 1996, trash burglars are taking newspapers and aluminum

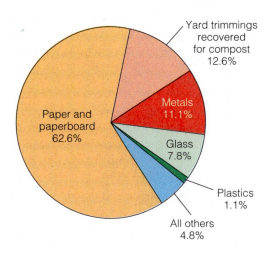

◉ **FIGURE 15.14** Materials recovered at landfills as percentage of the total weight recovered, 1990.

cans from curbside collection bins. This would be considered a blessing, but cities frown on it as it denies them their recycling revenues. One indicator of the anticipated value of trash is that glass and high-grade plastics have joined corn and soybeans as commodities that are bought and sold on the Chicago Board of Trade. Trash futures?

RECYCLING. An effective way to extend the lifetimes of landfills and natural resources is to recycle municipal wastes. Recycling includes such practices as composting, making concrete from incineration ash, and returning beverage bottles for reuse—the bottles average 15 to 20 fillings before breaking. Many cities have instituted regulations that

require residents to separate glass, metal, plastic, and paper products from other trash materials. Recycled paper, for example, is de-inked, shredded, pulped, and remade into newsprint, cardboard boxes, paper bags, and other products. Several kinds of plastic are being recycled by melting and extrusion processes to produce plastic "lumber," trash cans, furniture frames, and the like. Many states now require plastic objects to be stamped with a code indicating the type of plastic used. Seven kinds of plastic must be separated by hand in the recycling stream, which is expensive (■ Table 15.5). Only 2 percent of waste plastic was recycled in 1995. Ingenuity (see Chapter 13) plays a major role, however. For example, a small Massachusetts company quickly grew into a $1.3 million business by collecting spent plastic laser-printing cartridges from local businesses and replacing the spent parts. It sells the recycled cartridges to consumers and the worn plastic and metal components to recyclers.

Technology has been developed that separates out heavier materials such as metals, glass, and some plastics before the lighter, "wet" garbage is burned to produce energy. Electromagnets and various mechanical devices then separate metals and light objects from heavy glass along a conveyor belt. Metals are melted down and reformed, as is glass, which has first been separated by color. Automobiles are shredded, and their iron and steel removed by electromagnets. Their nonmagnetic metals, such as aluminum, copper, lead, and zinc, are separated out by a flotation method in a variable-density solution. Each metal sinks to the level in the liquid corresponding to its relative density.

Not easily solvable is the problem of what to do with old automobile tires. An estimated three billion tires are stockpiled in the United States today, and each year 240 million more are added. We discard three times more tires than we recycle. Resting in dumps, they add to environmental problems because they collect water, promoting the growth of mosquitoes, vermin, and other disease-spreading pests. Also, accidental tire fires can burn for weeks, giving off billows of irritating black smoke and fumes. No large-scale use for old tires has been found, although incidental small-scale uses are seen occasionally (◉ Figure 15.15). Chopped tires are mixed with asphalt to surface and repair highways and to fuel power plants in some places. In Modesto, California, for example, a specially built power plant burned five million tires each year to generate electricity. It was closed in 1996 for unexplained reasons. Because old tires are such an

■ TABLE 15.5 Coding System for Plastics

POLYETHYLENE TEREPHTHALATE (PET) a transparent plastic (may be colored) commonly used to make two-liter soda bottles and other containers such as peanut butter jars; is used in about 25% of all plastic bottles. Can be recycled into such items as strapping (for packaging), fiberfill for winter clothing, carpets, surfboards, and sailboat hulls.

HIGH-DENSITY POLYETHYLENE (HDPE) commonly used for plastic milk jugs, bleach and detergent bottles, motor-oil bottles, plastic bags, and other containers; is used in more than 50% of all plastic bottles. Can be recycled into trash cans, detergent bottles, drainage pipes, base cups for soda bottles, and other similar items.

VINYL (V) OR POLYVINYL CHLORIDE (PVC) used to make siding, pipes, hoses, shower curtains, and bottles for cooking-oil, shampoo, and household chemicals. Can be recycled into fencing, house siding, handrails, pipes, and similar items.

LOW-DENSITY POLYETHYLENE (LDPE) commonly used to make cellophane wrap, bread bags, and trash bags. Can be recycled into grocery and garbage bags.

POLYPROPYLENE (PP) a lightweight plastic commonly used for margarine and yogurt containers and certain types of lids and caps. Can be recycled into car-battery cases, bird feeders, and water pails.

POLYSTYRENE (Styrofoam™; PS) used to make coffee cups, plastic "peanuts" for packing, egg cartons, meat trays, plastic utensils, videocassettes, and other items. Has been recycled into tape dispensers and reusable cafeteria trays.

OTHER plastic resins other than the six categories listed above as well as mixtures of plastics. Some mixtures can be recycled into "plastic lumber" used for making benches, lawn furniture, picnic tables, marine pilings, and other outdoor equipment.

● FIGURE 15.15 Three thousand tires were recycled as children's playground equipment and landscaping material at this park.

enormous environmental problem, many engineers and scientists are working on a technological "fix" that will rid us of this blight.

Besides extending the life of our resources, recycling makes sense because it saves energy. The energy used in processing many scrap materials is less than that required for processing the raw material. Producing a pound of aluminum from scrap takes only 5 percent of the energy required to produce a pound from bauxite; scrap iron requires one-fourth the energy needed to make iron from ore; and producing a ton of paper from recycled stock uses only 30 percent of the energy used to make paper from trees. However, the costs of collection, transportation, and handling of recycled materials can be considerable, and recycling is still subsidized in most communities. As more curbside separation of garbage for recycling is mandated, and as more states enact container-bill legislation encouraging the return of beverage containers, the flow of recyclables increases. The "champion" recyclers (% recovered) in the United States are Minnesota (41%), New Jersey (39%), and Washington (35%). Taiwan is the world champion paper recycler; 98 percent of the paper used in that almost treeless country is recycled.

COMPOSTING. When decomposed by bacteria, biodegradable wastes, the so-called green garbage, can be used as a fertilizer and soil amendment. Yard cuttings, most

kitchen wastes, and organic wastes from commercial operations are acceptable compost material. Because yard wastes are second only to paper in landfill tonnage (see Figure 15.14), many states and municipalities have mandated separate collection for them. As a result, the number of composting facilities in the United States grew dramatically from 651 facilities in 1988 to more than 3,000 in 1993.

A state-of-the-art composting method, borrowed from the Netherlands, is used in Columbia County, Wisconsin. Leaves, grass, food scraps, and even road kill are placed in one end of gently sloping, rotating, 3½-meter × 50–meter drums. A week later, the oxygen, bacteria, and constant churning have produced crumbly, black, high-quality compost, which is removed at the low end of each drum and sold to farmers and home gardeners. Heat generated by oxidation and bacterial action kill any pathogens present in the material. Mecklenburg County, North Carolina, earned $100,000 in 1991 from its yard-waste composting program called "Curb It." Waste-management experts use flow charts such as ● Figure 15.16 for estimating waste recovery and landfill space needs.

WASTE-REDUCTION STRATEGIES. An integral component of waste management is reducing the amount of trash that is produced in the first place. Examples of waste-reduction strategies are now found in almost every office, manufacturing plant, school, and home in the United States. One company that now requires double-sided copying reduced its paper purchases by 77 million sheets a year. Some supermarkets sell their customers nylon-mesh grocery bags at cost. At least one furniture manufacturer ships its products in reusable padded quilts instead of heavy cardboard. Consumers can help by refusing to buy overpackaged items, patronizing firms that use some recycled materials, and returning beverage containers to the store to redeem the deposits. Many communities encourage waste reduction by charging residents for trash removal based upon the amount of waste they generate.

Multiple-Land-Use Solutions

Potential sites for sanitary landfills include canyons, abandoned rock quarries, and sand and gravel pits. Population pressure and resource conservation in the next century will require that we not only dispose of our waste efficiently, but

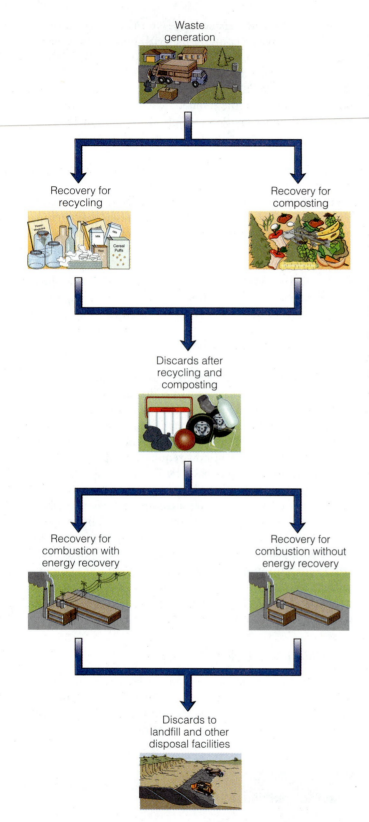

Waste generation

Recovery for recycling

Recovery for composting

Discards after recycling and composting

Recovery for combustion with energy recovery

Recovery for combustion without energy recovery

Discards to landfill and other disposal facilities

◉ **FIGURE 15.16 Waste-management flow chart.**

(a)

(b)

◉ **FIGURE 15.17 Palos Verdes, California, landfill** *(a)* **during filling operations and** *(b)* **reclaimed as South Coast Botanical Gardens, a tourist attraction and passive recreation area for local residents.**

will generate about a ton of trash a year. Thus it *makes sense* to place human-generated trash into the land openings created in extracting earth resources for meeting human needs. After a landfill is completed, methane can be recovered and converted to energy, and the site can be reclaimed as a golf course (see Figure 15.11), park, athletic field, public gardens (◉ Figure 15.17), drive-in theater, parking lot, lumberyard, and so forth. An ideal beneficial cycle of such multiple land uses is shown in ◉ Figure 15.18.

As explained earlier, structures placed directly on a landfill are subject to settlement and to methane emanations. A restaurant and shops built upon a thin landfill (<8 meters thick) south of Los Angeles were placed on piles driven through the fill to keep them from settling. The walkways around the buildings were not, however, and after a few years they settled and had to be built up to the base of the structures (◉ Figure 15.19). In the same commercial devel-

also that we gain some benefit from the stored trash and the disposal site. The average new home requires about 70 cubic yards of concrete (1,900 ft^3) in its construction and

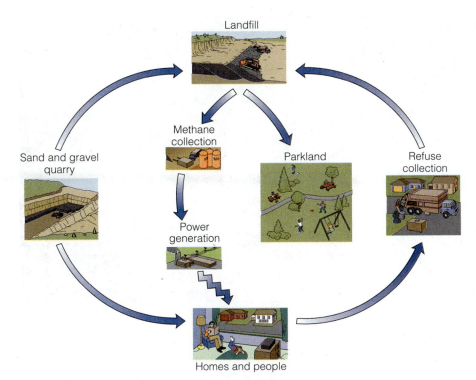

Landfill

Methane collection

Sand and gravel quarry

Parkland

Refuse collection

Power generation

Homes and people

opment, methane is extracted and used to illuminate exterior gas lanterns.

As cities expand and coalesce—such as in the New York-to-Boston corridor and the strip between Los Angeles and San Diego—waste-disposal-site selection must become very creative. Between 1940 and 1983 the Kaiser Corporation operated an open-pit iron mine at Eagle Mountain in the desert 115 kilometers (70 mi) east of Indio, California. The mining operation has now been abandoned, leaving an opening 3 kilometers (1.9 mi) long

● FIGURE 15.19 Commercial development built on a landfill. Whereas the buildings are founded on piles through the fill to bedrock, the pavement was placed directly on the fill. The lanterns are fueled by methane from the fill.

and more than 300 meters (about 1,000 ft) deep (● Figure 15.20). It is the proposed site for a landfill for all Southern California cities' household waste (no toxic or hazardous waste), which would come to it in closed railroad cars. The tracks that served the mine are still in place, and the site is known to be free from faults and potential for groundwater contamination. Far from any population center, the site could accommodate half of all the household trash generated in Southern California in the next 115 years! The open pit is in dense bedrock that would be lined with clay and high-density polyethylene plastic to protect ground water. The proposed site is acceptable geologically, and the only objections raised at public hearings regarding the proposal, from a technical point of view, seem to be entirely NIMBY, PIITBY, and NIMTOO (see Case Study 15.1). The Metropolitan Water District, which operates an open canal system 2 kilometers (1¼ mi) away that delivers millions of gallons of fresh water daily from the Colorado River to Southern California, has no objection to the project.

CONSIDER THIS . . .

During the past few decades recycling waste has evolved from a virtually nonexistent concept to a major component of waste management for many communities. Why do some environmental groups object to waste-problem solutions that emphasize recycling?

● FIGURE 15.20 Eagle Mountain Mine, site of the proposed Eagle Mountain landfill; Riverside County, California. As designed, the fill could accommodate up to 20,000 tons of trash per day, which would make it the largest-capacity landfill in the world. It would accept only nonhazardous waste and would have a recycling center.

HAZARDOUS WASTE DISPOSAL

Hazardous wastes are primarily such industrial products as sludges, solvents, acids, pesticides, and PCBs (polychlorinated biphenyls; highly toxic organic fluids used in plastics and electrical insulation). The threats that these and other pollutants pose to ground water are discussed in Chapter 10. The United States produces the largest amount of hazardous waste and the second-largest amount per unit area (see Table 15.1). EPA estimates are that U.S. industries were generating 260 million metric tons of hazardous waste per year in the mid 1990s. A typical rate at which liquid hazardous waste was generated in the early 1980s was 150 million metric tons per year, which can be thought of as 40 billion gallons. If that amount could be placed in 55-gallon drums and the drums placed end to end, the drums would encircle the earth 16 times (●Figure 15.21). In the mid 1990s this would be almost 28 times. Examples of haz-

ardous wastes individuals introduce to the environment include drain cleaners, paint products, used crankcase oil, and discarded fingernail polish (not trivial). The EPA has estimated that 90 percent of our hazardous waste, coming from about 750,000 sources, is disposed of improperly, and that this could leave problems that will last for generations. Fortunately, knowledge of how to dispose of these substances with a high degree of safety exists.

Hazardous-Waste Disposal Methods

Secure landfills are designed to totally isolate the received waste from the environment. Wastes are packaged and placed in an underground vault surrounded by a clay barrier, thick plastic liners, and a leachate removal system (●Figure 15.22). This is the *preferred* method of disposing of hazardous waste; unfortunately, few of the nation's existing 75,000 industrial landfills have plastic liners, clay barri-

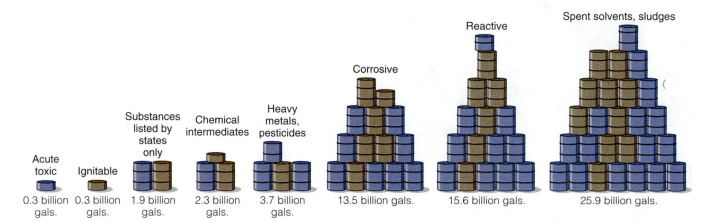

● FIGURE 15.21 Breakdown by category of the 40 billion gallons of toxic waste U.S. industries generated in 1981. The total of all categories is greater than 40 billion due to overlap; for example, a fluid may be both corrosive and reactive.

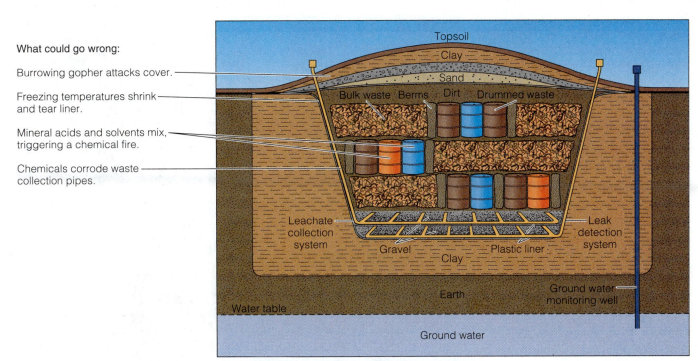

What could go wrong:

Burrowing gopher attacks cover.

Freezing temperatures shrink and tear liner.

Mineral acids and solvents mix, triggering a chemical fire.

Chemicals corrode waste collection pipes.

⊙ **FIGURE 15.22** A secure landfill for toxic waste. Note the liners, leachate collection system, leak detection system, and monitoring well.

ers, or leachate drains. Many older fills will either have to be retrofitted with modern environmental-protection systems or shut down. Additional protection is provided by existing regulations that mandate ground-water monitoring at disposal sites. An ideal secure landfill for hazardous waste would be a site in a dry climate with a very low water table. In humid climates where water tables are high, alternative means of safe disposal must be found.

Deep-well injection into permeable rocks far below fresh-water aquifers is another method of isolating toxic and other hazardous liquid substances (⊙ Figure 15.23). Before such a well is approved, thorough studies of subsurface geology must be made in order to determine the location of faults, the state of rock stresses at depth, and the impact of fluid pressures on the receiving formations (Case Study 15.4). The injection boreholes are lined with steel casing to prevent the hazardous fluids from leaking into fresh-water zones above the injection depth.

Disposal of hazardous wastes, both liquid and solid, is regulated by the Resource Conservation and Recovery Act of 1976 (RCRA). Exempt from the requirements of this act are businesses that generate less than a metric ton of waste per month. According to EPA estimates, about 700,000 of these small firms generate 90 percent of our hazardous waste. Consequently, most of these wastes end up in sanitary landfills.

Discharge into sealed pits is the least expensive way to dispose of large amounts of water containing relatively small amounts of hazardous substances. If the pit is well sealed and evaporation of water equals or exceeds the input of contaminated water, the pit may receive and hold hazardous waste

almost indefinitely. Problems can arise from leaky seals and overflow of holding ponds during heavy storms or floods.

Superfund

What about the thousands of old and mostly unrecorded sites where hazardous waste has been dumped improperly in the past? Hazardous substances have been disposed of indiscriminately since the eighteenth century in the United States. (■ Table 15.6 on page 476 gives some examples.) **Superfund,** the Comprehensive Environmental Response, Compensation, and Liability Act (CERCLA), was passed in 1980 to rectify past and present abuses from toxic-waste dumping. It authorizes the EPA to clean up a spill and send the bill to the responsible party. CERCLA also empowered the EPA to establish a National Priorities List (NPL) for cleaning up the worst of the 15,000 identified abandoned toxic and hazardous waste sites now known as *Superfund sites,* some of which date back many decades (Case Study 15.5). By 1996 1,225 NPL sites had been designated or were pending. Of those, 102 sites had been cleaned up, and work has been completed on 304 others. New Jersey has the dubious distinction of harboring 6 of the 15 worst Superfund sites as ranked by impact on the environment— 5 landfills and 1 industrial site. ■ Table 15.7, page 478, reports the number of Superfund sites in each state in 1990 and ranks the states by number of sites. (How does your state rank?)

Although the costs of cleaning them up are staggering, toxic dump sites do not just "go away" on their own. They need to be cleaned up. Some problems with Superfund also

● FIGURE 15.23 Deep-well injection of hazardous waste. This method presumes that wastes injected into strata of porous rock deep within the earth are isolated from the environment forever. What could go wrong? *(1)* Toxic spills may occur at the ground surface. *(2)* Corrosion of the casing may allow injected waste to leak into an aquifer. *(3)* Waste may migrate upward through rock fractures into the ground water.

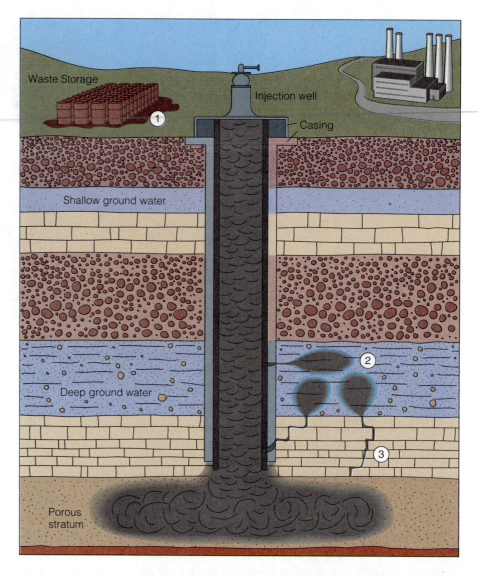

need to be cleaned up. The Superfund concept was well intended, but the program has been poorly executed. As a result, 40 percent of its funding has gone to litigation expenses. Superfund has become a "lawyers' money pit," as one journalist put it. A case in point is the Brio Superfund Site 18 miles south of Houston, Texas. Two waste-disposal plants once occupied the area that was developed as a subdivision of 2,800 people. Subsequently, water wells and soils there were discovered to be contaminated with chloroform, xylene, tar, TCE, and other health-endangering chemicals. High incidence of birth defects, heart disease, and other problems followed, and most residents fled their homes. By early 1996, more than a decade had passed since Brio was declared a Superfund site, and nearly a billion dollars of taxpayers' money had been paid in litigation fees and monetary damages to residents, but nothing had been done to remediate the site. In 1995 about $4 billion of the $20.4 billion spent on Superfund to date had been consumed by lawyers and filing fees, and legislation was underway to reduce litigation and settle liability issues more quickly.

CONSIDER THIS . . .

Perhaps no other recent environmental legislation has received as much public awareness as the act with the catchy nickname "Superfund." What is Superfund? Who manages it, and what legislation is behind it? From an environmental standpoint, what is its main problem?

 ## WASTEWATER TREATMENT

Sewage is required to be purified to a predetermined standard before it is delivered to a receiving stream or aquifer. An important measure of the purity of sewage effluent is **biological oxygen demand (BOD),** the amount of oxygen per unit volume of water—milligrams of O_2 per liter of water—required for the total aerobic decomposition of

Deep-Well Injection and Earthquakes

The U.S. Army's Rocky Mountain Arsenal in suburban Denver drilled a 3,764-meter (2⅓-mi) deep well for disposing of chemical-warfare wastes in 1961. The Tertiary and Cretaceous sedimentary rocks into which the well was drilled were cased off to prevent any pollution of these formations, and the lower 23 meters (75 ft) were open to Precambrian granite. About 40 kilometers (25 mi) to the west of the arsenal is the frontal fault system of the Rocky Mountains (◉Figure 1).

Injection began in 1962 at a rate of about 700 cubic meters per day (176,000 g/day). Soon after injection was started, numerous earthquakes were felt in the Denver area. The earthquake epicenters lay within 8 kilometers (5 mi) of the arsenal and plotted along a line northwest of the well. A local geologist attributed the quakes to the fluid injection at the arsenal. Although the army denied any cause-and-effect relationship, a graph of injection rates versus the number of earthquakes showed an almost direct correlation—especially during the quiet period when fluid injection was stopped (◉Figure 2).

The fractured granite into which the fluid was being injected was in a stressed condition, and the high fluid pressures at the bottom of the hole decreased the frictional resistance on the fracture surfaces. Earthquakes were generated along a fracture propagating northwest of the well as the rocks adjusted to relieve internal stresses.

Thus earthquakes *can* be "triggered" by human activities, and the Denver earthquakes initiated research into the possibility of modifying fault behavior by fluid injection; that is, relieving fault stresses by producing a large number of small earthquakes. Seismic studies conducted during a water-injection (see Chapter 13) program at the nearby Rangely Oil Field confirmed that earthquakes *could* be turned on or off at will, so to speak, in areas of geologic stress. To date no group or government entity has assumed responsibility for triggering earthquakes by fluid injection, for obvious reasons. What would be the consequences if, instead of triggering many small earthquakes, a major earthquake was triggered, accompanied by loss of life and property damage? More research is needed in this area.

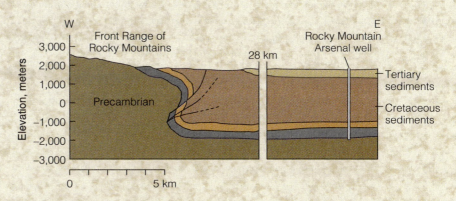

◉ FIGURE 1 Relationship between the Rocky Mountain Arsenal and the Rocky Mountain frontal fault system.

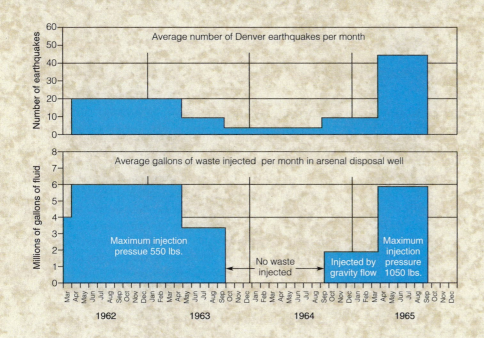

◉ FIGURE 2 Correlation of fluid waste injection and the number of earthquakes per month in the Denver area, 1962–1965.

■ **TABLE 15.6 Chronologic Identification of Selected Hazardous Substances in the United States**

DATE	SUBSTANCE	NATURE OF HAZARD	SOURCE
1860	Copper acetoarsenite (Paris Green)	serious health problems or death with chronic exposure	used against the Colorado potato beetle
1925	Ethylene, ethylene oxide, ethylene glycol	carcinogenic, liver disease, mutations, nervous disorders	antifreeze
1927	Polychlorinated biphenyls (PCBs)	carcinogenic, mutations, birth defects, liver disease	used for heat transfer in electrical transformers, motors, appliances
1940s	Chlorinated hydrocarbons (chlordane, toxaphene, dieldrin, aldrin, DDT) and organophosphate (parathion)	highly toxic or carcinogenic; persistent for years, low biodegradation, bioaccumulative in humans	insecticides
1955	Halogenated hydrocarbons	toxic	chemical warfare agents, pesticides, refrigerants; some are ozone scavengers
1968	Lead	human health hazard	paint and gasoline
1980s	Cyanide, pesticides, acids	toxic	aquifers at Swartz Creek, Michigan,★ contaminated by illegal burial of drums containing these substances
1980s	Dioxin-containing oil	carcinogenic, birth defects	road surfaces at Times Beach, Missouri★ (EPA bought the town and relocated 2,200 residents at cost of $33 million in 1985.)
1985	Paint thinner, pesticides, TV sets, chemical bleaches, batteries, fingernail polish	various	dumped with ordinary wastes at many landfills
1990s	Cyanide, mercury, lead, sulfur, hydrocarbons, etc.	various	abandoned mines and oil wells

★ Superfund site.

SOURCE: Adapted from Allen Hatheway, "Pre-RCRA History in Industrial Waste Management in Southern California," B. W. Pipkin and R. J. Proctor, eds., *Engineering Geology Practice in Southern California,* Association of Engineering Geologists Special Publication 4 (Belmont, Calif.: Star Publishing Company, 1992).

organic matter by microorganisms. Also referred to as *biochemical oxygen demand,* it indicates the degree of a water's organic pollution. BOD is determined by measuring the amount of oxygen in a closed liter bottle of treated water at intervals until uptake ceases. The total oxygen used is the BOD. Water that is totally free of organic matter and microorganisms has a BOD of zero. As many as three levels of sewage treatment may be required to bring BOD, water clarity, and chemical quality to within acceptable limits for a particular receiving water body (■ Table 15.8, page 480).

1. *Primary treatment* consists mostly of removing solids by gravity settling, screening, and then aerating the fluid portion to remove some organic material. There is considerable debris in raw sewage, including wood, rags,

paper, and sediment. (A group of students reported seeing a small bicycle on the screen during a field trip to a treatment facility!)

2. *Secondary treatment* reduces the BOD of the wastewater to a specified level. It introduces bacteria to digest organic matter and facilitates further sludge settling and clarification of the effluent. Secondary treatment requires the presence of microorganisms and a mechanism for replenishing oxygen in the wastewater. One method is to spray wastewater onto a trickling filter, where microbial films develop on the filter material. Another method uses *activated sludge,* whereby a microbial growth that consumes organic matter is suspended in the wastewater and settles to the floor of the reaction tank. Some of the settled sludge is recycled to the incoming wastewater to maintain the biomass required

CASE STUDY 15.5

Love Canal—No Love Lost Here

Love Canal, in the honeymoon city of Niagara Falls, New York, may sound very romantic, but it's not. Excavated by William T. Love in the 1890s but never finished, the canal is 1,000 meters long and 25 meters wide (3,300 ft × 80 ft). Unused and drained, Love Canal was bought by Hooker Chemical and Plastics Corporation in 1942 for use as a waste dump. In the next 11 years Hooker dumped 21,800 tons of toxic waste into the canal, which was believed to be impermeable and thus an ideal "grave" for hazardous substances. After the canal was filled and covered, Hooker Chemical sold the land to the Niagara Falls School Board for one dollar. A school was built on the landfill, and hundreds of homes and all the infrastructure needed to support a suburban community were built nearby.

In the early 1970s, after several years of heavy rainfall, water leaked through the "impermeable" clay cap and into basements and yards in the Love Canal area, and also into the local sewer system. Soon the new community was experiencing high rates of miscarriages, birth defects, liver cancer, and seizure-inducing diseases among children. Residents complained that toxic chemicals in the water were causing these adverse health effects.

In 1978 the New York State Health Commissioner requested the EPA's assistance in investigating the chemistry of fluids that were leaking into a few houses around the canal. The study revealed the presence of 82 toxic chemicals including benzene, chlorinated hydrocarbons, and dioxin, and the state commissioner declared Love Canal a health "emergency." Five days later President Carter declared it a federal disaster area, the first human-caused environmental problem to be so designated in the United States. The recommendation that pregnant women and children under the age of two be evacuated from the area followed, and New York State appropriated $22 million to buy homes and repair the leaks. In all, a thousand families were relocated (⦿ Figure 1).

The results of an EPA monitoring study were released to the public in 1982. An effective ground-water barrier and drain system were installed to intercept the toxic chemicals. The presence of relatively impermeable clays in the area suggests that future long-distance migration of pollutants from Love Canal is unlikely. Because the shallow and deep aquifers in the area are not hydraulically connected, there is little danger of contaminants migrating into deeper ground-water zones.

Love Canal was a precedent-setting case, because it spurred Congress to enact CERCLA, which provides federal money for toxic cleanup without long appeals. It also authorizes the EPA to sue polluters for the costs of cleanup and relocation of victims. Although Love Canal has been essentially contained, some chemicals still infect a nearby stream and schoolyard, and main-

(a)

(b)

⦿ FIGURE 1 Where the honeymoon ended and the nightmares began; Love Canal, Niagara Falls, New York.

tenance costs amount to a half-million dollars annually. The EPA no longer ranks Love Canal high among the nation's most dangerous waste sites, and the New York State Health Commissioner has announced that recent federal-state studies found four of the seven polluted Love Canal areas to be habitable. In spite of the commissioner's statement, there has been no stampede to reinhabit the area.

After 16 years of litigation, the legal battles over Love Canal ended in 1995. Occidental Chemical Corporation, which purchased Hooker Chemical in 1968, agreed to pay Superfund $102 million, and the Federal Emergency Management Agency (FEMA) agreed to pay $27 million. The State of New York had made a $98 million settlement with EPA in 1994. The cleanup is estimated to have cost $275 million.

TABLE 15.7 Hazardous Waste Sites on the National Priority List (Superfund) by State, 1990*

STATE	SITES	RANK	PERCENT DISTRIBUTION	STATE	SITES	RANK	PERCENT DISTRIBUTION
Alabama	12	27	1.0	Nebraska	6	43	0.5
Alaska	6	43	0.5	Nevada	1	50	0.1
Arizona	11	29	0.9	New Hampshire	16	22	1.3
Arkansas	10	34	0.8	New Jersey	109	1	9.1
California	88	3	7.4	New Mexico	10	34	0.8
Colorado	16	22	1.3	New York	83	4	6.9
Connecticut	15	24	1.3	North Carolina	22	17	1.8
Delaware	20	19	1.7	North Dakota	2	48	0.2
District of Columbia	0			Ohio	33	12	2.8
Florida	51	6	4.3	Oklahoma	11	29	0.9
Georgia	13	26	1.1	Oregon	8	40	0.7
Hawaii	7	42	0.6	Pennsylvania	95	2	7.9
Idaho	9	38	0.8	Rhode Island	11	29	0.9
Illinois	37	10	3.1	South Carolina	23	16	1.9
Indiana	35	11	2.9	South Dakota	3	46	0.3
Iowa	21	18	1.8	Tennessee	14	25	1.2
Kansas	11	29	0.9	Texas	28	13	2.3
Kentucky	17	21	1.4	Utah	12	27	1.0
Louisiana	11	29	0.9	Vermont	8	40	0.7
Maine	9	38	0.8	Virginia	20	19	1.7
Maryland	10	34	0.8	Washington	45	7	3.8
Massachusetts	25	14	2.1	West Virginia	5	45	0.4
Michigan	78	5	6.5	Wisconsin	39	9	3.3
Minnesota	42	8	3.5	Wyoming	3	46	0.3
Mississippi	2	48	0.2	**United States**	**1,197**		**100.0**
Missouri	24	15	2.0	Guam	1		
Montana	10	34	0.8	Puerto Rico	9		
				TOTAL	1,207		

* Includes both proposed and final sites listed on the National Priorities List for the Superfund program as authorized by the Comprehensive Environmental Response, Compensations, and Liability Act of 1980 and the Superfund Amendments and Reauthorization Act of 1986.
Source: *Statistical Abstract of United States, 1991*, p. 211.

for decomposition. Regardless of the level of wastewater treatment, a huge volume of solids remains that requires disposal. The alternative of dumping sludge at sea will soon be history. In landlocked areas and coastal areas where ocean dumping is prohibited, sludge must be dried and hauled to a landfill. Some is used commercially as a soil amendment for holding moisture and for fertilizer, although dried sludge has low nitrogen and little plant-nutrient value.

3. *Tertiary treatment* removes nitrogen and phosphorus compounds, reducing their levels to below those that cause eutrophication of the receiving waters. **Eutrophication** (from the Greek word, "to thrive") is a water problem that arises from nutrient enrichments and their associated explosive algal blooms. Sewage effluent is high in ammonia (NH_3), which is converted to nitrate $(NO_3)^-$, a plant fertilizer, by nitrifying bacteria. Ter-

tiary treatment is expensive, but it is essential where the discharge of partially treated sewage effluent would jeopardize the health of the receiving water body, such as at Lake Tahoe (Case Study 15.6). Tertiary-treated effluent is pure enough for human consumption.

Private Sewage Disposal

Modern household sewage-disposal systems utilize a septic tank, which separates solids from the effluent, and leach lines, which disseminate the liquids into the surrounding soil or rock (● Figure 15.24). A septic tank is constructed in such a way that all solid material stays within the tank and only liquid passes on into the leach field. A leach field consists of rows of tile pipe surrounded by a gravel filter through which the effluent percolates prior to percolating into the soil or underlying rock. Sewage is broken down by

The Cinder Cone Disposal Site—
Getting Your Sewage Together

Prior to 1965 sewage disposal in the north Lake Tahoe area was not much of a problem. Most households used septic tanks to dispose of their waste. With the area's population growth in the 1960s, however, it became clear that too much phosphate and nitrate were seeping from private septic-system leach lines into the lake. Nutrients in the effluent were stimulating algal growth, which in turn was clogging the shore zone and causing problems. It was decided to build a sewage disposal site outside the Tahoe basin to handle the 2.5 million gallons of sewage generated each day in the north Lake Tahoe area. This was to be an interim solution until a permanent tertiary-treatment facility could be built to serve the area.

Geologic studies indicated that a volcanic cinder cone near the Truckee River northwest of Tahoe City would be an ideal temporary disposal site (◉ Figure 1). It showed no evidence of recent activity, and its clinkery cinders and aa-type lava flows would be good percolation media for sewage effluent (◉ Figure 2). Furthermore, springs on the cone's flanks would facilitate chemical monitoring of the sewage effluent to determine its purity after percolation through the volcano. An interim primary-treatment plant was built in Tahoe City to separate the solid waste from sewage. The sludge from that plant was dried and trucked to a disposal site some 30 kilometers (19 mi) away, and the effluent was collected in ponds and pumped to several miles of percolation trenches in the central crater (◉ Figure 3). Winter freezing was not a problem, and chemical analyses indicated that the water emerging from the cone was in a relatively pure state.

The Cinder Cone disposal site dutifully took 2.8 million gallons of sewage each day during its operation, adequately serving the needs of the area until the permanent facility with a daily capacity of 30 million gallons was completed. Although the vol-

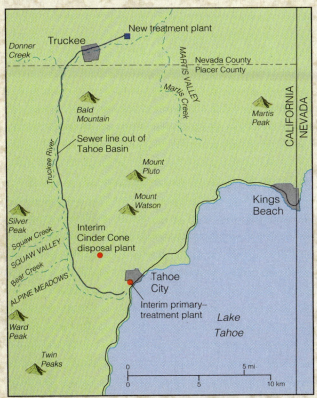

◉ **FIGURE 1** Temporary Cinder Cone disposal site and new water treatment plant; Tahoe City, California.

ume of effluent handled at Cinder Cone is impressive, it is very small in comparison to large cities' sewage plants, which handle several hundred million gallons of sewage per day. Nonetheless, this imaginative enterprise is notable in that it enlisted a natural geologic feature into the service of humankind.

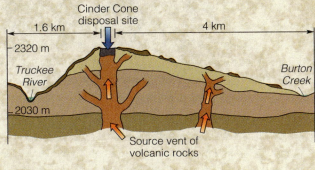

Pleistocene volcanics (younger)
Tertiary volcanics (older)
Scoria (volcanic cinder)
Older volcanics and granite (basement complex)

◉ **FIGURE 2** Geologic cross section of Cinder Cone disposal site (vertical scale exaggerated). The risk of polluting the Truckee River was a major consideration in geological studies of the proposed interim solution.

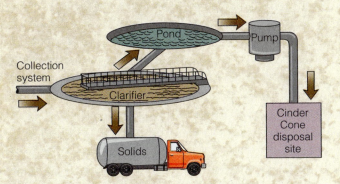

◉ **FIGURE 3** Solids were separated from the sewage in a clarifier and trucked to a Class II landfill near the California–Nevada border. Only the remaining liquid waste was pumped up to Cinder Cone disposal site.

Waste Disposal Then and Now

All humans generate solid waste, and disposing of it has always been a societal challenge. Safely disposing of containers of all kinds, old tires, large and small appliances, and just plain trash presents major problems today. As individuals we must recycle, demand recycled and recyclable containers, and actively assist in reducing the volume of waste (⊙ Figure 15.26). Historians point out that Ancient Rome was a filthy city; its cobblestones were noisy, its air was polluted from thousands of wood fires, and municipal sewage disposal was nonexistent. Medieval castles, monasteries, and convents had primitive waste-disposal systems, most of which were unhealthy and emanated noxious odors (⊙ Figure 15.27).

More recently, environmental paranoia has demanded the removal of all asbestos from all public buildings, and this has created another problem—how to get rid of asbestos-containing debris safely (⊙ Figure 15.28). Finally, certain activities at sanitary landfills appear to prove that some people have fun while "down in the dumps" (⊙ Figure 15.29).

⊙ **FIGURE 15.26** Norris McDonald organized a major cleanup of the Anacostia River in Washington, D.C., and adjacent Maryland.

● FIGURE 15.27 An adjacent stream was diverted to this medieval abbey in Scotland to create what was called a "great drain." All of the abbey's wastes went into the drain, which carried them to the stream, thus polluting its water for downstream users. Little wonder that lifespans were so short in those days; diseases such as cholera, typhoid fever, and dysentery and pestilence caused by the lack of sanitation took their toll. The correlation between stream pollution and disease was not widely recognized until the 1800s.

● FIGURE 15.28 Not an astronaut; a sanitation worker. Protective suits and masks are required for asbestos handlers at sanitary landfills.

● FIGURE 15.29 Caterpillar tractor operators play "king of the mountain" at a landfill. The 25-ton "Cats" are used to compact and flatten the debris to reduce its volume.

TABLE 15.8 What's Left after Primary and Secondary Sewage Treatment

	PERCENTAGE REMAINING AFTER TREATMENT	
WASTEWATER COMPONENT	Primary	Secondary
Biological oxygen demand (BOD) Dissolved oxygen necessary to decompose organic matter in wastewater	65	15
Nitrogen Nitrogen becomes a plant nutrient when it combines with hydrogen or oxygen to form ammonia, nitrate, or nitrite.	80–85	70
Phosphorus Phosphorus becomes a plant nutrient when it combines with oxygen to form phosphate.	80–85	70
Suspended solids Undissolved material suspended in water (sludge)	40	15

SOURCE: Data from T. M. Hawley, "Herculean Labors to Clean Wastewater," *Oceanus* 33 (1990), no. 2, p. 73.

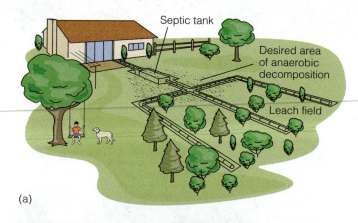

(a)

(b)

● FIGURE 15.24 *(a)* A septic tank near the house drains into a leach field of drain tiles that are laid in trenches and covered with filter gravel and soil. *(b)* A common but not recommended method of private sewage disposal.

anaerobic bacteria in the septic tank and the first few feet of the leach lines. As the effluent percolates into the soil or bedrock, it is filtered and oxidized, and *aerobic* conditions are established. Studies have shown that the movement of effluent through a few feet of unsaturated fine-grained soil under aerobic conditions reduces bacterial counts to almost nil. Fine-grained soils have been found to be the best filters of harmful pathogens, whereas coarse-grained soils allow pathogenic organisms and viruses to disperse over greater distances.

Under the best of conditions, leach lines have a life expectancy of 10–12 years. Factors that reduce infiltration rates and thus limit the life of a septic system are dispersion of clay particles by transfer of sodium ions (Na^+) in sewage to clay particles in the soil, plugging by solids, physical breakdown of the soil by wetting and drying, and biological "plugging." This plugging occurs when the leach system becomes overloaded with sewage and anaerobic conditions invade the leach lines, resulting in a growth of organisms that plug the filter materials.

After passing through the leach field, the effluent is still foul, smelly, and highly toxic—a harmful, pathogen-carrying fluid. It must be kept underground until oxidation and bacteria can purify it, making it indistinguishable from local

ground water. The geologic condition that most commonly causes a private sewage-disposal system to fail is a high water table that intersects or is immediately beneath the leach lines (● Figure 15.25, part a). This condition initiates the anaerobic growth that plugs filter materials and tile drains. Water tables may rise due to prolonged heavy precipitation or due to the addition of a number of septic tanks in a small area. One way to avoid exceeding the soil's ability to accept effluent is to design leach fields on the basis of percolation tests performed during wet periods. If necessary, a system's lifetime can be extended by constructing multiple leach fields whose use can be rotated regularly by switching a valve. Another problem is the improper placement of leach lines. This results in effluent seepage on hill slopes or stream valley walls (Figure 15.25, part b; see also Chapter 10). The proper horizontal distance of leach lines from a hill face is a function of *vertical permeability* and thus is best determined by a qualified professional. Although typical plumbing codes specify a

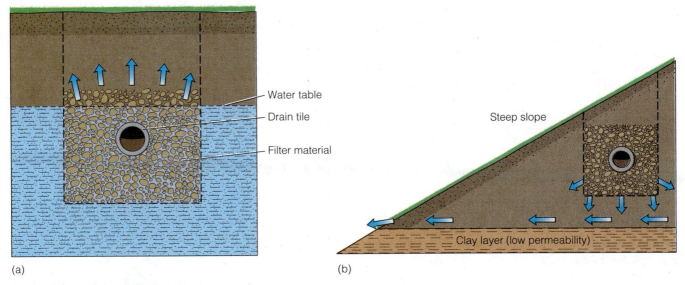

(a)   (b)

● **FIGURE 15.25** Conditions under which raw effluent can seep into the ground. *(a)* A drain tile and filter material intersect a high water table. When this occurs sewage rises, and the growth of anaerobic bacteria is promoted in the leach field. This leads to biological plugging of the filter material. *(b)* A leach line is underlain by low-permeability layers adjacent to a steep slope.

minimum of 5 meters (16 ft) horizontal distance from a hill face, this may not be sufficient if the vertical permeability is low and the hill slope is steep.

Nitrates in sewage effluent pose a eutrophication problem if households are built near a lake or pond. Eutrophication is a natural process in the life cycle of a lake, but cultural (human-caused) eutrophication accelerates the process and can convert an attractive, "real-estate" lake into a weed-choked bog in a short time. Nutrients from sewage systems and fertilized lawns must not be allowed to flow unchecked into bodies of water. Solving this problem requires the help of such experts in water flow and quality as aquatic biologists and hydrogeologists.

 **SUMMARY**

Waste Management

MANAGEMENT COMPONENTS
1. Waste disposal
2. Source reduction
3. Waste recovery
4. Recycling

RATIONALE Extending the lifetimes of landfills and natural resources.

CATEGORIES OF WASTE-DISPOSAL METHODS Isolation and attenuation.

Municipal Waste Disposal

SANITARY LANDFILL Trash is covered with clean soil each day.

OCEAN DUMPING Favored in the past for disposal of sewage, sludge, dredgings, and canistered nuclear wastes. Now regulated by Marine Protection Reserve and Sanctuary Act (Ocean Dumping Act).

LANDFILL PROBLEMS Leachate, settlement, gas generation, visual blight.

RESOURCE RECOVERY Amount of waste delivered to landfills can be reduced through recycling programs, composting, and implementing waste-reduction strategies.

MULTIPLE LAND USE CONCEPT A mine or quarry becomes a landfill, which when full may be converted to a park or other recreational area. Although some shallow landfills have been built upon, settlement is always a problem.

Hazardous-Waste Disposal

SECURE LANDFILL Lined with plastic and impermeable clay seal. Built with leachate drains.

DEEP-WELL INJECTION Into dry permeable rock body sealed from aquifers above and below. Has triggered earthquakes in Denver area.

SUPERFUND Comprehensive Environmental Response, Compensation, and Liability Act (CERCLA) created in 1980 to correct and stop abuses from irresponsible toxic-waste dumping.

Wastewater Treatment

PURPOSE To clarify and purify water. The measure of purification is biological oxygen demand (BOD).

LEVELS
1. *Primary*—Solids are separated by settling and some aeration to remove organic materials.

2. *Secondary*—BOD is reduced by bacterial action and aeration.

3. *Tertiary*—Inorganic nutrients such as nitrates and phosphates are removed.

PRIVATE SEWAGE DISPOSAL Requires permeability and porosity and can lead to local pollution.

KEY TERMS

attenuation (waste disposal)
biological oxygen demand (BOD)
composting
deep-well injection
EPA
eutrophication
isolation (waste disposal)
leachate

NEPA
RCRA
recycling
resource recovery
sanitary landfill
secure landfill
Superfund
waste management

STUDY QUESTIONS

1. Explain the federal legislation known as NEPA and RCRA and describe their impacts on the environmental safety of landfills.

2. What is leachate, and what methods are used to prevent landfill leachates from entering bodies of water?

3. Explain the "cell" method of placing trash in a landfill.

4. How much solid waste does the average American generate each day in kilograms? In pounds?

5. What problems are associated with operating a sanitary landfill?

6. What resources are found in a landfill, and how can they be recovered? What can individuals do to assist in recov-

ering and extending the lifetimes of certain limited natural resources?

7. Describe two common methods of isolating hazardous wastes from the environment. Explain the steps that are taken to keep these wastes from polluting underground water.

8. What is eutrophication, and what can be done in sewage treatment to minimize this threat to lakes and rivers?

FURTHER INFORMATION

BOOKS AND PERIODICALS

Alexander, J. H. 1993. *In defense of garbage.* New York: Praeger.

Brunner, D. R., and D. K. Keller. 1972. *Sanitary landfill design and operations.* Washington, D.C.: U.S. Environmental Protection Agency.

Grove, Noel. 1994. Recycling. *National geographic* 186, no. 1 (July).

Lewin, David. 1991. Solid waste: Bury, burn, or reuse. *Mechanical engineering,* August, p. 75.

McKinney, Michael, and Robert Schoch. 1996. *Environmental science.* St. Paul, Minn.: West Educational Publishing.

Rathje, William L. 1991. Once and future landfills. *National geographic,* May, pp. 114–134.

Rathje, William L., and Cullen Murphy. 1992. *Rubbish! The archeology of garbage.* New York: Harper Collins.

U.S. Environmental Protection Agency. 1992. *Characterization of municipal solid waste in the United States: Final report.*

White, Peter T. 1983. The wonderful world of trash. *National geographic,* April.

World Resources Institute. 1992. *World resources, 1991–1992.* New York: Oxford University Press.

1 THE PERIODIC TABLE

Period	IA																	Noble Gases
1	1 **H** 1.008	IIA										IIIA	IVA	VA	VIA	VIIA		2 **He** 4.003
2	3 **Li** 6.941	4 **Be** 9.012											5 **B** 10.81	6 **C** 12.01	7 **N** 14.01	8 **O** 16.00	9 **F** 19.00	10 **Ne** 20.18
3	11 **Na** 22.99	12 **Mg** 24.31	IIIB	IVB	VB	VIB	VIIB	VIII		IB	IIB		13 **Al** 26.98	14 **Si** 28.09	15 **P** 30.97	16 **S** 32.06	17 **Cl** 35.45	18 **Ar** 39.95
4	19 **K** 39.10	20 **Ca** 40.08	21 **Sc** 44.96	22 **Ti** 47.90	23 **V** 50.94	24 **Cr** 52.00	25 **Mn** 54.94	26 **Fe** 55.85	27 **Co** 58.93	28 **Ni** 58.71	29 **Cu** 63.85	30 **Zn** 65.37	31 **Ga** 69.72	32 **Ge** 72.59	33 **As** 74.92	34 **Se** 78.96	35 **Br** 79.90	36 **Kr** 83.80
5	37 **Rb** 85.47	38 **Sr** 87.62	39 **Y** 88.91	40 **Zr** 91.22	41 **Nb** 92.91	42 **Mo** 95.94	43 **Tc**^{x} 98.91	44 **Ru** 101.1	45 **Rh** 102.9	46 **Pd** 106.4	47 **Ag** 107.9	48 **Cd** 112.4	49 **In** 114.8	50 **Sn** 118.7	51 **Sb** 121.8	52 **Te** 127.6	53 **I** 126.9	54 **Xe** 131.3
6	55 **Cs** 132.9	56 **Ba** 137.3	57 **La** 138.9	72 **Hf** 178.5	73 **Ta** 180.9	74 **W** 183.9	75 **Re** 186.2	76 **Os** 190.2	77 **Ir** 192.2	78 **Pt** 195.1	79 **Au** 197.0	80 **Hg** 200.6	81 **Tl** 204.4	82 **Pb** 207.2	83 **Bi** 209.0	84 **Po**^{x} (210)	85 **At**^{x} (210)	86 **Rn**^{x} (222)
7	87 **Fr**^{x} (223)	88 **Ra**^{x} 226.0	89 **Ac**^{x} (227)															

Atomic Number — 1
Symbol of Element — H
Atomic Mass — 1.008
(rounded to four significant figures)

● Lanthanide series

58 **Ce** 140.1	59 **Pr** 140.9	60 **Nd** 144.2	61 **Pm**^{x} (147)	62 **Sm** 150.4	63 **Eu** 152.0	64 **Gd** 157.3	65 **Tb** 158.9	66 **Dy** 162.5	67 **Ho** 164.9	68 **Er** 167.3	69 **Tm** 168.9	70 **Yb** 173.0	71 **Lu** 175.0
90 **Th**^{x} 232.0	91 **Pa**^{x} 231.0	92 **U**^{x} 238.0	93 **Np**^{x} 237.0	94 **Pu**^{x} (244)	95 **Am**^{x} (243)	96 **Cm**^{x} (247)	97 **Bk**^{x} (247)	98 **Cf**^{x} (251)	99 **Es**^{x} (254)	100 **Fm**^{x} (257)	101 **Md**^{x} (258)	102 **No**^{x} (255)	103 **Lr**^{x} (256)

■ Actinide series

x: All isotopes are radioactive.

( ) indicates mass number of isotope with longest known half-life.

The Elements

ELEMENT	SYMBOL	ATOMIC NUMBER	ELEMENT	SYMBOL	ATOMIC NUMBER
Actinium	Ac	89	Mercury	Hg	80
Aluminum	Al	13	Molybdenum	Mo	42
Americium	Am	95	Neodymium	Nd	60
Antimony	Sb	51	Neon	Ne	10
Argon	Ar	18	Neptunium	Np	93
Arsenic	As	33	Nickel	Ni	28
Astatine	At	85	Niobium	Nb	41
Barium	Ba	56	Nitrogen	N	7
Berkelium	Bk	97	Nobelium	No	102
Beryllium	Be	4	Osmium	Os	76
Bismuth	Bi	83	Oxygen	O	8
Boron	B	5	Palladium	Pd	46
Bromine	Br	35	Phosphorus	P	15
Cadmium	Cd	48	Platinum	Pt	78
Calcium	Ca	20	Plutonium	Pu	94
Californium	Cf	98	Polonium	Po	84
Carbon	C	6	Potassium	K	19
Cerium	Ce	58	Praseodymium	Pr	59
Cesium	Cs	55	Promethium	Pm	61
Chlorine	Cl	17	Protactinium	Pa	91
Chromium	Cr	24	Radium	Ra	88
Cobalt	Co	27	Radon	Rn	86
Copper	Cu	29	Rhenium	Re	75
Curium	Cm	96	Rhodium	Rh	45
Dysprosium	Dy	66	Rubidium	Rb	37
Einsteinium	Es	99	Ruthenium	Ru	44
Erbium	Er	68	Samarium	Sm	62
Europium	Eu	63	Scandium	Sc	21
Fermium	Fm	100	Selenium	Se	34
Fluorine	F	9	Silicon	Si	14
Francium	Fr	87	Silver	Ag	47
Gadolinium	Gd	64	Sodium	Na	11
Gallium	Ga	31	Strontium	Sr	38
Germanium	Ge	32	Sulfur	S	16
Gold	Au	79	Tantalum	Ta	73
Hafnium	Hf	72	Technetium	Tc	43
Helium	He	2	Tellurium	Te	52
Holmium	Ho	67	Terbium	Tb	65
Hydrogen	H	1	Thallium	Tl	81
Indium	In	49	Thorium	Th	90
Iodine	I	53	Thulium	Tm	69
Iridium	Ir	77	Tin	Sn	50
Iron	Fe	26	Titanium	Ti	22
Krypton	Kr	36	Tungsten	W	74
Lanthanum	La	57	Uranium	U	92
Lawrencium	Lr	103	Vanadium	V	23
Lead	Pb	82	Xenon	Xe	54
Lithium	Li	3	Ytterbium	Yb	70
Lutetium	Lu	71	Yttrium	Y	39
Magnesium	Mg	12	Zinc	Zn	30
Manganese	Mn	25	Zirconium	Zr	40
Medelevium	Md	101			

2 MINERAL TABLE

	MINERAL	FORMULA	COLOR	HARDNESS	OTHER CHARACTERISTICS	PRIMARY OCCURRENCE
1	amphibole (e.g., hornblende) (hydrous)	Na, Ca, Mg, Fe Al silicate	green, blue brown, black	5 to 6	often forms needlelike crystals; two good cleavages forming 60°–120°-degree angle crystals hexagonal in cross-section	Igneous and Metamorphic rocks; Minor amounts in all rock types
2	apatite (F, Cl, OH)	$Ca_5(PO_4)_3$	blue-green cross-section	5 rocks	crystals hexagonal in	Minor amounts, all
3	biotite (a mica)	Hydrous K, Mg, Fe silicate	brown to black	2 ½–3 thin sheets	excellent cleavage into	All rock types
4	calcite	$CaCO_3$	variable; colorless if pure	3	effervesces in weak acid; cleaves into rhombohedral fragments	Sedimentary and metamorphic rocks
5	corundum	Al_2O_3	variable; colorless in pure form	9	most readily identified by its hardness	Metamorphic rocks
6	dolomite	$CaMg(CO_3)_2$	white or pink	3½ to 4	powdered mineral effervesces in acid	Sedimentary rocks
7	galena	PbS	silver-gray	2½	metallic luster; cleaves into cubes; high specific gravity	Sulfide veins
8	garnet	Ca, Mg, Fe, Al silicate	variable; often dark red	7	glassy luster	Metamorphic rocks
9	graphite	C	dark gray	1 to 2	streaks like pencil lead	Metamorphic rocks
10	gypsum	$CaSO_4 \cdot 2H_2O$	colorless	2	2 directions of cleavage	Sedimentary rocks
11	halite	NaCl	colorless	2½	salty taste; cleaves into cubes	Sedimentary rocks
12	hematite	Fe_2O_3	red or dark gray	5½ to 6½	red-brown streak regardless of color	Oxidation of iron-bearing minerals
13	kaolinite	Hydrous Al silicate	white feel to the touch	2	earthy luster, smooth	Sedimentary rocks
14	limonite	$Fe_2O_3 \cdot 3H_2O$	yellow-brown	2 to 3	earthy luster; yellow-brown streak	Hydration of iron oxides
15	magnetite	Fe_3O_4	black rock types	6	strongly magnetic	Minor amounts, all
16	muscovite	K, Al, Si, silicate, hydrated	colorless to light brown	2 to 2 ½	excellent cleavage into thin sheets	All rock types
17	olivine	Fe, Mg silicate	yellow-green metamorphic rocks	6 ½ to 7	glassy luster	Igneous and
18	plagioclase feldspar	Na, Ca, Al silicate	wite to gray; some compositions iridescent blue-gray	6	2 cleavages at 90°, striations on cleavage surfaces	All rock types

MINERAL	FORMULA	COLOR	HARDNESS	OTHER CHARACTERISTICS	PRIMARY OCCURRENCE
19 orthoclase feldspar	K, Al silicate	white; or colored pink or aqua	6	two good cleavages forming a 90-degree angle; no striations	All rock types
20 pyrite	FeS_2	brassy yellow	6 to 6½	metallic luster; black streak	All rock types
21 pyroxene (e.g., augite)	Na, Ca, Mg, Fe, All silicate	usually green or black	5 to 7	two good cleavages forming a 90-degree angle	Igneous and metamorhpic rocks
22 quartz	SiO_2	variable; commonly colorless or white	7	glassy luster; conchoidal fracture; crystals hexagonal in cross section	All rock types

3 PLANES IN SPACE— DIP AND STRIKE

Geologists describe the orientation of a plane surface by the plane's "strike" and "dip." Geologists measure such tangible planes as those made by sedimentary strata, foliation, and faults, because it allows them to anticipate hazards during excavations for such things as tunnels, dam foundations, and roadcuts. Accurate surface measurements of planes also enable geologists to predict the underground locations of valuable resources such as oil reservoirs, ore deposits, and water-bearing strata.

Strike is the direction of the trace of an inclined plane on the horizontal plane, and the direction is referenced to north. In the illustrated example, the strike is 30° west of north. In geologic notation, this is written *N30°W*.

Dip is the vertical angle that the same inclined plane makes with the horizontal plane. Dip is assessed perpendicular to the strike, because any other angle will give an "apparent dip" that is less than the true dip. Because a plane that strikes northwest may be inclined to either the southwest or the northeast, the general direction of dip must be indicated. In our example, the dip is 60° to the northeast, which is notated *60°NE*.

Thus, a geologist's notes for the bedding planes in this example would describe their orientation as "strike N30°W, dip 60°NE." All other geologists would understand that the plane has a strike that is 30° west of north and a dip that is 60° to the northeast.

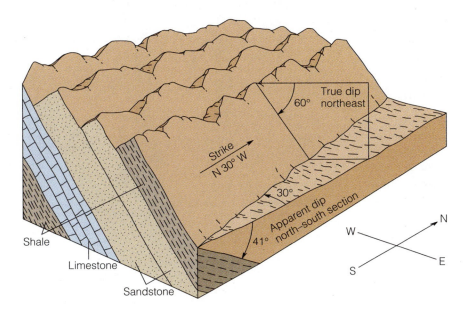

Block diagram showing sedimentary rock that is striking N30°W and dipping 60° northeast. The apparent dip is shown on the north–south cross section at the lower front of the diagram.

APPENDIX

4 PORES, FLOW, AND FLOATING ROCKS

Porosity is the ratio of the volume of pore spaces in a rock or other solid to the material's total volume. Expressed as a percentage, it is

$$P = \frac{V_p}{V_t} \times 100$$

where P = porosity
V_p = pore volume
V_t = total volume

If an aquifer has a porosity of 33 percent, a cubic meter of the aquifer is composed of ⅔ cubic meter of solid material and ⅓ cubic meter of pore space, which may or may not be filled with water. Porosities of 20 to 40 percent are not uncommon in clastic sedimentary rocks.

Permeability is a measure of how fast fluids can move through a porous medium and is thus expressed in a unit of velocity, such as meters per day. Knowing the permeability of an aquifer allows us to determine in advance how much water a well will produce and how far apart wells should be spaced. To determine permeability we use **Darcy's Law,** which states that the rate of flow (Q in cubic meters) is the product of hydraulic conductivity (K), the **hydraulic gradient,** or the slope of the water table (I in meters per meter), and the cross-sectional area (A in square meters). Thus,

Q	=	K	×	I	×	A
rate of flow		hydraulic conductivity		hydraulic gradient		cross-sectional area

Hydraulic conductivity, K, is a measure of the permeability of a rock or a sediment, its water-transmitting characteristics. It is defined as the quantity of water that will flow in a unit of time under a unit hydraulic gradient through a unit of area measured perpendicular to the flow direction. K is expressed in distance of flow with time, usually meters per day.

If we use a unit hydraulic gradient, then Darcy's Law reduces to:

$$Q = K \times A$$

Unit hydraulic gradient is the slope of 1-unit vertical to 1-unit horizontal ($I = 1/1 = 45°$), and it allows us to compare the water-transmitting ability of aquifers composed of materials with different permeabilities. Flow rates may be less than a millimeter per day for clays, many meters per day for well-sorted sands and clean gravels. In general, ground water moves at a snail's pace.

Pumice provides an interesting example. It is a porous but almost impermeable rock. It has porosities as high as 90 percent but few connected pore spaces. Pumice will float for a time because it is light-weight and its low permeability prevents immediate saturation. A cubic meter of pumice may weigh as little as 500 pounds, a fifth of the weight of solid rock of the same composition.

The unit hydraulic gradient (I) and unit prism (1 m²) of a sandstone aquifer. When the hydraulic conductivity (K) is known, then the discharge (Q) in liters or cubic meters per day per square meter of aquifer can be calculated.

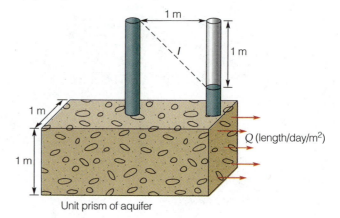

Unit prism of aquifer

5 INTERNET SOURCES OF INFORMATION

The World Wide Web (www) is part of a network of large interlinking computers around the globe. It is an information retrieval system that allows you to access information on computers connected to the Internet using an interactive user interface called a Browser. The Web has the capacity to present graphics and is an ideal place to search for photographs and figures. Searching (or surfing) the Web can be time consuming, particularly if your computer is not particularly fast. Therefore, we present here some dependable sites that are almost always on-line (available), are easy to access, and inevitably lead to exploring other subject areas. Nevertheless, virtually all sites on the Internet are subject to change. If you have difficulty trying to connect with any of the following sites listed, look for a change of address file. If no change of address file is listed try connecting again at a later time.

Here are a few definitions to help you if you are not an experienced surfer on the Internet.

- *URL*—Uniform Resource Locator is the Internet address for documents located on the Internet.
- *HTML*—Hypertext Markup Language is a format used for creating hypertext documents on the World Wide Web.
- *HTTP*—Hypertext Transport Protocol is an information retrieval mechanism for HTML documents.
- *FTP*—File Transfer Protocol is the standard for downloading files from one computer to another.

Opening Web addresses leads to home pages (cover sheets) of organizations that supply information. The Web sites begin with (http://www.) and this portion is followed by initials of the organization, such as (http://www.usgs.gov), which is the address of the U.S. Geological Survey. The parentheses are used here only to isolate the address from the adjacent text and are not part of the address. For example, the U.S.G.S. Web site is probably your best bet for general environmental geologic data and information. The U.S.G.S. offers four "Theme" sites which are Hazards, Natural Resources, Environment, and Information Management. After clicking on Themes, click on Hazards and you get a choice of earthquakes, floods, landslides, coastal storms and tsunamis, volcanoes, and further information. As you investigate more deeply into a subject the offerings become more specific—for example, you can explore a particular earthquake, volcano, landslide, or flood.

While surfing the Internet can be fun, it is also a bit like taking a cross-country car trip without a road map or a travel guide. To be sure, there will be signs along the way, but you may prefer to have a more general overview of some of the possible routes or a description of a few of the most interesting things to see. For such Internet users, West Publishing offers an excellent guide for exploring some of the geology-related offramps on the Internet. *Geology and the Environment—An Internet-Based Resource Guide* by John C. Butler (University of Houston) is a disk-based Internet exploration tool containing files that correspond to the chapters in this text, as well as several appendices on other general topics in geology. The files have been prepared for viewing using the Netscape browser, although an attempt has been made to limit the code to those features (such as tables) that are supported by other commonly used browsers. The Guide contains both internal links between the chapter files and external links to outside Web sites. Many chapters contain exercises related to exploration of environmental geology-related sites on the Internet.

We have listed the following sites that we consider to be very good, but we are sure you can find many more sites as you surf the web:

GENERAL INFORMATION. The U.S.G.S. site (http://www.usgs.gov) is excellent for basic geologic information.

Yahoo (http://www.yahoo.com/) is an electronic index that attempts to collect and organize scattered earth science information.

The site of the National Geophysical Data Center (http://www.ngdc.noaa.gov/) offers stunning images and photo collections.

EARTHQUAKES. The National Earthquake Information Center (http://gldfs.cr.usgs.gov) has a real-time listing of earthquakes worldwide over the past few weeks. In addition, there is a listing of the world's most damaging earthquakes and an explanation of plate tectonics.

The U.S.G.S. (http://quake.wr.usgs.gov) site also contains very useful earthquake information.

For the state of California try the Southern California Earthquake Center (http://scec.gps.caltech.edu).

Alaskans might find the great maps useful at (http://giseis.alaska.edu/Seis/).

TSUNAMIS. As part of a global research project try this site for tsunami information. (http://www.geophys.washington.edu/tsunami/welcome.html)

VOLCANOES. Volcano World (http://volcano.und.nodak.edu/) is both an educational and entertaining site. It is maintained by the University of North Dakota and has an "Ask a Volcanologist" feature.

Michigan Technological Institute (http://www.geo.mtu.edu/volcanoes/) lists continuing and renewed volcanic activity.

A site known as Virtually Hawaii (http://www.satlab.hawaii.edu/space/hawaii/) has "virtual" field trips by land or air to Hawaii's great volcanoes. A fun site.

LANDSLIDES. Try the U.S.G.S. site (http://www.usgs.gov), click on Themes, Hazards and then Landslides.

RIVERS. The U.S.G.S. site (http://www.usgs.gov), click on Themes, Hazards and then Floods.

UNDERGROUND WATER. Try the U.S.G.S. site (http://www.usgs.gov), click on Themes, Resources and then Water Resources.

COASTAL PROCESSES. The U.S.G.S. home page (http://www.usgs.gov), click on Themes, then click on Hazards, and coastal floods and tsunami.

ENERGY. The U.S. Department of Energy (http://www.eren.doe.gov) maintains a site known as EREN that has all kinds of information on renewable energy sources. EREN is an acronym for "energy renewable."

Try the U.S.G.S. site (http://www.usgs.gov), click on Themes, Resources, and finally Energy.

MINERAL RESOURCES. Open the U.S.G.S. site (http://www.usgs.gov) and click on Themes, Resources, and then on Minerals.

GLOSSARY

aa A basaltic lava that has a rough, blocky surface.

ablation All processes by which snow or ice is lost from a glacier, floating ice, or snow cover. These processes include melting, evaporation, and calving.

ablation zone The part of a glacier or a snowfield in which ablation exceeds accumulation over a year's time; the region below the firn line.

abrasion The mechanical grinding, scraping, wearing, or rubbing away of rock surfaces by friction or impact. Abrasion agents include wind, running water, waves, and ice.

absolute age dating Assigning units of time, usually years, to fossils, rocks, and geologic features or events. The method utilizes mostly the decay of radioactive elements, but ages may also be obtained by counting tree rings or annual sedimentary layers.

accretion The mechanism by which the earth is believed to have formed from a small nucleus by additions of solid bodies, such as meteorites, asteroids, or planetesimals. Also used in plate tectonics to indicate the addition of terranes to a continent.

accumulation zone The part of a glacier or snowfield in which accumulation exceeds ablation over a year's time; the region above the firn line. Also the B-horizon of a soil.

acid mine drainage (AMD) The acid runoff from a mine or mine waste.

acid rain Corrosive precipitation due to the solution of atmospheric sulfuric acid in rain, which makes it acidic. Although it is generally considered human-made, it can also be natural, as from volcanic activity.

active layer The zone in permafrost that is subject to seasonal freezing and thawing.

active sea cliff A cliff whose formation is dominated by wave erosion.

adsorption The attraction of ions or molecules to the surface of particles, in contrast to *absorption*, the process by which a substance is taken into a clay-mineral structure.

aftershocks A series of earthquakes following the first or main shock. They are usually of lower magnitude than the main shock.

air fall *See* ashfall.

alluvial fan A gently sloping, fan-shaped mass of alluvium deposited in arid or semiarid regions where a stream flows from a narrow canyon onto a plain or valley floor.

alluvial placer A placer deposit that is concentrated in a streambed.

alluvium Stream-deposited sediment.

alpha decay The transformation that occurs when the nucleus of an atom spontaneously emits a helium atom, reducing its atomic mass by 4 units and its atomic number by 2; for example, $^{238}_{92}U \rightarrow \,^{234}_{90}Th + \,^{4}_{2}a$.

amorphous Descriptive term applied to glasses and other substances that lack an orderly internal crystal structure.

andesite line The petrographic boundary between basalts and andesites. It runs roughly around the Pacific basin margin. It also marks the boundary between continental (explosive) and oceanic (Hawaiian-type) volcanoes in the Pacific Ocean.

angle of repose The maximum stable slope angle that granular, cohesionless material can assume. For dry sand this is about 34°.

anomaly A deviation from the average or expected value. In paleomagnetism, a "positive" anomaly is a stronger-than-average magnetic field at the earth's surface.

anticline A generally convex-upward fold that has older rocks in its core. (*Contrast with* syncline.)

aphanitic A fine-grained igneous-rock texture resulting from rapid cooling. Aphanites are fine-grained igneous rocks.

A.P.I. gravity (American Petroleum Institute) The weight of crude oil arbitrarily compared to that of water at 60° F, which is assigned a value of 10°. Most crude oils range from 5° to 30° A.P.I.

aquiclude A layer of low-permeability rock that hampers ground-water movement. Often applied to the confining bed in a confined aquifer.

aquifer A water-bearing body of rock or sediment that will yield water to a well or spring in usable quantities.

area mining The mining of coal in flat terrain (strip mining).

artificial levee An embankment constructed to contain a stream at flood stage. *See* levee.

ashfall (air fall) A "rain" of airborne volcanic ash.

ash flow (nuée ardente) A highly heated, dense cloud of volcanic gases and pyroclastic material that travels with great velocity down the slope of a volcano; produced by explosive disintegration of viscous lava in a vent.

ash Very fine, gritty volcanic dust: particles are less than 4 mm in diameter.

asperity Rough spot or "knot" along a fault that stops movement along that section for a period of time. The bigger the asperity, the greater the stress buildup and the larger the earthquake that results. (*See also* seismic gap.)

asthenosphere The *plastic* layer of the earth, at whose upper boundary the brittle overlying lithosphere is detached from the underlying mantle.

atom The smallest particle of an element that can exist either alone or in combination with similar particles of the same or a different element.

atomic number The number of protons (positively-charged particles) in an atomic nucleus. Atoms with the same atomic number belong to the same chemical element.

atomic mass The sum of the number of protons and neutrons (uncharged particles) in the nucleus of an atom.

attenuation (waste disposal) To dilute or spread out the waste material.

average crustal abundance The percentage of a particular element in the composition of the earth's crust.

a-zone The upper soil horizon (topsoil) capped by a layer of organic matter. It is the zone of leaching of soluble soil salts.

back slope The gentler slope of a dune, scarp, or fault block.

ball mill A large rotating drum containing steel balls that grinds ore to a fine powder. Batteries of ball mills are used in modern milling operations.

banded-iron formation An iron deposit consisting essentially of iron oxides and chert occurring in prominent layers or bands of brown or red and black.

barchan dune An isolated, moving, crescent-shaped sand dune lying transverse to the direction of the prevailing wind. A gently sloping, convex side faces the wind, and the horns of the crescent point downwind.

barrier beach (barrier island) A narrow sand ridge rising slightly above hightide level that is oriented generally parallel with the coast and separated from the coast by a lagoon or a tidal marsh.

basalt A dark-colored, fine-grained volcanic rock that is commonly called *lava*. Basaltic lavas are quite fluid.

base level The level below which a stream cannot erode its bed. The ultimate base level is sea level.

base shear Due to a rapid horizontal ground acceleration during an earthquake. It can cause a building to part from its foundation; thus the term.

batholith A large, discordant intrusive body greater than 100 km^2 (40 mi^2) in surface area with no known bottom. Batholiths are granitic in composition.

bauxite An off-white, grayish, brown, or reddish brown rock composed of a mixture of various hydrous aluminum oxides and aluminum hydroxides.

beach cusps A series of regularly spaced points of beach sand or gravel separated by crescent-shaped troughs.

bedding *See* stratification.

bedding planes The divisions or surfaces that separate successive layers of sedimentary rocks from the layers above and below them.

bench placer Remnant of an ancient alluvial placer deposit that was concentrated beneath a former stream and that is now pre-

served as a bench on a hillside.

Benioff zone A zone of earthquake epicenters beneath oceanic trenches that extends from near the surface to a depth of 700 km (430 mi). The zone is named after Hugo Benioff and is thought to indicate active subduction.

bentonite A soft, plastic, light-colored clay that swells to many times its volume when placed in water. Named for outcrops near Fort Benton, Wyoming, bentonite is formed by chemical alteration of volcanic ash.

beta decay The radioactive emission of a nuclear electron (beta ray, β^-) that converts a neutron to a proton. The element is changed to the element of next-highest atomic number; for example, $^{14}_{6}C \rightarrow ^{14}_{7}N + \beta^-$.

big bang The hypothesis that the currently observed expansion of the universe may be traced back to an exploding compact cosmic fireball.

biogenic sedimentary rock Rocks resulting directly from the activities of living organisms, such as a coral reef, shell limestone, or coal.

biological oxygen demand (B.O.D.) Indicator of organic pollutants (microbial) in an effluent measured as the amount of oxygen required to support them. The greater the B.O.D. (ppm or mg/liter) the greater the pollution and less oxygen available for higher aquatic organisms.

biomass fuels Fuels that are manufactured from plant materials; e.g., grain alcohol (ethanol) is made from corn.

bioremediation The use of microbes to clean hydrocarbon residues from contaminated soils.

black smoker A submarine hydrothermal vent associated with a mid-oceanic spreading center that has emitted or is emitting a "smoke" of metallic sulfide particles.

block (volcanology) A mass of cold country rock greater than 32 mm (1¼ in) in size that is ripped from the vent and ejected from a volcano.

block glide (translational slide) A landslide in which movement occurs along a well-defined plane surface or surfaces, such as bedding planes, foliation planes, or faults.

body wave Earthquake waves that travel through the body of the earth.

bomb (volcanology) A molten mass greater than 32 mm (1¼ in) in diameter that is ejected from a volcano and cools sufficiently before striking the ground to retain a rounded, streamlined shape.

breakwater An offshore rock or concrete structure, attached at one end or parallel to the shoreline, intended to create quiet water for boat anchorage purposes.

breeder reactor A reactor that produces more fissile (split- table) nuclei than it consumes.

British thermal unit (Btu) The amount of heat, in calories, needed to raise the temperature of 1 pound of water 1° F.

caldera Large crater caused by explosive eruption and/or subsidence of the cone into the magma chamber, usually more than 1.5 km (1 mi) in diameter.

caliche In arid regions, a sediment or soil that is cemented by calcium carbonate ($CaCO_3$).

calving The breaking away of a block of ice, usually from the front of a tidewater glacier, that produces an iceberg.

caprock An impermeable stratum that caps an oil reservoir and prevents oil and gas from escaping to the ground surface.

capillary fringe The moist zone in an aquifer above the water table. The water is held in it by capillary action.

capillary water Water that is held in tiny openings in rock or soil by capillary forces.

carrying capacity The number of creatures a given tract of land can adequately support with food, water, and other necessities of life.

central-core meltdown The melting of an atomic-reactor fuel containment vessel as a result of runaway atomic fission that generates extreme heat.

CERCLA The Comprehensive Environmental Response, Compensation and Liability Act of 1980. *See* Superfund.

chain reaction A self-sustaining nuclear reaction that occurs when atomic nuclei undergo fission. Free neutrons are released that split other nuclei, causing more fissions and the release of more neutrons, which split other nuclei, and so on.

chemical sedimentary rocks Rock resulting from precipitation of chemical compounds from a water solution. *See* evaporite.

chlorofluorocarbons (CFCs) A group of compounds that, on escape to the atmosphere, break down, releasing chlorine atoms that destroy ozone molecules; e.g., Freon, which is widely used in refrigeration, air-conditioning, and formerly was a propellant in aerosol-spray cans.

cinder cone A small, straight-sided volcanic cone consisting mostly of pyroclastic material (usually basaltic).

cinders Glassy and vesicular (containing gas holes) volcanic fragments, only a few centimeters across, that are thrown from a volcano.

clastic Describing rock or sediment composed primarily of detritus of preexisting rocks or minerals.

clay minerals Hydrous aluminum silicates that have a layered atomic structure. They are very fine grained and become plastic (moldable) when wet. Most belong to one of three clay groups: kaolinite, illite, and smectite, the last of which are expansive when they absorb water between layers.

claypan A dense, relatively impervious subsoil layer that owes its character to a high clay-mineral content due to concentration by downward-percolating waters.

cleavage The breaking of minerals along certain crystallographic planes of weakness that reflects their internal structure.

coastal desert A desert on the western edge of continents in tropical latitudes; i.e., near the Tropic of Capricorn or Cancer. The daily and annual temperature fluctuations are much less than in inland tropical deserts.

columnar jointing Vertically oriented polygonal columns in a solid lava flow or shallow intrusion formed by contraction upon cooling. They are usually formed in basaltic lavas.

compaction The reduction in bulk volume of fine-grained sediments due to increased overburden weight or to tighter packing of grains. Clay is more compressible than is sand and therefore compacts more.

complex landslide When both rotational and translational sliding are found within one landslide mass.

composting Bacterial recycling of organic material.

concentrate Enriched ore, often obtained by flotation, produced at a mill.

concentration factor The enrichment of a deposit of an element, expressed as the ratio of the element's abundance in the deposit to its average crustal abundance.

cone of depression The cone shape formed by the water table around a pumping well when pumping exceeds the flow of water to the well.

confined aquifer A water-bearing formation bounded above and below by impermeable beds or beds of distinctly lower permeability.

contact metamorphism A change in the mineral composition of rocks in contact with an invading magma from which fluid constituents are carried out to combine with some of the country-rock constituents to form a new suite of minerals.

continental divide A drainage divide that separates streams flowing toward opposite sides of a continent.

continental drift The term applied by American geologists to Wegener's hypothesis of long-distance horizontal movement of continents he called "continental displacement".

contour mining The mining of coal by cutting into and following the coal seam around the perimeter of the hill.

convection The circulatory movement within a body of nonuniformly heated fluid. Warmer material rises in the center, and cooler material sinks at the outer boundaries. It has been proposed as the mechanism of plate motion in plate tectonic theory.

convergent boundary Tectonic boundary where two plates collide. Where a tectonic plate sinks beneath another plate and is destroyed, this boundary is known as a *subduction zone*. Subduction zones are marked by deep-focus seismic activity, strong earthquakes, and violent volcanic eruptions.

core The central interior of the earth beginning at a depth of 2,900 km (1,800 mi). It probably consists of an iron–nickel alloy, with an outer liquid part and a solid interior.

Coriolis effect A condition peculiar to motion on a rotating body such as the earth. It appears to deflect moving objects in the Northern Hemisphere to the right and objects in the Southern Hemisphere to the left, to an observer on the rotating body.

craton A part of the earth's continental crust that has attained stability—that has not been deformed for a long period of time. The term is restricted to continents and includes their most stable areas, the continental shields.

creep (soil) The imperceptibly slow downslope movement of rock and soil particles by gravity.

cross-bedding Arrangement of strata, greater than 1 cm thick, inclined at an angle to the main stratification.

crust The outermost layer of the earth consisting of granitic continental crust and basaltic oceanic crust; that part of the earth above the Mohorovičić discontinuity.

crystal settling In a magma, the sinking of crystals because of their greater density, sometimes aided by magmatic convection. (*See* gravity separation).

cubic feet per second (cfs) Unit of water flow or stream discharge. One cubic foot per second is 7.48 gallons passing a given point in one second.

cutoff A new channel that is formed when a stream cuts through a very tight meander. (*See* oxbow lake).

cyanide heap leaching A process of using cyanide to dissolve and recover gold and silver from ore.

c-zone Soil zone of weathered parent material leading downward to fresh bedrock.

Darcy's Law A derived formula for the flow of fluids in a porous medium.

debris avalanche A sudden, rapid movement of a water–soil–rock mixture down a steep slope.

debris flow A moving mass of a water, soil, and rock mixture. More than half of the soil and rock particles are coarser than sand, and the mass has the consistency of wet concrete.

deflation The lifting and removal of loose, dry, fine-grained particles by the action of wind.

delta The nearly flat land formed where a stream empties into a body of standing water, such as a lake or the ocean.

desert An arid region of low rainfall, usually less than 25 cm (10 in) annually, and of high evaporation or extreme cold. Deserts are generally unsuited for human occupation under natural conditions.

desert pavement An interlocking cobble mosaic that remains on the earth surface in a dry region after winds have removed all fine-grained sediments.

desertification The process by which semiarid grasslands are converted to desert.

detrital *See* clastic.

detrital sedimentary rocks Rocks composed of fragments of minerals or other rock materials.

dilatancy–diffusion model A model for explaining certain earthquake precursory phenomena, for example changes in P-wave velocities.

dip It is the angle that an inclined geological planar surface (sedimentary bedding plane, fault, joint) makes with the horizontal plane.

discharge (Q) The volume of water, usually expressed in cfs, that passes a given point within a given period of time.

discontinuity In solid earth geophysics this means a chemical of phase change in materials with depth. At discontinuities the earth materials become denser and earthquake waves change velocity.

disseminated deposit A mineral deposit, especially of a metal, in which the minerals occur as scattered particles in the rock, but in sufficient quantity to make the deposit a commercially worthwhile ore.

divergent boundary Found at mid-ocean ridge or spreading center where new crust is created as plates move apart. For this reason they are sometime referred to as *constructive* boundaries and are the sites of weak shallow-focus earthquakes and volcanic action.

doldrums The area of warm rising air, calm winds, and heavy precipitation between 5° north latitude and 5° south latitude. It is where the northeast and southeast tradewinds meet.

doubling time The number of years required for a population to double in size.

drainage basin The tract of land that contributes water to a particular stream, lake, reservoir, or other body of surface water.

drainage divide The boundary between two drainage basins (*see* continental divide).

drawdown The lowering of the water table immediately adjacent to a pumping well.

dredging The excavation of earth material from the bottom of a body of water by a floating barge or raft equipped to scoop up, discharge by conveyors, and process or transport materials.

drift A general term for all rock material that is transported and deposited by a glacier or by running water emanating from a glacier.

dripstone A general term for calcite or other minerals deposited in caves by dripping water.

driving force (landslide) Gravity.

dune A mound, bank, ridge, or hill of loose, windblown granular material (generally sand), either bare or covered with vegetation. It is capable of movement, but maintains its characteristic shape.

Dust Veil Index (DVI) A measure of the impact of a volcanic eruption on incoming solar radiation and climate.

eccentricity The deviation of the earth's path around the sun from a perfect circle.

effective stress The average normal force per unit area (stress) that is transmitted directly across grain-to-grain boundaries in a sediment or rock mass.

effluent stream A stream that receives water from the zone of saturation.

elastic Said of a body in which strains are totally recoverable, as in a rubber band. (*Contrast with* plastic).

Elastic Rebound Theory The theory that movement along a fault and the resulting seismicity is the result of an abrupt release of stored elastic strain energy between two rock masses on either side of the fault.

element A substance that cannot be separated into different substances by usual chemical means.

end moraine (terminal moraine) A moraine that is produced at the front of an actively flowing glacier; a moraine that has been deposited at the lower end of a glacier.

engineering geology The application of the science of geology to human engineering works.

ENSO Abbreviation for the periodic meteorological event known as El Niño–Southern Oscillation.

environmental geology The relationship between humans and their geological environment.

eolian Of, produced by, or carried by wind.

epicenter The point at the surface of earth directly above the focus of an earthquake.

equatorial low The belt of low-pressure air masses near the equator. (*See* doldrums.)

erosion The weathering and transportation of the materials of the earth's surface.

erratic A rock fragment that has been carried by glacial ice or floating ice and deposited when the ice melted at some distance from the outcrop from which it was derived.

estuary A semienclosed coastal body where outflowing river water meets tidal sea water.

eutrophication The increase in nitrogen, phosphorous and other plant nutrients in the aging of an aquatic system. Blooms of algae develop preventing light penetration, and causing reduction of oxygen needed in a healthy system.

evaporite A nonclastic sedimentary rock composed primarily of minerals produced when a saline solution becomes concentrated by evaporation of the water; especially a deposit of salt that precipitated from a restricted or enclosed body of seawater or from the water of a salt lake.

evapotranspiration That portion of precipitation that is returned to the air through evaporation and transpiration, the latter being the escape of water from the leaves of plants.

exfoliation The process whereby slabs of rock bounded by sheet joints peel off the host rock, usually a granite or sandstone.

exfoliation dome A large rounded dome resulting from exfoliation; e.g., Half Dome in Yosemite National Park.

exotic terranes Fault-bounded bodies of rock that have been transported some distance from their place of origin and that are unrelated to adjacent rock bodies or terranes.

expansive soils Clayey soils that expand when they absorb water and shrink when they dry out.

extrusive rocks Igneous rocks that formed at or near the surface of the earth. Because they cooled rapidly, they generally have an *aphanitic* texture.

fault creep (tectonic) The gradual slip or motion along a fault without an earthquake.

fault Fracture in the earth along which there has been displacement.

felsic Descriptive of magma or rock with abundant light-colored minerals and a high silica content. The term is derived from *fel*dspar + *si*lica + *c*.

fetch The unobstructed stretch of sea over which the wind blows to create wind waves.

fiord (fjord) A narrow, steep-walled inlet of the sea between cliffs or steep slopes, formed by the passage of a glacier. A drowned glacial valley.

firn A transitional material between snow and glacial ice, being older and denser than snow, but not yet transformed into glacial ice. Snow becomes firn after surviving one summer melt season; firn becomes ice when its permeability to liquid water becomes zero.

fissile isotopes Isotopes with nuclei that are capable of being split into other elements (*see* fission).

fission The rupture of the nucleus of an element into lighter elements (fission products) and free neutrons spontaneously or by absorption of a neutron.

fjord *See* fiord.

flood frequency The average time interval between floods that are equal to or greater than a specified discharge.

floodplain The portion of a river valley adjacent to the channel that is built of sediments deposited during times when the river overflows its banks at flood stage.

floodwall A heavily reinforced wall designed to contain a stream at flood stage.

floodway Floodplain area under federal regulation for flood insurance purposes.

flotation The process of concentrating minerals with distinct non-wettable properties by floating them in liquids containing soapy frothing agents such as pine oil.

fluidization The process whereby granular solids (corn, wheat, volcanic ash and lapilli) under high gas pressures become fluid-like and flow downslope or can be pumped.

fluvial Pertaining to rivers.

focus (hypocenter) Point within the earth where an earthquake originates.

foliation The planar or wavy structure that results from the flattened growth of minerals in a metamorphic rock.

foreshore slope The zone of a beach that is regularly covered and uncovered by the tide.

fossil fuels Coal, oil, natural gas, and all other solid or liquid hydrocarbon fuels.

fracture Break in a rock caused by tensional, compressional or shearing forces (*see also* joints).

frost wedging The opening of joints and cracks by the freezing and thawing of water.

fusion The combination of two nuclei to form a single heavier nucleus accompanied by a loss of some mass that is converted to heat.

Gaia The controversial hypothesis that proposes that life has a controlling influence on the oceans and atmosphere. It states that the earth is a giant self-regulating body with close connections between living and non-living components.

geologic agents All geologic processes—e.g., wind, running water, glaciers, waves, mass wasting—that erode, move, and deposit earth materials.

geological repository (nuclear waste) An underground vault in terrane area free from geologic hazards. Repositories in salt mines, granite, and welded tuff have been studied.

geothermal energy Energy derived from circulating hot water and steam; usually associated with cooling magma or hot rocks.

gigawatt (GW) A thousand million (a billion) watts; 1,000 megawatts.

glacial outwash Deposits of stratified sand, gravel, and silt that have been removed from a glacier by meltwater streams.

glacier A large mass of ice formed, at least in part, on land by the compaction and recrystallation of snow. It moves slowly downslope by creep or outward in all directions due to the stress of its weight, and it survives from year to year.

global warming The general term for the steady rise in the average global termperatures over the last 100 years.

Gondwana, Gondwanaland The southern part of the Permo-Triassic drift landmass of Pangaea.

graben An elongate flat-floored valley bounded by faults on each side.

graded stream A stream that is in equilibrium, showing a balance of erosion, transporting capacity, and material supplied to it. Graded streams have a smooth profile.

gradient (stream) The slope of a streambed usually expressed as the amount of drop per horizontal distance in meters per kilometer (m/km) or in feet per mile (ft/mi).

gravity separation *See* crystal settling.

greenhouse effect The warming of the atmosphere by an increase in the amount of certain gases in the atmosphere, primarily CO_2 and CH_4, that increase the retention of heat that has been radiated by the earth.

groin A structure of rock, wood, or concrete built roughly perpendicular to a beach to trap sand.

ground water That part of subsurface water that is in the zone of saturation (below the water table).

gullying The cutting of channels into the landscape by running water. When extreme, it renders farmland useless.

half life The period of time during which one-half of a given number of atoms of a radioactive element or isotope will disintegrate.

hard rock (*a*) A term loosely used for an igneous or metamorphic rock, as distinguished from a sedimentary rock. (*b*) In mining, a rock that requires drilling and blasting for economical removal.

hardpan Impervious layer just below the land surface produced by calcium carbonate ($CaCO_3$) in the B horizon (*see* claypan).

high-grade deposit A mineral deposit with a high concentration of a desired element.

high-level nuclear waste Mostly radioactive elements in spent fuel with long half lives.

horse latitudes Oceanic regions in the *subtropical highs* about 30° north and south of the equator that are characterized by light winds or calms, heat, and dryness.

hot spot A point (or area) on the lithosphere over a plume of lava rising from the mantle.

hydration A process whereby anhydrous minerals combine with water.

hydraulic conductivity A ground-water unit expressed as the volume of water that will move in a unit of time through a unit of area measured perpendicular to the flow direction. Commonly

called *permeability,* it is expressed as the particular aquifer's $m^3/day/m^2$ (or m/day) or $ft^3/day/ft^2$ (or ft/day).

hydraulic gradient The slope or vertical change (ft/ft or m/m) in water-pressure head with horizontal distance in an aquifer.

hydraulic mining, hydraulicking A mining technique by which high-pressure jets of water are used to dislodge unconsolidated rock or sediment so that it can be processed.

hydrocarbon One of the many chemical compounds solely of hydrogen and carbon atoms; may be solid, liquid, or gaseous.

hydrocompaction The compaction of dry, low-density soils due to the heavy application of water.

hydrogeologist A person who studies the geology and management of underground and related aspects of surface waters.

hydrograph A graph of the stage (height), discharge, velocity, or some other characteristic of a body of water over time.

hydrologic cycle (water cycle) The constant circulation of water from the sea to the atmosphere, to the land, and eventually back to the sea. The cycle is driven by solar energy.

hydrology The study of liquid and solid water on, under, and above the earth's surface, including economic and environmental aspects.

hydrolysis The chemical reaction between hydrogen ions in water with a mineral, commonly a silicate, usually forming clay minerals.

hydrothermal Of or pertaining to heated water, the action of heated water, or a product of the action of heated water—such as a mineral deposit that precipitated from a hot aqueous solution.

hypocenter *See* focus.

igneous rocks Rocks that crystallize from molten material at the surface of the earth (volcanic) or within the earth (plutonic).

inactive sea cliff A cliff whose erosion is dominated by subaerial processes rather than wave action.

indicated reserves The hydrocarbons that can only be recovered by secondary recovery techniques such as fluid injection. These reserves are in addition to the reserves directly recoverable from a known reservoir.

influent stream A stream or reach of a stream that contributes water to the zone of saturation.

intensity scale An earthquake rating scale (I–XII) based upon subjective reports of human reactions to ground shaking and upon the damage caused by an earthquake.

intrusive rocks Igneous rocks that have "intruded" into the crust, hence they are slowly cooled and generally have phaneritic texture.

ion An atom with a positive or negative electrical charge because it has gained or lost one or more electrons.

isolation (waste disposal) Buried or otherwise sequestered waste.

isoseismals Lines on a map that enclose areas of equal earthquake shaking based upon an intensity scale.

isostasy The condition of equilibrium, comparable to floating, of large units of lithosphere above the asthenosphere. Isostatic compensation demands that as mountain ranges are eroded, they rise—just as a block of wood floating in water would rise if a part of it above the waterline were removed.

isostatic rebound Uplift of the crust of the earth that results from unloading such as results from the melting of ice sheets.

isotope One of two or more forms of an element that have the same atomic number but different atomic masses.

J-curve The shape of a graphing of geometric (exponential) growth. Growth is slow at first, and the rate of growth increases with time.

jetties Structures built perpendicular to the shoreline to improve harbor inlets or river outlets.

joint Separation or parting in a rock that has not been displaced. Joints usually occur in groups (``sets''), the members of a set having a common orientation.

karst topography, karst terrane Area underlain by soluble limestone or dolomite and riddled with caves, caverns, sinkholes, lakes, and disappearing streams.

kerogen A carbonaceous residue in sediments that has survived bacterial metabolism. It consists of large molecules from which hydrocarbons and other compounds are released on heating.

kettle A bowl-shaped depression without surface drainage in glacial-drift deposits, often containing a lake, believed to have formed by the melting of a large detached block of stagnant ice left behind by a retreating glacier.

lahar A debris flow or mudflow consisting of volcanic material.

landslide The downslope movement of rock and/or soil as a semicoherent mass on a discrete slide surface or plane (*see also* mass wasting).

lapilli (cinders) Fragments only a few centimeters in size that are ejected from a volcano.

lateral spreading The horizontal movement on nearly level slopes of soil and mineral particles due to liquefaction of quick clays.

laterite A highly weathered brick-red soil characteristic of tropical and subtropical rainy climates. Laterites are rich in oxides or iron and aluminum and have some clay minerals and silica. Bauxite is an Al-rich deposit of a similar origin.

Laurasia The northern part of the Permo-Triassic landmass of Pangaea. (*See* Gondwana)

lava Molten material at the surface of the earth.

lava dome (volcanic dome) Steep-sided protrusion of viscous, glassy lava, sometimes within the crater of a larger volcano (e.g. Mount St. Helens) and sometimes free-standing (e.g. Mono Craters).

lava plateau A broad elevated tableland, thousands of square kilometers in extent, underlain by a thick sequence of lava flows.

leachate The noxious fluids that percolate through sanitary landfills. If released to the surface or underground environment, leachate can pollute surface and underground water.

levees (natural) Embankments of sand or silt built by a stream along both banks of its channel. They are deposited during floods, when waters overflowing the stream banks are forced to deposit sediment.

liquefaction *See* spontaneous liquefaction.

lithification The conversion of sediment into solid rock through processes of compaction, cementation, and crystallization.

lithified To change into stone as in the transformation of loose sand to sandstone.

lithosphere The rigid outer layer from the earth's surface down to the asthenosphere. It includes the crust and part of the upper mantle and is of the order of 100 km (60 mi) thick. The "plates" in plate tectonic theory.

littoral drift Sediment (sand, gravel, silt) that is moved parallel to the shore by longshore currents.

loam A rich, permeable soil composed of a mixture of clay, silt, sand, and organic matter.

lode A mineral deposit consisting of a zone of veins in consolidated rock, as opposed to a *placer deposit.*

loess A blanket deposit of buff-colored calcareous silt that is porous and shows little or no stratification. It covers wide areas in Europe, eastern China, and the Mississippi valley. It is generally considered to be windblown dust of Pleistocene age.

longitudinal wave (P-wave) A type of seismic wave involving particle motion that is alternating expansion and compression in the direction of wave propagation. Also known as *compressional waves,* they resemble sound waves in their motion.

longshore current (littoral current) The current adjacent and parallel to a shoreline that is generated by waves striking the shoreline at an angle.

low-grade deposit A mineral deposit in which the mineral content is minimal but still exploitable at a profit.

low-level nuclear wastes Wastes generated in research labs, hospitals, and industry. Shallow burial in containers is sufficient to sequester this waste.

luster The manner in which a mineral reflects light using such terms as metallic, resinous, silky, and glassy.

L-wave Seismic waves that travel at the surface of the earth. Surface waves are generated by the unreflected energy of P- and S-waves striking the earth's surface. They are the most damaging of earthquake waves.

M-discontinuity (Moho) The contact between the earth's crust and the mantle. There is a sharp increase in earthquake P-wave velocity across this boundary.

maar A low-relief, broad volcanic crater formed by multiple shallow explosive eruptions. Maars commonly contain a lake and are surrounded by a low rampart or ring of ejected material. They typically form from *phreatic eruptions.*

mafic Descriptive of a magma or rock rich in iron and magnesium. The mnemonic term is derived from *ma*gnesium + *f*erric + *ic.*

magma Molten material within the earth that is capable of intrusion or extrusion and from which igneous rocks form.

magnitude (earthquake) A measure of the strength or the strain energy released by an earthquake at its source. (*See* moment magnitude, Richter magnitude.)

manganese nodule A small irregular black to brown, laminated concretionary mass consisting primarily of manganese minerals with some iron oxides and traces of other metallic minerals, abundant on the floors of the world's oceans.

mantle The zone of the earth below the crust and above the core, generally thought to be composed of rocky material.

mass wasting A general term for all downslope movements of soil and rock material under the direct influence of gravity.

massive sulfide deposit Usually volcanogenic, often rich in zinc and sometimes in lead.

meander One of a series of sinuous curves in the course of a stream.

megawatt (MW) A million watts; 1,000 kilowatts.

meltdown See central-core meltdown.

metamorphic rocks Preexisting rocks that have been altered by heat, pressure, or chemically active fluids.

mid-latitude desert A desert area occurring within latitudes 30°–50° north and south of the equator in the deep interior of a continent, usually on the lee side of a mountain range that blocks the path of prevailing winds, and commonly characterized by a cold, dry climate (see rain-shadow desert).

millirem A measure of radioactive emissions in the environment.

mineral Naturally occurring crystalline substance with well-defined physical properties and a definite range of chemical composition.

mineral deposit A localized concentration of naturally occurring mineral material (e.g., a metallic ore or a nonmetallic mineral), usually of economic value, without regard to its mode of origin.

mineral reserves Known mineral deposits that are recoverable under present conditions but as yet undeveloped. The term excludes *potential ore*.

mineral resources The valuable minerals of an area that are presently legally recoverable or that may be so in the future; include both the known ore bodies *(mineral reserves)* and the *potential ores* of a region.

miscible Soluble; capable of mixing.

Modified Mercalli Scale Earthquake intensity scale from I (not felt) to XII (total destruction) based upon damage and reports of human reactions.

Mohorovičić discontinuity See M-discontinuity.

Moh's hardness scale A scale that indicates the relative hardness of minerals on a scale of 1 (talc, very soft) to 10 (diamond, very hard).

moment magnitude (M_w or M) A scale of seismic energy released by an earthquake based on the product of the rock rigidity along the fault, the area of rupture on the fault plane, and the amount of slip.

moraine A mound, ridge, hill, or other distinct accumulation of unsorted, unstratified glacial drift, predominantly till, deposited chiefly by direct action of glacial ice in a variety of landforms.

mother lode (a) A main mineralized unit that may not be economically valuable in itself but to which workable deposits are related; e.g., the Mother Lode of California. (b) An ore deposit from which a placer is derived; the *mother rock* of a placer.

mountain-top mining The mining of coal that underlies the tops of mountains.

natural levee See levee.

neutron An electronically neutral particle (zero charge) of an atomic nucleus with an atomic mass of approximately 1.

nucleus The positively charged central core of an atom containing protons and neutrons that provide its mass.

nuée ardente See ash flow.

NWPA Nuclear Waste Policy Act of 1982 mandating selection of a repository for nuclear waste.

obliquity The tilt of the earth's axis.

open-pit mining Mining from open excavations, most commonly for low-grade copper and iron deposits, and coal.

ore deposit The same as a *mineral reserve* except that it refers only to a metal-bearing deposit.

ore mineral The part of an ore, usually metallic, that is economically desirable.

oxbow lake A crescent-shaped lake formed along a stream course when a tight meander is cut off and abandoned.

pahoehoe Basaltic lava typified by smooth, billowy, or ropy surfaces.

paleomagnetism Study of the remanent magnetism in rocks, which indicates the strength and direction of the earth's magnetic field in ancient landmasses.

paleoseismicity The rock record of past earthquake events in displaced beds and liquefaction features in trenches or natural outcrops.

Pangaea A supercontinent that existed from about 300–200 Ma, which included all the continents we know today.

pedalfer Soil of humid regions characterized by an organic-rich A horizon and clays and iron oxides in the B horizon.

pedocal Soil of arid or semiarid regions that is rich in calcium carbonate.

pedologist (Greek *pedo*, "soil," and *logos*, "knowledge.") A person who studies soils.

pedology The study of soils.

pegmatite An exceptionally coarse-grained igneous rock with interlocking crystals, usually found as irregular dikes, lenses, or veins, especially at the margins of batholiths.

perched water table The upper surface of a body of ground water held up by a discontinuous impermeable layer above the static water table.

permafrost Permanently frozen ground, with or without water, occurring in arctic, subarctic, and alpine regions.

permafrost table The upper limit of permafrost.

permeability The degree of ease with which fluids flow through a porous medium (see hydraulic conductivity).

phaneritic texture A coarse-grained igneous-rock texture resulting from slow cooling. Phanerites are coarse-grained igneous rocks.

photochemical smog Atmospheric haze that forms when automobile-exhaust emissions are activated by ultraviolet radiation from the sun to produce highly reactive oxidants known to be health hazards.

phreatic eruption Volcanic eruption, mostly steam, caused by the interaction of hot magma with underground water, lakes, or seawater. Where significant amounts of new (magmatic) material are ejected in addition to steam, the eruptions are described as *phreatomagmatic.*

phytoremediation The use of plants to remove soil contaminants, particularly excess salt.

piping Subsurface erosion in sandy materials caused by the percolation of water under pressure. A somewhat similar effect is obtained by placing a garden-hose nozzle in a sandbox and jetting an opening through the sand.

placer deposit (pronounced "plass-er") A surficial mineral deposit formed by settling from streams of mineral particles from weathered debris. (*See also* lode.)

planetesimals Space debris, up to a few hundred kilometers in diameter, that first condensed from the solar nebula. Their accretion is believed to have formed the planets in our solar system.

plastic Capable of being deformed continuously and permanently in any direction without rupture.

Pleistocene The geologic Epoch of the Quaternary Period known as the Ice Age, falling by definition between 1.6 million and 12,000 years ago. Extensive glaciation is known as far back as the Pliocene Epoch 2.5 million years ago. The exact boundaries shift in time as new data are revealed.

Plinian eruption (volcanology) An explosive eruption in which a steady, turbulent, nearly vertical column of ash and steam is released from a vent at high velocity.

plutonic Pertaining to rock formed by any process at depth, usually with a phaneritic texture.

point bar A deposit of sand and gravel found on the inside curve of a *meander.*

polar desert The bitterly cold, arid region centered at the south pole that receives very little precipitation.

polar front The boundary in the atmosphere where cold polar surface air and mild-temperature surface air converge.

polar high The region of cold, high-density air that exists in the polar regions.

polarity The magnetic positive (north) or negative (south) character of a magnetic pole.

population dynamics The study of how populations grow.

population growth rate The number of live births less deaths per 1,000 people, usually expressed as a percentage; a growth rate of 40 would represent 4.0% growth.

pore pressure The stress exerted by the fluids that fill the voids between particles of rock or soil.

porosity A material's ability to contain fluid. The ratio of the volume of pore spaces in a rock or sediment to its total volume, usually expressed as a percentage.

porphyry copper A copper deposit, usually of low grade, in which the copper-bearing minerals occur in disseminated grains and/or in veinlets through a large volume of rock.

potable (water) Drinkable without ill effects.

potential ore As yet undiscovered mineral deposits and also known mineral deposits for which recovery is not yet economically feasible.

potentiometric surface The hydrostatic head of ground water and the level to which the water will rise in a well. The water table is the potentiometric surface in an unconfined aquifer. Also called *piezometric surface.*

precession The slow gyration of the Earth's axis, analogous to that of a spinning top, which causes the slow change in the orientation of the earth's axis relative to the sun that accounts for the reversal of the seasons of winter and summer in the Northern and Southern Hemispheres every 11,500 years.

precipitation The separation of a solid substance from a solution by a chemical reaction.

precursors (of earthquakes) Observable phenomena that occur before an earthquake and indicate that an event is soon to occur.

pregnant pond A catchment basin that holds a gold- and silver-bearing cyanide solution.

proglacial lake An ice-marginal lake formed just beyond the frontal moraine of a retreating glacier and generally in direct contact with the ice.

proton A particle in an atomic nucleus with a positive electrical charge and an atomic mass of approximately one.

protostar A star in the process of formation that has entered the slow gravitational contraction phase.

pumice Frothy-appearing rock composed of natural glass ejected from high-silica, gas-charged magmas. It is the only rock that floats.

P-wave *See* longitudinal wave.

pyroclastic ("fire broken") Descriptive of the fragmental material, ash, cinders, blocks, and bombs ejected from a volcano. (*See also* tephra.)

quad A quadrillion British thermal units (10^{15} Btu).

quick clay (sensitive clay) A clay possessing a "house-of-cards" sedimentary structure that collapses when it is disturbed by an earthquake or other shock.

radioactivity The property possessed by certain elements that spontaneously emit energy as the nucleus of the atom changes its proton-to-neutron ratio.

radiometric dating *See* absolute age dating.

radon A radioactive gas emitted in the breakdown of uranium.

rain-shadow desert A desert on the lee side of a mountain or a mountain range (*see* mid-latitude desert).

rank (coal) A coal's carbon content depending upon its degree of metamorphism.

recessional moraine One of a series of nested end moraines that record the stepwise retreat of a glacier at the end of an ice age.

recrystallization The formation in the solid state of new mineral grains in a metamorphic rock. The new grains are generally larger than the original mineral grains.

recurrence interval The return period of an event, such as a flood or earthquake, of a given magnitude. For flooding, it is the average interval of time within which a given flood will be equalled or exceeded by the annual maximum discharge.

regional metamorphism A general term for metamorphism that affects an extensive area.

regolith Unconsolidated rock and mineral fragments at the surface of the earth.

Regulatory Floodplain The part of a floodplain that is subject to federal government regulation for insurance purposes.

relative age dating The chronologic ordering of geologic strata, features, fossils, or events with respect to the geological time scale without reference to their absolute age.

replacement rate The birth rate at which just enough offsprings are born to replace their parents.

reserves Those portions of an identified resource that can be recovered economically.

reservoir rock A permeable, porous geologic formation that will yield oil or natural gas.

residence time The average length of time a substance (atom, ion, molecule) remains in a given reservoir.

residual soil Soil formed in place by decomposition of the rocks upon which it lies.

resisting force (landslide) Friction along the slide plane and normal forces across the plane.

resonance The tendency of a system (a structure) to vibrate with maximum amplitude when the frequency of the applied force (seismic waves) is the same as the vibrating body's natural frequency.

revetment A rock or concrete structure built landward of a beach to protect coastal property.

Richter Magnitude Scale (earthquakes) The logarithm of the maximum trace amplitude of a particular seismic wave on a seismogram, corrected for distance to epicenter and type of seismometer. A measure of the energy released by an earthquake.

rift zone (volcanology) A linear zone of weakness on the flank of a shield volcano that is the site of frequent flank eruptions.

rill erosion The carving of small channels, up to 25 cm (10 in) deep, in soil by running water.

rock cycle The cycle or sequence of events involving the formation, alteration, destruction, and reformation of rocks as a result of processes such as erosion, transportation, lithification, magmatism, and metamorphism.

rocks Aggregates of minerals or rock fragments.

rotational landslide (slump) A slope failure in which sliding occurs on a well-defined, concave-upward, curved surface, producing a backward rotation of the slide mass.

runoff That part of precipitation falling on land that runs into surface streams.

runup The elevation to which a wind wave or tsunami advances onto the land.

salt dome A column or plug of rock salt that rises from depth because of its low density and pierces overlying sediments.

sanitary landfill A method of burying waste where each day's dumping is covered with soil to protect the surrounding area. This is also monitored to insure protection of local underground water.

sea-floor spreading A hypothesis that oceanic crust forms by convective upwelling of lava at mid-ocean ridges and moves laterally to trenches where it is destroyed.

seawall A rock or concrete structure built landward of a beach to protect the land from wave action. (*See also* revetment.)

secondary enrichment Processes of near-surface mineral deposition, in which oxidation produces acidic solutions that leach metals, carry them downward, and reprecipitate them, thus enriching sulfide minerals already present.

sedimentary rock A layer of rock deposited from water, ice, or air and subsequently lithified to form a coherent rock.

seif (longitudinal) dune A very large, sharp-crested, tapering chain of sand dunes; commonly found in the Arabian Desert.

seismic gap The segment along an active fault that has seismically been quiet relative to segments at either end. It is the part of the fault most likely to rupture and generate the next earthquake. (*See also* asperity.)

seismogram The recording from a seismograph.

seismograph An instrument used to measure and record seismic waves from earthquakes.

sensitive clay *See* quick clay.

settlement This occurs when the applied load of a structure's foundation is greater than the bearing strength of the foundation material (soil or rock).

sheet erosion The removal of thin layers of surface rock or soil from an area of gently sloping land by broad continuous sheets of running water (sheet flow), rather than by channelized streams.

sheet joints (*See* joint) Cracks more or less parallel to the ground surface that result from expansion due to deep erosion and the unloading of overburden pressure.

shield volcano (cone) The largest type of volcanoes, composed of piles of lava in a convex-upward slope. Found mainly in Hawaii and Iceland.

sinkhole Circular depression formed by the collapse of a shallow cavern in limestone.

slant-drilling (whipstocking) Purposely deflecting a drill hole from the vertical in order to tap a reservoir not directly below the drill site. The deflecting tool is a whipstock.

slide plane (slip plane) A curved or planar surface along which a landslide moves.

slip face The steeper, lee side of a dune, standing at or near the angle of repose of loose sand and advancing downwind by a succession of slides whenever that angle is exceeded.

slump *See* rotational landslide.

smelter An industrial plant that mechanically and chemically produces metals from their ores.

SMCRA The Surface Mining Control and Reclamation Act of 1977.

soil (*a*) Loose material at the surface of the earth that supports the growth of plants (pedological definition). (*b*) All loose surficial earth material resting on coherent bedrock (engineering definition).

soil horizon A layer of soil that is distinguishable from adjacent layers by properties such as color, texture, structure, or chemical composition.

soil profile A vertical section of soil that exposes all of its horizons or zones.

solution (weathering) The dissolving of rocks and minerals by natural acids or water.

source rock The geologic formation in which oil and/or gas originate.

specific yield The ratio of the volume of water drained from an aquifer by gravity to the total volume considered.

speleology The study of caves and caverns.

spheroidal weathering Weathering of rock surfaces that creates a rounded or spherical shape as the corners of the rock mass are weathered faster than its flat faces.

spontaneous liquefaction Process whereby water-saturated sands, clays, or artificial fill suddenly become fluid upon shaking, as in an earthquake.

spreading center *See* divergent boundary.

stack A pillarlike rocky island that has become detached from a sea cliff or headland by wave erosion.

stage The height of flood waters in feet or meters above an established datum plane.

stalactite Conical deposit of mineral matter, usually calcite, that has developed downward from the ceiling of a limestone cave.

stalagmite Conical deposit of mineral matter, usually calcite, developed upward from the floor of a cavern or cave.

steppe An extensive treeless grassland in the semiarid mid-latitudes of Asia and southeastern Europe. It is drier than the *prairie* that develops in the mid-latitudes of the United States.

storm surge The sudden rise of sea level on an open coast due to a storm. Storm surge is caused by strong onshore winds and results in water being piled up against the shore. It may be enhanced by high tides and low atmospheric pressure.

strain Deformation resulting from an applied stress (=force per unit area). It may be elastic (recoverable) or ductile (nonreversible).

stratification (bedding) The arrangement of sedimentary rocks in strata or layers.

stratigraphic trap An accumulation of oil that results from a change in the character (permeability) of the reservoir rock, rather than from structural deformation.

stratovolcano A large volcano that is shaped slightly concave upward and is composed of layers of ash and lava.

stress The force applied on an object per unit area, expressed in pounds per square foot (lb/ft^2) or kilograms per square meter (kg/m^2).

strike The direction west or east of north of the trace of the inclined plane (sedimentary bedding plane, fault, or joint surface) on the horizontal plane.

strip-mining Surficial mining, in which the resource is exposed by removing the overburden.

subduction The sinking of one lithospheric plate beneath another at a convergent plate margin. (*See* convergent boundary.)

subsidence Sinking or downward settling of the earth's surface due to solution, compaction, withdrawal of underground fluids, or to cooling of the hot lithosphere or to loading with sediment or ice.

subtropical desert A hot, dry desert occurring between latitudes 15°–30° north and south of the equator where subtropical air masses prevail. (*See* trade-wind desert.)

subtropical highs The climatic belts at latitudes of approximately 30°–35° north and south of the equator characterized by semipermanent high-pressure air masses, moderate seasonal rainfall, summer heat, dryness, and generally clear skies.

Superfund A spinoff of CERCLA insuring financial accountability for environmental polluters.

supernova The explosive destruction of a massive star that occurs when the star's nuclear fuel has been depleted; the star catastrophically collapses inward.

surcharge Weight added to the top of a natural or artificial slope by sedimentation or by artificial fill.

surging glacier A glacier that alternates periodically between brief periods (usually 1–4 years) of very rapid flow, called *surges*, and longer periods (usually 10–100 years) of near-stagnation.

suspension (*a*) A mode of sediment transport in which the upward currents in eddies of turbulent flow are capable of supporting the weight of sediment particles and keeping them held indefinitely in the surrounding fluid (as silt in water or dust in air). (*b*) The state of a substance in such a mode of transport.

sustainable society A population capable of maintaining itself while leaving sufficient resources for future generations.

sustained yield (hydrology) The amount of water an aquifer can yield on a daily basis over a long period of time.

S-wave *See* transverse wave.

syncline A concave-upward fold that contains younger rocks in its core. (*Contrast with* anticline.)

synthetic fuels (synfuels) A manufactured fuel; for example, a fuel derived by liquefaction or gasification of coal.

tailings The worthless rock material discarded from mining operations.

talus Coarse angular rock fragments lying at the base of a cliff or steep slope from which they were derived.

tectonism (adj. tectonic) Deformation of the earth's crust by natural processes leading to the formation of ocean basins, continents, mountain systems, and other earth features.

tephra A collective term for all *pyroclastic* material ejected from a volcano.

terranes (*See* exotic terannes.)

texture The size, shape, and arrangement of the component particles or crystals in a rock.

till Unsorted and unstratified drift, generally unconsolidated, that is deposited directly by and beneath a glacier without subsequent reworking by water from the glacier.

tombolo A sandbar that connects an island with the mainland or another island.

total fertility rate Average number of children born to women of child-bearing age.

trace element An element that occurs in minute quantities in rocks or plant or animal tissue. Some are essential for human health.

trade winds Steady winds blowing from areas of high pressure at 30° north and south latitudes toward the area of lower pressure at the equator. This pressure differential produces the northeast and southeast trade winds in the Northern and Southern Hemispheres, respectively.

trade wind desert *See* subtropical desert.

transform boundary *See* transform fault.

transform faults Plate boundaries with mostly horizontal (lateral) movement that connect spreading centers to each other or to subduction zones.

translational slide *See* block glide.

transported soils Soil developed on *regolith* transported and deposited by a geological agent. Hence the term *glacial* soils, alluvial soils, *eolian* soils, *volcanic* soils, and so forth.

transuranic elements (TRU) Unstable (radioactive) elements of atomic numbers greater than 92.

transverse dune A strongly asymmetrical sand dune that is elongated perpendicularly to the direction of the prevailing winds. It has a gentle windward slope and a steep leeward slope that stands at or near the angle of repose of sand.

transverse wave (S-wave) A seismic wave propagated by a shearing motion that involves oscillation perpendicular to the direction of travel. It travels only in solids.

troposphere That part of the atmosphere next to the earth's surface, in which temperature generally decreases rapidly with altitude and clouds and convection are active. Where the weather occurs.

tsunami Large sea wave produced by a submarine earthquake, a volcanic eruption, or a landslide. Incorrectly known as a tidal wave.

tuff A coherent rock formed from volcanic ash, cinders, or other *pyroclastic* material.

tundra (soil) A treeless plain characteristic of arctic regions with organic rich, poorly drained tundra soils and permanently frozen ground.

unconfined aquifer Underground body of water that has a free (static) water table; that is, water that is not confined under pressure beneath an aquiclude.

upwelling A process whereby cold, nutrient-rich water is brought to the surface. Coastal upwelling occurs mostly in trade-wind belts.

vapor extraction The use of vacuum pumps placed in shallow bore holes to remove vaporized organic contaminants in the soil, such as gasoline or TCE.

vein deposit A thin, sheetlike igneous intrusion into a crevice.

ventifact Any stone or pebble that has been shaped, worn, faceted, cut, or polished by the abrasive or sandblast action of windblown sand, generally under arid conditions.

Volcanic Explosivity Index (VEI) A scale for rating volcanic eruptions according to the volume of the ejecta, the height to which the ejecta rise, and the duration of the eruption.

volcanogenic deposit Of volcanic origin; e.g., volcanogenic sediments or ore deposits.

water cycle *See* hydrologic cycle.

water table The contact between the zone of aeration and the zone of saturation.

wave base The depth at which water waves no longer move sediment, equal to approximately half the wavelength.

wavelength The distance between two equivalent points on two consecutive waves, for example, crest to crest or trough to trough.

wave of oscillation A water wave in which water particles follow a circular path with little forward motion.

wave of translation A water wave in which the individual particles of water move in the direction of wave travel, as in broken waves (surf).

wave period (T) The time (in seconds) between passage of equivalent points on two consecutive waveforms.

wave refraction The bending of wave crests as they move into shallow water.

weathering The physical and chemical breakdown of materials of the earth's crust by interaction with the atmosphere and biosphere.

welded tuff A glassy pyroclastic rock that has been made hard by the welding together of its particles under the action of retained heat.

wetted perimeter (stream) The length of stream bed and bank that are in contact with running water in the stream in a given cross-section.

wind The movement of air from a region of high pressure to a region of low pressure.

zero population growth A population growth rate of 0; that is, the number of deaths in a population is equal to the number of live births on average.

zone of aeration (vadose zone) The zone of soil or rock through which water infiltrates by gravity to the water table. Pore spaces between rock and mineral grains in this zone are filled with air; hence the name.

zone of flowage The lower part of a glacier that deforms due to the glacier's own weight.

zone of saturation The ground-water zone where voids in rock or sediment are filled with water.

INDEX

CREDITS

KEY TO ABBREVIATIONS:

AAPG   American Association of Petroleum Geologists
AIPG   American Institute of Professional Geologists (Arvada, Colorado)
API   American Petroleum Institute
ARCO   Atlantic Richfield Co.
CDMG   California Division of Mines and Geology
CIT   California Institute of Technology
DOE   United States Department of Energy
EPA   Environmental Protection Agency
FEMA   Federal Emergency Management Agency
GSA   Geological Society of America (Boulder, Colorado)
JPL   Jet Propulsion Laboratory
NAGT   National Association of Geology Teachers
NASA   National Aeronautics and Space Administration
NGDC   National Geophysical Data Center
NOAA   National Oceanic and Atmospheric Administration
NPS   National Park Service
NRC   National Research Council
SCEC   Southern California Earthquake Center
TVA   Tennessee Valley Authority
USAF   United States Air Force
USC   University of Southern California
USDA   United States Department of Agriculture
USGS   United States Geological Survey
USN   United States Navy

CHAPTER 1

Opener, Breck, P. Kent, JLM Visuals; **1.2** After George A. Kiersch, *The heritage of engineering geology: First hundred years, 1888–1988,* vol. 3 (GSA, 1991), p. 4; **1.3** After George A. Kiersch, *The heritage of engineering geology: First hundred years, 1888–1988,* vol. 3 (GSA, 1991), p. 3; **1.4** USGS; **1.5a, b** B. Pipkin; **c** Thomas L. Holzer, USGS; **1.6** Jim Mori, USGS; **1.8** After Donald Mann, *Negative Population Growth newsletter* (Teaneck, N.J., 1995); **1.9** Blue Planet Group data; **1.10** Blue Planet Group data; **1.11** NOAA data; **1.12** United Nations Food and Agriculture Organization data; **1.13** B. Pipkin; **1.14** Cathlyn Melloan/Tony Stone Images; **1.15** Owen Franken/Sygma; **1.16** Reproduced by permission of Earth Observation Satellite Corp., Lanham, MD, U.S.A.; **Case Study 1.1a** B. Pipkin; **b, c** D. D. Trent

CHAPTER 2

Opener, D. D. Trent; **2.5** Adapted from Larry P. Gough, compiler, Understanding our fragile environment, USGS circular 1105, 1993; **2.6** After NASA, Earth system science report (1986), in NRC, *Solid earth sciences and society* (Washington, D.C.: National Academy Press, 1993); **2.10a, b** B. Pipkin; **c, d** JLM Visuals; **2.14** B. Pipkin; **2.16** B. Pipkin; **2.18** B. Pipkin; **2.19** D. D. Trent; **2.20** D. D. Trent; **2.22** B. Pipkin; **2.23** B. Pipkin; **2.24** D. D. Trent; **2.26** Adapted with permission from A. R. Palmer, The decade of North American geology, 1983 geologic time scale, *Geology* (GSA, 1983), p. 504; **2.31a** © David Austen/Tony Stone Images; **b** Monroe/Wicander, The changing earth, West Pub. Co., © 1994; **2.32** Monroe/Wicander, The changing earth, West Pub. Co., © 1994; **2.33** Theodore Roosevelt Collection, Harvard College Library; **2.34** B. Pipkin; **Case Study 2.1 1, 2** Janet Blabaum, Highland Geotechnical, courtesy AIPG; **Case Study 2.2 1a, d** B. Pipkin; **1b** Martin Land, Science Photo Library/Photo Researchers; **1c** Carl Frank, Photo Researchers; **2** after Christopher B. Gunn, A description catalog of the drift diamonds of the Great Lakes region, North America, *Gems and gemology* 12, nos. 10 and 11 (1968), pp. 297–303, 333–334. © Gemological Institute of America

CHAPTER 3

Opener, D. D. Trent; **3.1** After Alfred Wegener, *Continents and ocean basins* (Dover Publications, 1996); **3.2a, b** Patricia G. Gensel, Univ. of North Carolina; **3.3c** Scott Katz; **3.5** Carlyn Iverson; **3.8** NGDC; **3.12** Data from J. B. Minster and T. H. Jordan, Present-day plate motions, *Journal of geophysical research* 83 (1978), pp. 5331–51; **3.14b** H. E. Enlow, USGS; **d** USGS; **3.16b inset** Marie Tharp; **3.18** Bildarchiv Preussischer Kulturbesitz; **3.19a** Courtesy Rodney Supple; **b** D. D. Trent; **Case Study 3.1 2** After I. W. D. Dalziel, Allan Cox, and D. Engebretson, GSA special paper 207, 1987

CHAPTER 4

Opener, © Michael Edwards, *Los Angeles Times;* **4.1b, e** Keery E. Sieh, Seismological Laboratory, CIT; **c** Peter W. Weigand; **d** B. Pipkin; **4.2b** D. D. Trent; **d** B. Bradley, Univ. of Colorado; **f** NOAA; **4.3e** USGS; **4.6** B. Pipkin; **4.7** SCEC data; **4.10** SCEC data; **4.12** B. Pipkin; **4.13b** Dorothy Stout, Cypress College; **d** Dennis Fox; **4.14** James C. Anderson, USC; **4.15** C. J. Langer, USGS; **4.16c** B. Pipkin; **4.17b** B. Bradley, Univ. of Colorado; **4.18a** NOAA; **b** courtesy Richard Hilton; **4.19** NGDC; **4.21a** Courtesy Dave Johnson; **b** USGS, Menlo Park; **4.22** Martin E. Klimek, *Marin independent journal;* **4.23** Katsushika Hokusai, *The great wave off Kanagawa.* The Metropolitan Museum of Art, Bequest of Mrs. H. O. Havemeyer, 1929; The H. O. Havemeyer Collection (JP 1847); **4.25** USGS data; **4.26a** Data from USGS; **b** NGDC; **4.27** USGS data; **4.28** USN; **4.29** Courtesy Bishop Museum; **4.30a, b, c** Courtesy Henry Helbush; **4.32** USGS data; **4.33** Robert Sydnor, CDMG; **4.34** Gregory A. Davis, USC; **4.35** Grant Marshall, USGS; **4.36a** University of California, San Diego, after C. Arnold, *Earthquake disaster prevention planning in Japan* BSD (Palo Alto, Calif.: Buildings System Development, 1982); planning density example by R. Eisner and H. Shapiro; **b** University of California, San Diego, after C. Arnold, *Earthquake disaster prevention planning in Japan* BSD (Palo Alto, Calif.: Buildings System Development, 1982); planning density example by R. Eisner and H. Shapiro; **4.37a** Carol Prentice, USGS; courtesy GSA; **b** C. Rojahn, Applied Technology Council; courtesy GSA; **c** Thomas L. Holzer, USGS; **4.38** After T. Fujiwara, Y. Suzuki, M. Iwai

Nakashima, A. S. Kitahara, and M. Bruneau, Overview of building damage from the 1995 Great Hanshin Earthquake, *Newsletter of the Disaster Prevention Institute, Kyoto University,* Preliminary report on the Great Hanshin Earthquake of January 17, 1995, February 1995 special issue, pp. 13–20; **4.39** University of California, San Diego, after C. Arnold, *Earthquake disaster prevention planning in Japan* BSD (Palo Alto, Calif.: Buildings System Development, 1982); planning density example by R. Eisner and H. Shapiro; **4.41b** E. V. Leyendecker, National Bureau of Standards (NBS); **4.42a** USGS data; **4.44** Dennis Fox; **4.45** Frederick Larson, *San Francisco chronicle;* **4.46** American Geophysical Union, *EOS;* **4.47** M. L. Blair and W. W. Spangle, USGS prof. paper 941-B, 1979; **4.48b** J. K. Hillers, USGS Photo Library; **4.49** USGS; **4.50** Oregon Dept. of Geology and Mineral Industries; **4.54** After Sykes and Seeber, Great earthquakes and great asperities, San Andreas fault, California, *Geology* 13 (Dec. 1985), pp. 835–838; **4.55** Kerry E. Sieh, Seismological Laboratory, CIT; **4.56** FEMA; **4.57** FEMA; **4.58** CDMG; **4.59a, b** USC Earth Sciences Dept.; **c** USGS; **Case Study 4.2 1** Kerry E. Sieh, Seismological Laboratory, CIT; **2** redrawn courtesy Edwin Harp and Randall Jibson, USGS; **Case Study 4.3 1** SCEC; **2** Mark Peterson, The January 17, 1994, Northridge earthquake, *California geology,* March/April 1994, pp. 40–45; **Case Study 4.4 1, 2** After *The Los Angeles times;* **Case Study 4.5 1** B. Pipkin

CHAPTER 5

Opener, Courtesy Edwin D. W. Belcher, Wellington, New Zealand; **5.1b** B. Pipkin; **c** Ric Hazleff, Pomona College; **5.2** After R. I. Tilling, C. Heliker, and T. L. Wright, *Eruptions of Hawaiian volcanoes: Past, present, and future* (USGS, 1987); **5.3** After Hays, USGS prof. paper 1240-B, 1981; data compiled by Donald Mullineaux, USGS; **5.5** After Simpkin and Siebert, *Volcanoes of the world,* 2d ed. (Tucson, Ariz.: Geoscience Press, 1994); **5.6** Data from T. Simpkin et al., *Volcanoes of the world* (Dowden, Hutchinson, and Ross); **5.8** B. Pipkin; **5.9** Eros Data Center; **5.10a** USGS; **b** J. D. Griggs, Hawaiian Volcano Observatory; **c** D. D. Trent; **5.11** J. B. Stokes, USGS; **5.12** B. Pipkin; **5.13b** B. Pipkin; **5.14** B. Pipkin; **5.15a** Björn Bölstad, Photo Researchers; **b** Tau Rho Alpha, USGS; **5.16** B. Pipkin; **5.17a** B. Pipkin; **b** Tau Rho Alpha, USGS; **5.18** USGS; **5.19** B. Pipkin; **5.20** B. Pipkin; **5.21a** B. Pipkin; **b** USGS; Hawaiian Volcano Observatory; **5.22** Jim Mori, USGS; **5.23** NOAA; **5.24a** After Tau Rho Alpha, USGS; **b** K. Staufel; **c** after Tau Rho Alpha, USGS; **d** Keith Ronnholm; **e** B. Pipkin; **5.25** USGS; **5.26** B. Pipkin; **5.27** USGS data; **5.28** Matthew Pipkin; **5.29** Jim Lockwood, USGS; **5.30** USGS; **5.31** USGS data; **5.32** USGS data; **5.33** USGS data; **5.34a** Tracy Knauer, Photo Researchers; **b** NAGT; **5.35** John Mead, Science Photo Library/Photo Researchers; **Case Study 5.1 1b** David M. Bailey, Grant Co. Airport, Moses Lake, Washington; **2** USGS data; **3** NOAA; **4** NASA; **Case Study 5.2 2** Jim Mori, USGS; **3** NOAA; **Case Study 5.3 1a** Karen Williams, *Volcanoes of the southwind: A field guide to the volcanoes and landscape of Tongariro National Park* (Tongariro Natural History Society); **b** B. Pipkin; **2a** Karen Williams, *Volcanoes of the southwind: A field guide to the volcanoes and landscape of Tongariro National Park* (Tongariro Natural History Society); **2b** Karen Williams, *Volcanoes of the southwind: A field guide to the volcanoes and landscape of Tongariro National Park* (Tongariro Natural History Society); **3a** B. Pipkin; **3b** Edwin D. W. Belcher, Wellington, New Zealand; **Case Study 5.4 2a** Mike Tuttle, USGS; **2b** T. Orban/Sygma

CHAPTER 6

Opener, © David Schultz/Tony Stone Images; **6.2** B. Pipkin; **6.3** John Shelton; **6.5** B. Pipkin; **6.6** USGS; **6.7** B. Pipkin; **6.8** Bruce Hayes, Photo Researchers; **6.10** B. Pipkin; **6.12** Rodman Snead, JLM Visuals; **6.14** Douglas J. Lathwell, Cornell Univ.; **6.15** B. Pipkin; **6.16** B. Pipkin; **6.18** Monroe/Wicander Physical Geology, 2d ed. West Pub. Co., © 1995; **6.19** USDA; **6.21** Howard G. Wilshire, USGS; **6.22** Doug Burbank, USC; **6.23** Visuals Unlimited; **6.25** Courtesy Edward B. Nuhfer and AIPG; **6.26** Hays, USGS prof. paper 1240-B, 1981; **6.29** Troy L. Pewe, USGS; **6.30** Troy L. Pewe, USGS; **6.31a** R. A. Combellick, Alaska Div. of Geological and Geophysical Sciences; **b** Troy Pewe, Arizona State Univ.; **6.32** B. Pipkin; **6.33** ARCO; **6.35** Douglas J. Lathwell, Cornell Univ.; **6.36** Alain Nogues/Sygma; **6.37** Robert Thayer/Projections; **6.38** B. Pipkin; **6.39** J. P. Laffont/Sygma; **Case Study 6.1 1** Visuals Unlimited; **Case Study 6.2 1** UPI/Bettmann; **3** The Bettmann Archive; **Case Study 6.3 1** J. M. Spencer, *Texas journal of science* 21 (1970); **2** Dr. Verle Bohman, Dept. of Animal Sciences, Univ. of Nevada–Reno; **Case Study 6.4 1** Al Robb III

CHAPTER 7

Opener, Courtesy Siang Tan, CDMG; **7.1** J. P. Krohn and J. E. Slosson, Landslide potential in the United States, *California geology* 29, no. 10 (1976), pp. 224–231; **7.3b** B. Pipkin; **7.4** USGS; **7.5b** Randall Jibson, USGS; **7.6** *California geology* Data from CDMG; **7.7** B. Pipkin; **7.9a** Courtesy Burt Amundson; **b** after M. G. Smelser, *California geology,* March 1983; **7.10c** B. Pipkin; **d** D. D. Trent; **7.11** B. Pipkin; **7.12** Randall W. Jibson, USGS; **7.18b** NGDC; **7.19** Canadian Air Force; **7.21** USGS; **7.23** D. D. Trent; **7.24** Pam Irvine, CDMG; **7.25** Siang Tan, CDMG; **7.26b** B. Pipkin; **7.28** B. Pipkin; **7.30a** Bob Hollingsworth; **b** D. D. Trent; **7.31** B. Pipkin; **7.32** Redrawn courtesy M. Natland; **7.33** Perry Ehlig, California State Univ.–Los Angeles; **7.34** Redrawn courtesy Perry Ehlig, California State Univ.–Los Angeles; **7.35** Redrawn courtesy Perry Ehlig, California State Univ.–Los Angeles; **7.36** Rick Giardino, Texas A&M Univ.; **7.37** Rick Giardino, Texas A&M Univ.; **7.38** Dedman's Photo Shop, Skagway, Alaska; **7.39** B. Pipkin; **7.40** USGS; **7.41** B. Pipkin; **7.42** B. Pipkin; **7.43** B. Pipkin; **Case Study 7.1 1** Joan Van Velsor, California Dept. of Transportation; **2** after M. G. Smelser, *California geology,* 1987; **3** redrawn courtesy of Joan Van Velsor, California Dept. of Transportation; **4a** B. Pipkin; **4b** Courtesy Paul Trent; **Case Study 7.2 1** Robert Schuster, USGS; **2** redrawn courtesy Lynn Highland, *Earthquakes and volcanoes* 24, no. 3 (1993); **3** redrawn courtesy Lynn Highland, USGS

CHAPTER 8

Opener, Porterfield–Chickering, Photo Researchers; **8.1** USGS data; **8.2** After USGS prof. paper 1240-6; **8.4** Courtesy Xenophon Colazas, City of Long Beach; **8.5** City of Long Beach, Dept. of Oil Properties; **8.6** City of Long Beach, Dept. of Oil Properties data; **8.7** Anne Bryant, USAF; **8.8** USGS Photo Library; **8.9** Ronny Jacques, Photo Researchers; **8.10** Adapted from P. Gatto and L. Carbognin, The lagoon of Venice, *Hydrological sciences bulletin* 26, no. 4 (1981); **8.11** Redrawn from 1967 GSA annual meeting (Mexico City) *Field trip guide;* **8.12** R. V. Dietrich; **8.13a** State of California data; **b** James Rhul Enterprises, Fullerton, Calif.; **8.14a** USGS; **b** Frank Kujawa, Univ. of Central Florida, Geophoto Pub. Co.; **c** B. Pipkin; **8.16** USGS; **8.18** USGS data; **8.20** USGS; **8.21** James Farrell; **8.22** Scott Fee, American Speleological Society; **8.23** From O. J. Snowden and J. B. Rucker, Subsidence in New Orleans: A photographic case history of the Lakeview Subdivision, *Transactions, Gulf Coast Association of Geological Societies* 42 (1992), p. 857; **Case Study 8.1 2** California Dept. of Water Resources data; **3** *Los Angeles times;* **Case**

Study 8.2 1a After R. T. Saucier and Jesse Snowden, Engineering geology of the New Orleans area, 1995 GSA annual meeting *Field trip guide book no. 6;* **1b** EROS; **c** B. Pipkin; **2** B. Pipkin; **3** redrawn courtesy Jesse O. Snowden, Dept. of Geosciences, Southeast Missouri State Univ., from Snowden, Simmons, Traughber, and Stephens, *Transactions, Gulf Coast Association of Geological Societies* 27; **4** B. Pipkin

CHAPTER 9

Opener, © Frank Oberle/Tony Stone Images; **9.2** USGS data; **9.3** USGS data; **9.4** State of California data; **9.5** USGS data; **9.7** USGS data; **9.8** USGS data; **9.10a** Monroe/Wicander Physical Geology, 2d ed., West Pub. Co., © 1995; **9.11c** W. L. Fisher et al., *Delta systems in the exploration for oil and gas—A research colloquium,* 1969; **9.12** USN, courtesy Robert Stevenson; **9.13** NASA/JPL; **9.14b** Victor R. Baker, Univ. of Arizona; **9.15e** See R. C. Scott, Introduction to Physical Geography, 2d ed., West Pub. Co., © 1996; **9.17** Chris Stewart, Black Star; **9.18** Courtesy Ed Nuhfer and AIPG; **9.22** Modified from W. D. Hays, USGS prof. paper 1240-B, 1981; **9.24a** NRC data; **b** data from North Dakota Geological Survey miscellaneous series 35; **9.26a** Data from North Dakota Geological Survey miscellaneous series 35; **b** USGS data; **9.27** Redrawn from USGS 1993 Yearbook; **9.28** Landsat imagery courtesy The Earth Observation Satellite Co., Lanham, Maryland; **9.29a, b** B. Pipkin; **c** courtesy Ed Nuhfer and AIPG; **9.30** B. Pipkin; **9.31** A. O. Waananen et al., USGS; **9.32** San Diego Co. Engineers Office prof. paper 942, 1977; **9.34** Bradford B. Van Diver, *Imprints of time* (Missoula, Mont.: Mountain Pass Pub. Co., 1988); **9.36** B. Pipkin; **9.37** B. Pipkin; **Case Study 9.1 2** NRC data; **Case Study 9.3 1, 3** Security Pacific National Bank Historical Collection, Los Angeles Co. Library

CHAPTER 10

Opener, Jim Richardson; **10.1** After W. B. Solley, R. R. Pierce, and Howard Perelman, Estimated use of water in the United States in 1990, USGS circular 1081, 1993; **10.6** USGS data; **10.10c** J. R. Stacey, USGS; **10.13** J. B. Weeks et al., USGS prof. paper 1400-A, 1988; **10.14** Kevin McGillicuddy; **10.16** Data from AIPG, *Ground water: Issues and answers,* 1985; **10.18** Monroe/Wicander Physical Geology, 2d ed., West Pub. Co., © 1995; **10.20** D. D. Trent; **10.21** Monroe/Wicander Physical Geology, 2d ed., West Pub. Co., © 1995; **10.22** Visuals Unlimited; **10.23** Martin Land, Science Photo Library/Photo Researchers; **Case Study 10.1 1** AIPG, *Ground water: Issues and answers,* 1985; **Case Study 10.2 1, 2, 3** After B. L. Foxworthy, USGS prof. paper 950, 1978; **4** courtesy Nassau Co. Dept. of Public Works; **Case Study 10.3 1** Courtesy K. W. Brown, Texas A&M Univ.; **2** After Stephen Testa in Pipkin and Proctor, *Engineering geology practice in southern California* (Belmont, Calif.: Star, 1993)

CHAPTER 11

Opener, Ruth Flanagan, © 1993, Kalmbach Pub. Co., *Earth* magazine; **11.3b** Tom Servais, *Surfer* magazine; **11.4b** Tom Servais, *Surfer* magazine; **11.5b** B. Pipkin; **11.7b** B. Pipkin; **11.8** John S. Shelton; **11.10** B. Pipkin; **11.11** B. Pipkin; **11.12a, b** B. Pipkin; **c** Craig Everts; **d** USAF; **11.14** NOAA; **11.15** Data from USGS; **11.16** After Robert Dolan et al., 1985, Coastal erosion and accretion, USGS National Atlas; **11.18** USGS data; **11.19a** Craig H. Everts; **b** J. S. Williams et al., Coasts in crisis, USGS circular 1075, 1990; **11.20** B. Pipkin; **11.21** B. Pipkin; **11.22** B. Pipkin; **11.24a** NAGT; **b** GeoPhoto Pub. Co.; **11.25** Courtesy Orrin Pilkey, Duke Univ.; **11.26** B. Pipkin; **11.27** B. Pipkin; **11.28a** B. Pipkin; **b** U.S. Army

Corps of Engineers; **11.29** After *California geology,* Sept. 1988; **11.30** B. Pipkin; **11.31** Paul Moser, courtesy Ed Nuhfer and AIPG; **11.33** P. Godfrey; **11.35b** NOAA; **11.38** Data from various sources; **11.39a, b** National Weather Service (NOAA), National Hurricane Center, courtesy Ed Nuhfer and AIPG; **11.40** B. Pipkin; **11.41** Suzanne and Nick Geary, Tony Stone Worldwide; **11.42** National Weather Service (NOAA), Hurricane Information Center, courtesy Ed Nuhfer and AIPG; **11.43** B. Pipkin; **Case Study 11.1 1, 3** Courtesy Orrin Pilkey, Duke Univ.; **2** after Orrin H. Pilkey et al., The concept of shoreface profile of equilibrium: A critical review, *Journal of coastal research* 9 (1993), no. 1, p. 255; **Case Study 11.2 1** Coastal Weather Research Station, Univ. South Alabama

CHAPTER 12

Opener, D. D. Trent; **12.9** D. D. Trent; **12.10** D. D. Trent; **12.11** D. D. Trent; **12.12** D. D. Trent; **12.13a, b, d** D. D. Trent; **c** Joseph Allen; **12.16** Wilfried Haeberli; **12.17** NPS data; **12.18** John S. Shelton; **12.19** Richard P. Jacobs/JLM Visuals; **12.21** D. D. Trent; **12.22** D. D. Trent; **12.23** B. Pipkin; **12.24** C. Hillaire-Marcel; **12.25** After Dott and Batten, *Evolution on the earth,* 3d ed. (McGraw-Hill, 1981), fig. 18.6, p. 514; **12.26** Foundation for Glacier and Environmental Research data; **12.27** Rolf Å. Larsson; **12.28** After Weiss and Newman, The channeled scablands of eastern Washington, USGS pamphlet, 1973; and USGS prof. paper 424-B, 1961; **12.29** D. D. Trent; **12.30** D. D. Trent; **12.31** John S. Shelton; **12.34** Data from Baggeroer and Munk, Heard Island feasibility test. *Physics today* 45, no. 9 (Sept. 1992), fig. 2, p. 24; **12.35** After Nansen and Lebedeff, "Global trends of measured surface-air temperatures," *Journal of geophysical research* 29, no. D11 (20 Nov. 1987), fig. 15, p. 13370; **12.36** After Radford Byerly, Jr., The policy dynamics of global change, *Earth-Quest,* Spring 1989 (Univ. Corp. for Atmospheric Research, Office of Interdisciplinary Earth Sciences, Boulder, Colo.), pp. 11–24; **12.37** D. D. Trent; **12.38** D. D. Trent; **12.39** D. D. Trent; **12.40** D. D. Trent; **12.41** Austin Post, USGS; **Case Study 12.2 2** British War Museum; **Case Study 12.3 1** Jürg Alean; **Case Study 12.4 2** D. D. Trent; **3a** D. D. Trent; **Case Study 12.5 2** D. D. Trent; **Case Study 12.6 1** Data from Wallace S. Broecker, Chaotic climate, *Scientific American* 273, no. 5 (Nov. 1995), pp. 62–68; **Case Study 12.7 1** Data from Greenland Ice Core Project (GRIP), Climate instability during the last interglacial period recorded in the GRIP ice core. *Nature* 364, no. 6434 (15 July 1993), pp. 203–207

CHAPTER 13

Opener, © John Zoiner/International Stock; **13.1** USGS data; **13.7** Steven Dutch; **13.10** API data; **13.11** After Charles D. Masters, Energy realities of the world: A projected role for fossil fuels, in L. M. H. Carter, *Energy and the environment: Application of geosciences to decision-making,* USGS circular 1108, 1995, pp. 7–8; **13.12** After AAPG *Explorer,* March 1991, p. 39; **13.13** USGS data; **13.14** After DOE *Annual energy review, 1992,* courtesy Herb Adams, California State Univ.–Northridge; **13.15** Visuals Unlimited; **13.16** USGS data; **13.17** Peggy Foster; **13.18** Courtesy Hobart King, Mansfield Univ.; **13.19** After A. C. Stern, ed., *Air pollution,* vol. I (San Diego: Academic Press); **13.20** Richard Keeler, Southern California Edison Co.; **13.22** D. D. Trent; **13.23** After Carl J. Wienberg and Robert Williams, *Energy from the sun.* Copyright © 1990 Scientific American. All rights reserved; **13.25** B. Pipkin; **13.26** Data from Wendell Duffield et al., USGS circular 1125, 1994; **13.27** B. Pipkin; **13.28** After Wendell Duffield et al., USGS circular 1125, 1994; **13.33** DOE data; **13.34** DOE data, Office of Environmental Management, August 1994; **13.35** DOE data; **13.36a** DOE data; **b** DOE; **13.39**